THE PHYSICAL
GEOGRAPHY OF THE SEA

and its Meteorology

THE PHYSICAL
GEOGRAPHY OF THE SEA
and its Meteorology

MATTHEW FONTAINE MAURY

Edited by
JOHN LEIGHLY

DOVER PUBLICATIONS, INC.
Mineola, New York

Bibliographical Note

This Dover edition, first published in 2003, is an unabridged republication of the work published in 1963 by The Belknap Press of Harvard University Press, Cambridge, Massachusetts. Maury's text was originally published in 1855.

Library of Congress Cataloging-in-Publication Data

Maury, Matthew Fontaine, 1806–1873.
    The physical geography of the sea and its meteorology / Matthew Fontaine Maury ; edited by John Leighly.
        p. cm.
    "... an unabridged republication of the work published in 1963 by the Belknap Press of Harvard University Press, Cambridge, Massachusetts. Maury's text was originally published in 1855"—T.p. verso.
    Includes index.
    ISBN 0-486-43248-3 (pbk.)
    1. Oceanography. 2. Marine meteorology. I. Title.

GC11.2.M38 2003
551.46—dc22

2003055663

Manufactured in the United States of America
Dover Publications, Inc., 31 East 2nd Street, Mineola, N.Y. 11501

CONTENTS

TABLES

FIGURES

PLATES

following page 432

INTRODUCTION

In one respect the fate of Matthew Fontaine Maury's *Physical Geography of the Sea* has been an unusual one. Few books with scientific pretensions can have enjoyed as much popularity among the general reading public. Its publishers, both in the United States and in England, kept it in print for twenty years after its first appearance in 1855. It was translated into a half-dozen Continental languages. It never received the approval of those contemporary scientists who were best able to judge it, however, and can not be looked upon as representative of the best scientific thought of the eighteen-fifties. Yet in the twentieth century, long after the general reading public had forgotten the book, adepts of the sciences of the sea usually have mentioned it with respect as a pioneering contribution to their field of investigation. Only when they have had occasion to examine particular points in Maury's exposition have they recognized his shortcomings.[1] Maury had, indeed, an abundance of factual information to impart, but his presentation of it was constantly obscured by his eagerness to organize it by means of hastily formulated and poorly grounded hypotheses. The history of the application of physical theory to the sea and the atmosphere would have been little different if the book had never been written. Its value as a document in our intellectual history lies in a more general sphere than in the contribution it made to scientific insight.

A work that falls within the continuous sequence of investigations of natural phenomena is likely to have a somewhat impersonal quality; such a work often gives the impression that if a particular author had not written it someone else equally qualified would have done so. Maury's *Physical Geography of the Sea* is, on the contrary, a highly individual book; it could have been written by no one else. It is not a segment of the highroad of marine science, but a bypath, interesting enough with its unexpected turnings and unfamiliar views, but leading in a direction different from that of the main road. The organization of knowledge of the seas had in Maury's time a long history; he mentions a few general works, but it does not appear that he was acquainted with the greater part of the pertinent literature. He borrowed facts readily enough from writings he knew, but no scheme of organization.[2] *The Physical Geog-*

---

[1] For example, Henry Stommel, *The Gulf Stream; a Physical and Dynamical Description* (Berkeley and Los Angeles, 1960), pp. 7–8.

[2] One exemplary organization of knowledge of the sea that was available is in

raphy of the Sea is to an unusual degree the product of one man's mind, shaped by his unique experience.

The story of Maury's life has often been told, and only the briefest sketch of it is required here.[3] He was born near Fredericksburg, Virginia, on January 14, 1806. In 1810 his family migrated westward across the mountains, settling near Franklin, Tennessee. Here Maury spent his early youth, attending local elementary schools and Harpeth Academy, then situated near Franklin. He was appointed midshipman in the United States Navy in 1825 and immediately entered on service afloat, since there was at the time no academy for the training of American naval officers. The instruction imparted to midshipmen, of which he held a low opinion, completed his formal education.

Maury spent most of the period 1825–1834 on cruises across the Atlantic, about South America, and westward around the earth. After his return from this circumnavigation he saw no further service on the high seas. During an extended leave he began writing for publication; his first published articles, one a discussion of problems of navigation in the vicinity of Cape Horn based on his experiences in that region in 1831, and the other on an instrument for making certain computations in nautical astronomy, appeared in 1834.[4] Far more important was a textbook in navigation he wrote expressly for use in the training of future officers of the United States Navy, which in successive editions was used for that purpose.[5]

A little later Maury began a series of pseudonymous articles in newspapers and other periodicals, concerned mainly with shortcomings in the Navy, particularly in the training of officers. In writing these articles, published in the late eighteen-thirties, he acquired the habit of hortatory and polemic utterance that remained with him throughout his life and that finds expression in many passages of *The Physical Geography of the Sea*.

The event that determined Maury's later career was a severe injury to

---

Heinrich Berghaus, "Umriss der Hydrographie, erste Abtheilung, von der allgemeinen Wasserhülle der Erde oder dem Ozean," *Allgemeine Länder- und Völkerkunde*, I (Stuttgart, 1837), 402–640.

[3] Besides the accounts given in the usual reference works, several separate biographies of Maury have been published. The first of these, and the source of most of the personal information about Maury used by later authors, is by one of his daughters: Diana Fontaine Maury Corbin, *A Life of Matthew Fontaine Maury* . . . (London, 1888). The best of the later biographies is Charles Lee Lewis, *Matthew Fontaine Maury, the Pathfinder of the Seas* (Annapolis, 1927). Lewis gives a fuller account of Maury's public career than Mrs. Corbin does.

[4] "On the Navigation of Cape Horn," *American Journal of Science*, XXVI(1834), 54–63; "Plan of an Instrument for Finding the True Lunar Distance," *ibid.*, pp. 63–65.

[5] *A New Theoretical and Practical Treatise on Navigation* . . . (Philadelphia, 1836).

his right leg, suffered in the overturning of a stage coach in October 1839, when he was on his way from Tennessee to report for duty in New York. Having never recovered the full use of his leg, he was not again assigned to duty at sea. During the enforced idleness following this injury he continued his pseudonymous writing on naval matters and other questions of public policy. When he was recalled to duty on July 1, 1842, it was as superintendent of the Navy's Depôt of Charts and Instruments, the agency that was renamed National, or Naval, Observatory when it took possession of new quarters and new equipment in 1844. (From 1854 to 1866 it was called "Naval Observatory and Hydrographical Office"; in 1866 the Hydrographic Office was separated from the Naval Observatory.) It was in this position, which he held until the outbreak of the Civil War, when he resigned his commission to serve the cause of the Confederacy, that Maury achieved worldwide renown. In his early years at the Observatory his time was occupied principally in astronomic work, but he soon turned his attention from the heavens to the sea and the atmosphere. As has often been stated, his most significant contribution was the compilation, for use in the navigation of sailing ships, of observations of wind and weather made as a routine part of the operation of vessels of both the military and the merchant marine. The source of the first information of this kind compiled under his direction was the accumulation of logbooks, stored at the Observatory, of ships of the United States Navy. The first sheet of his *Wind and Current Charts*, the primary vehicle by which the accumulated information was disseminated, appeared in 1847; the early sheets were based entirely on material "quarried" from the logbooks. Soon, however, Maury prepared special charts and forms for recording observations, which were used first on vessels of the Navy and then on American merchant ships. As more information was received and compiled, more charts, and the *Explanations and Sailing Directions to Accompany the Wind and Current Charts*, were prepared and circulated, and earlier charts revised.

The culmination of Maury's work in organizing the collection of observations at sea was an international Maritime Conference held at Brussels in 1853, in which he played the leading part; he gives a brief account of the conference in the "Introduction to the First Edition. — 1855," prefaced as well to all later editions of *The Physical Geography of the Sea*. His role in the work of this conference brought him into international prominence and into communication with the world of science.

Maury's published writings about the sea, beyond the compilation of the *Wind and Current Charts*, began to appear frequently only about 1850. Of the previously published papers, the content of which was later incorporated into *The Physical Geography of the Sea*, only one besides the Cape Horn article of 1834 appeared before 1850: one on the Gulf

Stream published in 1844.[6] This paper became the nucleus of the chapter on the Gulf Stream in all editions of *The Physical Geography*. In 1850 Maury read three papers before the American Association for the Advancement of Science, meeting in Charleston, South Carolina.[7] Two of these, "On the Influence Arising from the Discovery of the Gulf Stream on the Commerce of Charleston" and "On the General Circulation of the Atmosphere," became chapters in *The Physical Geography*; in revised form they constitute, respectively, parts of Chapter II and Chapter III of the final edition. Two further articles that eventually became chapters in the book appeared in 1851.[8]

An important step in Maury's career as author was the publication in 1851 of the *Explanations and Sailing Directions to Accompany the Wind and Current Charts*. Successive editions of *Sailing Directions* appeared during the eighteen-fifties, each bulkier than its predecessors; the last edition, the eighth, was published in two generous volumes in 1858 and 1859.[9] This work was distributed with the *Wind and Current Charts* to the seamen who contributed the observations compiled at the Observatory. It is, in all editions, much more than what its title promises. Into it Maury poured, without much order, such material as he considered of use or instruction to the navigator: tables of sailing times in different parts of the ocean, extracts from logs of ships, and correspondence between himself and his collaborators. Each edition contains a number of charts, some of which became plates in *The Physical Geography*. In addition, Maury put into *Sailing Directions* material he had published elsewhere. Thus the edition of 1851 contains sections headed "The Gulf Stream and the Trade of Charleston," "Currents of the Sea," "General Circulation of the Atmosphere," and "The Equatorial Cloud Ring," all of which titles can be identified with titles of articles already published and with chapter headings to appear in *The Physical Geography*. Other material from *Sailing Directions* found its way into the succeeding work, though not always as distinct chapters. In the sixth edition of *Sailing Directions*, 1854, general

---

[6] "Remarks on the Gulf Stream and Currents of the Sea," *Amer. Jour. Sci.*, XLVII(1844), 161–181; "Paper on the Gulf Stream and Currents of the Sea," *Southern Literary Messenger*, X(1844) [393]–409.

[7] "On the Influence Arising from the Discovery of the Gulf Stream upon the Commerce of Charleston," *Proceedings of the American Association for the Advancement of Science*, III(1850), 17–20; "On the Currents of the Atlantic Ocean," *ibid.*, pp. 74–79; "On the General Circulation of the Atmosphere," *ibid.*, pp. 126–147.

[8] "On the Geological Agency of the Winds," *Proc. Amer. Assoc. Adv. Sci.*, IV(1851), 277–296; "On the Probable Relation between Magnetism and the Circulation of the Atmosphere, Especially the Trade Winds," *Edinburgh New Philosophical Journal*, LI(1851), 271–292 (also as "Appendix" to *Washington Astronomical Observations for 1846*, Washington, 1851).

[9] In numbering later editions of *Sailing Directions*, Maury counted that of 1851 as the third. Other editions bear the following dates: fourth, 1852; fifth, 1853; sixth, 1854; seventh, 1855; eighth, I, 1858, II, 1859.

didactic matter is separated, more distinctly than in the earlier editions, from information designed for the immediate use of the navigator. The seventh edition (1855) was compiled after publication of the first edition of *The Physical Geography*; it contains thirteen general chapters bearing headings identical with those of thirteen of the eighteen chapters in *The Physical Geography*. The content of the first volume of the eighth edition of *Sailing Directions* is nearly the same as that of the contemporary edition (1858) of *The Physical Geography*.

The general didactic part of *Sailing Directions* was thus the immediate precursor of *The Physical Geography*: almost all of the content of the first edition of *The Physical Geography* was drawn from the sixth edition of *Sailing Directions*. What precipitated the publication of the new volume, according to Maury's biographer C. L. Lewis,[10] was a suggestion made by a member of the firm E. C. and J. Biddle of Philadelphia, publishers of the sixth and seventh editions of *Sailing Directions* and Maury's textbook of navigation. The publisher called attention to the fact that Maury's writings on the sea and the atmosphere were not protected by copyright and might easily be appropriated and published by anyone else. Maury acted immediately by preparing the copy for the first edition of *The Physical Geography* for publication by Harper and Brothers early in 1855. Maury credits the expression "physical geography of the sea" to Alexander von Humboldt; his first use of it, to my knowledge, is in the sixth edition of *Sailing Directions*, 1854, where it appears as the heading of a section on investigations of the Atlantic Ocean made under Maury's direction on certain vessels of the United States Navy detailed for the purpose. (The inappropriate use of the expression as the title of the map of the Antarctic regions called in the present volume Plate X is inexplicable at this late date.)

Maury subjected *The Physical Geography* to the same kind of hasty revision for repeated printings as he had in preparing *Sailing Directions*. I am unable to give a complete bibliography of the work, even of the American editions. The list given by the bibliographer of Maury's writings, R. M. Brown,[11] is probably imperfect; a more detailed collation than he provides would be required to bring order into the confusion of successive printings. Indeed, a definitive bibliography of the American editions and printings would require examination and collation of a larger number of individual copies than are apparently held by any one library. "Editions" often differ from their immediate predecessors only in supplementary material appended after the numbered chapters, material

---

[10] *Matthew Fontaine Maury*, pp. 67–68.
[11] Ralph Minthorne Brown, "Bibliography of Commander Matthew Fontaine Maury," 2nd ed., *Bulletin of the Virginia Polytechnic Institute*, XXXVII(no. 12, 1944), 23–24.

that was gradually incorporated into formal chapters. I record the following information, not in the pretense of supplying a definitive bibliography, but in order to show how tangled a skein the bibliography of *The Physical Geography* is, and how highly the work was valued by publishers as a literary property.

The editions of *The Physical Geography* following the first were not all numbered. Editions numbered second and third and, according to Sabin's *Dictionary of Books Relating to America* (XI, 509), a fifth edition (following an unnumbered fourth?) appeared in 1855. Later editions until the eighth bear no numbers on the title pages but, instead, such identifications as "an entirely new edition" and "an entirely new edition, with addenda." Not all printings bearing these respective designations are identical. No edition is numbered "sixth" on the title page, but printings from 1856 to 1859 contain an "Introduction to the sixth edition" dated April 1856; the eighth edition carries this material as "Introduction to the revised edition. — 1856." There were only two general revisions, in the accepted sense of the term, of the content of the book: one in 1856 and the more thorough one made in 1860 that yielded the final, eighth edition, published early in 1861.

The authorized British publishers of *The Physical Geography* were Sampson Low, Son, & Co., London, who, it appears, issued there most, if not all, of the editions published by Harper in New York. I have seen one copy printed by Harper evidently intended for distribution in Britain ("an entirely new edition, with addenda," 1858). It is identical with the simultaneous American edition except for the dedication, as in all the English editions, to Lord (John) Wrottesley (1797–1867), a distinguished amateur astronomer who had befriended Maury in England when he visited Europe in connection with the Brussels conference of 1853. The corresponding American edition, as all previous to the eighth, was dedicated to George Manning of New York, who as agent distributed the *Wind and Current Charts* and *Sailing Directions* to the seafaring world. The other copies from Sampson Low that I have seen, printed in England, differ in format from the one Harper used. Designations of the British editions do not always agree with those of the corresponding American editions. Sampson Low continued to give new "edition" numbers to printings even after Maury's death in 1873, although there is no evidence of any revision of the book after the 1861 edition corresponding to the eighth American. Printing errors, for example, more numerous in the British than in the American final edition, recur in successive printings. In some later editions Sampson Low included an index, which Harper never provided, though Harper kept the book in print until 1876.

The appearance of unauthorized editions in England introduces a complication that does not afflict the bibliography of the American printings.

At least two firms other than Sampson Low published *The Physical Geography* in England. R. M. Brown lists "a new edition, with improvements, additions, and revised charts by the author," London, T. Nelson & Sons, 1859. Sampson Low's edition of 1859 is described thus on the title page, but I have not seen such a copy published by Nelson. I have, however, seen one issued by Nelson in 1859 with no identification of the edition on the title page; of all copies examined it contains the handsomest maps, printed in two colors. In addition to another Nelson copy, of 1860 "with the author's latest additions," I have seen a copy issued in London by George Philip and Son, 1863, a firm not mentioned by Brown among the publishers of the book. Despite the date and the fact that Sampson Low then had in print Maury's final edition, the content of this copy is the same as that of the American "entirely new edition, with addenda," of 1859. Maury protected his final British edition by sailing to London in November 1860 for its publication. This London edition differs from the American eighth by greater fullness in some passages: the internal evidence is that Maury continued to work on the manuscript after leaving the United States. Thus the authorized publishers issued printing after printing; and in England two firms, both reputable, published unauthorized editions. Brown lists Dutch, French, German, and Italian translations; I can add to his list a (somewhat abridged) Norwegian and a Spanish translation.[12]

The eighth and last American edition republished here, *The Physical Geography of the Sea, and its Meteorology*, was kept in print without revision for the longest time. The revision that produced it was, moreover, more extensive than any earlier one. Seven chapters, VIII, XI, XII, XIII, XVII, XVIII, and XX, are changed only slightly from the "entirely new edition, with addenda," of 1859, which contained 389 pages of text as compared with 474 in the eighth edition. One chapter, "The Submarine Telegraph of the Atlantic," is omitted. Three chapters are new, the introductory "The Sea and the Atmosphere," and the closing two on "The Antarctic Regions and Their Climatology," and "The Actinometry of the Sea." Chapter VI combines two short chapters of the edition of 1859. The material retained in Chapters IV and V formed a single long chapter, "The Atmosphere," in the preceding edition. Chapters XV and XVI are a result of the division of another long chapter, "The Winds." The remaining chapters were extensively rewritten, and new material was added to them. Most of the figures in the text are new.

No edition of the book is systematically organized. The work began as

---

[12] *Havets physiske Geografi og Meteorologi* i Uddrag og Bearbeidelse efter M. F. Maury's "The Physical Geography of the Sea and its Meteorology" af H. J. Müller (Christiania, 1865), vii + 332 pp.; M. F. Maury, *Geografía física del mar.* Traducida al castellano de la quinta edicion, par el brigadier de la armada D. Juan Nepomuceno de Vizcarrondo (Madrid, 1860), 316 pp.

a collection of papers already published, augmented by some new matter. The first chapter of the eighth edition, "The Sea and the Atmosphere," represents one step toward an improved organization, but a short one. Earlier editions had all commenced with Chapter II, "The Gulf Stream," and its startling opening sentence, "There is a river in the ocean." [13] But in general it is difficult to find any logical order in the sequence of chapters. There is, moreover, a good deal of repetition of facts and arguments. Some chapters combine unrelated topics, notably Chapter VII, "The Easting of the Trade-winds, the Crossing at the Calm Belts, and the Magnetism of the Atmosphere."

The attentive reader will recognize a certain lack of homogeneity in Maury's interpretation of the circulation of the atmosphere, to which he recurs in several chapters. In the matter held over from earlier editions he presents the mechanism of the circulation as a simple convection dependent upon differences in temperature of the air (Chapters II and IV). In material added later, for example in Chapters XV, XVI, and XIX, he gives more attention to pressure gradients in the atmosphere. In § 691 he uses the expression "barometric declivity," meaning precisely "pressure gradient." He undoubtedly owed this belated awareness of the role of pressure in producing motion in the atmosphere to the work of contemporaries, especially James P. Espy. But this awareness did not penetrate the material written earlier.

In order to avoid anachronism I shall so far as is practicable let the contemporaries express their own judgments of *The Physical Geography*. Comments cited do not all refer to the same edition, but since the character of the work did not change they may be taken as appropriate to all editions.

On the title page of a copy of the eighth edition of *The Physical Geography* in the library of the University of California, Berkeley, its original owner wrote: "This is a great book." Equivalent judgments appear in published reviews, for example: "On the whole, this treatise . . . may be considered as occupying the first rank in its appropriate department, and Lieutenant Maury has erected for himself a statue that is imperishable"; [14] ". . . hereafter Lieutenant Maury will be numbered among the great scientific men of the age and the benefactors of mankind." [15] The most

---

[13] An expression not forgotten in the middle of the twentieth century. It echoes in the title of Hans Leip, *Der grosse Fluss im Meer* (München, 1954), and in less happy paraphrase in that of Henry Chapin and F. G. Walton Smith, *The Ocean River* (New York, 1952).

[14] "The Physical Geography of the Sea" [review], *De Bow's Review*, XIX(1855), 72–78, esp., p. 78.

[15] "From *The Economist*," *Littell's Living Age*, XLV(1855), 655–658, esp., p. 657.

uncritical praise of the book is found in reviews published in general and literary periodicals, usually written by persons having little or no scientific competence, who judged the work from the viewpoint of the ordinary reader. The extracts just given may suffice; most such comments may be passed over in silence, since they express little beyond general approval on ill-defined grounds. Some notices in the literary journals were, however, written by competent scientists. One of these notices, unsigned but attributed by R. M. Brown to Sir David Brewster, appeared in *The North British Review* in 1858.[16] Brewster takes issue with Maury on several points, for example, Maury's ascription of the blueness of the water of the Gulf Stream to its saltness (§ 71 of the eighth edition) and his interpretation of the significance of cold water underneath the Gulf Stream for the conveyance of heat to western Europe (§ 143). Another such review, attributed by Brown to Sir Henry Holland, contains the following general criticism:

[Maury] theorizes too largely and hazardously, and does not clearly separate the *known* from the *unknown*. His volume is replete with valuable and ingenious suggestions, but they are not methodized enough for the common reader, who will probably rise from the chapters on winds and atmospheric currents, his head confused by a whirl of facts and theories and questions, as fleeting as the very air of which he has been reading.[17]

The leading American scientific journal received the first edition of *The Physical Geography* briefly and coolly: "While the work contains much instruction, we cannot adopt some of its theories, believing them unsustained by facts." [18]

There is little evidence in the eighth edition of the book of any appreciable modification of Maury's views as a consequence of adverse criticism of his earlier editions. Rather, he stoutly defends some of his hypotheses against published objections. In §§ 79ff. he takes to the field against no less formidable an opponent than Sir John Herschel, who in *The Encyclopædia Britannica* had rejected Maury's interpretation of the origin of the Gulf Stream as what would today be called a thermohaline circulation, that is, a circulation arising from horizontal gradients of density owing to differences in temperature and salinity within a continuous body of water. The more generally accepted interpretation, to which Herschel adhered, was that the current represents the escape northward and eastward of water impelled westward by the trade winds of the Atlantic. In his rebuttal Maury is less than fair in ascribing to Herschel an explanation of the Gulf Stream as a rebound of water from

---

[16] "The Geography of the Sea," *North British Review*, XXVIII(1858), 403–436, esp., p. 408, n. 1.
[17] *Edinburgh Review*, CV(1857), 360–390, esp., p. 380.
[18] *Amer. Jour. Sci.*, ser. 2, XIX(1855), 449.

the eastern shores of the American continent "as billiard balls are [deflected] from the cushions of the table" (§ 80). Now the only reference Herschel makes to billiard balls is in an argument that a circulation may exist in a body of water, the surface of which is level:

A circulation in a closed area, produced by an impulse acting horizontally on the surface water, may perfectly well co-exist with a truly level course of each molecule. A billiard ball runs along a level table by an impulse from the cue quite as naturally as if it rolled on an inclined plane by its weight.[19]

In a different article, "Meteorology," in the *Britannica*,[20] Herschel does ascribe the origin of the Gulf Stream to a deflection of the north equatorial current of the Atlantic Ocean by the shores of the Caribbean Sea and the Gulf of Mexico, but without mentioning billiard balls. Maury, indifferent to the niceties of citation, may have confused the two passages in his mind and used his combination of them as a target for his derision.

With the possible exception of Maury's appeal to magnetism as a control of the wind, nothing in *The Physical Geography* evoked more unanimous opposition in the world of science than his scheme of the circulation of the atmosphere, presented graphically in Plate I, described in §§ 204ff., and returned to repeatedly in such later passages as §§ 288ff., 330, 355–369, and 540ff. The features of this scheme that no competent critic would accept are (a) low atmospheric pressure at the surface of the earth in high latitudes and the associated convergence of wind toward the poles, and (b) the "crossing" of currents of air in vertical motion at the tropical high-pressure belts and near the equator. Item (a) could be refuted by citing observations made in high latitudes. There was no observational evidence regarding (b), but all that was known about the behavior of fluids in motion made it highly improbable. Maury first published this scheme [21] in 1850 and maintained it — how insistently is realized only when it is encountered in many passages in the book — through this last edition of *The Physical Geography*. Herschel expressly rejected it.[22] On this side of the Atlantic, Espy in 1857 publicly and courteously besought Maury to abandon it:

Nor do I doubt, from his love of truth, that he will abandon the unintelligible paradox of supposing that the whole of the southeast trade wind passes to the north, and the whole of the northeast trade wind passes to the south, each permeating the other as they pass through the region of the calms . . .[23]

[19] Sir John Herschel, "Physical Geography," *Encyclopædia Britannica*, 8th ed., XVII(Boston, 1859), 569–647, esp., p. 579.

[20] In XIV (Boston, 1857), 652.

[21] "On the General Circulation of the Atmosphere," *Proceedings*, AAAS, III, 126–147.

[22] "Physical Geography," *Encyclop. Britan.*, XVII, 615.

[23] James P. Espy, *Fourth Meteorological Report* (34th Cong., 3rd sess., Sen. Ex. Doc. no. 65, 1857), p. 159.

In short, Maury's insistence, from 1850 to 1861, on his scheme of circulation of the atmosphere succeeded in convincing no one competent to judge it.

Neither Herschel nor Espy offered a system of general circulation of the atmosphere as an alternative to Maury's. But Maury's scheme called forth one reaction that was positive and in the end highly fruitful. In 1856, after the appearance of the first edition of *The Physical Geography*, William Ferrel published an article provoked by his reading of the interpretation Maury presents there.[24] In refuting Maury, Ferrel first cites observational evidence that the surface winds near the poles have a component of motion away from the poles instead of toward them, as Maury insisted. Second, Ferrel introduces an effect of the earth's rotation in deflecting moving bodies on the earth's surface away from a path along a great circle, whatever the direction in which the bodies are moving, an effect that had not earlier been taken into account in discussions of the circulation of the atmosphere or the oceans.[25] By invoking this new "force," Ferrel was able to account for the existence of belts of winds on the surface of the earth much more logically than Maury had done. He rejected even more emphatically than Herschel or Espy the "crossing" of winds at what Maury called the "nodes" of the circulation. Moreover, he formulated a corresponding theory of the motions of the surface water of the oceans, taking into account the new "force" he had defined. Ferrel's article was published obscurely, but he distributed it so widely in reprint form that it may be inferred that Maury read it. But Maury nowhere mentions it in his published writings.

In 1859 and 1860, in good time to affect the last revision of *The Physical Geography*, Ferrel returned to the question of the circulation of the atmosphere with a fuller and more rigorous treatment[26] than he had formulated in 1856. It may confidently be presumed that Maury could not understand Ferrel's mathematics, but there is more than enough plain English in his series of articles to give the non-mathematical reader a conception of the circulation of both the atmosphere and the oceans much clearer and more nearly in accordance with observation and physical theory than any Maury had offered. Again, there is no trace in the eighth edition of any influence of Ferrel's reasoning. None of Maury's theoretical

[24] "An Essay on the Winds and Currents of the Ocean," *Nashville Journal of Medicine and Surgery*, vol. XI(1856). The article is usually cited from a later, more accessible source: William Ferrel, "Popular Essays on the Movements of the Atmosphere," *Professional Papers of the Signal Service*, no. XII(1882), [7]–19.

[25] With the single exception, to my knowledge, of a clear statement in Charles Tracy, "On the Rotary Action of Storms," *Amer. Jour. Sci.*, XLV(1843), 65–72, which for some reason remained long unnoticed.

[26] "The Motions of Fluids and Solids Relative to the Earth's Surface," *Mathematical Monthly*, I(1859), 140–148, 210–216, 300–307, 366–373; II(1860), 85–97, 339–346, 374–382.

generalizations approaches in significance the effect of his scheme of atmospheric circulation on Ferrel, who was sufficiently outraged by it to apply his knowledge of theoretical mechanics to the motions of the atmosphere, a field of investigation that he cultivated for many years with most valuable results.

If one charitably overlooks Maury's hypothesis in Chapter VII that the circulation of the atmosphere is controlled by terrestrial magnetism, which did not convince even his lay critics, the most sweeping generalizations he made were those just cited: his denial of any significant role of winds in producing currents in the ocean and his scheme of atmospheric circulation, with the manifold consequences he derived from it. (The fallacious notion that the Arctic Ocean is free of ice, which he expounds in Chapter IX, was shared by many responsible contemporaries.) In defense of his views concerning the motions of the water of the sea he was willing to enter the lists against Sir John Herschel, but he insists on his scheme of circulation of the atmosphere without mentioning the names of any opponents of it or the objections they had expressed.

In 1863, too late to influence the content of the eighth edition of *The Physical Geography*, there appeared in France the most thorough criticism ever directed against Maury's scheme of circulation of the atmosphere. It was written by a distinguished French naval officer, Captain (later Vice-Admiral) Siméon Bourgois.[27] Using the information Maury had compiled in his *Wind and Current Charts* but also observations made in higher latitudes than are included in them, Bourgois demolishes, first, Maury's notion that winds with a poleward component of motion prevail all the way to the poles in both hemispheres; and second, Maury's system of belts of winds and calms that extend all the way about the earth, from which mistaken conception he derived many erroneous conclusions regarding the climates of the lands. Bourgois shows, instead, that the general pattern of mean circulation of the air over each of the ocean basins, from the equator to latitudes 45° or 50°, is a great anticyclonic whirl.

Since *The Physical Geography* was not revised after the appearance of Bourgois' devastating criticism, it is impossible to know whether Maury would have dealt with it as confidently as he did with Herschel's objections to his explanation of the Gulf Stream. It may be worth noting, however, that in *Physical Geography*, a textbook Maury completed just before his death, there is no "diagram of the winds" such as Plate I. Instead, there is a world map that presents a generalization of the circulation of the atmosphere in which zonation is not unduly insisted on, and in which winds having a component of motion away from the pole are indicated in high northern latitudes. Maury repeats in this book, however, an error

---

[27] M. J. Bourgois, *Réfutation du système des vents de M. Maury* (Paris, 1863); repr. from *Revue Maritime et Coloniale*. (Initials as per title page.)

that runs through all editions of *The Physical Geography* (§ 34 and elsewhere in the eighth edition): he ascribes the explanation of the deviation of the trade winds from meridional directions to directions having components toward the west to "Halley," [28] whereas the explanation he offers (an erroneous one, as Ferrel showed) is that of George Hadley [29] (1685–1768), published much later than Halley's. By way of extenuation of this slip, Gustav Hellmann's remark may be cited, that confusion of Hadley with the better-known Halley was common until the middle of the nineteenth century.[30] I have not observed that any of the contemporary reviewers of *The Physical Geography* called attention to Maury's error.

It is evident from what has been cited that as a scientific treatise *The Physical Geography* fell below the level of the best contemporary knowledge. Maury did not know enough physical theory to recognize the difficulty of the problems he disposes of so confidently in his discussions of the motions of the sea and the atmosphere. It was the valor of ignorance, not confidence born of superior knowledge, that guided his pen in defense of his interpretation of the mechanics of the Gulf Stream against Sir John Herschel and impelled him to insist so repeatedly on his scheme of atmospheric circulation. (He had a firmer basis for his contradiction of Sir Charles Lyell in §§ 386–389.) It was not the approval of the scientific world that induced publishers to keep the book in print for two decades. The modern scientific reader finds it more difficult to account for its popularity than to assess its limited value. In any event, that popularity could owe nothing to Maury's insistence on his dubious hypotheses.

Yet *The Physical Geography* did exert a scientific influence of a sort. Let the reader examine Chapters I to V and XVII and note how many popularly held conceptions of the motions of the atmosphere and the sea and ideas about weather and climate are presented in them. There, for example, he will find the misuse of the term "Gulf Stream" to designate most of the poleward movement of warm water in the North Atlantic Ocean, and the attribution of the effects of the ocean on the climates of the lands primarily to currents. There is found also the simple concept of belts of wind continuous about the earth, over land and sea alike, which migrate northward and southward with the sun's declination, bringing alternating dry and rainy seasons to the lands over which the boundaries

---

[28] Edmund Halley, "An Historical Account of the Trade-Winds and Monsoons, observable in the Seas between and near the Tropicks, with an attempt to assign the Phisical cause of the said Winds," *Philosophical Transactions*, XVI(1688), 153–158. Halley (1656–1742) was Astronomer Royal from 1721 to 1742.

[29] "Concerning the Cause of the General Trade-Winds," *Philos. Trans.*, XXXIX(1735), 58–62.

[30] "George Hadley, Concerning the Cause of the General Trade-Winds," *Neudrucke von Schriften und Karten über Meteorologie und Erdmagnetismus*, no. 6 (Berlin, 1896), p. 5.

of the belts pendulate. These notions and others, few of which were original with Maury but most of which were obsolete or gravely questioned in his day, have lived on in the underworld of science: in popular literature and school textbooks. These physically untenable notions have apparently persisted in nearly complete independence of the results of a century's active investigation of the atmosphere and the ocean. They seem to be continuous not only with *The Physical Geography* but also with the textbooks for elementary and secondary schools that Maury wrote in his last years.

Since the reviewers who praised *The Physical Geography* with few or no reservations were less specific than the critical scientists who called attention to its weaknesses, the qualities of the book that evoked their uncritical praise are not easy to identify. Evidently the style in which the book is written recommended it. "Here is a subject, in the abstract hopelessly dry, treated in a manner that, from the opening of the book to its close, never tires." [31] The anonymous author of an obituary notice of Maury called it "one of the most charming books in the English language." [32] Maury's characteristic style is emphatically not that appropriate to a sober scientific treatise. The reviewer in *Blackwood's* calls it "nervously-eloquent." It is the style of the pulpit and the lecture platform in Maury's day and obviously owes much to the habits he formed while composing the polemic articles published in his earlier years. His daughter, Mrs. Corbin, tells in her biography how, in the years when the later revisions of the book were being written, Maury was accustomed to dictate to her and a sister while pacing the floor.[33] The more carefully formulated passages are aimed at the ear of a listener, rolling off the tongue with the rhythm of oratory. Maury liked striking sentences at the beginning of chapters. Besides the one that introduces the chapter on the Gulf Stream, Chapters I, V, XI, and XVI begin with strong and terse assertions. Other chapters, however, such as IV, IX, XV, and XVII, begin weakly. In spite of the judgment of the reviewer in *Blackwood's*, Maury's lively style is not sustained; some passages, especially in the later chapters — parts of Chapter XIX, for example — are tedious. Long quotations, such as those from George Buist in Chapter I and from M. H. Jansen in Chapters VI and XVI, give the reader the uncomfortable feeling that Maury is parading too long in borrowed finery. These passages and the one from M. J. Schleiden that introduces Chapter XIII are all couched in exuberant language. Since a paraphrase in less florid words would have sufficed to convey the information they contain, it must have been the

---

[31] "The Physical Geography of the Sea" [review], *Blackwood's Magazine*, LXXXIX(1861), 277–287, esp., p. 278.

[32] "Captain Maury," *Temple Bar*, XXXVIII(1873), 58–65, esp., p. 60.

[33] *A Life of Matthew Fontaine Maury*, pp. 150–151.

style of these passages that led Maury to quote them verbatim, a style he evidently admired and was eager to emulate. In his "Introduction to the revised edition. – 1856," he speaks of Jansen's "very delightful and pleasant manner." Sir David Brewster, on the contrary, called Jansen's language "too poetical for science." [34]

Those who found Maury's prose "charming" were presumably persons whose literary tastes were in accord with the prevailing style of the time. Maury evidently sought to satisfy their tastes by loading his exposition with words calculated to evoke emotional responses. Repeatedly, starting from some fact of observation, he spins out a long train of reflections called forth from his lively imagination. He uses rhetorical questions in an effort to involve the reader in his own profound concern with the ideas he promulgates. He compares the physical processes of the sea and the atmosphere with the workings of familiar industrial and domestic mechanisms, as in the opening paragraph of Chapter III, in § 315, and in § 489, in the last of which his simile is wholly unconvincing. In one passage, at the beginning of Chapter VI, he composes a high-flown description of a scene remembered from his years at sea.[35]

All these rhetorical devices, which Maury uses skillfully, are familiar from popular writing. In places, however, his prose is embarrassingly overinflated; there is in it, unfortunately, not a little of the romantic bombast that has often been identified and deplored in the literature of the ante bellum South. Some reviewers reproached Maury for the exuberance of his language; Sir Henry Holland, for example, in the notice in the *Edinburgh Review* cited earlier, wrote:

We are unwilling to be hypercritical where there is so much real merit, but it is impossible not to see in his work a desultory desire for novelty, occasionally going beyond the bounds of true inductive science; and venting itself in a phraseology which loses its force and effect by being to sedulous to attain them. With a little more constraint upon his speculations, and a clearer separa-

---

[34] "The Geography of the Sea," *North British Review*, XXVIII, 416. The "language" quoted in *The Physical Geography* is of course not Jansen's, but that of a translator identified in earlier editions as "Mrs. Dr. Breed, Washington." Comparison of the translations with the original shows that Mrs. Breed's knowledge of the Dutch language was inadequate to her task. But though there are many mistakes in her translation, she did preserve Jansen's style.

[35] Readers concerned with the use of the English language in the United States may find some interesting archaisms in Maury's writing. For example: "lither sky," § 239, a Shakespearean expression; "ley," § 461 (but "lye" in § 43); "lichen caves," § 591, meaningless if "lichen" is the word intended (a corruption of *licham*, "corpse," and the expression thus signifying "ossuaries"?); "eat," § 596, as past participle. *The Oxford English Dictionary* (s. vbb. "lither," "lye") cites Maury for its latest examples of the use of "lither," and of "lye," "ley" in the sense *lixivium*. As for Maury's Latin, his use of the erroneous plural *animalculæ* (e. g., §§ 499, 590, 594, 618) — "das ganz barbarische Wort" — made the German translator of *The Physical Geography* wince. (*Die physische Geographie des Meeres*, deutsch bearbeitet von Dr. C. Böttger, 2nd ed., Leipzig, 1858, p. 204, n.)

tion of fact and hypothesis, he would be a valuable scientific writer; with somewhat less intention of fine writing, he would be an eloquent one.[36]

The reading public was obviously not only less critical of Maury's scientific ideas than Holland, but also less fastidious about his style.

Besides his inflated literary style, two other mutually related qualities of Maury's writing strike the reader: his frequent use of the "argument from design" prevalent in the popular natural theology of the first half of the nineteenth century, and his more than liberal borrowing of quotations from Scripture. Maury expounds the "argument from design" in its most elementary form in §§ 165 and 201, and urges it in many other passages, notably §§ 289, 302–303, and 351. In § 370 he identifies the "laws of Nature" with the divine design and insists further upon "design" in §§ 453, 455–456, 496, 550, and 837. More or less perfunctory obeisance to the notion of "design" in nature is a commonplace in popular scientific writings from Maury's time; but except in overt works on natural theology it is seldom appealed to with such frequency or made so essential a part of an exposition as in *The Physical Geography*. Maury shows his acquaintance with some of these works by alluding in § 495 to William Whewell, "the learned author of the 'Plurality of Worlds,'" and by mentioning in § 498 Thomas Chalmers, author of the first of the *Bridgewater Treatises*. He even composes a passage in imitation of Chalmers. The more widely read works of natural theology consisted, of course, of little more than tautologies: every new discovery of a structure or a process in nature could be claimed as the identification of an additional marvel of divine ingenuity and foresight and made the occasion of pious reflections. Maury freely indulges in this kind of tautology. All coherence and balance, all of the smooth working of the manifold processes on earth were, according to him, carefully thought out and adjusted once for all by an infinitely competent cosmic watchmaker. Maury assumes a priori that the whole mechanism of the earth in its place in the solar system is in equilibrium; balance and mutual compensation of its component parts and processes do not need to be established inductively (e. g., see §§ 302–303, 465, 487).

It is perhaps a reflection of the general acceptance of the argument from design in Maury's time that critics of his book did not expressly object to his derivation of equilibrium and compensation in the atmosphere and the ocean from a hypothetical divine plan. Most of those who commented on *The Physical Geography* remarked, however, on its religious tone. Some nonscientific reviewers praised the book for this quality:

. . . we have not met for a long period with a book which is at once so minute

---

[36] In CV, 366.

and profound in research, so plain, manly, and eloquent in expression. At almost every page there are proofs that Lieutenant Maury is as pious as he is learned. He sees the Almighty everywhere and in all things.[37]

Few works of science receive laudatory reviews in the religious press, as *The Physical Geography* did at least once:

We have selected this volume for comment, not only as interesting in itself and honorable to American science, but because of its connection with religion . . . thankful for the evidence it affords of the scientific attainments of the officers, both in our Navy and commercial marine; and still more that in their noble daring, and ingenious research, they recognize with a devout spirit the proofs spread all around them of the wisdom and greatness of the Creator.[38]

Contemporary notices contain numerous comments on Maury's inordinate resort to biblical passages in support of his hypotheses. Perhaps the extreme example of such use — misuse in the eyes of his more severe critics — is his argument concerning the origin of the salt of the sea (§ 491). Sonorous citations from the King James Bible for literary effect are a convention so firmly established in English literature as scarcely to excite remark, but Maury's dependence on scripture for part of his scientific argument repeatedly evoked adverse comment. One rather amusing instance of such criticism was occasioned by his quotation about the Pleiades from the Book of Job, in § 216 of the eighth edition. In earlier versions Maury had cited the passage as "Canst thou tell [sic] the sweet influences of the Pleiades?" Sir Henry Holland reproved him for misquoting the Bible and remarked, with a quotation in the Greek of the Septuagint, that the passage is in any event disputed and hence unclear. He pointed out, too, that Maury's acceptance of the hypothesis that the pole of rotation of the sidereal system lies in the Pleiades is based on the conclusions of only one person, J. H. Maedler, and is not generally accepted by astronomers.[39] The translator of *The Physical Geography* into German also called attention to Maury's mistake, citing the passage from Job in six languages.[40] Maury corrected the quotation, but left the verb "tell" later in the paragraph, in a sentence where it has little relation to its context, "Who but the astronomer, then, could tell their 'sweet influences'? "

In more general terms, Holland wrote:

It is with reluctance that we advert to another characteristic of this volume:— we mean the very frequent and incautious reference to passages in Scripture; not solely for illustration, but even as authority for physical truths, or arguments

---

[37] *Littell's Living Age*, XLV, 657.

[38] S[tephen] G[reenleaf] B[ulfinch], "The Physical Geography of the Sea" [review], *The Christian Examiner*, LXI(1856), 11–18, esp., pp. 12, 18.

[39] *Edinburgh Review*, CV, 367.

[40] *Die physische Geographie des Meeres*, p. 59, n.

for hypotheses still unproved. Lieut. Maury is evidently a man of strong and sincere religious feelings, and we honour the earnestness with which he expresses them. But he unhappily does not see that in forcing Scripture to the interpretation of physical facts, he is mistaking the whole purport of the sacred Books, misappropriating their language, and discrediting their evidence on matters of deep concern by applying it to objects and cases of totally different nature.[41]

A little later Sir David Brewster expressed essentially the same sentiments:

It is now, we think, almost universally admitted, and certainly by men of the soundest faith . . . that the Bible was not intended to teach us the truths of science. The geologist has sought in vain for geological truth in the inspirations of Moses, and the astronomer has equally failed to discover in Scripture the facts and laws of his science. Our author, however, seems to think otherwise, and has taken the opposite side, in the unfortunate controversy which still rages between the divine and the philosopher.[42]

Maury's French fellow officer, Captain Bourgois, is more forthright than the learned Scots:

Scientific investigations can certainly have no nobler purpose than the discovery of the laws of harmony that the Creator has imposed upon matter; and the knowledge of these laws is well adapted to imbue the mind with a profound admiration for the infinite wisdom from which they emanate. But the author who seeks to awake in us so laudable a sentiment should carefully avoid offering as the object of this admiration some ephemeral system born of his own imagination in place of the immutable laws of Providence.

Maury does not seem to have avoided this dangerous reef. He may even be reproached with having too often used sacred texts to support his theoretical concepts for lack of precise facts and rigorous arguments.[43]

Maury provides some indications of the way in which he viewed his own work. In the several editions of *Sailing Directions* he enumerates the general results obtained at the National Observatory from the compilation and study of observations of wind and currents made at sea. He divides these results into two groups. The first group, consisting of 15 items, contains empirical statements concerning winds, temperature and density of the surface water of the sea, the occurrence of dry and rainy seasons, and the distribution of whales. These are rather immediate generalizations of the data of observation. The second group, "probable results, which these investigations encourage us to anticipate, or induce us to inquire for," includes the broad generalizations about the circulation of the atmosphere and the sources of water vapor precipitated on the lands embodied in *The Physical Geography* and received by contemporary scientists with the skepticism illustrated in the passages already cited.

[41] *Edinburgh Review*, CV, 366.
[42] *North British Review*, XXVIII, 434–435.
[43] *Réfutation du système* . . . p. 6 (my translation).

In *Sailing Directions* Maury presents the statements of the second group quite tentatively:

I do not wish to be understood as claiming this catalogue of phenomena as actual results already derived from the investigation of log-books; nor do I intend, by this enumeration of them, to commit myself with regard to them, farther than I have done in the body of this work.[44]

No such disclaimer appears in *The Physical Geography*. In the introduction to the (sixth) revised edition of 1856 Maury does try to disarm his critics: ". . . in accounting for the various phenomena that present themselves, I am wedded to no theories, and do not advocate the doctrines of any particular school." The reader of the book finds, however, little exemplification of such restraint.

Maury's division of his conclusions into two groups, positive results and hypothetical generalizations, finds approximate expression in his introduction to the first edition of *The Physical Geography*, 1855, in the distinction he draws between "what has been accomplished by these researches in the way of shortening passages and lessening the dangers of the sea" and "a good of higher value . . . yet to come out of the moral, the educational influence which they are calculated to exert upon the seafaring community of the world." The words he chooses to designate these accomplishments leave no doubt about the relative weight Maury attached to them. He seems to be more eager to enforce the homily on the goodness and wisdom of the Creator preached throughout the book than with achieving objective scientific results. In the judgment of his contemporaries, and especially in retrospect, the distinction Maury makes is between the *Wind and Current Charts* and those parts of *Sailing Directions* directly applicable to navigation on the one hand, and a large fraction of the content of *The Physical Geography* on the other. A generation after the appearance of the last edition of *The Physical Geography* these two aspects of Maury's work were clearly distinguished. Thus, an American meteorologist could write: "Maury showed his strength by collecting and mapping the normal winds of the ocean; but shows his weakness in speculating on a philosophy of their origin." [45] In the same spirit, two judicious European scientists wrote of "the celebrated navigator who contributed so much to both science and navigation," but also of "his theoretical investigations, which have always left many doubts in the minds of meteorologists." [46]

Sharper words than those of Waldo or of Hildebrandsson and Teis-

[44] *Sailing Directions*, 6th ed. (Philadelphia, 1854), pp. 102–104.
[45] Frank Waldo, "Some Results on Theoretical Meteorology in the United States," *Report of the International Meteorological Congress held at Chicago, Ill., August 21– 24, 1893* . . . (U. S. Weather Bureau, Bull. no. 11, 1894), pp. 317–325, esp., p. 319.
[46] H. Hildebrand Hildebrandsson and Léon Teisserenc de Bort, *Les bases de la météorologie dynamique* . . . I(Paris, 1902), 14–15 (my translation).

serenc de Bort were written at the turn of the century by the French physicist Marcel Brillouin.[47] These words are, however, carefully weighed, and may serve as something approaching a final judgment of Maury's place in science. By 1900 Brillouin needed to do no more than cite Maury on the circulation of the atmosphere in order to refute him. He reminded his readers, however, of the extent to which Maury's notions had been accepted and speculated on the reasons for their acceptance:

> On every occasion his religious optimism breaks forth in lyrical paragraphs of a rather naive inspiration but of fine literary form. In addition to his qualities as a writer, Maury had a practical, if not a scientific, mind. Moreover, from the charts in which the observations collected under his direction were summarized he derived rules of navigation under sail that shortened some passages by half. This success, a direct result of the observations collected and having no relation whatever to Maury's views concerning the circulation of the atmosphere, must have created the illusion that the latter were valid; for they have been disseminated everywhere, in textbooks of geography and in all books written for popular consumption . . .
>
> Maury's own ideas are pure fantasy, but everyone cites them; little by little his assertions have been introduced everywhere, and repetitions of these assertions are presented as rational arguments.

Yet Brillouin wonders whether everyone who has speculated on the circulation of the atmosphere has not been influenced by Maury, "perhaps without his combination of literary talent and practical mind." We can do no better than accept Brillouin's diagnosis of the reasons why *The Physical Geography* enjoyed so favorable a reception and contributed so much to popular interpretations of the phenomena of the sea and the atmosphere: "literary talent" for the immediate impact of Maury's writings on the reader, and the products of his "practical mind" as recommendation to the unwary — recommendation even of Maury's "pure fantasy." The scientific world, however, completely reversed Maury's estimate of the relative value of his empirical and his theoretical contributions.

By referring to large generalizations more than to details I have perhaps overemphasized the scientific weaknesses of *The Physical Geography*. There is in fact a great deal of sound information in the book, but the reader now, as a hundred years ago, must weigh each paragraph and sort out its content of objective fact, of material selected to support Maury's interpretations, and of fantasy. The most valuable parts are those in which he describes investigations done under his supervision, such as his account of deep-sea soundings in Chapter XIII. We can not attach importance to all the "results" Maury reports from his work. Some are false, some are answers to the wrong questions. But his book contains two unquestioned "firsts": the map of the temperature of the surface water of

---

[47] *Mémoires originaux sur la circulation générale de l'atmosphère* (Paris, 1900), pp. 12–13, 17 (my translation).

the Atlantic (Figure 10), and the map of the relief of the bottom of the
same ocean (Plate VII, first published as Plate XIV of the 6th edition of
Sailing Directions, 1854). Both of these are compilations of observations
collected at Maury's observatory in Washington and in large part made at
his instigation. The information in Chapter XIV on the sediments of the
deep sea, based on samples brought up by Brooke's sounding apparatus,
also represents an important contribution to knowledge. As is to be expected, Maury is at his best when he sticks to his last. His apparently
irrepressible urge to speculate led him, undisciplined as he was in rigorous
thinking, into untenable hypotheses on his all-too-frequent departures
from the solid ground of fact. But occasionally his active imagination
directed him toward valuable perceptions. The last chapter in the eighth
edition of *The Physical Geography*, for example, "The Actinometry of the
Sea," contains, in addition to worthless speculation, some ideas on the
heat economy of the ocean that anticipate interpretations accepted today.
I refer in particular to Maury's comment on the "office" of the waves.
Here he was on the threshold of modern ideas concerning the role of
turbulence set up by the wind in the surface water of the ocean in distributing vertically the heat derived from absorption of solar radiation.

The *Physical Geography* went out of print at about the time when the
first results of the *Challenger* expedition of 1873–1876 were appearing.
That expedition initiated a new era in methods by which the sea is
investigated and put into both scientific and popular literature a flood of
fresh observations. Among its minor results it gave currency in English
to the term "oceanography" as a designation for what Humboldt and
Maury had called "the physical geography of the sea." The obsolescence
of Maury's title is, however, much less significant than the overshadowing
of all earlier literature about the sea by the immense mass of new facts the
Challenger expedition provided. We read *The Physical Geography* today
not for Maury's insights into the mechanisms of the ocean and the atmosphere, which were dim in the light of his own day. We scarcely read it
as a datum from which to measure a century's advance in understanding.
Rather, we read it as the highly personal testament of an energetic and
self-assertive man unacquainted with the rigorous methods practiced in
the academies, but possessing first-hand knowledge of ships and the sea,
boundless self-confidence, and a pen well exercised in persuasive writing.

Maury wrote at a time when sailing ships were in the full glory of
technical adaptation to their purpose, when on the high seas they were
still unchallenged by steam. Especially were the exploits of the American
clipper ships, aided in their navigation by Maury's charting of the best
sailing routes, a source of justifiable pride among his countrymen. The
popularity of *The Physical Geography* abroad is evidence that the inhabitants of other countries shared the interests of their American con-

temporaries. Plausible reasons why a readable book about the sea found an appreciative audience a century ago are easy to formulate. But he would be bold indeed who would try to explain the distinctive qualities that Maury gave to his writings, or who would pretend to understand how a man of high integrity might be beguiled into mistaking "ephemeral system[s] born of his own imagination" for "the immutable laws of Providence."

John Leighly

University of California,
Berkeley

A NOTE ON THE TEXT

In preparing *The Physical Geography of the Sea, and its Meteorology* for this edition I have used the eighth edition, published in 1861, the development of which from earlier versions I have explained in the Introduction. I have left untouched the text organization that Maury gave it, except for placing the "Explanation of the Plates" at the end of the chapters instead of immediately after the table of contents. The reader will note, for example, that the numbering of sections is usually not continuous from chapter to chapter; there is in most instances a gap in section numbering between chapters, which does not signify that anything has been left out. Having discussed the content of Maury's introductions to the sixth and eighth editions in my Introduction, I have left in place only his introduction to the first edition.

I have supplied numbers to tables in the text to facilitate reference to them; I have also rearranged the illustrations, giving numbers to Maury's unnumbered text figures, and changing the designations of some illustrations from plates to text figures. This rearrangement has involved a renumbering of the plates at the end of the book, and corresponding changes in references in the text to some of the illustrations. Using as far as possible words from Maury's text, I have supplied in brackets legends to all of the figures, and titles to those tables which lacked them. Minor stylistic adjustments have made headings in the book more consistent for today's reading.

In the last sentence of § 248 I have supplied my estimate of something accidentally left out in printing the passage. In § 492 I have omitted nearly a page of the original text, since it is repeated, with insignificant changes in wording, from §§ 161 and 162. In §§ 855 and 856 I have likewise omitted two tables of the mean height of the barometer in different latitudes and longitudes, repeated from §§ 362 and 670.

In footnotes containing Maury's references to literature I have completed as many of his citations as I could and have given them a consistent form. My additions are in square brackets. In a few instances, because Maury does not give enough information or because I have not had access to the works referred to, I have had to leave references as he gave them. I have also silently corrected such errors in dates and in the spelling of names of persons as I have detected.

J. L.

THE PHYSICAL
GEOGRAPHY OF THE SEA
and its Meteorology

[William Cornelius Hasbrouck (1800–1870) had just graduated
from Union College at the time he was one of Maury's teachers at
Harpeth Academy. He soon returned North, and became a prominent
lawyer in Newburgh. He and Maury remained close friends until
Hasbrouck's death; Maury entrusted to Hasbrouck the management
of his business interests in the North during the Civil War.]

INTRODUCTION TO THE FIRST EDITION – 1855

The primary object of "The Wind and Current Charts," out of which has grown this Treatise on the Physical Geography of the Sea, was to collect the experience of every navigator as to the winds and currents of the ocean, to discuss his observations upon them, and then to present the world with the results on charts for the improvement of commerce and navigation.

By putting down on a chart the tracks of many vessels on the same voyage, but at different times, in different years, and during all seasons, and by projecting along each track the winds and currents daily encountered, it was plain that navigators hereafter, by consulting this chart, would have for their guide the results of the combined experience of all whose tracks were thus pointed out.

Perhaps it might be the first voyage of a young navigator to the given port, when his own personal experience of the winds to be expected, the currents to be encountered by the way, would itself be blank. If so, there would be the wind and current chart. It would spread out before him the tracks of a thousand vessels that had preceded him on the same voyage, wherever it might be, and that, too, at the same season of the year. Such a chart, it was held, would show him not only the tracks of the vessels, but the experience also of each master as to the winds and currents by the way, the temperature of the ocean, and the variation of the needle. All this could be taken in at a glance, and thus the young mariner, instead of groping his way along until the lights of experience should come to him by the slow teachings of the dearest of all schools, would here find, at once, that he had already the experience of a thousand navigators to guide him on his voyage. He might, therefore, set out upon his first voyage with as much confidence in his knowledge as to the winds and currents he might expect to meet with, as though he himself had already been that way a thousand times before.

Such a chart could not fail to commend itself to intelligent shipmasters, and such a chart was constructed for them. They took it to sea, they tried it, and to their surprise and delight they found that, with the knowledge it afforded, the remote corners of the earth were brought closer together, in some instances, by many days' sail. The passage hence to the equator alone was shortened ten days. Before the commencement of this undertaking, the average passage to California was 183 days; but with these charts for their guide, navigators have reduced that average, and brought it down to 135 days.

Between England and Australia, the average time going, without these charts, is ascertained to be 124 days, and coming, about the same; making the round voyage one of about 250 days on the average.

These charts, and the system of research to which they have given rise, bid fair to bring that colony and the mother country nearer by many days, reducing, in no small measure, the average duration of the round voyage.[1]

At the meeting of the British Association of 1853, it was stated by a distinguished member — and the statement was again repeated at its meeting in 1854 — that in Bombay, whence he came, it was estimated that this system of research, if extended to the Indian Ocean, and embodied in a set of charts for that sea, such as I have been describing, would produce an annual saving to British commerce, in those waters alone, of one or two millions of dollars; [2] and in all seas, of ten millions.[3]

A system of philosophical research which is so rich with fruits and abundant with promise could not fail to attract the attention and commend itself to the consideration of the seafaring community of the whole civilized world. It was founded on observation; it was the result of the experience of many observant men, now brought together for the first time and patiently discussed. The results tended to increase human knowledge with regard to the sea and its wonders, and therefore the system of research could not be wanting in attractions to right-minded men.

---

[1] The outward passage, it has since been ascertained, has been reduced to 97 days on the average, and the homeward passage has been made in 63.

[2] The Earl of Harrowby, "Address [of the President]," *British Association* [*for the Advancement of Science, Report of*] *24th Meeting, Liverpool, 1854* [(London, 1855) pp. lv–lxxii, esp., pp. lx, lxii].

[3] . . . "Now let us make a calculation of the annual saving to the commerce of the United States effected by those charts and sailing directions. According to Mr. Maury, the average freight from the United States to Rio Janeiro is 17.7 cts. per ton per day; to Australia, 20 cts.; to California, also, about 20 cts. The mean of this is a little over 19 cts. per ton per day; but to be within the mark, we will take it at 15, and include all the ports of South America, China, and the East Indies.

"The sailing directions have shortened the passages to California 30 days, to Australia 20, to Rio Janeiro 10. The mean of this is 20, but we will take it at 15, and also include the above-named ports of South America, China, and the East Indies.

"We estimate the tonnage of the United States engaged in trade with these places at 1,000,000 tons per annum.

"With these data, we see that there has been effected a saving for each one of these tons of 15 cents per day for a period of 15 days, which will give an aggregate of $2,250,000 saved per annum. This is on the outward voyage alone, and the tonnage trading with all other parts of the world is also left out of the calculation. Take these into consideration, and also the fact that there is a vast amount of foreign tonnage trading between these places and the United States, and it will be seen that the annual sum saved will swell to an enormous amount." — *Hunt's Merchant's Magazine*, [XXX] (1854) [531–547, esp., pp. 546–547].

The results of the first chart, however, though meagre and unsatisfactory, were brought to the notice of navigators; their attention was called to the blank spaces, and the importance of more and better observations than the old sea-logs generally contained was urged upon them.

They were told that if each one would agree to co-operate in a general plan of observations at sea, and would send regularly, at the end of every cruise, an abstract log of their voyage to the National Observatory at Washington, he should, for so doing, be furnished, free of cost, with a copy of the charts and sailing directions that might be founded upon those observations.

The quick, practical mind of the American ship-master took hold of the proposition at once. To him the field was inviting, for he saw in it the promise of a rich harvest and of many useful results.

So, in a little while, there were more than a thousand navigators engaged day and night, and in all parts of the ocean, in making and recording observations according to a uniform plan, and in furthering this attempt to increase our knowledge as to the winds and currents of the sea, and other phenomena that relate to its safe navigation and physical geography.

To enlist the service of such a large corps of observers, and to have the attention of so many clever and observant men directed to the same subject, was a great point gained: it was a giant stride in the advancement of knowledge, and a great step toward its spread upon the waters.

Important results soon followed, and great discoveries were made. These attracted the attention of the commercial world, and did not escape the notice of philosophers every where.

The field was immense, the harvest was plenteous, and there was both need and room for more laborers. Whatever the reapers should gather, or the merest gleaner collect, was to inure to the benefit of commerce and navigation — the increase of knowledge — the good of all.

Therefore, all who use the sea were equally interested in the undertaking. The government of the United States, so considering the matter, proposed a uniform system of observations at sea, and invited all the maritime states of Christendom to a conference upon the subject.

This conference, consisting of representatives from France, England, and Russia, from Sweden and Norway, Holland, Denmark, Belgium, Portugal, and the United States, met in Brussels, August 23, 1853, and recommended a plan of observations which should be followed on board the vessels of all friendly nations, and especially of those there present in the persons of their representatives.

Prussia, Spain, Sardinia, the Holy See, the free city of Hamburg, the republics of Bremen and Chili, and the empires of Austria and Brazil, have since offered their co-operation also in the same plan.

Thus the sea has been brought regularly within the domains of philosophical research, and crowded with observers.

In peace and in war these observations are to be carried on, and, in case any of the vessels on board of which they are conducted may be captured, the abstract log — as the journal which contains these observations is called — is to be held sacred.

Baron Humboldt is of opinion that the results already obtained from this system of research are sufficient to give rise to a new department of science, which he has called the PHYSICAL GEOGRAPHY OF THE SEA. If so much have already been accomplished by one nation, what may we not expect in the course of a few years from the joint co-operation of so many?

Rarely before has there been such a sublime spectacle presented to the scientific world: all nations agreeing to unite and co-operate in carrying out one system of philosophical research with regard to the sea. Though they may be enemies in all else, here they are to be friends. Every ship that navigates the high seas with these charts and blank abstract logs on board may henceforth be regarded as a floating observatory, a temple of science. The instruments used by every co-operating vessel are to be compared with standards that are common to all; so that an observation that is made any where and in any ship may be referred to and compared with all similar observations by all other ships in all parts of the world.

But these meteorological observations which this extensive and admirable system includes will relate only to the sea. This is not enough. The plan should include the land also, and be universal. Other great interests of society are to be benefited by such extension no less than commerce and navigation have been. A series of systematic observations, directed over large districts of country, nay, over continents, to the improvement of agricultural and sanitary meteorology, would, I have no doubt, tend to a development of many interesting, important, and valuable results.

With proper encouragement, this plan of research is capable of great expansion. With the aid of the magnetic telegraph, and by simply a properly devised system of daily reports through it, sentinels upon the weather may be so posted that we may have warning in advance of every storm that traverses the country.

The agricultural societies of many states of the Union have addressed memorials to the American Congress, asking for such extension; and it is hoped that that enlightened body will not fail favorably to respond.

This plan contemplates the co-operation of all the states of Christendom, at least so far as the form, method, subjects of observations, time of making them, and the interchange of results are concerned. I hope that my fellow-citizens will not fail to second and co-operate in such a humane, wise, and noble scheme. The Secretary of the Navy, taking the enlarged and enlightened views which do honor to great statesmen, has officially recommended the adoption of such a system, and the President has asked the favorable consideration thereof by Congress. These researches for the land look not only to the advancement of the great interests of sanitary and agricultural meteorology, but they involve also a study of the laws which regulate the atmosphere, and a careful investigation of all its phenomena.

Another beautiful feature in this system is, that it costs nothing additional. The instruments that these observations at sea call for are such as are already in use on board of every well-conditioned ship, and the observations that are required are precisely those which are necessary for her safe and proper navigation.

As great as is the value attached to what has been accomplished by these researches in the way of shortening passages and lessening the dangers of the sea, a good of higher value is, in the opinion of many seamen, yet to come out of the moral, the educational influence which they are calculated to exert upon the seafaring community of the world. A very clever English ship-master, speaking recently of the advantages of educational influences among those who intend to follow the sea, remarks:

To the cultivated lad there is a new world spread out when he enters on his first voyage. As his education has fitted, so will he perceive, year by year, that his profession makes him acquainted with things new and instructive. His intelligence will enable him to appreciate the contrasts of each country in its general aspect, manners, and productions, and in modes of navigation adapted to the character of coast, climate, and rivers. He will dwell with interest on the phases of the ocean, the storm, the calm, and the breeze, and will look for traces of the laws which regulate them. All this will induce a serious earnestness in his work, and teach him to view lightly those irksome and often offensive duties incident to the beginner.[4]

And that these researches do have such an effect many noble-hearted mariners have testified. Captain Phinney, of the American ship Gertrude, writing from Callao, January, 1855, thus expresses himself:

"Having to proceed from this to the Chincha Islands and remain three months, I avail myself of the present opportunity to forward to you abstracts of my two passages over your southern routes, although not required to do so until my own return to the United States next summer; knowing that you are less amply supplied with abstracts of voyages over these regions than of many other parts of the ocean, and, such as it is, I am happy to contribute my mite toward furnishing you with material to

[4] Robert Methuen, *The Log of a Merchant Officer* (London, 1854).

work out still farther toward perfection your great and glorious task, not only of pointing out the most speedy routes for ships to follow over the ocean, but also of teaching us sailors to look about us, and see by what wonderful manifestations of the wisdom and goodness of the great God we are continually surrounded.

"For myself, I am free to confess that for many years I commanded a ship, and, although never insensible to the beauties of nature upon the sea or land, I yet feel that, until I took up your work, I had been traversing the ocean blindfolded. I did not think; I did not know the amazing and beautiful combination of all the works of Him whom you so beautifully term 'the Great First Thought.'

"I feel that, aside from any pecuniary profit to myself from your labors, you have done me good as a man. You have taught me to look above, around, and beneath me, and recognize God's hand in every element by which I am surrounded. I am grateful for this personal benefit. Your remarks on this subject, so frequently made in your work, cause in me feelings of the greatest admiration, although my capacity to comprehend your beautiful theory is very limited.

"The man of such sentiments as you express will not be displeased with, or, at least, will know how to excuse, so much of what (in a letter of this kind) might be termed irrelevant matter. I have therefore spoken as I feel, and with sentiments of the greatest respect." Sentiments like these can not fail to meet with a hearty response from all good men, whether ashore or afloat.

Never before has such a corps of observers been enlisted in the cause of any department of physical science as is that which is now about to be engaged in advancing our knowledge of the Physical Geography of the Sea, and never before have men felt such an interest with regard to this knowledge.

Under this term will be included a philosophical account of the winds and currents of the sea; of the circulation of the atmosphere and ocean; of the temperature and depth of the sea; of the wonders that lie hidden in its depths; and of the phenomena that display themselves at its surface. In short, I shall treat of the economy of the sea and its adaptations — of its salts, its waters, its climates, and its inhabitants, and of whatever there may be of general interest in its commercial uses or industrial pursuits, for all such things pertain to its PHYSICAL GEOGRAPHY.

The object of this little book, moreover, is to show the present state, and, from time to time, the progress of this new and beautiful system of research, as well as of this interesting department of science; and the aim of the author is to present the gleanings from this new field in a manner that may be interesting and instructive to all, whether old or young, ashore or afloat, who desire a closer look into "the wonders of the great deep," or a better knowledge as to its winds, its adaptations, or its Physical Geography.[5]

[5] There is an old and very rare book which treats upon some of the subjects to which this little work relates. It is by Count L. F. MARSIGLI, an Italian, and is called NATURAL DESCRIPTION OF THE SEAS. The copy to which I refer was translated into Dutch by Boerhaave in 1786. [Louis Ferdinand, comte de Marsilli, *Histoire physique de la mer* (Amsterdam, 1725); *Natuurkundige beschrijving der zeeën* ('s Gravenhage, 1786)]. The learned count made his observations along the coast of Provence and Languedoc. The description only relates to that part of the Mediterranean. The book is divided into four chapters: the first, on the bottom and shape of the sea; the second, of sea water; the third, on the movements of sea water; and the fourth, of sea plants.

He divides sea water into surface and deep-sea water; because, when he makes salt from surface water (not more than half a foot below the upper strata), this salt will give a red color to blue paper; whereas the salt from deep-sea water will not alter the colors at all. The blue paper can only change its color by the action of an acid. The reason why this acid (iodine?) is found in surface and not in deep-sea water is, it is derived from the air; but he supposes that the saltpetre that is found in sea water, by the action of the sun's rays and the motion of the waves, is deprived of its coarse parts, and, by evaporation, embodied in the air, to be conveyed to beasts or plants for their existence, or deposited upon the earth's crust, as it occurs on the plains of Hungary, where the earth absorbs so much of this saltpetre vapor.

Donati, also, was a valuable laborer in this field. His inquiries enabled Mr. Trembley to conclude that there are, "at the bottom of the water, mountains, plains, valleys, and caverns, just as upon the land." [Vitaliano Donati, *Della storia naturale marina dell' Adriatico* (Venice, 1750); *Essai sur l'histoire naturelle de la mer Adriatique* (The Hague, 1758). Abraham Trembley, "An Account of a Work published in Italian by Vitaliano Donati, M. D. containing, An Essay towards a natural History of the Adriatic Sea," *Phil. Trans.*, XLIX(1756), 281–284, esp., p. 282.]

But by far the most interesting and valuable book touching the physical geography of the Mediterranean is Admiral Smyth's last work: William Henry Smyth, *The Mediterranean; a Memoir, Physical, Historical, and Nautical* (London, 1854).

CONTENTS

CHAPTER III

INFLUENCE OF THE GULF STREAM UPON CLIMATES AND COMMERCE

CHAPTER IV

THE ATMOSPHERE

CHAPTER V

RAINS AND RIVERS

CHAPTER VI

RED FOGS AND SEA BREEZES

CHAPTER VII

THE EASTING OF THE TRADE-WINDS, THE CROSSING AT THE CALM BELTS, AND THE MAGNETISM OF THE ATMOSPHERE

CHAPTER VIII

CURRENTS OF THE SEA

CHAPTER IX

THE SPECIFIC GRAVITY OF THE SEA, AND THE OPEN WATER IN THE ARCTIC OCEAN

CHAPTER X

THE SALTS OF THE SEA

CHAPTER XI

THE CLOUD REGION, THE EQUATORIAL CLOUD-RING, AND SEA FOGS

CHAPTER XII

THE GEOLOGICAL AGENCY OF THE WINDS

CHAPTER XIII

THE DEPTHS OF THE OCEAN

CHAPTER XIV

THE BASIN AND BED OF THE ATLANTIC

CHAPTER XV

SEA ROUTES, CALM BELTS, AND VARIABLE WINDS

CHAPTER XVI

MONSOONS

CHAPTER XVII

THE CLIMATES OF THE SEA

CHAPTER XVIII

TIDE-RIPS AND SEA DRIFT

CHAPTER XIX

STORMS, HURRICANES, AND TYPHOONS

CHAPTER XX

THE WINDS OF THE SOUTHERN HEMISPHERE

CHAPTER XXI

THE ANTARCTIC REGIONS AND THEIR CLIMATOLOGY

CHAPTER XXII

THE ACTINOMETRY OF THE SEA

CHAPTER I

§ 1–55. THE SEA AND THE ATMOSPHERE

§1. *The two oceans of air and water.* Our planet is invested with two great oceans; one visible, the other invisible; one underfoot, the other overhead; one entirely envelops it, the other covers about two thirds of its surface. All the water of the one weighs about 400 times as much as all the air of the other.

2. *Their meeting.* It is at the bottom of this lighter ocean where the forces which we are about to study are brought into play; to treat of the Physical Geography of the Sea, we must necessarily refer to the phenomena which are displayed at the meeting of these two oceans. Let us, therefore, before entering either of these fields for study, proceed first to consider each one in some of its most striking characteristics. They are both in a state of what is called unstable equilibrium; hence the currents of one and the winds of the other.

3. *Their depth.* As to their depth, we know very little more of the one than of the other; but the conjecture that the average depth of the sea does not much exceed four miles is probably as near the truth as is the commonly received opinion that the height of the atmosphere does not exceed 50 miles. If the air were, like water, non-elastic, and not more compressible than this non-elastic fluid, we could sound out the atmospherical ocean with the barometer and gauge it by its pressure. The mean height of the barometer in the torrid and temperate zones, at the level of the sea, is about 30 inches — rather under than over. Now, it has been ascertained that, if we place a barometer 87 feet above the level of the sea, its average height will be reduced from 30.00 in. to 29.90 in.; that is, it will be diminished one tenth of an inch, or the three hundredth part of the whole; consequently, by going up $300 \times 87 (= 26{,}100)$ feet, the barometer, were the air non-elastic, would stand at 0. It would then be at the top of the atmosphere. The height of 26,100 feet is just five miles lacking 300 feet.

4. *Weight of the atmosphere.* But the air is elastic, and very unlike water. That at the bottom is weighed down by the superincumbent air with the force of about 15 pounds to the square inch, while that at the top is inconceivably light. If we imagine the lightest down, in layers of equal weight and ten feet thick, to be carded into a pit several miles deep, we can readily imagine that the bottom layer, though it might

have been ten feet thick when it first fell, yet with the weight of the accumulated and superincumbent mass, it might now, the pit being full, be compressed into a layer of only a few inches in thickness, while the top layer of all, being uncompressed, would be exceedingly light, and still ten feet thick; so that a person ascending from the bottom of the pit would find the layers of equal weight thicker and thicker until he reached the top. So it is with the barometer and the atmosphere; when it is carried up in the air through several strata of 87 feet, the observer does not find that it falls a tenth of an inch for every successive 87 feet upward through which he may carry it. To get it to fall a tenth of an inch, he must carry it higher and higher for every successive layer.

5. *Three fourths below the mountain-tops.* More than three fourths of the entire atmosphere is below the level of the highest mountains; the other fourth is rarefied and expanded in consequence of the diminished pressure, until the height of many miles be attained. From the reflection of the sun's rays after he has set, or before he rises above the horizon, it is calculated that this upper fourth part must extend at least forty or forty-five miles higher.

6. *Its height.* At the height of 26,000 miles from the earth, the centrifugal force would counteract gravity; consequently, all ponderable matter that the earth carries with it in its diurnal revolution must be within that distance, and consequently the atmosphere can not extend beyond that. This limit, however, has been greatly reduced, for Sir John Herschel has shown, by balloon observations,[1] that at the height of 80 or 90 miles there is a vacuum far more complete than any which we can produce by any air-pump. In 1783 a large meteor, computed to be half a mile in diameter and fifty miles from the earth, was heard to explode. As sound can not travel through vacuum, it was inferred that the explosion took place within the limits of the atmosphere. Herschel concludes that the aerial ocean is at least 50 miles deep.

7. *Data conjectural.* The data from which we deduce our estimate, both as to the mean height of the atmosphere and average depth of the ocean, are, to some extent, conjectural; consequently, the estimates themselves must be regarded as approximations, but sufficiently close, nevertheless, for the present purposes of this work.

8. *Analysis of air.* Chemists who have made the analysis tell us that, out of 100 parts of atmospheric air, 99.5 consist of oxygen and nitrogen, mixed in the proportion of 21 of oxygen to 79 of nitrogen by volume, and of 23 to 77 by weight. The remaining *half of a part* consists of .05 of carbonic acid and .45 of aqueous vapor.

9. *Information respecting the depth of the ocean.* The average depth of the ocean has been variously computed by astronomers, from such

[1] Those of Mr. Welsh, in his ascent from Kew.

arguments as the science affords, to be from 26 to 11 miles. About ten years ago I was permitted to organize and set on foot in the navy a plan for "sounding out" the ocean with the plummet.[2] Other navies, especially the English, have done not a little in furtherance of that object. Suffice it to say that, within this brief period, though the undertaking has been by no means completed — no, not even to the tenth part — yet more knowledge has been gained concerning the depths and bottom of the *deep* sea, than all the world had before acquired in all previous time.

10. *Its probable depth.* The system of deep-sea soundings thus inaugurated does not thus far authorize the conclusion that the average depth of ocean water is more than three or four miles (§ 3), nor have any reliable soundings yet been made in water over five miles deep.

11. *Relation between the depth and the waves of the sea.* In very shallow pools, where the water is not more than a few inches deep, the ripples or waves, as all of us, when children, have observed, are small; their motion, also, is slow. But when the water is deep, the waves are larger and more rapid in their progress, thus indicating the existence of a numerical relation between their breadth and their velocity, and the depth of the water. It may be inferred, therefore, that if we knew the size and velocity of certain waves, we could compute the depth of the ocean.

12. *Airy's tables.* Such a computation has been made, and we have the authority of Mr. Airy,[3] the Astronomer Royal of England, that waves of given breadths will travel in water of certain depths with the velocities as per Table 1.

13. *The earthquake of Simoda.* Accident has afforded us an opportunity of giving a *quasi* practical application to Mr. Airy's formulæ. On the 23d of December, 1854, at 9:45 A.M.,[4] the first shocks of an earthquake were felt on board the Russian frigate "Diana," as she lay at anchor in the harbor of Simoda, not far from Jeddo, in Japan. In fifteen minutes afterward (10 o'clock), a large wave was observed rolling into the harbor, and the water on the beach to be rapidly rising. The town, as seen from the frigate, appeared to be sinking. This wave was followed by

---

[2] "*And be it further enacted,* That the Secretary of the Navy be directed to detail three suitable vessels of the navy in testing new routes and perfecting the discoveries made by Lieut. Maury in the course of his investigations of the winds and currents of the ocean; and to cause the vessels of the navy to co-operate in procuring materials for such investigations, in so far as said co-operation may not be incompatible with the public interests." — From *Naval Appropriation Bill,* approved March 3, 1849.

[3] [George Biddell] Airy ["Tides and Waves"], *Encyclop. Metropol.* [V(London, 1845), 241–396; table on p. 291. The form in which Maury cites the table indicates that he took it from one in which the units are recomputed, as given in Sir John Herschel, "Physical Geography," *Encyclop. Britan.,* 8th ed., XVII(Boston, 1859), 569–647, esp., p. 582].

[4] "Notes of a Russian Officer," *Nautical Magazine,* XXV (1856), 97.

Table 1

[Relation of breadth and velocity of surface waves to depth of water]

Depth of the Water in Feet	Breadth of the Wave in Feet							
	1	10	100	1000	10,000	100,000	1,000,000	10,000,000
	Corresponding Velocity of Wave per Hour in Miles							
1	1.54	3.81	3.86	3.86	3.86	3.86	3.86	3.86
10	1.54	4.87	11.51	12.21	12.22	12.22	12.22	12.22
100	1.54	4.87	15.18	36.40	38.64	38.66	38.66	38.66
1,000	1.54	4.87	15.18	48.77	115.11	122.18	122.27	122.27
10,000	1.54	4.87	15.18	48.77	154.25	364.02	386.40	386.66
100,000	1.54	4.87	15.18	48.77	154.25	487.79	1151.11	1222.70

another, and when the two receded — which was at 10h. 15m. — there was not a house, save an unfinished temple, left standing in the village. These waves continued to come and go until 2.30 P.M., during which time the frigate was thrown on her beam ends five times. A piece of her keel 81 feet long was torn off, holes were knocked in her by striking on the bottom, and she was reduced to a wreck. In the course of five minutes the water in the harbor fell, it is said, from 23 to 3 feet, and the anchors of the ship were laid bare. There was a great loss of life; many houses were washed into the sea, and many junks carried up — one two miles — and dashed to pieces on the shore. The day was beautifully fine, and no warning was given of the approaching convulsion; the barometer standing at 29.87 in., thermometer 58°; the sea perfectly smooth when its surface was broken by the first wave. It was calm in the morning, and the wind continued light all day.

14. *The propagation of its waves.* In a few hours afterward, at San Francisco and San Diego, the tide-gauges showed that several well-marked and extraordinary waves had arrived off the coast of California.[5] The origin of these waves, and those which destroyed the town of Simoda, in Japan, and wrecked the Diana, was doubtless the same. But where was their birthplace? Supposing it to be near the coasts of Japan, we may, with the tide-gauge observations in California and Mr. Airy's formulæ, calculate the average depth of the sea along the path of the wave from Simoda to both San Francisco and San Diego.

15. *Their breadth and velocity.* Supposing the waves to have taken up their line of march from some point along the coast of Japan, the San Francisco wave, having a breadth of 256 miles, had a velocity of 438 miles

[5] ["Notice of Earthquake Waves on the Western Coast of the United States, on the 23d and 25th December, 1854," *Rept. of the Superintendent of the Coast Survey, 1855* (1856), pp. 342–346. The results Maury cites are also given in the article by Sir John Herschel cited above (n. 3), on p. 582.]

an hour; while the breadth of the San Diego wave was 221 miles, and its rate of travel 427 miles an hour.

16. *Average depth of the North Pacific.* Admitting these premises — which are partly assumed — to be correct, then, according to Airy's formulæ, the average depth of the North Pacific between Japan and California is, by the path of the San Francisco wave, 2149 fathoms, by the San Diego, 2034 (say 2½ miles).

17. *Specific gravity of sea-water.* At the temperature of 60°, the specific gravity of average sea-water [6] is 1.0272, and the weight of a cubic foot is 64.003 lbs.

18. *Of air.* With the barometer at 30 in. and the thermometer at 32°, the weight of a cubic foot of dry atmospheric air is 1.291 oz., and its specific gravity .00129. Such is the difference in weight between the two elements, the phenomena of which give the physical geography of the sea its charms as a study.

19. *Unequal distribution of light, etc.* There is in the northern hemisphere more land, less sea, more fresh water, and more atmospheric air, than there is in the southern.

20. *The sun longer in northern declination.* The sun dispenses light and heat unequally also between them. In his annual round he tarries a week (7¾ days) longer on the north than he does on the south side of the equator, and consequently the antarctic night and its winter are both longer, and should, if there were no interferences, be more intense than the polar winter and night of the arctic regions. The southern hemisphere is said also to be cooler, and this is true to a certain degree. In the summer of the southern hemisphere the sun is in perigee, and during the course of a diurnal revolution there the southern half of our planet receives more heat than the northern half during the same period of our summer. This difference, Sir John Herschel thinks, is compensated by the longer duration of the northern summer. Admitting this to be so — and it is susceptible of mathematical demonstration — there remains outstanding the longer winter for the southern hemisphere, which may radiate more heat than the northern hemisphere can do in its short winter. But the southern hemisphere is the cooler chiefly on account of the latent heat which is brought thence by vapor, and set free here by condensation.

21. *England about the pole of hemisphere with most land.* Within the torrid zone the land is nearly equally divided north and south of the equator, the proportion being as 5 to 4. In the temperate zones, however, the north with its land is thirteen times in excess of the south. Indeed, such is the inequality in the distribution of land over the surface of the globe that the world may be divided into hemispheres consisting, the one

---

[6] *Sailing Directions* [8th ed.], I[(1858), 259]. Sir John Herschel quotes it at 1.0275 for 62°.

with almost all the land in it, except Australia and a slip of America lying south of a line drawn from the Desert of Atacama to Uruguay; England is the centre of this hemisphere. The other, an aqueous hemisphere, contains all the great waters except the Atlantic Ocean; New Zealand is the nearest land to its centre.

22. *Effects of inequality in distribution of land and water.* This unequal distribution of land, air, and water is suggestive. To it we owe the different climates of the earth. Were it different, they would be different also; were it not for the winds, the vapors that rise from the sea would from the clouds be returned in showers back to the places in the sea whence they came; we should have neither green pastures, still waters, nor running brooks to beautify the landscape on an earth where no winds blow. Were there no currents in the sea, the seasons might change, but climates would be a simple affair, depending solely on the declination of the sun in the sky.

23. *Quantity of fresh water in American lakes.* About two thirds of all the fresh water on the surface of the earth is contained in the great American lakes; and though there be in the northern, as compared with the southern hemisphere, so much less sea surface to yield vapor, so much more land to swallow up rain, and so many more plants to drink it in, yet the fresh-water courses are far more numerous and copious on the north than they are on the south side of the equator.

24. *Southern seas the boiler, and northern lands the condenser.* These facts have suggested the comparison in which the southern hemisphere has been likened to the boiler and the northern to the condenser of the steam-engine. This vast amount of steam or vapor rising up in the extra-tropical regions of the south, expels the air thence, causing the barometer to show a much less weight of atmosphere on the polar side of 40° S. than we find in corresponding latitudes north.

25. *Offices of the atmosphere.* The offices of the atmosphere are many, marvelous, and various. Though many of them are past finding out, yet, beautiful to contemplate, they afford most instructive and profitable themes for meditation.

26. *Dr. Buist.* When this system of research touching the physics of the sea first began — when friends were timid and co-laborers few, the learned Dr. Buist stood up as its friend and champion in India; and by the services he thus rendered entitled himself to the gratitude of all who with me take delight in the results which have been obtained. The field which it was proposed to occupy — the firstlings of which were gathered in this little book [7] — was described by him in glowing terms, colored to

---

[7] [George Buist, *Trans. Bombay Geogr. Soc.,* IX(1850), lxxxi–lxxxv. Although Maury reproduces this material as continuous discourse, it is a selection of passages that make up only about half of Buist's text.]

nature. They are apropos, and it is a pleasure to repeat them, with such alterations only as experience shall approve:

27. *The sea and the atmosphere contrasted.* "The weight of the atmosphere is equal to that of a solid globe of lead sixty miles in diameter. Its principal elements are oxygen and nitrogen gases, with a vast quantity of water suspended in them in the shape of vapor, and commingled with these a quantity of carbon in the shape of fixed air, equal to restore from its mass many fold the coal that now exists in the world. In common with all substances, the ocean and the air are increased in bulk, and, consequently, diminished in weight, by heat; like all fluids, they are mobile, tending to extend themselves equally in all directions, and to fill up depressions wherever vacant space will admit them; hence, in these respects, the resemblance betwixt their movements. Water is not compressible or elastic, and it may be solidified into ice or vaporized into steam; the air is elastic; it may be condensed to any extent by pressure, or expanded to an indefinite degree of tenuity by pressure being removed from it; it is not liable to undergo any change in its constitution beyond these, by any of the ordinary influences by which it is affected.

28. *Influence of the sun.* "These facts are few and simple enough; let us see what results arise from them: As the constant exposure of the equatorial regions of the earth to the sun must necessarily there engender a vast amount of heat, and as his absence from the polar regions must in like manner promote an infinite accumulation of cold, to fit the entire earth for a habitation to similar races of beings, a constant interchange and communion betwixt the heat of the one and the cold of the other must be carried on. The ease and simplicity with which this is effected surpass all description. The air, heated near the equator by the overpowering influences of the sun, is expanded and lightened; it ascends into upper space, leaving a partial vacuum at the surface to be supplied from the regions adjoining. Two currents from the poles toward the equator are thus established at the surface, while the sublimated air, diffusing itself by its mobility, flows in the upper regions of space from the equator toward the poles. Two vast whirlpools are thus established, constantly carrying away the heat from the torrid toward the icy regions, and, there becoming cold by contact with the ice, they carry back their gelid freight to refresh the torrid zone.

29. *Of diurnal rotation.* "Did the earth, as was long believed, stand still while the sun circled around it, we should have had directly from north and south two sets of meridional currents blowing at the surface of the earth toward the equator; in the upper regions we should have had them flowing back again to the place whence they came. On the other hand, were the heating and cooling influences just referred to to cease, and the earth to fail in impressing its own motion on the atmos-

phere, we should have a furious hurricane rushing round the globe at the rate of 1000 miles an hour — tornadoes of ten times the speed of the most violent now known to us, sweeping every thing before them. A combination of the two influences, modified by the friction of the earth, which tends to draw the air after it, gives us the trade-winds, which, at the speed of from ten to twenty miles an hour, sweep round the equatorial region of the globe unceasingly.

30. *Currents.* "Impressed with the motion of the air, constantly sweeping its surface in one direction, and obeying the same laws of motion, the great sea itself would be excited into currents similar to those of the air, were it not walled in by continents and subjected to other control. As it is, there are constant currents flowing from the torrid toward the frigid zone to supply the vast amount of vapor there drained off, while other whirlpools and currents, such as the gigantic Gulf Stream, come to perform their part in the same stupendous drama. The waters of this vast ocean river are, to the north of the tropic, greatly warmer than those around; the climate of every country it approaches is improved by it, and the Laplander is enabled by its means to live and cultivate his barley in a latitude which, every where else throughout the world, is condemned to perpetual sterility. There are other laws which the great sea obeys which peculiarly adapt it as the vehicle of interchange of heat and cold betwixt those regions where either exists in excess.

31. *Icebergs.* "In obedience to these laws water warmer than ice attacks the basis and saps the foundations of the icebergs — themselves gigantic glaciers, which have fallen from the mountains into the sea, or which have grown to their present size in the shelter of bays and estuaries, and by accumulations from above. Once forced from their anchorage, the first storm that arises drifts them to sea, where the beautiful law which renders ice lighter than the warmest water enables it to float, and drifts southward a vast magazine of cold to cool the tepid water which bears it along — the evaporation at the equator causing a deficit, the melting and accumulation of the ice in the frigid zone giving rise to an excess of accumulation, which tends, along with the action of the air and other causes, to institute and maintain the transporting current. These stupendous masses, which have been seen at sea in the form of church spires, and gothic towers, and minarets, rising to the height of from 300 to 600 feet, and extending over an area of not less than six square miles, the masses above water being only one tenth of the whole, are often to be found within the tropics.

32. *Mountain ranges.* "But these, though among the most regular and magnificent, are but a small number of the contrivances by which the vast and beneficent ends of nature are brought about. Ascent from the surface of the earth produces the same change, in point of climate, as an

approach to the poles; even under the torrid zone mountains reach the line of perpetual congelation at nearly a third less altitude than the extreme elevation which they sometimes attain. At the poles snow is perpetual on the ground, and at the different intervening latitudes reaches some intermediate point of congelation betwixt one and 20,000 feet. In America, from the line south to the tropics, as also, as there is now every reason to believe, in Africa within similar latitudes, vast ridges of mountains, covered with perpetual snow, run northward and southward in the line of the meridian right across the path of the trade-winds. A similar ridge, though of less magnificent dimensions, traverses the peninsula of Hindoostan, increasing in altitude as it approaches the line, attaining an elevation of 8500 feet at Dodabetta, and above 6000 in Ceylon. The Alps in Europe, and the gigantic chain of the Himalayas in Asia, both far south in the temperate zone, stretch from east to west, and intercept the aerial current from the north. Others of lesser note, in the equatorial or meridional, or some intermediate direction, cross the paths of the atmospherical currents in every direction, imparting to them fresh supplies of cold, as they themselves obtain from them warmth in exchange; in strictness the two operations are the same.

33. *Water.* "Magnificent and stupendous as are the effects and results of the water and of air acting independently on each other, in equalizing the temperature of the globe, they are still more so when combined. One cubic inch of water, when invested with a sufficiency of heat, will form one cubic foot of steam — the water before its evaporation, and the vapor which it forms, being exactly of the same temperature; though in reality, in the process of conversion, 1100 degrees of heat have been absorbed or carried away from the vicinage, and rendered latent or imperceptible; this heat is returned in a sensible and perceptible form the moment the vapor is converted once more into water. The general fact is the same in the case of vapor carried off by dry air at any temperature that may be imagined; for, down far below the freezing point, evaporation proceeds uninterruptedly.

34. *Latent heat.* "The air, heated and dried as it sweeps over the arid surface of the soil, drinks up by day myriads of tons of moisture from the sea — as much, indeed, as would, were no moisture restored to it, depress its whole surface at the rate of eight or ten feet annually. The quantity of heat thus converted from a sensible or perceptible to an insensible or latent state is almost incredible. The action equally goes on, and with the like results, over the surface of the earth, where there is moisture to be withdrawn. But night and the seasons of the year come around, and the surplus temperature, thus withdrawn and stored away at the time it might have proved superfluous or inconvenient, is rendered back so soon as it is required; thus the cold of night and the rigor of winter are modi-

fied by the heat given out at the point of condensation by dew, rain, hail, and snow.

35. *Effects upon the earth.* "The earth is a bad conductor of heat; the rays of the sun, which enter its surface and raise the temperature to 100° or 150°, scarcely penetrate a foot into the ground; a few feet down, the warmth of the ground is nearly the same night and day. The moisture which is there preserved free from the influence of currents of air is never raised into vapor; so soon as the upper stratum of earth becomes thoroughly dried, capillary action, by means of which all excess of water was withdrawn, ceases; so that, even under the heats of the tropics, the soil two feet down will be found, on the approach of the rains, sufficiently moist for the nourishment of plants. The splendid flowers and vigorous foliage which burst forth in May, when the parched soil would lead us to look for nothing but sterility, need in no way surprise us; fountains of water, boundless in extent and limited in depth only by the thickness of the soil which contains them, have been set aside and sealed up for their use, beyond the reach of those thirsty winds or burning rays which are suffered to carry off only the water which is superfluous, and would be pernicious. They remove it to other lands, where its agency is required, or treasure it up, as the material of clouds and dew, in the crystal vault of the firmament, the source, when the fitting season comes round again, of those deluges of rain which provide for the wants of the year. Such are some of the examples which may be supplied of general laws operating over nearly the whole surface of the terraqueous globe. Among the local provisions ancillary to these are the monsoons of India, and the land and sea breezes prevalent throughout the tropical coasts.

36. *The tides.* "We have not noticed the tides, which, obedient to the sun and moon, daily convey two vast masses of water round the globe, and which twice a month, rising to an unusual height, visit elevations which otherwise are dry. During one half of the year the highest tides visit us by day, the other half by night; and at Bombay, at spring tide, the depths of the two differ by two or three feet from each other. The tides simply rise and fall, in the open ocean, to an elevation of two or three feet in all; along our shores, and up gulfs and estuaries, they sweep with the violence of a torrent, having a general range of ten or twelve feet — sometimes, as at Fundy, in America, at Brest and Milford Haven, in Europe, to a height of from forty to sixty feet. The tides sweep our shores from filth, and purify our rivers and inlets, affording to the residents of our islands and continents the benefits of a bi-diurnal ablution, and giving a health, and freshness, and purity wherever they appear. Obedient to the influence of bodies many millions of miles removed from them, their subjection is not the less complete; the vast volume of water, capable of crushing by its weight the most stupendous barriers that can

be opposed to it, and bearing on its bosom the navies of the world, impetuously rushing against our shores, gently stops at a given line, and flows back again to its place when the word goes forth, 'Thus far shalt thou go, and no farther;' and that which no human power or contrivance could have repelled, returns at its appointed time so regularly and surely that the hour of its approach, and measure of its mass, may be predicted with unerring certainty centuries beforehand.

37. *Hurricanes.* "The hurricanes which whirl with such fearful violence over the surface, raising the waters of the sea to enormous elevations, and submerging coasts and islands, attended as they are by the fearful attributes of thunder and deluges of rain, seem requisite to deflagrate the noxious gases which have accumulated, to commingle in one healthful mass the polluted elements of the air, and restore it fitted for the ends designed for it. We have hitherto dealt with the sea and air — the one the medium through which the commerce of all nations is transported, the other the means by which it is moved along — as themselves the great vehicles of moisture, heat, and cold throughout the regions of the world — the means of securing the interchange of these inestimable commodities, so that excess may be removed to where deficiency exists, deficiency substituted for excess, to the unbounded advantage of all. This group of illustrations has been selected because they are the most obvious, the most simple, and the most intelligible and beautiful that could be chosen.

38. *Powers of the air.* "We have already said that the atmosphere forms a spherical shell, surrounding the earth to a depth which is unknown to us, by reason of its growing tenuity, as it is released from the pressure of its own superincumbent mass. Its upper surface can not be nearer to us than fifty, and can scarcely be more remote than five hundred miles. It surrounds us on all sides, yet we see it not; it presses on us with a load of fifteen pounds on every square inch of surface of our bodies, or from seventy to one hundred tons on us in all, yet we do not so much as feel its weight. Softer than the finest down, more impalpable than the finest gossamer, it leaves the cobweb undisturbed, and scarcely stirs the lightest flower that feeds on the dew it supplies; yet it bears the fleets of nations on its wings around the world, and crushes the most refractory substances with its weight. When in motion, its force is sufficient to level with the earth the most stately forests and stable buildings, to raise the waters of the ocean into ridges like mountains, and dash the strongest ships to pieces like toys. It warms and cools by turns the earth and the living creatures that inhabit it. It draws up vapors from the sea and land, retains them dissolved in itself or suspended in cisterns of clouds, and throws them down again, as rain or dew, when they are required. It bends the rays of the sun from their path to give us the aurora of the

morning and twilight of evening; it disperses and refracts their various tints to beautify the approach and the retreat of the orb of day. But for the atmosphere, sunshine would burst on us in a moment and fail us in the twinkling of an eye, removing us in an instant from midnight darkness to the blaze of noon. We should have no twilight to soften and beautify the landscape, no clouds to shade us from the scorching heat; but the bald earth, as it revolved on its axis, would turn its tanned and weakened front to the full and unmitigated rays of the lord of day.

39. *Its functions.* "The atmosphere affords the gas which vivifies and warms our frames; it receives into itself that which has been polluted by use, and is thrown off as noxious. It feeds the flame of life exactly as it does that of the fire. It is in both cases consumed, in both cases it affords the food of consumption, and in both cases it becomes combined with charcoal, which requires it for combustion, and which removes it when combustion is over. It is the girdling encircling air that makes the whole world kin. The carbonic acid with which to-day our breathing fills the air, to-morrow seeks its way round the world. The date-trees that grow round the falls of the Nile will drink it in by their leaves; the cedars of Lebanon will take of it to add to their stature; the cocoa-nuts of Tahiti will grow rapidly upon it; and the palms and bananas of Japan will change it into flowers. The oxygen we are breathing was distilled for us some short time ago by the magnolias of the Susquehanna and the great trees that skirt the Orinoco and the Amazon; the giant rhododendrons of the Himalayas contributed to it, and the roses and myrtles of Cashmere, the cinnamon-tree of Ceylon, and the forest, older than the flood, that lies buried deep in the heart of Africa, far behind the Mountains of the Moon, gave it out. The rain we see descending was thawed for us out of the icebergs which have watched the Polar Star for ages, or it came from snows that rested on the summits of the Alps, but which the lotus lilies have soaked up from the Nile, and exhaled as vapor again into the ever-present air."

40. *The operations of water.* There are processes no less interesting going on in other parts of this magnificent field of research. Water is nature's carrier. With its currents it conveys heat away from the torrid zone and ice from the frigid; or, bottling the caloric away in the vesicles of its vapor, it first makes it impalpable, and then conveys it, by unknown paths, to the most distant parts of the earth. The materials of which the coral builds the island, and the sea-conch its shell, are gathered by this restless leveler from mountains, rocks, and valleys in all latitudes. Some it washes down from the Mountains of the Moon, or out of the gold-fields of Australia, or from the mines of Potosi, others from the battle-fields of Europe, or from the marble quarries of ancient Greece and Rome. These materials, thus collected and carried over falls or down rapids, are transported from river to sea, and delivered by the obedient waters to each

insect and to every plant in the ocean at the right time and temperature, in proper form, and in due quantity.

41. *Its marvelous powers.* Treating the rocks less gently, it grinds them into dust, or pounds them into sand, or rolls them in pebbles, rubble, or boulders: the sand and shingle on the sea-shore are monuments of the abrading, triturating power of water. By water the soil has been brought down from the hills and spread out into valleys, plains, and fields for man's use. Saving the rocks on which the everlasting hills are established, every thing on the surface of our planet seems to have been removed from its original foundation and lodged in its present place by water. Protean in shape, benignant in office, water, whether fresh or salt, solid, fluid, or gaseous, is marvelous in its powers.

42. *It caters on land for insects of the sea.* It is one of the chief agents in the manifold workshops in which and by which the earth has been made a habitation fit for man. Circulating in veins below the surface, it pervades the solid crust of the earth in the fulfillment of its offices; passing under the mountains, it runs among the hills and down through the valley in search of pabulum for the moving creatures that have life in the sea. In rivers and in rain it gathers up by ceaseless lixiviation food for the creatures that wait upon it. It carries off from the land whatever of solid matter the sea in its economy requires.

43. *Leaching.* The waters which dash against the shore, which the running streams pour into the flood, or with which the tides and currents scour the bottom of their channel ways, have soaked from the soil, or leached out of the disintegrated materials which strew the beach or line the shores, portions of every soluble ingredient known in nature. Thus impregnated, the laughing, dancing waters come down from the mountains, turning wheels, driving machinery, and serving the manifold purposes of man. At last they find their way into the sea, and so make the lye of the earth brine for the ocean.

44. *Solid ingredients.* Iron, lime, silver, sulphur, and copper, silex, soda, magnesia, potash, chlorine, iodine, bromine, ammonia, are all found in sea-water; some of them in quantities too minute for the nicest appliances of the best chemists to detect, but which, nevertheless, are elaborated therefrom by physical processes.

45. *Quantity of silver in the sea.* By examining the copper that had been a great while on the bottom of a ship in Valparaiso, the presence of silver, which it obtained from the sea, was detected in it. It was in such quantities as to form the basis of a calculation, by which it would appear that there is held in solution by the sea a quantity of silver sufficient to weigh no less than two hundred million tons, could it all, by any process, be precipitated and collected into bars.

46. *Its inhabitants — their offices.* The salts of the sea, as its solid

ingredients may be called, can neither be precipitated on the bottom, nor taken up by the vapors, nor returned again by the rains to the land; and, but for the presence in the sea of certain agents to which has been assigned the task of collecting these ingredients again, in the sea they would have to remain. There, accumulating in its waters, they would alter the quality of the brine, injure the health of its inhabitants, retard evaporation, change climates, and work endless mischief upon the fauna and the flora of both sea, earth, and air. But in the oceanic machinery all this is prevented by compensations the most beautiful and adjustments the most exquisite. As in the atmosphere the plants are charged with the office of purifying the air by elaborating into vegetable tissue and fibre the impurities which the animals are continually casting into it, so also to the mollusks, to the madrepores, and insects of the sea, has been assigned the office of taking out of its waters and making solid again all this lixiviated matter as fast as the dripping streams and searching rains discharge it into the ocean.

47. *Monuments of their industry.* As to the extent and magnitude of this endless task some idea may be formed from the coral islands, the marl beds, the shell banks, the chalk cliffs, and other marine deposits which deck the sea or strew the land.

48. *Analysis of sea-water.* Fresh water is composed of oxygen and hydrogen gas in the proportion by weight of 1 to 8; and the principal ingredients which chemists, by treating small samples of sea-water in the laboratory, have found in a thousand grains are,

Water	962.0 grains
Chloride of Sodium	27.1 "
Chloride of Magnesium	5.4 "
Chloride of Potassium	0.4 "
Bromide of Magnesia	0.1 "
Sulphate of Magnesia	1.2 "
Sulphate of Lime	0.8 "
Carbonate of Lime	0.1 "
Leaving a residuum of	2.9 " = 1000,

consisting of sulphureted hydrogen gas, hydrochlorate of ammonia, etc., etc., in various quantities and proportions, according to the locality of the specimen.

49. *Proportion of water to the mass of the earth.* If we imagine the whole mass of the earth to be divided into 1786 equal parts by weight, then the weight of all the water in the sea would, according to an estimate by Sir John Herschel, be equivalent to one of such parts. Such is the quantity, and such some of the qualities of that delightful fluid to which, in the laboratories and workshops of nature, such mighty tasks, such important offices, such manifold and multitudinous powers have been assigned.

50. *The three great oceans.* This volume of water, that outweighs the

atmosphere about 400 times, is divided into three great oceans, the Atlantic, the Pacific, and the Arctic; for in the rapid survey which in this chapter we are taking of the field before us, the Indian and Pacific oceans may be regarded as one.

51. *The Atlantic.* The Atlantic Ocean, with its arms, is supposed to extend from the Arctic to the Antarctic — perhaps from pole to pole; but, measuring from the icy barrier of the north to that of the south, it is about 9000 miles in length, with a mean breadth of 2700 miles. It covers an area of about 25,000,000 square miles. It lies between the Old World and the New; passing beyond the "stormy capes," there is no longer any barrier, but only an imaginary line to separate its waters from that great southern waste in which the tides are cradled.

52. *Its tides.* The young tidal wave, rising in the circumpolar seas of the south, rolls thence into the Atlantic, and in 12 hours after passing the parallel of Cape Horn, it is found pouring its flood into the Bay of Fundy.

53. *Its depths.* The Atlantic is a deep ocean, and the middle its deepest part, therefore the more favorable (§ 13) to the propagation of this wave.

54. *Contrasted with the Pacific.* The Atlantic Ocean contrasts very strikingly with the Pacific. The greatest length of one lies east and west; of the other, north and south. The currents of the Pacific are broad and sluggish, those of the Atlantic swift and contracted. The Mozambique current, as it is called, has been found by navigators in the South Pacific to be upward of 1600 miles wide — nearly as broad as the Gulf Stream is long. The principal currents in the Atlantic run to and fro between the equator and the Northern Ocean. In the Pacific they run between the equator and the southern seas. In the Atlantic the tides are high, in the Pacific they are low. The Pacific feeds the clouds with vapors, and the rains feed the Atlantic with rivers. If the volume of rain which is discharged into the Pacific and on its slopes be represented by 1, that discharged upon the hydrographical basin of the Atlantic would be represented by 5. The Atlantic is crossed daily by steamers, the Pacific not once a year. The Atlantic washes the shores of the most powerful, intelligent, and Christian nations; but a pagan or a heathen people in the countries to which the Pacific gives drainage are like the sands upon its shores for multitude. The Atlantic is the most stormy sea in the world, the Pacific the most tranquil.

55. *The Telegraphic Plateau.* Among the many valuable discoveries to which these researches touching the physics of the sea have led, none perhaps is more interesting than the Telegraphic Plateau of the Atlantic, and the fact that the bottom of the deep sea is lined with its own dead, whose microscopic remains are protected from the abrading action of its currents and the violence of its waves by cushions of still water.

CHAPTER II

§ 70–143. THE GULF STREAM

70. *Its color.* There is a river in the ocean: in the severest droughts it never fails, and in the mightiest floods it never overflows; its banks and its bottom are of cold water, while its current is of warm; the Gulf of Mexico is its fountain, and its mouth is in the Arctic Seas. It is the Gulf Stream. There is in the world no other such majestic flow of waters. Its current is more rapid than the Mississippi or the Amazon, and its volume more than a thousand times greater. Its waters, as far out from the Gulf as the Carolina coasts, are of an indigo blue. They are so distinctly marked that their line of junction with the common sea-water may be traced by the eye. Often one half of the vessel may be perceived floating in Gulf Stream water, while the other half is in common water of the sea — so sharp is the line, and such the want of affinity between those waters, and such, too, the reluctance, so to speak, on the part of those of the Gulf Stream to mingle with the littoral waters of the sea.

71. *How caused.* At the salt-works in France, and along the shores of the Adriatic, where the *"salines"* are carried on by the process of solar evaporation, there is a series of vats or pools through which the water is passed as it comes from the sea, and is reduced to the briny state. The longer it is exposed to evaporation, the salter it grows, and the deeper is the hue of its blue, until crystallization is about to commence, when the now deep blue water puts on a reddish tint. Now the waters of the Gulf Stream are salter (§ 102) than the littoral waters of the sea through which they flow, and hence we can account for the deep indigo blue which all navigators observe off the Carolina coasts. The salt-makers are in the habit of judging of the richness of the seawater in salt by its color — the greener the hue, the fresher the water. We have in this, perhaps, an explanation of the contrasts which the waters of the Gulf Stream present with those of the Atlantic, as well as of the light green of the North Sea and other Polar waters; also of the dark blue of the trade-wind regions, and especially of the Indian Ocean, which poets have described as the "black waters."

72. *Speculations concerning the Gulf Stream.* What is the cause of the Gulf Stream has always puzzled philosophers. Many are the theories and numerous the speculations that have been advanced with regard to it. Modern investigations and examinations are beginning to throw some

light upon the subject, though all is not yet entirely clear. But they seem to encourage the opinion that this stream, as well as all the *constant* currents of the sea, is due *mainly* to the *constant* difference produced by temperature and saltness in the specific gravity of water in certain parts of the ocean. Such difference of specific gravity is inconsistent with aqueous equilibrium, and to maintain this equilibrium these great currents are set in motion. The agents which derange equilibrium in the waters of the sea, by altering specific gravity, reach from the equator to the poles, and in their operations they are as ceaseless as heat and cold, consequently they call for a system of perpetual currents to undo their perpetual work.

73. *Agencies concerned.* These agents, however, are not the *sole* cause of currents. The winds *help* to make currents by pressing upon the waves and drifting before them the water of the sea; so do the rains, by raising its level here and there; and so does the atmosphere, by pressing with more or less superincumbent force upon different parts of the ocean at the same moment, and as indicated by the changes of the barometric column. But when the winds and the rains cease, and the barometer is stationary, the currents that were the consequence also cease. The currents thus created are therefore *ephemeral.* But the changes of temperature and of saltness, and the work of other agents which affect the specific gravity of sea-water and derange its equilibrium, are as ceaseless in their operations as the sun in his course, and in their effects they are as endless. Philosophy points to them as the *chief* cause of the Gulf Stream and of all the *constant* currents of the sea.

74. *Early writers.* Early writers, however, maintained that the Mississippi River was the father of the Gulf Stream. Its floods, they said, produce it; for the velocity of this river in the sea might, it was held, be computed by the rate of the current of the river on the land.

75. *Objection to the fresh-water theory.* Captain Livingston overturned this hypothesis by showing that the volume of water which the Mississippi River empties into the Gulf of Mexico is not equal to the three thousandth part of that which escapes from it through the Gulf Stream. Moreover, the water of the Gulf Stream is salt — that of the Mississippi, fresh; and the advocates of this fresh-water theory (§ 74) forgot that just as much salt as escapes from the Gulf of Mexico through this stream, must enter the Gulf through some other channel from the main ocean; for, if it did not, the Gulf of Mexico, in process of time, unless it had a salt bed at the bottom, or was fed with salt springs from below — neither of which is probable — would become a fresh-water basin.

76. *Livingston's hypothesis.* The above quoted argument of Captain Livingston, however, was held to be conclusive; and upon the remains of the hypothesis which he had so completely overturned, he set up an-

other, which, in turn, has also been upset. In it he ascribed the velocity of the Gulf Stream as depending "on the motion of the sun in the ecliptic, and the influence he has on the waters of the Atlantic."

77. *Franklin's theory.* But the opinion that came to be most generally received and deep-rooted in the mind of seafaring people was the one repeated by Dr. Franklin, and which held that the Gulf Stream is the escaping of the waters that have been *forced* into the Caribbean Sea by the trade-winds, and that it is the pressure of those winds upon the water which drives up into that sea a head, as it were, for this stream.

78. *Objections to it.* We know of instances in which waters have been accumulated on one side of a lake, or in one end of a canal, at the expense of the other. The pressure of the trade-winds may *assist* to give the Gulf Stream its initial velocity, but are they of themselves adequate to such an effect? Examination shows that they are not. With the view of ascertaining the average number of days during the year that the N.E. trade-winds of the Atlantic operate upon the currents between 25° N. and the equator, log-books [1] containing no less than 380,284 observations on the force and direction of the wind in that ocean were examined. The data thus afforded were carefully compared and discussed. The results show that within those latitudes, and on the average, the wind from the N.E. quadrant is in excess of the winds from the S.W. only 111 days out of the 365. During the rest of the year the S.W. counteract the effect of the N.E. winds upon the currents. Now can the N.E. trades, by blowing for less than one third of the time, cause the Gulf Stream to run all the time, and without varying its velocity either to their force or their prevalence?

79. *Herschel's explanation.* Sir John Herschel maintains [2] that they can; that the trade-winds are the *sole cause* [3] of the Gulf Stream; not, indeed, by causing "a head of water" in the West Indian seas, but by rolling particles of water before them somewhat as billiard balls are rolled over the table. He denies to evaporation, temperature, salts, and sea-shells, any effective influence whatever upon the circulation of the waters in the ocean. According to him, the winds are the supreme current-producing power in the sea. [4]

80. *Objections to it.* This theory would require all the currents of the sea to set with the winds, or, when deflected, to be deflected from the shore, as billiard balls are from the cushions of the table, making the

---

[1] [Maury, "The Winds at Sea; their Mean Direction and Annual Average Duration from each of the Four Quarters,"] *Nautical Monographs,* no. 1 (Washington [1859]).

[2] "Physical Geography," [*Encyclop. Britan.,* XVII, 578].

[3] "The dynamics of the Gulf Stream have of late, in the work of Lieut. Maury, already mentioned, been made the subject of much (we can not but think misplaced) wonder, as if there could be any possible ground for doubting that it owes its origin *entirely* to the trade-winds." — [*Ibid.*]

[4] [*Ibid.,* pp. 579–580.]

littoral angles of incidence and reflection equal. Now, so far from this being the case, *not* ONE of the *constant* currents of the sea either makes such a rebound or sets with the winds. The Gulf Stream sets, as it comes out of the Gulf of Mexico, and for hundreds of miles after it enters the Atlantic, against the trade-winds; for a part of the way it runs right in the "wind's eye." The Japan current, "the Gulf Stream of the Pacific," does the same. The Mozambique current runs to the south, against the S.E. trade-winds, and it changes not with the monsoons. The ice-bearing currents of the north oppose the winds in their course. Humboldt's current has its genesis in the ex-tropical regions of the south, where the "brave west winds" blow with almost if not with quite the regularity of the trades, but with double their force. And this current, instead of setting to the S.E. before these winds, flows north in spite of them. These are the main and constant currents of the sea — the great arteries and jugulars through which its circulation is conducted. In every instance, and regardless of winds, those currents that are warm flow toward the poles, those that are cold set toward the equator. And this they do, not by the force of the winds, but in spite of them, and by the force of those very agencies that make the winds to blow. They flow thus by virtue of those efforts which the sea is continually making to restore that equilibrium to its waters which heat and cold, the forces of evaporation, and the secretion of its inhabitants are everlastingly destroying.

81. *The supremacy of the winds disputed.* If the winds make the *upper*, what makes the *under* and counter currents? This question is of itself enough to impeach that supremacy of the winds upon the currents, which the renowned philosopher, with whom I am so unfortunate as to differ, traveled so far out of his way to vindicate.[5]

82. *The Bonifaccio current.* That the winds do make currents in the sea no one will have the hardihood to deny; but currents that are born of the winds are as unstable as the winds; uncertain as to time, place, and direction, they are sporadic and ephemeral; they are not the constant currents such as have been already enumerated. Admiral Smyth, in his valuable memoir on the Mediterranean (p. 162), mentions that a continuance in the Sea of Tuscany of *"gusty gales"* from the southwest has been known to raise its surface no less than twelve feet above its ordinary level. This, he says, occasions a strong surface drift through the Strait of Bonifaccio. But in this we have nothing like the Gulf Stream; no deep and narrow channel-way to conduct these waters off like a miniature river even

[5] "We have, perhaps, been more diffuse on the subject of oceanic currents than the nature of this article may seem to justify; but some such detail seemed necessary to vindicate to the winds their supremacy in the production of currents, without calling in the feeble and ineffective aid of heated water, or the still more insignificant influence of insect secretion, which has been pressed into the service as a cause of buoyancy in the regions occupied by coral formations." — [*Ibid.*]

in that sea, but a mere surface flow, such as usually follows the piling up of water in any pond or gulf above the ordinary level. The Bonifaccio current does not flow like a "river in the sea" across the Mediterranean, but it spreads itself out as soon as it passes the Straits, and, like a circle on the water, loses itself by broad spreading as soon as it finds sea room. As soon as the force that begets it expends itself, the current is done.

83. *The bed of the Gulf Stream an ascending plane.* Supposing, with Franklin, and those of his school, that the pressure of the waters that are *forced* into the Caribbean Sea by the trade-winds is the *sole cause* of the Gulf Stream, that sea and the Mexican Gulf should have a much higher level than the Atlantic. Accordingly, the advocates of this theory require for its support "a great degree of elevation." Major Rennell likens the stream to "an immense river descending from a higher level into a plain." Now we know very nearly the average breadth and velocity of the Gulf Stream in the Florida Pass. We also know, with a like degree of approximation, the velocity and breadth of the same waters off Cape Hatteras. Their breadth here is about seventy-five miles against thirty-two in the "Narrows" of the Straits, and their mean velocity is three knots off Hatteras against four in the "Narrows." This being the case, it is easy to show that the depth of the Gulf Stream off Hatteras is not so great as it is in the "Narrows" of Bemini by nearly 50 per cent., and that, consequently, instead of *descending*, its bed represents the surface of an inclined plane — inclined from the north toward the south — *up* which plane the lower depths of the stream *must* ascend. If we assume its depth off Bemini [6] to be two hundred fathoms, which are thought to be within limits, the above rates of breadth and velocity will give one hundred and fourteen fathoms for its depth off Hatteras. The waters, therefore, which in the Straits are below the level of the Hatteras depth, so far from *descending*, are actually forced up an inclined plane, whose submarine ascent is not less than ten inches to the mile.

84. *The Niagara.* The Niagara is an "immense river descending into a plain." But instead of preserving its character in Lake Ontario as a distinct and well-defined stream for several hundred miles, it spreads itself out, and its waters are immediately lost in those of the lake. Why should not the Gulf Stream do the same? It gradually enlarges itself, it is true; but, instead of mingling with the ocean by broad spreading, as the "immense rivers" descending into the northern lakes do, its waters, like a stream of oil in the ocean, preserve a distinctive character for more than three thousand miles.

85. *A current counter to the Gulf Stream.* Moreover, while the Gulf

[6] The superintendent reports that the officers of the Coast Survey have sounded with the deep-sea lead, and ascertained its depth here to be 370 fathoms (January, 1856).

Stream is running to the north from its supposed elevated level at the south, there is a cold current coming down from the north; meeting the warm waters of the Gulf midway the ocean, it divides itself, and runs by the side of them right back into those very reservoirs at the south, to which theory gives an elevation sufficient to send out entirely across the Atlantic a jet of warm water said to be more than three thousand times greater in volume than the Mississippi River. This current from Baffin's Bay has not only no trade-winds to give it a head, but the prevailing winds are unfavorable to it, and for a great part of the way it is below the surface, and far beyond the propelling reach of any wind. And there is every reason to believe that this, with other polar currents, is quite equal in volume to the Gulf Stream. Are they not the effects of like causes? If so, what have the trade-winds to do with the one more than the other?

86. *Bottle chart.* It is a custom often practiced by sea-faring people to throw a bottle overboard, with a paper, stating the time and place at which it is done. In the absence of other information as to currents, that afforded by these mute little navigators is of great value. They leave no tracks behind them, it is true, and their routes can not be ascertained. But knowing where they were cast, and seeing where they are found, some idea may be formed as to their course. Straight lines may at least be drawn, showing the shortest distance from the beginning to the end of their voyage, with the time elapsed. Captain Becher, R. N., has prepared a chart representing in this way the tracks of more than one hundred bottles. From it it appears that the waters from every quarter of the Atlantic tend toward the Gulf of Mexico and its stream. Bottles cast into the sea midway between the Old and the New Worlds, near the coasts of Europe, Africa, and America, at the extreme north or farthest south, have been found either in the West Indies, on the British Isles, or within the well-known range of Gulf Stream waters.

87. *Their drift.* Of two cast out together in south latitude on the coast of Africa, one was found on the island of Trinidad; the other on Guernsey, in the English Channel. In the absence of positive information on the subject, the circumstantial evidence that the latter performed the tour of the Gulf is all but conclusive. And there is reason to suppose that some of the bottles of the gallant captain's chart have also performed the tour of the Gulf Stream; then, without being cast ashore, have returned with the drift along the coast of Africa into the intertropical region; thence through the Caribbean Sea, and so on with the Gulf Stream again. (Plate III.) Another bottle, said to be thrown over off Cape Horn by an American ship-master in 1837, was afterward picked up on the coast of Ireland. An inspection of the chart, and of the drift of the other bottles, seems to *force* the conclusion that this bottle too went even from that remote region to the so-called *higher* level of the Gulf Stream reservoir.

88. *The Sargasso Sea.* Midway the Atlantic, in the triangular space between the Azores, Canaries, and the Cape de Verd Islands, is the great Sargasso Sea. (Plate III.) Covering an area equal in extent to the Mississippi Valley, it is so thickly matted over with Gulf weeds (*fucus natans*) that the speed of vessels passing through it is often much retarded. When the companions of Columbus saw it, they thought it marked the limits of navigation, and became alarmed. To the eye, at a little distance, it seems substantial enough to walk upon. Patches of the weed are always to be seen floating along the outer edge of the Gulf Stream. Now, if bits of cork or chaff, or any floating substance, be put into a basin, and a circular motion be given to the water, all the light substances will be found crowding together near the centre of the pool, where there is the least motion. Just such a basin is the Atlantic Ocean to the Gulf Stream, and the Sargasso Sea is the centre of the whirl. Columbus first found this weedy sea in his voyage of discovery; there it has remained to this day, moving up and down, and changing its position, like the calms of Cancer, according to the seasons, the storms, and the winds. Exact observations as to its limits and their range, extending back for fifty years, assure us that its mean position has not been altered since that time. This indication of a circular motion by the Gulf Stream is corroborated by the bottle chart, by Plate III, and other sources of information. If, therefore, this be so, why give the endless current a higher level in one part of its course than another?

89. *A bifurcation.* Nay, more; at the very season of the year when the Gulf Stream is rushing in greatest volume through the Straits of Florida, and hastening to the north with the greatest rapidity, there is a cold stream from Baffin's Bay, Labrador, and the coasts of the north, running to the south with equal velocity. Where is the trade-wind that gives the higher level to Baffin's Bay, or that even presses upon, or assists to put this current in motion? The agency of winds in producing currents in the deep sea must be very partial. These two currents meet off the Grand Banks, where the latter is divided. One part of it underruns the Gulf Stream, as is shown by the icebergs which are carried in a direction tending across its course. The probability is, that this "fork" flows on toward the south, and runs into the Caribbean Sea, for the temperature of the water at a little depth there has been found far below the mean temperature of the earth's crust, and quite as cold as at a corresponding depth off the Arctic shores of Spitzbergen.

90. *Winds exercise but little influence upon constant currents.* More water can not run from the equator or the pole than to it. If we make the trade-winds to cause the Gulf Stream, we ought to have some other wind to produce the Polar flow; but these currents, for the most part, and for great distances, are *submarine*, and therefore beyond the influence of winds. Hence it should appear that *winds* have little to do with the

general system of aqueous circulation in the ocean. The other "fork" runs between our shores and the Gulf Stream to the south, as already described. As far as it has been traced, it warrants the belief that it, too, runs *up* to seek the so-called *higher* level of the Mexican Gulf.

91. *Effects of diurnal rotation upon the Gulf Stream.* The power necessary to overcome the resistance opposed to such a body of water as that of the Gulf Stream, running several thousand miles without any renewal of impulse from the forces of gravitation or any other known cause, is truly surprising. It so happens that we have an argument for determining, with considerable accuracy, the resistance which the waters of this stream meet with in their motion toward the east. Owing to the diurnal rotation, they are carried around with the earth on its axis *toward the east* with an hourly velocity of one hundred and fifty-seven [7] miles greater when they enter the Atlantic than when they arrive off the Banks of Newfoundland; for in consequence of the difference of latitude between the parallels of these two places, their rate of motion around the axis of the earth is reduced from nine hundred and fifteen [8] to seven hundred and fifty-eight miles the hour. Hence this immense volume of water would, if we suppose it to pass from the Bahamas to the Grand Banks in an hour, meet with an opposing force in the shape of resistance sufficient, in the aggregate, to retard it two miles and a half the minute in its eastwardly rate. If the actual resistance be calculated according to received laws, it will be found equal to several atmospheres. And by analogy, how inadequate must the pressure of the gentle trade-winds be to such resistance, and to the effect assigned them!

92. *The Gulf Stream can not be accounted for by a higher level.* If, therefore, in the proposed inquiry, we search for a propelling power nowhere but in the *higher level* of the Gulf, or in the "billiard-ball" rebound from its shores, we must admit, in the head of water there, the existence of a force capable of putting in motion, and of driving over a plain at the rate of four miles the hour, all the waters, as fast as they can be brought down by three thousand (§ 75) such streams as the Mississippi River — a power, at least, sufficient to overcome the resistance required to reduce from two miles and a half to a few feet per minute the velocity of a stream that keeps in perpetual motion one fourth of all the waters in the Atlantic Ocean.

93. *Nor by the trade-wind theory.* The advocates of the trade-wind theory, whether, with Franklin (§ 77), they make the propelling power to be derived from a "*head of water*" in the Gulf, or, with Herschel (§ 79),

[7] In this calculation the earth is treated as a perfect sphere, with a diameter of 7925.56 miles.

[8] Or, 915.26 to 758.60. On the latter parallel the current has an east set of about one and a half miles the hour, making the true velocity to the east, and on the axis of the earth, about seven hundred and sixty miles an hour at the Grand Banks.

from the rebound, *à la* billiard-balls, against its shores, require that the impulse then and there communicated to the waters of the Gulf Stream should be sufficient to send them entirely across the Ocean; for in neither case does their theory provide for any renewal of the propelling power by the wayside. Can this be? Can water flow on any more than cannon balls can continue their flight after the propelling force has been expended?

94. *Illustration.* When we inject water into a pool, be the force never so great, the jet is soon overcome, broken up, and made to disappear. In this illustration the Gulf Stream may be likened to the jet, and the Atlantic to the pool. We remember to have observed as children how soon the mill-tail loses its current in the pool below; or we may see at any time, and on a larger scale, how soon the Niagara, current and all, is swallowed up in the lake.

95. *Gulf Stream the effect of some constantly operating power.* Nothing but a continually-acting power can keep currents in the sea, any more than cannon balls in the air or rivers on the land, in motion. But for the forces of gravitation the waters of the Mississippi would remain at its fountain, and but for *difference* of specific gravity the waters of the Gulf Stream would remain in the caldron, as the intertropical parts of the Atlantic Ocean may be called.

96. *The production of currents without wind.* For the sake of further illustration, let us suppose a globe of the earth's size, and with a solid nucleus to be covered all over with water two hundred fathoms deep, and that every source of heat and cause of radiation be removed, so that its fluid temperature becomes constant and uniform throughout. On such a globe, the equilibrium remaining undisturbed, there would be neither wind nor current. Let us now suppose that all the water within the tropics, to the depth of one hundred fathoms, suddenly becomes oil. The aqueous equilibrium of the planet would thereby be disturbed, and a general system of currents and counter currents would be immediately commenced — the oil, in an unbroken sheet on the surface, running toward the poles, and the water, in an under current, toward the equator. The oil is supposed, as it reaches the polar basin, to be reconverted into water, and the water to become oil as it crosses Cancer and Capricorn, rising to the surface in the intertropical regions and returning as before. Thus, *without wind,* we should have a perpetual and uniform system of tropical and polar currents, though *without wind,* Sir John Herschel maintains,[9] we should have no currents whatever in the sea. In consequence of the diurnal rotation of the planet on its axis, each particle of oil, were resistance small, would approach the poles on a spiral turning to the east with

---

[9] "If there were no atmosphere, there would be no Gulf Stream or any other considerable oceanic current (as distinguished from a mere surface drift) whatever." — "Physical Geography," [*Encyclop. Britan.,* XVII, 578.]

a relative velocity greater and greater, until, finally, it would reach the pole, and whirl about it at the rate of nearly a thousand miles the hour. Becoming water and losing its velocity, it would approach the tropics by a similar, but reversed spiral, turning toward the west. Owing to the principle here alluded to, all currents from the equator to the poles should have an eastward tendency, and all from the poles toward the equator a westward. Let us now suppose the solid nucleus of this hypothetical globe to assume the exact form and shape of the bottom of our seas, and in all respects, as to figure and size, to represent the shoals and islands of the sea, as well as the coast lines and continents of the earth. The uniform system of currents just described would now be interrupted by obstructions and local causes of various kinds, such as unequal depth of water, contour of shore lines, &c.; and we should have at certain places currents greater in volume and velocity than at others. But still there would be a system of currents and counter currents to and from either pole and the equator. Now, do not the cold waters of the north, and the warm waters of the Gulf, made specifically lighter by tropical heat, and which we see actually preserving such a system of counter currents, hold, at least in some degree, the relation of the supposed water and oil?

97. *Warm currents flow toward the pole, cold toward the equator.* In obedience to the laws here hinted at, there is a constant tendency (Plate VI) of polar waters toward the tropics and of tropical waters toward the poles. Captain Wilkes, of the United States Exploring Expedition, crossed one of these hyperborean under currents two hundred miles in breadth at the equator.

98. *Edges of the Gulf Stream a striking feature.* No feature of the Gulf Stream excites remark among seamen more frequently than the sharpness of its edges, particularly along its inner borders. There, it is a streak on the water. As high up as the Carolinas this streak may be seen, like a greenish edging to a blue border — the bright indigo of the tropical contrasting finely (§ 70) with the dirty green of the littoral waters. It is this apparent reluctance of the warm waters of the stream to mix with the cool of the ocean that excites wonder and calls forth remark. But have we not, so to speak, a similar reluctance manifested by all fluids, only upon a smaller scale, or under circumstances less calculated to attract attention or excite remark?

99. *Illustrations.* The water, hot and cold, as it is let into the tub for a warm bath, generally arranges itself in layers or sections, according to temperature; it requires violent stirring to break them up, mix, and bring the whole to an even temperature. The jet of air from the blow-pipe, or of gas from the burner, presents the phenomenon still more familiarly; here we have, as with the Gulf Stream, the dividing line between fluids in motion and fluids at rest finely presented. There is a like reluctance for mix-

ing between streams of clear and muddy water. This is very marked between the red waters of the Missouri and the inky waters of the upper Mississippi; here the waters of each may be distinguished for the distance of several miles after these two rivers come together. It requires force to inject, as it were, the particles of one of these waters among those of the other, for mere *vis inertia* tends to maintain in their *statu quo* fluids that have already arranged themselves in layers, streaks, or aggregations.

100. *How the water of the Gulf Stream differs from the littoral waters.* In the ocean we have the continual heaving of the sea and agitation of the waves to overcome this *vis inertia,* and the marvel is, that they in their violence do not, by mingling the Gulf and littoral waters together (§ 70), sooner break up and obliterate all marks of a division between them. But the waters of the Gulf Stream differ from the in-shore waters not only in color, transparency, and temperature, but in specific gravity, in saltness (§ 102), and in other properties, I conjecture, also. Therefore they *may* have a peculiar viscosity, or molecular arrangement of their own, which further tends to prevent mixture, and so preserve their line of demarkation.

101. *Action on copper.* Observations made for the purpose in the navy show that ships cruising in the West Indies suffer in their copper sheathing more than they do in any other seas. This would indicate that the waters of the Caribbean Sea and Gulf of Mexico, from which the Gulf Stream is fed, have some peculiar property or other which makes them so destructive upon the copper of cruisers.

102. *Saltness of the Gulf Stream.* The story told by the copper and the blue color (§ 71) indicates a higher point of saturation with salts than sea water generally has; and the salometer confirms it. Dr. Thomassy, a French *savant,* who has been extensively engaged in the manufacture of salt by solar evaporation, informs me that on his passage to the United States he tried the saltness of the water with a most delicate instrument: he found it in the Bay of Biscay to contain 3½ per cent. of salt; in the trade-wind region, 4⁴⁄₁₀ per cent.; and in the Gulf Stream, off Charleston, 4 per cent., notwithstanding the Amazon and the Mississippi, with all the intermediate rivers, and the clouds of the West Indies, had lent their fresh water to dilute the saltness of this basin.

103. *Agents concerned.* Now the question may be asked, What should make the waters of the Mexican Gulf and Caribbean Sea salter than the waters in those parts of the ocean through which the Gulf Stream flows? There are physical agents that are known to be at work in different parts of the ocean, the tendency of which is to make the waters in one part of the ocean salter and heavier, and in another part lighter and less salt than the average of sea water. These agents are those employed by sea-

shells in secreting solid matter for their structures; they are also heat [10] and radiation, evaporation and precipitation. In the trade-wind regions at sea (Plate V), evaporation is generally in excess of precipitation, while in the extra-tropical regions the reverse is the case; that is, the clouds let down more water there than the winds take up again; and these are the regions in which the Gulf Stream enters the Atlantic. Along the shores of India, where observations have been carefully made, the evaporation from the sea is said to amount to three fourths of an inch daily. Suppose it in the trade-wind region of the Atlantic to amount to only half an inch, that would give an annual evaporation of fifteen feet. In the process of evaporation from the sea, fresh water only is taken up; the salts are left behind. Now a layer of sea water fifteen feet deep, and as broad as the trade-wind belts of the Atlantic, and reaching across the ocean, contains an immense amount of salts. The great equatorial current (Plate III) which often sweeps from the shores of Africa across the Atlantic into the Caribbean Sea is a surface current; and may it not bear into that sea a large portion of those waters that have satisfied the thirsty trade-winds with saltless vapor? If so — and it probably does — have we not detected here the foot-prints of an agent that does tend to make the waters of the Caribbean Sea salter, and therefore heavier, than the average of sea water at a given temperature?

104. *Evaporation and precipitation.* It is immaterial, so far as the correctness of the principle upon which this reasoning depends is concerned, whether the annual evaporation from the trade-wind regions of the Atlantic be fifteen, ten, or five feet. The layer of water, whatever be its thickness, that is evaporated from this part of the ocean, is not all poured back by the clouds upon the same spot whence it came. But they take it and pour it down in showers upon the extra-tropical regions of the earth — on the land as well as in the sea — and on the land more water is let down than is taken up into the clouds again. The rest sinks down through the soil to feed the springs, and return through the rivers to the sea. Suppose the excess of precipitation in these extra-tropical regions of the sea to amount to but twelve inches, or even to but two — it is twelve inches or two inches, as the case may be, of fresh water added to the sea in those parts, and which therefore tends to lessen the specific gravity of sea water there to that extent, and to produce a double dynamical effect, for the simple reason that what is taken from one scale, by being put into the other, doubles the difference.

[10] According to Dr. Marcet, sea water contracts down to 28°; my own to about 25.6. [Alexander Marcet, "On the Specific Gravity, and Temperature of Sea Waters, in Different Parts of the Ocean, and in Particular Seas; with Some Account of their Saline Contents," *Philos. Trans.*, CIX(1819), 161–208, esp., pp. 182, 183, 188. Maury's results: *Sailing Directions*, 8th ed., I(1858), 250.]

105. *Current into the Caribbean Sea.* Now that we may form some idea as to the influence which the salts left by the vapor that the trade-winds, northeast and southeast, take up from sea water, is calculated to exert in creating currents, let us make a partial calculation to show how much salt this vapor held in solution before it was taken up, and, of course, while it was yet in the state of sea water. The northeast trade-wind regions of the Atlantic embrace an area of at least three million square miles, and the yearly evaporation from it is (§ 103), we will suppose, fifteen feet. The salt that is contained in a mass of sea water covering to the depth of fifteen feet an area of three million square miles in superficial extent, would be sufficient to cover the British islands to the depth of fourteen feet. As this water supplies the trade-winds with vapor, it therefore becomes salter, and as it becomes salter, the forces of aggregation among its particles are increased, as we may infer from the fact (§ 98) that the waters of the Gulf Stream are reluctant to mix with those of the ocean.

106. *Amount of salt left by evaporation.* Whatever be the cause that enables these trade-wind waters to remain on the surface, whether it be from the fact just stated, and in consequence of which the waters of the Gulf Stream are held together in their channel; or whether it be from the fact that the expansion from the heat of the torrid zone is sufficient to compensate for this increased saltness; or whether it be from the low temperature and high saturation of the submarine waters of the inter-tropical ocean; or whether it be owing to all of these influences together that these waters are kept on the surface, suffice it to say, we do know that they go into the Caribbean Sea (§ 103) as a surface current. On their passage to and through it, they intermingle with the fresh waters that are emptied into the sea from the Amazon, the Orinoco, and the Mississippi, and from the clouds, and the rivers of the coasts round about. An immense volume of fresh water is supplied from these sources. It tends to make the sea water, that the trade-winds have been playing upon and driving along, less briny, warmer, and lighter; for the waters of these large intertropical streams are warmer than sea water. This admixture of fresh water still leaves the Gulf Stream a brine stronger than that of the extra-tropical sea generally, but not quite so strong (§ 102) as that of the trade-wind regions.

107. *Currents created by storms.* The dynamics of the sea confess the power of the winds in those tremendous currents which storms are sometimes known to create; and that even the gentle trade-winds may have influence and effect upon the currents of the sea has not been denied (§ 82). But the effect of moderate winds, as the trades are, is to cause what may be called the *drift* of the sea rather than a current. Drift is confined to surface waters, and the trade-winds of the Atlantic may *assist* in

creating the Gulf Stream by drifting the waters which have supplied them with vapor toward the Caribbean Sea. But admit never so much of the water which the trade-winds have played upon to be drifted into the Caribbean Sea, what should make it flow thence with the Gulf Stream to the shores of Europe? It is because of the difference in the specific gravity of sea water in an intertropical sea on one side, as compared with the specific gravity of water in northern seas and frozen oceans on the other, that they so flow.

108. *The dynamical force that calls forth the Gulf Stream to be found in the difference as to specific gravity of intertropical and polar waters.* The dynamical forces which are expressed by the Gulf Stream may with as much propriety be said to reside in those northern waters as in the West India seas; for on one side we have the Caribbean Sea and Gulf of Mexico, with their waters of brine; on the other, the great Polar basin, the Baltic, and the North Sea, the two latter with waters that are but little more than brackish.[11] In one set of these sea-basins the water is heavy; in the other it is light. Between them the ocean intervenes; but water is bound to seek and to maintain its level; and here, therefore, we unmask one of the agents concerned in causing the Gulf Stream. What is the power of this agent — is it greater than that of other agents, and how much? We can not say; we only know it is one of the chief agents concerned. Moreover, speculate as we may as to all the agencies concerned in collecting these waters, that have supplied the trade-winds with vapor, into the Caribbean Sea, and then in driving them across the Atlantic — we are forced to conclude that the salt which the trade-wind vapor leaves behind in the tropics has to be conveyed away from the trade-wind region, to be mixed up again in due proportion with the other water of the sea — the Baltic Sea and the Arctic Ocean included — and that these are some of the waters, at least, which we see running off through the Gulf Stream. To convey them away is doubtless one of the offices which, in the economy of the ocean, has been assigned to it. But as for the seat of the forces which put and keep the Gulf Stream in motion, theorists may place them *exclusively* on one side of the ocean with as much philosophical propriety as on the other. Its waters find their way into the North Sea and the Arctic Ocean by virtue of their specific gravity, while water thence, to take their place, is, by virtue of its specific gravity and by counter currents, carried back into the Gulf. The dynamical force which causes the Gulf Stream may therefore be said to reside both in the polar and in the intertropical waters of the Atlantic.

[11] The Polar basin has a known water area of 3,000,000 square miles, and an unexplored area, including land and water, of 1,500,000 square miles. Whether the water in this basin be more or less salt than that of the intertropical seas, we know it is quite different in temperature, and difference of temperature will beget currents quite as readily as difference in saltness, for change in specific gravity follows either.

109. *Winter temperature of the Gulf Stream.* As to the temperature of the Gulf Stream, there is, in a winter's day, off Hatteras, and even as high up as the Grand Banks of Newfoundland in mid-ocean, a difference between its waters and those of the ocean near by of 20°, and even 30°. Water, we know, expands by heat, and here the difference of temperature may more than compensate for the difference in saltness, and leave, therefore, the waters of the Gulf Stream, though salter, yet lighter by reason of their warmth.

110. *Top of Gulf Stream roof-shaped.* If they be lighter, they should therefore occupy a higher level than those through which they flow. Assuming the depth off Hatteras to be one hundred and fourteen fathoms, and allowing the usual rates of expansion for sea water, figures show that the middle or axis of the Gulf Stream there should be nearly two feet higher than the contiguous waters of the Atlantic. Hence the surface of the stream should present a double inclined plane, from which the water would be running down on either side as from the roof of a house. As this runs off at the top, the same weight of colder water runs in at the bottom, and so raises up the cold-water bed of the Gulf Stream, and causes it to become shallower and shallower as it goes north. That the Gulf Stream is therefore roof-shaped, causing the waters on its surface to flow off to either side from the middle, we have not only circumstantial evidence to show, but observations to prove. Navigators, while drifting along with the Gulf Stream, have lowered a boat to try the surface current. In such cases, the boat would drift either to the east or to the west, as it happened to be on one side or the other of the axis of the stream, while the vessel herself would drift along with the stream in the direction of its course; thus showing the existence of a shallow roof-current from the middle toward either edge, which would carry the boat along, but which, being superficial, does not extend deep enough to affect the drift of the vessel.

111. *Drift matter sloughed off to the right.* That such is the case (§ 110) is also indicated by the circumstance that the sea-weed and driftwood which are found in such large quantities along the outer edge of the Gulf Stream, are rarely, even with the prevalence of easterly winds, found along its inner edge — and for the simple reason that to cross the Gulf Stream, and to pass over from that side to this, they would have to drift up an inclined plane, as it were; that is, they would have to stem this roof-current until they reached the middle of the stream. We rarely hear of planks, or wrecks, or of any floating substance which is cast into the sea on the other side of the Gulf Stream being found along the coast of the United States. Drift-wood, trees, and seeds from the West India islands, are often cast up on the shores of Europe, but rarely on the Atlantic shores of this country.

112. *Why so sloughed off.* We are treating now of the effects of physical causes. The question to which I ask attention is, Why does the Gulf Stream slough off and cast upon its outer edge, sea-weed, drift-wood, and all other solid bodies that are found floating upon it? One cause has been shown to be in its roof-shaped current; but there is another which tends to produce the same effect; and because it is a physical agent, it should not, in a treatise of this kind, be overlooked, be its action never so slight. I allude now to the effects produced upon the drift matter of the stream by the diurnal rotation of the earth.

113. *Illustration.* Take, for illustration, a railroad that lies north and south in our hemisphere. It is well known to engineers that when the cars are going north on such a road, their tendency is to run off on the east side; but when the train is going south, their tendency is to run off on the west side of the track — *i. e.*, always on the right-hand side. Whether the road be one mile or one hundred miles in length, the effect of diurnal rotation is the same; and, whether the road be long or short, the tendency to run off, as you cross a given parallel at a stated rate of speed, is the same; for the tendency to fly off the track is in proportion to the speed of the train, and not at all in proportion to the length of the road. Now, *vis inertiæ* and velocity being taken into the account, the tendency to obey the force of this diurnal rotation, and to trend to the right, is proportionably as great in the case of a patch of sea-weed as it drifts along the Gulf Stream, as it is in the case of the train of cars as they speed to the north along the iron track of the Hudson River railway, or any other railway that lies north and south. The rails restrain the cars and prevent them from flying off; but there are no rails to restrain the sea-weed, and nothing to prevent the drift-matter of the Gulf Stream from going off in obedience to this force. The slightest impulse tending to turn aside bodies moving freely in water is immediately felt and implicitly obeyed.

114. *Drift-wood on the Mississippi.* It is in consequence of this diurnal rotation that drift-wood coming down the Mississippi is so very apt to be cast upon the west or right bank. This is the reverse of what obtains upon the Gulf Stream, for it flows to the north; it therefore sloughs off (§ 111) to the east.

115. *Effect of diurnal rotation upon.* The effect of diurnal rotation upon the winds and upon the currents of the sea is admitted by all — the trade-winds derive their *easting* from it — it must, therefore, extend to all the matter which these currents bear with them, to the largest iceberg as well as to the smallest spire of grass that floats upon the waters, or the minutest organism that the most powerful microscope can detect among the impalpable particles of sea-dust. This effect of diurnal rotation upon drift will be frequently alluded to in the pages of this work.

116. *Formation of the Grand Banks.* In its course to the north, the

Gulf Stream gradually trends more and more to the eastward, until it arrives off the Banks of Newfoundland, where its course becomes nearly due east. These banks, it has been thought, deflect it from its proper course, and cause it to take this turn. Examination will prove, I think, that they are an effect, certainly not the cause. It is here that the frigid current already spoken of (§ 85), and its icebergs from the north, are met and melted by the warm waters of the Gulf. Of course the loads of earth, stones, and gravel brought down upon these bergs are here deposited. Captain Scoresby, far away in the north, counted at one time five hundred icebergs setting out from the same vicinity upon this cold current for the south. Many of them, loaded with earth, have been seen aground on the Banks. This process of transferring deposits from the north for these shoals, and of snowing down upon them the infusoria and the corpses of "living creatures" that are brought forth so abundantly in the warm waters of the Gulf Stream, and delivered in myriads for burial where the conflict between it and the great Polar current (§ 89) takes place, is everlastingly going on. These agencies, with time, seem altogether adequate to the formation of extensive bars or banks.

117. *Deep water near.* The deep-sea soundings that have been made by vessels of the navy (Plate VII) tend to confirm this view as to the formation of these Banks. The greatest contrast in the bottom of the Atlantic is just to the south of these Banks. Nowhere in the open sea has the water been found to deepen so suddenly as here. Coming from the north, the bottom of the sea is shelving; but suddenly, after passing these Banks, it dips down by a precipitous descent to unknown depths — thus indicating that the debris which forms the Grand Banks comes from the north.

118. *The Gulf Stream describes in its course the path of a trajectory.* From the Straits of Bemini the course of the Gulf Stream (Plate III) describes (as far as it can be traced over toward the British Islands which are in the midst of its waters) the arc of a great circle nearly. Such a course as the Gulf Stream takes is very nearly the course that a cannon ball, could it be shot from these straits to those islands, would follow.

119. *Its path from Bemini to Ireland.* If it were possible to see Ireland from Bemini, and to get a cannon that would reach that far, the person standing on Bemini and taking aim, intending to shoot at Ireland as a target, would, if the earth were at rest, sight direct, and make no allowance for difference of motion between marksman and target. Its path would lie in the plane of a great circle. But there *is* diurnal rotation; the earth *does* revolve on its axis; and since Bemini is nearer to the equator than Ireland is, the gun would be moving in diurnal rotation (§ 91) faster than the target, and therefore the marksman, taking aim point blank at his target, would miss. He would find, on examination, that he had shot south — that is, to the *right* (§ 103) of his mark. In other words, that the

path actually described by the ball would be a resultant arising from this difference in the rate of rotation and the trajectile force. Like a ray of light from the stars, the ball would be affected by aberration. The ball so shot presents the case of the passenger in the railroad car throwing an apple, as the train sweeps by, to a boy standing by the way-side. If he throw straight at the boy, he will miss, for the apple, partaking of the motion of the cars, will go ahead of the boy, and for the very reason that the shot will pass in advance of the target, for both the marksman and the passenger are going faster than the object at which they aim.

120. *Tendency of all currents both in the sea and air to move in great circles a physical law.* Hence we may assume it as a law, that the natural tendency of all currents in the sea, like the natural tendency of all projectiles through the air, is to describe each its curve of flight very nearly in the plane of a great circle. The natural tendency of all matter, when put in motion, is to go from point to point by the shortest distance, and it requires force to overcome this tendency. Light, heat, and electricity, the howling wind, running water, and all substances, whether ponderable or imponderable, seek, when in motion, to obey this law. Electricity may be turned aside from its course, and so may the cannon ball or running water; but remove every obstruction, and leave the current or the shot free to continue on in the direction of the first impulse, or to turn aside of its own volition, so to speak, and straight it will go, and continue to go — if on a plane, in a straight line; if about a sphere, in the arc of a great circle — thus showing that it has no volition except to obey impulse; and that impulse comes from the physical requirements upon it to take the shortest way to its point of destination.

121. *This law recognized by the Gulf Stream.* The waters of the Gulf Stream, as they escape from the Gulf, are bound for the British Islands, to the North Sea, and Frozen Ocean (Plate VI). Accordingly, they take (§ 118), in obedience to this physical law, the most direct course by which nature will permit them to reach their destination. And this course, as already remarked, is nearly that of the great circle, and of the supposed cannon ball.

122. *Shoals of Nantucket do not control its course.* Many philosophers have expressed the opinion — indeed, the belief (§ 116) is common among mariners — that the coasts of the United States and the Shoals of Nantucket turn the Gulf Stream toward the east; but if the view I have been endeavoring to make clear be correct, it would appear that the course of the Gulf Stream is fixed and prescribed by exactly the same laws that require the planets to revolve in orbits, the planes of which shall pass through the centre of the sun; and that, were the Nantucket Shoals not in existence, the course of the Gulf Stream, in the main, would be exactly as it is and where it is. The Gulf Stream is bound over to the North Sea

and Bay of Biscay partly for the reason, perhaps, that the waters there are lighter than those of the Mexican Gulf; and if the Shoals of Nantucket were not in existence, it could not pursue a more direct route. The Grand Banks, however, are encroaching (§ 116), and cold currents from the north come down upon it: they may, and probably do, assist now and then to turn it aside.

123. *Herschel's theory not consistent with known facts.* Now if this explanation as to the *course* of the Gulf Stream and its eastward tendency hold good, a current setting from the north toward the south should (§ 103) have a westward tendency. It should also move in a circle of trajection, or such as would be described by a trajectile moving through the air without resistance and for a great distance. Accordingly, and in obedience to the propelling powers derived from the rate at which different parallels are whirled around in diurnal motion (§ 91), we find the current from the north, which meets the Gulf Stream on the Grand Banks (Plate VI), taking a south*westwardly* direction, as already described (§ 114). It runs down to the tropics by the side of the Gulf Stream, and stretches as far to the west as our own shores will allow. Yet, in the face of these facts, and in spite of this force, both Major Rennell and M. Arago would make the coasts of the United States and the Shoals of Nantucket to turn the Gulf Stream toward the east; and Sir John Herschel (§ 79) makes the trade-winds, which blow *from* the eastward, drive the stream *to* the eastward!

124. *The channel of the Gulf Stream shifts with the season.* But there are other forces operating upon the Gulf Stream. They are derived from the effect of changes in the waters of the whole ocean, as produced by changes in their temperature from time to time. As the Gulf Stream leaves the coasts of the United States, it begins to vary its position according to the seasons; the limit of its northern edge, as it passes the meridian of Cape Race (Plate III), being in winter about latitude 40–41°, and in September, when the sea is hottest, about latitude 45–46°. The trough of the Gulf Stream, therefore, may be supposed to waver about in the ocean not unlike a pennon in the breeze. Its head is confined between the shoals of the Bahamas and the Carolinas; but that part of it which stretches over toward the Grand Banks of Newfoundland is, as the temperature of the waters of the ocean changes, first pressed down toward the south, and then again up toward the north, according to the season of the year.

125. *The phenomenon thermal in its character.* To appreciate the extent of the force by which it is so pressed, let us imagine the waters of the Gulf Stream to extend all the way to the bottom of the sea, so as completely to separate, by an impenetrable liquid wall, if you please, the waters of the ocean on the right from the waters in the ocean on the left

of the stream. It is the height of summer: the waters of the sea on either hand are for the most part in a liquid state, and the Gulf Stream, let it be supposed, has assumed a normal condition between the two divisions, adjusting itself to the pressure on either side so as to balance them exactly and be in equilibrium. Now, again, it is the dead of winter, and the temperature of the waters over an area of millions of square miles in the North Atlantic has been changed many degrees, and this change of temperature has been followed likewise by a change in volume of those waters, amounting, no doubt, in the aggregate, to many hundred millions of tons, over the whole ocean; for sea water, unlike fresh (§ 103), contracts to freezing, and below. Now is it probable that, in passing from their summer to their winter temperature, the sea waters to the right of the Gulf Stream should change their specific gravity exactly as much in the aggregate as do the waters in the whole ocean to the left of it? If not, the difference must be compensated by some means. Sparks are not more prone to fly upward, nor water to seek its level, than Nature is sure with her efforts to restore equilibrium in both sea and air whenever, whereever, and by whatever it be disturbed. Therefore, though the waters of the Gulf Stream do not extend to the bottom, and though they be not impenetrable to the waters on either hand, yet, seeing that they have a waste of waters on the right and a waste of waters on the left, to which (§ 70) they offer a sort of resisting permeability, we are enabled to comprehend how the waters on either hand, as their specific gravity is increased or diminished, will impart to the trough of this stream a vibratory motion, pressing it now to the right, now to the left, according to the seasons and the consequent changes of temperature in the sea.

126. *Limits of the Gulf Stream in March and September.* Plate III shows the limits of the Gulf Stream for March and September. The reason for this change of position is obvious. The banks of the Gulf Stream (§ 70) are cold water. In winter the volume of cold water on the American, or left side of the stream, is greatly increased. It must have room, and gains it by pressing the warmer waters of the stream farther to the south, or right. In September, the temperature of these cold waters is modified; there is not such an extent of them, and then the warmer waters, in turn, press them back, and so the pendulum-like motion is preserved.

127. *Reluctance of layers or patches to mingle.* In the offings of the Balize, sometimes as far out as a hundred miles or more from the land, puddles or patches of Mississippi water may be observed on the surface of the sea with little or none of its brine mixed with it. This *anti-mixing* property in water has already (§ 98) been remarked upon. It may be observed, from the gutters in the street to the rivers in the ocean, and every where, wherever two bodies of water that differ in color are found

in juxtaposition. The patches of white, black, green, yellow, and reddish waters so often met with at sea are striking and familiar examples. We have seen, also, that a like proclivity exists (§ 99) between bodies or streams of water that differ in temperature or velocity. This peculiarity is often so strikingly developed in the neighborhood of the Gulf Stream that persons have been led to suppose that the Gulf Stream has forks in the sea, and that these are they.

128. *Streaks of warm and cool.* Now, if any vessel will take up her position a little to the northward of Bermuda, and steering thence for the Capes of Virginia, will try the water-thermometer all the way at short intervals, she will find its readings to be now higher, now lower; and the observer will discover that he has been crossing streak after streak of warm and cool water in regular alternations. He will then cease to regard them as bifurcations of the Gulf Stream, and view them rather in the light of thermal streaks of water which have, in the plan of oceanic circulation and in the system of unequal heating and cooling, been brought together.

129. *Waters of the ocean kept in motion by thermo-dynamical means.* The waters of the Gulf Stream form by no means the only body of warm water that the thermo-dynamical forces of the ocean keep in motion. Nearly all that portion of the Atlantic which lies between the Gulf Stream and the island of Bermuda has its surface covered with water which a tropical sun and tropical winds have played upon — with water, the specific gravity of which has been altered by their action, and which is now drifting to more northern climes in the endless search after lost equilibrium. This water, moreover, as well as that of the Gulf Stream, cools unequally. It would be surprising if it did not; for by being spread out over such a large area, and then drifting for so great a distance, and through such a diversity of climates, it is not probable that all parts of it should have been exposed to like vicissitudes by the way, or even to the same thermal conditions; therefore all of the water over such a surface can not be heated alike; radiation here, sunshine there; clouds and rain one day, and storms the next; the unequal depths; the breaking up of the fountains below, and the bringing their cooler waters to the surface by the violence of the waves, may all be expected, and are well calculated to produce unequal heating in the torrid and unequal cooling in the temperate zone; the natural result of which would be streaks and patches of water differing in temperature. Hence it would be surprising if, in crossing this drift and stream (Plate III) with the water-thermometer, the observer should find the water all of one temperature. By the time it has reached the parallel of Bermuda or "the Capes" of the Chesapeake, some of this water has been ten days, some ten weeks, and some perhaps longer on its way from the "caldron" at the south. It has consequently had ample

time to arrange itself into those differently-tempered streaks and layers (§ 127) which are so familiar to navigators, and which have been mistaken for "forks of the Gulf Stream."

130. *Fig. A, Plate III.* Curves showing some of these variations of temperature have been projected by the Coast Survey on a chart of engraved squares (Fig. A, Plate III). These curves show how these waters have arranged themselves off the Capes of Virginia into a series of thermal elevations and depressions.

131. *The high temperature and drift in the western half of North Atlantic and Pacific Oceans.* In studying the Gulf Stream, the high temperature and drift of the waters to the east of it are worthy of consideration. The Japan current (§ 80) has a like drift of warm water to the east of it also (Plates III and VI). In the western half, reaching up from the equator to the Gulf Stream, both of the North Atlantic and North Pacific, the water is warmer, parallel for parallel, than it is in the eastern half. On the west side, where the water is warm, the flow is to the north; on the east side, where the temperature is lower, the flow is to the south — making good the remark (§ 80) that, when the waters of the sea meet in currents, the tendency of the warm is to seek cooler latitudes, and of the cool, warmer.

132. *A Gulf Stream in each.* The Gulf Stream of each has its genesis on the west side, and in its course skirts the coast along; leaving the coast, it strikes off to the eastward in each case, losing velocity and spreading out. Between each of these Gulf Streams and its coasts there is a current of cool water setting to the south. On the outside of each stream, and coming up from the tropics, is a broad sheet of warm water; it covers an area of thousands of square miles, and its drift is to the north. Between the northern drift on the one side and the southern set on the other, there is in each ocean a sargasso (§ 88), into which all drift matter, such as wood and weeds, finds its way. In both oceans the Gulf Streams sweep across to the eastern shores, and so, bounding these seas, interpose a barrier between them and the higher parallels of latitude, which this drift matter can not pass. Such are the points of resemblance between the two oceans and in the circulation of their waters.

133. *Their connection with the Arctic Ocean.* A prominent point for contrast is afforded by the channels or water-ways between the Arctic and these two oceans. With the Atlantic they are divers and large; with the Pacific there is but one, and it is both narrow and shallow. In comparison with that of the Atlantic, the Gulf Stream of the Pacific is sluggish, ill defined, and irregular. Were the water-ways between the Atlantic and the Arctic Ocean no larger than Behring's Straits, our Gulf Stream would fall far below that of the Pacific in majesty and grandeur.

134. *The sargassos show the feeble power of the trade-winds upon*

currents. Here I am reminded to turn aside and call attention to another fact that militates against the vast current-begetting power that has been given by theory to the gentle trade-winds. In both oceans these weedy seas lie partly within the trade-wind region; but in neither do these winds give rise to any current. The weeds are partly out of water, and the wind has therefore more power upon them than it has upon the water itself; and if the supreme power over the currents of the sea reside in the winds, as Sir John Herschel would have it, then, of all places in the trade-wind region, we should have here the strongest currents. Had there been currents here, these weeds would have been borne away long ago; but, so far from it, we simply know that they have been in the Sargasso Sea (§ 88) of the Atlantic since the first voyage of Columbus. But to take up the broken thread.

135. *The drift matter confined to sargassos by currents.* The water that is drifting north, on the outside of the Gulf Stream, turns, with the Gulf Stream, to the east also. It can not reach the high latitudes (§ 80), for it can not cross the Gulf Stream. Two streams of water can not cross each other, unless one dip down and underrun the other; and if this drift water do dip down, as it may, it can not carry with it its floating matter, which, like its weeds, is too light to sink. They, therefore, are cut off from a passage into higher latitudes.

136. *Theory as to the formation of sargassos.* According to this view, there ought to be a sargasso sea somewhere in the sort of middle ground between the grand equatorial flow and reflow which is performed by the waters of all the great oceans. The place where the drift matter of each sea would naturally collect would be in this sort of pool, into which every current, as it goes from the equator, and again as it returns, would slough off its drift matter. The forces of diurnal rotation would require this collection of drift to be, in the northern hemisphere, on the right-hand side of the current, and, in the southern, to be on the left. (See Chap. XVIII and Plate VI.)

137. *Sargassos of southern seas to the left of the southern, to the right of the great polar and equatorial flow and reflow.* Thus, with the Gulf Stream of the Atlantic, and the "Black Stream" of the Pacific, their sargassos are on the right, as they are also on the right of the returning and cooler currents on the eastern side of each one of those northern oceans. So, also, with the Mozambique current, which runs south along the east coast of Africa from the Indian Ocean, and with the cooler current setting to the north on the Australian side of the same sea. Between these there is a sargasso on the left.

138. *Their position conforms to the theory.* Again, there is in the South Pacific a flow of equatorial waters to the Antarctic on the east of Australia, and of Antarctic waters (Humboldt's current) to the north,

along the western shores of South America; and, according to this principle, there ought to be another sargasso somewhere between New Zealand and the coast of Chili. (See Plate VI.)

139. *The discovery of a new sargasso.* To test the correctness of this view, I requested Lieut. Warley to overhaul our sea-journals for notices of kelp and drift matter on the passage from Australia to Cape Horn and the Chincha Islands. He did so, and found it abounding in small patches, with "many birds about," between the parallels of 40° and 50° south, the meridians of 140° and 178° west. This sargasso is directly south of the Georgian Islands, and is, perhaps, less abundantly supplied with drift matter, less distinct in outline, and less permanent in position than any one of the others.

140. *One in the South Atlantic.* There is no warm current, or if one, a very feeble one, flowing out of the South Atlantic. Most of the drift matter borne upon the ice-bearing current into that sea finds its way to the equator, and then into the veins which give volume to the Gulf Stream and supply the sargasso of the North Atlantic with extra quantities of drift. The sargasso of the South Atlantic is therefore small. The formations and physical relations of sargassos will be again alluded to in Chapter XVIII.

141. *The large volume of warm water outside of the Gulf Stream.* Let us return (§ 129) to this great expanse of warm water which, coming from the torrid zone on the southwestern side of the Atlantic, drifts along to the north on the outside of the Gulf Stream. Its velocity is slow, not sufficient to give it the name of current; it is a drift, or what sailors call a "set." By the time this water reaches the parallel of 35° or 40° it has parted with a good deal of its intertropical heat; consequent upon this change in temperature is a change in specific gravity also, and by reason of this change as well as by the difficulties of crossing the Gulf Stream, its progress to the north is arrested. It now turns to the east with the Gulf Stream, and, yielding to the force of the westerly winds of this latitude, is by them *slowly drifted* along; losing temperature by the way, these waters reach the southwardly flow on the east side with their specific gravity so altered that, disregarding the gentle forces of the wind, they heed the voice of the sea, and proceed to unite with this cool flow, and to set south in obedience to those dynamical laws that derive their force in the sea from differing specific gravity.

142. *The resemblance between the currents in the North Atlantic and the North Pacific.* The Thermal Charts of the North Atlantic afford for these views other illustrations which, when compared with the charts of the North Pacific now in the process of construction, will make still more striking the resemblance of the two oceans in the general features of their systems of circulation. We see how, in accordance with this principle

(§ 132), the currents necessary for the formation of thickly-set sargassos are generally wanting in southern oceans. How closely these two seas of the north resemble each other; and how, on account of the large openings between the Atlantic and the Frozen Oceans, the flow of warm waters to the north and of cold waters to the south is so much more active in the Atlantic than it is in the Pacific. Ought it not so to be?

143. *A cushion of cool water protects the bottom of the deep sea from abrasion by its currents.* As a rule, the hottest water of the Gulf Stream is at or near the surface; and as the deep-sea thermometer is sent down, it shows that these waters, though still far warmer than the water on either side at corresponding depths, gradually become less and less warm until the bottom of the current is reached. There is reason to believe that the warm waters of the Gulf Stream are nowhere permitted, in the oceanic economy, to touch the bottom of the sea. There is every where a cushion of cool water between them and the solid parts of the earth's crust. This arrangement is suggestive, and strikingly beautiful. One of the benign offices of the Gulf Stream is to convey heat from the Gulf of Mexico, where otherwise it would become excessive, and to dispense it in regions beyond the Atlantic for the amelioration of the climates of the British Islands and of all Western Europe. Now cold water is one of the best non-conductors of heat, and if the warm water of the Gulf Stream was sent across the Atlantic in contact with the solid crust of the earth — comparatively a good conductor of heat — instead of being sent across, as it is, in contact with a cold, non-conducting cushion of cool water to fend it from the bottom, much of its heat would be lost in the first part of the way, and the soft climates of both France and England would be, as that of Labrador, severe in the extreme, ice-bound, and bitterly cold.

CHAPTER III

§ 150–191. INFLUENCE OF THE GULF STREAM UPON CLIMATES AND COMMERCE

150. *How the Washington Observatory is warmed.* Modern ingenuity has suggested a beautiful mode of warming houses in winter. It is done by means of hot water. The furnace and the caldron are sometimes placed at a distance from the apartments to be warmed. It is so at the Observatory. In this case, pipes are used to conduct the heated water from the caldron under the superintendent's dwelling over into one of the basement rooms of the Observatory, a distance of one hundred feet. These pipes are then flared out so as to present a large cooling surface; after which they are united into one again, through which the water, being now cooled, returns of its own accord to the caldron. Thus cool water is returning all the time and flowing in at the bottom of the caldron, while hot water is continually flowing out at the top. The ventilation of the Observatory is so arranged that the circulation of the atmosphere through it is led from this basement room, where the pipes are, to all other parts of the building; and in the process of this circulation, the warmth conveyed by the water to the basement is taken thence by the air and distributed over all the rooms. Now, to compare small things with great, we have, in the warm waters which are confined in the Gulf of Mexico, just such a heating apparatus for Great Britain, the North Atlantic, and Western Europe.

151. *An analogy showing how the Gulf Stream raises temperature in Europe.* The furnace is the torrid zone; the Mexican Gulf and Caribbean Sea are the caldrons; the Gulf Stream is the conducting pipe. From the Grand Banks of Newfoundland to the shores of Europe is the basement — the hot-air chamber — in which this pipe is flared out so as to present a large cooling surface. Here the circulation of the atmosphere is arranged by nature; it is from west to east; consequently it is such that the warmth thus conveyed into this warm-air chamber of mid-ocean is taken up by the genial west winds, and dispensed, in the most benign manner, throughout Great Britain and the west of Europe. The mean temperature of the water-heated air-chamber of the Observatory is about 90°. The maximum temperature of the Gulf Stream is 86°, or about 9° above the ocean temperature due the latitude. Increasing its latitude 10°, it loses but 2° of temperature; and, after having run three thousand miles toward

the north, it still preserves, even in winter, the heat of summer. With this temperature, it crosses the 40th degree of north latitude, and there, overflowing its liquid banks, it spreads itself out for thousands of square leagues over the cold waters around, covering the ocean with a mantle of warmth that serves so much to mitigate in Europe the rigors of winter. Moving now more slowly, but dispensing its genial influences more freely, it finally meets the British Islands. By these it is divided (Plate VI), one part going into the polar basin of Spitzbergen, the other entering the Bay of Biscay, but each with a warmth considerably above the ocean temperature. Such an immense volume of heated water can not fail to carry with it beyond the seas a mild and moist atmosphere. And this it is which so much softens climate there.

152. *Depth and temperature.* We know not, except approximately in a few places, what the depth or the under temperature of the Gulf Stream may be; but *assuming* the temperature and velocity at the depth of two hundred fathoms to be those of the surface, and taking the well-known difference between the capacity of air and of water for specific heat as the argument, a simple calculation will show that the quantity of heat discharged over the Atlantic from the waters of the Gulf Stream in a winter's day would be sufficient to raise the whole column of atmosphere that rests upon France and the British Islands from the freezing point to summer heat.

153. *Contrasts of climates in the same latitudes.* Every west wind that blows crosses the stream on its way to Europe, and carries with it a portion of this heat to temper there the northern winds of winter. It is the influence of this stream upon climate that makes Erin the "Emerald Isle of the Sea," and that clothes the shores of Albion in evergreen robes; while in the same latitude, on this side, the coasts of Labrador are fast bound in fetters of ice. In a valuable paper on currents, Mr. Redfield [1] states that in 1831 the harbor of St. John's, Newfoundland, was closed with ice as late as the month of June; yet who ever heard of the port of Liverpool, on the other side, though 2° farther north, being closed with ice, even in the dead of winter?

154. *Mildness of an Orkney winter.* The Thermal Chart (Fig. 9, p. 346) shows this. The isothermal lines of 60°, 50°, etc., starting off from the parallel of 40° near the coasts of the United States, run off in a northeastwardly direction, showing the same oceanic temperature on the European side of the Atlantic in latitude 55° or 60°, that we have on the western side in latitude 40°. Scott, in one of his beautiful novels, tells us that the ponds in the Orkneys (latitude near 60°) are not frozen in winter. The people there owe their soft climate to this grand heating ap-

[1] [W. C. Redfield, "Remarks on Tides and the Prevailing Currents of the Ocean and Atmosphere,"] *Amer. Jour. Sci.*, ser. 2, XIV[(1843)] 293[-300, esp., p. 298].

paratus, for driftwood from the West Indies is occasionally cast ashore there by the Gulf Stream.

155. *Amount of heat daily escaping through the Gulf Stream.* Nor do the beneficial influences of this stream upon climate end here. The West Indian Archipelago is encompassed on one side by its chain of islands, and on the other by the Cordilleras of the Andes, contracting with the Isthmus of Darien, and stretching themselves out over the plains of Central America and Mexico. Beginning on the summit of this range, we leave the regions of perpetual snow, and descend first into the *tierra témplada,* and then into the *tierra caliente,* or burning land. Descending still lower, we reach both the level and the surface of the Mexican seas, where, were it not for this beautiful and benign system of aqueous circulation, the peculiar features of the surrounding country assure us we should have the hottest, if not the most pestilential climate in the world. As the waters in these two caldrons become heated, they are borne off by the Gulf Stream, and are replaced by cooler currents through the Caribbean Sea; the surface water, as it enters here, being 3° or 4°, and that in depth even 40° cooler than when it escapes from the Gulf.[2] Taking only this difference in surface temperature as an index of the heat accumulated there, a simple calculation will show that the quantity of heat daily carried off by the Gulf Stream from those regions, and discharged over the Atlantic, is sufficient to raise mountains of iron from zero to the melting point, and to keep in flow from them a molten stream of metal greater in volume than the waters daily discharged from the Mississippi River.

156. *Its benign influences.* Who, therefore, can calculate the benign influence of this wonderful current upon the climate of the South? In the pursuit of this subject, the mind is led from nature up to the great Architect of nature; and what mind will the study of this subject not fill with profitable emotions? Unchanged and unchanging alone, of all created things, the ocean is the great emblem of its everlasting Creator. "He treadeth upon the waves of the sea," and is seen in the wonders of the deep. Yea, "He calleth for its waters, and poureth them out upon the face of the earth." In obedience to this call, the aqueous portion of our planet preserves its beautiful system of circulation. By it heat and warmth are dispensed to the extra-tropical regions; clouds and rain are sent to refresh the dry land; and by it cooling streams are brought from Polar Seas to temper the heat of the torrid zone. At the depth of two hundred and forty fathoms, the temperature of the currents setting into the Caribbean Sea has been found as low as 48°, while that of the sur-

---

[2] Temperature of the Caribbean Sea (from the journals of Mr. Dunsterville): Surface temperature: 83°, September; 84°, July; 83°–86½°, Mosquito Coast. Temperature in depth: 48°, 240 fathoms; 43°, 386 fathoms; 42°, 450 fathoms; 43°, 500 fathoms.

face was 85°. Another cast with three hundred and eighty-six fathoms gave 43° below against 83° at the surface. The hurricanes of those regions agitate the sea to great depths; that of 1780 tore rocks up from the bottom seven fathoms deep, and cast them ashore. They therefore can not fail to bring to the surface portions of the cooler water below.

157. *Cold water at the bottom of the Gulf Stream.* At the very bottom of the Gulf Stream, when its surface temperature was 80°, the deep-sea thermometer of the Coast Survey has recorded a temperature as low as 35° Fahrenheit. These cold waters doubtless come down from the north to replace the warm water sent through the Gulf Stream to moderate the cold of Spitzbergen; for within the Arctic Circle the temperature at corresponding depths off the shores of that island is said to be only one degree colder than in the Caribbean Sea, while on the coasts of Labrador and in the Polar Seas the temperature of the water beneath the ice was invariably found by Lieutenant De Haven at 28°, or 4° below the melting point of fresh-water ice. Captain Scoresby relates, that on the coast of Greenland, in latitude 72°, the temperature of the air was 42°; of the water, 34°; and 29° at the depth of one hundred and eighteen fathoms. He there found a surface current setting to the south, and bearing with it this extremely cold water, with vast numbers of icebergs, whose centres, perhaps, were far below zero. It would be curious to ascertain the routes of these under currents on their way to the tropical regions, which they are intended to cool. One has been found at the equator (§ 97) two hundred miles broad and 23° colder than the surface water. Unless the land or shoals intervene, it no doubt comes down in a spiral curve (§ 96), approaching in its course the great circle route.

158. *Fish and currents.* Perhaps the best indication as to these cold currents may be derived from the fish of the sea. The whales, by avoiding its warm waters, pointed out to the fisherman the existence of the Gulf Stream. Along our own coasts, all those delicate animals and marine productions which delight in warmer waters are wanting; thus indicating, by their absence, the prevalence of the cold current from the north now known to exist there. In the genial warmth of the sea about the Bermudas on one hand, and Africa on the other, we find, in great abundance, those delicate shell-fish and coral formations which are altogether wanting in the same latitudes along the shores of South Carolina. The same obtains in the west coast of South America; for there the immense flow of polar waters known as Humboldt's Current almost reaches the line before the first sprig of coral is found to grow. A few years ago, great numbers of bonita and albercore — tropical fish — following the Gulf Stream, entered the English Channel, and alarmed the fishermen of Cornwall and Devonshire by the havoc which they created among the pilchards there. It may well be questioned if our Atlantic cities and towns do not owe

their excellent fish-markets, as well as our watering-places their refreshing sea-bathing in summer, to this littoral stream of cold water. The temperature of the Mediterranean is 4° or 5° above the ocean temperature of the same latitude, and the fish there are, for the most part, very indifferent. On the other hand, the temperature along our coast is several degrees below that of the ocean, and from Maine to Florida our tables are supplied with the most excellent of fish. The sheep's-head of this cold current, so much esteemed in Virginia and the Carolinas, loses its flavor, and is held in no esteem, when taken on the warm coral banks of the Bahamas. The same is the case with other fish: when taken in the cold water of that coast, they have a delicious flavor and are highly esteemed; but when taken in the warm water on the other edge of the Gulf Stream, though but a few miles distant, their flesh is soft and unfit for the table. The temperature of the water at the Balize reaches 90°. The fish taken there are not to be compared with those of the same latitude in this cold stream. New Orleans, therefore, resorts to the cool waters on the Florida coasts for her choicest fish. The same is the case in the Pacific. A current of cold water (§ 398) from the south sweeps the shores of Chili, Peru, and Columbia, and reaches the Gallipagos Islands under the equator. Throughout this whole distance, the world does not afford a more abundant or excellent supply of fish. Yet out in the Pacific, at the Society Islands, where coral abounds, and the water preserves a higher temperature, the fish, though they vie in gorgeousness of coloring with the birds, and plants, and insects of the tropics, are held in no esteem as an article of food. I have known sailors, even after long voyages, still to prefer their salt beef and pork to a mess of fish taken there. The few facts which we have bearing upon this subject seem to suggest it as a point of the inquiry to be made, whether the habitat of certain fish does not indicate the temperature of the water; and whether these cold and warm currents of the ocean do not constitute the great highways through which migratory fishes travel from one region to another. Why should not fish be as much the creatures of climate as plants, or as birds and other animals of land, sea, and air? Indeed, we know that some kinds of fish are found only in certain climates. In other words, they live where the temperature of the water ranges between certain degrees.

159. *A shoal of sea-nettles.* Navigators have often met with vast numbers of young sea-nettles (*medusæ*) drifting along with the Gulf Stream. They are known to constitute the principal food for the whale; but whither bound by this route has caused much curious speculation, for it is well known that the habits of the right whale are averse to the warm waters of this stream. An intelligent sea-captain informs me that, several years ago, in the Gulf Stream off the coast of Florida, he fell in with such a "school of young sea-nettles as had never before been heard of." The

sea was covered with them for many leagues. He likened them, as they appeared on near inspection in the water, to acorns floating on a stream; but they were so thick as to completely cover the sea, giving it the appearance, in the distance, of a boundless meadow in the yellow leaf. He was bound to England, and was five or six days in sailing through them. In about sixty days afterward, on his return, he fell in with the same school off the Western Islands, and here he was three or four days in passing them again. He recognized them as the same, for he had never before seen any like them; and on both occasions he frequently hauled up buckets full and examined them.

160. *Food for whales.* Now the Western Islands is the great place of resort for whales; and at first there is something curious to us in the idea that the Gulf of Mexico is the harvest field, and the Gulf Stream the gleaner which collects the fruitage planted there, and conveys it thousands of miles off to the hungry whale at sea. But how perfectly in unison is it with the kind and providential care of that great and good Being which caters for the sparrow, and feeds the young ravens when they cry!

161. *Piazzi Smyth's description.* Piazzi Smyth, the Astronomer Royal of Edinburgh, when bound to Teneriffe on his celebrated astronomical expedition of 1856, fell in with the annual harvest of these creatures. They were in the form of hollow gelatinous lobes, arranged in groups of five or nine — each lobe having an orange vein down the centre. Thus each animal was formed of an aggregation of lobes, with an orange-colored vein, or stomach, in every lobe. "Examining," says he, "in the microscope a portion of one of the orange veins, apparently the stomach of the creature, it was found to be extraordinarily rich in diatomes, and of the most bizarre forms, as stars, Maltese crosses, embossed circles, semicircles, and spirals. The whole stomach could hardly have contained less than seven hundred thousand; and when we multiply them by the number of lobes, and then by the number of groups, we shall have some idea of the countless millions of diatomes that go to make a feast for the medusæ — some of the softest things in the world thus confounding and devouring the hardest — the flinty-shelled diatomacæ." [3] Each of these "sea-nettles," as the sailors call them, had in his nine stomachs not less, according to this computation, than five or six millions of these mites of flinty shells, the materials of which their inhabitants had collected from the silicious matter which the rains washed out from the valleys, and which the rivers are continually rolling down to the sea.

162. *The waters of the sea bring forth — oh how abundantly!* The medusæ have the power of sucking in the sea-water slowly, and of ejecting it again with more or less force. Thus they derive both food and the power of locomotion, for, in the passage of the water, they strain it

[3] [C. Piazzi Smyth, *Teneriffe: An Astronomer's Experiment* (London, 1858), p. 6.]

and collect the little diatomes. Imagine, now, how many medusæ-mouthfuls of water there must be in the sea, which, though loaded with diatomes, are never filtered through the stomachs of these creatures; imagine how many medusæ the whale must gulp down with every mouthful; imagine how deep and thickly the bottom of the sea must, during the process of ages, have become covered with the flinty remains of these little organisms; now call to mind the command which was given to the waters of the sea on the fifth day of creation; and then the powers of the imagination are silenced by the emotions of wonder, love, and praise.

163. *Contrasts between the climates of land and sea.* The sea has its climates as well as the land. They both change with the latitude; but one varies with the elevation above, the other with the depression below the sea level. The climates in each are regulated by circulation; but the regulators are, on the one hand, winds; on the other, currents.

164. *Order and design.* The inhabitants of the ocean are as much the creatures of climate as are those of the dry land; for the same Almighty hand which decked the lily and cares for the sparrow, fashioned also the pearl and feeds the great whale; He adapted each to the physical conditions by which his providence has surrounded it. Whether of the land or the sea, the inhabitants are all his creatures, subjects of his laws, and agents in his economy. The sea, therefore, we may safely infer, has its offices and duties to perform; so, may we infer, have its currents, and so, too, its inhabitants; consequently, he who undertakes to study its phenomena must cease to regard it as a waste of waters. He must look upon it as a part of that exquisite machinery by which the harmonies of nature are preserved, and then he will begin to perceive the developments of order and the evidences of design; viewed in this light, it becomes a most beautiful and interesting subject for contemplation.

165. *Terrestrial adaptations.* To one who has never studied the mechanism of a watch, its main-spring or the balance-wheel is a mere piece of metal. He may have looked at the face of the watch, and, while he admires the motion of its hands, and the time it keeps, or the tune it plays, he may have wondered in idle amazement as to the character of the machinery which is concealed within. Take it to pieces, and show him each part separately; he will recognize neither design, nor adaptation, nor relation between them; but put them together, set them to work, point out the offices of each spring, wheel, and cog, explain their movements, and then show him the result; now he perceives that it is all *one* design; that, notwithstanding the number of parts, their diverse forms and various offices, and the agents concerned, the whole piece is of *one* thought, the expression of *one* idea. He now rightly concludes that when the main-spring was fashioned and tempered, its relation to all the other parts must have been considered; that the cogs on this wheel are cut and

regulated — *adapted* — to the rachets on that, &c.; and his final conclusion will be, that such a piece of mechanism could not have been produced by chance; for the adaptation of the parts is such as to show it to be according to design, and obedient to the will of *one* intelligence. So, too, when one looks out upon the face of this beautiful world, he may admire its lovely scenery, but his admiration can never grow into adoration unless he will take the trouble to look behind and study, in some of its details at least, the exquisite system of machinery by which such beautiful results are brought about. To him who does this, the sea, with its physical geography, becomes as the main-spring of a watch; its waters, and its currents, and its salts, and its inhabitants, with their adaptations, as balance-wheels, cogs, and pinions, and jewels in the terrestrial mechanism. Thus he perceives that they too are according to design — parts of the physical machinery that are the expression of One Thought, a unity with harmonies which One Intelligence, and One Intelligence alone, could utter. And when he has arrived at this point, then he feels that the study of the sea, in its physical aspects, is truly sublime. It elevates the mind and ennobles the man. The Gulf Stream is now no longer, therefore, to be regarded by such an one merely as an immense current of warm water running across the ocean, but as a balance-wheel — a part of that grand machinery by which air and water are adapted to each other, and by which this earth itself is adapted to the well-being of its inhabitants — of the flora which decks, and the fauna which enlivens its surface.

166. *Meteorology of the Sea: Gulf Stream the weather-breeder — its storms — the great hurricane of 1780.* Let us now consider the *Influence of the Gulf Stream upon the Meteorology of the Ocean.* To use a sailor expression, the Gulf Stream is the great "weather-breeder" of the North Atlantic Ocean. The most furious gales of wind sweep along with it; and the fogs of Newfoundland, which so much endanger navigation in spring and summer, doubtless owe their existence to the presence, in that cold sea, of immense volumes of warm water brought by the Gulf Stream. Sir Philip Brooke found the air on each side of it at the freezing point, while that of its waters was 80°. "The heavy, warm, damp air over the current produced great irregularities in his chronometers." The excess of heat daily brought into such a region by the waters of the Gulf Stream would, if suddenly stricken from them, be sufficient to make the column of superincumbent atmosphere hotter than melted iron. With such an element of atmospherical disturbance in its bosom, we might expect storms of the most violent kind to accompany it in its course. Accordingly, the most terrific that rage on the ocean have been known to spend their fury within or near its borders. Of all storms, the hurricanes of the West Indies and the typhoons of the China seas cause the most ships to founder. The stoutest men-of-war go down before them, and seldom, indeed, is any one of the crew left to tell the tale. Of this the Hornet, the

Albany, and the Grampus, armed cruisers in the American navy, all are memorable and melancholy examples. Our nautical works tell us of a West India hurricane so violent that it forced the Gulf Stream back to its sources, and piled up the water in the Gulf to the height of thirty feet. The Ledbury Snow attempted to ride it out. When it abated, she found herself high up on the dry land, and discovered that she had let go her anchor among the tree-tops on Elliott's Key. The Florida Keys were inundated many feet, and, it is said, the scene presented in the Gulf Stream was never surpassed in awful sublimity on the ocean. The water thus dammed up is said to have rushed out with frightful velocity against the fury of the gale, producing a sea that beggared description. The "great hurricane" of 1780 commenced at Barbadoes. In it the bark was blown from the trees, and the fruits of the earth destroyed; the very bottom and depths of the sea were uprooted, and the waves rose to such a height that forts and castles were washed away, and their great guns carried about in the air like chaff; houses were razed, ships were wrecked, and the bodies of men and beasts lifted up in the air and dashed to pieces in the storm. At the different islands, not less than twenty thousand persons lost their lives on shore, while farther to the north, the "Sterling Castle" and the "Dover Castle," men-of-war, went down at sea, and fifty sail were driven on shore at the Bermudas.

167. *Inquiries instituted by the Admiralty.* Several years ago the British Admiralty set on foot inquiries as to the cause of the storms in certain parts of the Atlantic, which so often rage with disastrous effects to navigation. The result may be summed up in the conclusion to which the investigation led: that they are occasioned by the irregularity between the temperature of the Gulf Stream and of the neighboring regions, both in the air and water.

168. *The most stormy sea.* The southern points of South America and Africa have won for themselves, among seamen, the name of "the stormy capes;" but investigations carried on in that mine of sea-lore contained in the log-books at the National Observatory have shown that there is not a storm-fiend in the wide ocean can out-top that which rages along the Atlantic coasts of North America. The China seas and the North Pacific may vie in the fury of their gales with this part of the Atlantic, but Cape Horn and the Cape of Good Hope can not equal them, certainly, in frequency, nor do I believe in fury.

169. *Northern seas more boisterous than southern.* In the ex-tropical regions of the south we lack those contrasts which the mountains, the deserts, the plains, the continents, and the seas of the north afford for the production of atmospherical disturbances. Neither have we in the southern seas such contrasts of hot and cold currents. The flow of warm water toward the pole, and of polar water toward the equator is as great — perhaps if measured according to volume, is greater in the southern

hemisphere. But in the southern hemisphere the currents are broad and sluggish; in the northern, narrow, sharp, and strong. Then we have in the north other climatic contrasts for which we may search southern seas in vain. Hence, without further investigation, we may infer southern seas to be less boisterous than northern.

170. *Storms in the North Atlantic and Pacific.* By a like reasoning we may judge the North Pacific to be less boisterous than the North Atlantic; for, though we have continental climates on either side of each, and a Gulf Stream in both, yet the Pacific is a very much wider sea, and its Gulf Stream is (§ 54) not so warm, nor so sharp, nor so rapid; therefore the broad Pacific does not, on the whole, present the elements of atmospherical disturbance in that compactness which is so striking in the narrow North Atlantic.

171. *Storms along their western shores.* Nevertheless, though the North Pacific generally may not be so stormy as the North Atlantic, we have reason to believe that meteorological agents of nearly equal power are clustered along the western shores of each ocean. Though the Gulf Stream of the Pacific is not so hot, nor the cool littoral currents so cold as those of our ocean are, yet they lave the shores of a broader continent, and hug them quite as closely as ours do. Moreover, the Japan Current, with its neighboring seas, is some 500 miles nearer to the pole of maximum cold than the Gulf Stream of the Atlantic is.

172. *Position of the poles of maximum cold, and their influence upon the meteorology of these two oceans.* Some philosophers hold that there are in the northern hemisphere two poles of maximum cold: The Asiatic, near the intersection of the parallel of 80° with the meridian of 120° E., and the American, near lat. 79° and long. 100° W. The Asiatic pole is the colder. The distance between it and the Japan Current is about 1500 miles; the distance between the other pole and the Gulf Stream is about 2000 miles. The bringing of the heat of summer, as these two streams do, in such close juxtaposition with the cold of winter, can not fail to produce violent commotions in the atmosphere. These commotions, as indicated by the storms, are far more frequent and violent in winter, when the contrasts between the warm and cool places are greater, than they are in summer, when those contrasts are least. Moreover, each of these poles is to the northwest of its ocean, the quarter whence come the most terrific gales of winter. Whatever be the exact degree of influence which future research may show to be exercised by these cool places, and the heat dispensed so near them by these mighty streams of tepid water, there is reason to believe that they do act and react upon each other with no inconsiderable meteorological power. In winter the Gulf Stream carries the temperature of summer as far north as the Grand Banks of Newfoundland.

173. *Climates of England and silver fogs of Newfoundland.* The

habitual dampness of the climate of the British Islands, as well as the occasional dampness of that along the Atlantic coasts of the United States when easterly winds prevail, is attributable also to the Gulf Stream. These winds come to us loaded with vapors gathered from its warm and smoking waters. The Gulf Stream carries the temperature of summer, even in the dead of winter, as far north as the Grand Banks of Newfoundland, and there maintains it in the midst of the severest frosts. It is the presence of this warm water and a cold atmosphere in juxtaposition there which gives rise to the "silver fogs" of Newfoundland, one of the most beautiful phenomena to be seen any where among the treasures of the frost-king.

174. *Influences upon storms.* The Gulf Stream exercises a powerful influence upon the storms of the North Atlantic, especially those which take their rise within the tropics, even as far over as the coast of Africa, and also upon those which, though not intertropical in their origin, are known to visit the offings of our own coasts. These gales, in whatever part of the ocean east of the Gulf Stream they take their rise, march to the northwest until they join it, when they "recurvate," as the phrase is, and take up their line of march to the northeast along with it. Gales of wind have been traced from latitude 10° N. on the other side of the Atlantic to the Gulf Stream on this, and then with it back again to the other side, off the shores of Europe. By examining the log-books of ships, the tracks of storms have been traced out and followed for a week or ten days. Their path is marked by wreck and disaster. At a meeting of the American Association for the Advancement of Science, in 1854, Mr. Redfield mentioned one which he had traced out, and in which no less than seventy odd vessels had been wrecked, dismasted, or damaged.[4]

175. *More observations in and about the Gulf Stream a desideratum.* Now, what should attract these storms to the Gulf Stream, is a question which yet remains to be satisfactorily answered. A good series of simultaneous barometric observations within and on either side of the Gulf Stream is a great desideratum in the meteorology of the Atlantic. At the equator, where the trade-winds meet and ascend, where the air is loaded with moisture, and the vapor from the warm waters below condensed into the equatorial cloud-ring above, we have a low barometer.

176. *Certain storms make for it and follow it.* How is it with the Gulf Stream when these storms from right and left burst in upon it, and, turning about, course along with it? Its waters are warm; they give off vapor rapidly; and, were this vapor visible to an observer in the moon, he no doubt would, on a winter's day especially, be able to trace out by the mist in the air the path of the Gulf Stream through the sea.

---

[4] [W. C. Redfield, "On the First Hurricane of September 1853, in the Atlantic, with a Chart; and Notices of Other Storms, *Amer. Jour. Sci.*, ser. 2, XVIII(1854), 1–18, 176–190, esp., p. 9, n.]

177. *How aqueous vapor assists in producing winds.* Let us consider the effect of vapor upon winds, and then the importance of the observations proposed (§ 175) will perhaps be better appreciated. Aqueous vapor assists in at least five, perhaps six ways to put air in motion and produce winds. (1.) By evaporation the air is cooled; by cooling its specific gravity is changed, and, consequently, here is one cause of movement in the air, as is manifest in the tendency of the cooled air to flow off, and of warmer and lighter to take its place. (2.) Excepting hydrogen and ammonia, there is no gas so light as aqueous vapor, its weight being to common air in the proportion of nearly 5 to 8; consequently, as soon as it is formed it commences to rise; and, as each vesicle of vapor may be likened, in the movements which it produces in the air, to a balloon as it rises, it will be readily perceived how these vaporous particles, as they ascend, become entangled with those of the air, and so, carrying them along, upward currents are produced: thus the wind is called on to rush in below, that the supply for the upward movement may be kept up. (3.) The vapor, being lighter than air, presses it out, and, as it were, takes its place, causing the barometer to fall: thus again an in-rush or wind is called for below. (4.) Arrived in the cloud-region, this vapor, being condensed, liberates the latent heat which it borrowed from the air and water below; which heat being now set free and made sensible, raises the temperature of the surrounding air, causing it to expand and ascend still higher; and so winds are again called for. Ever ready, they come; thus we have a fourth way. (5.) Innumerable rain-drops now begin to fall, and in their descent, as in a heavy shower, they displace and press the air out below with great force. To this cause Espy ascribes the gusts of wind which are often found to blow outward from the centre, as it were, of sudden and violent thunder-showers. (6.) Probably, and especially in thunder-storms, electricity may assist in creating movements in the atmosphere, and so make claim to be regarded as a wind-producing agent. But the winds are supposed to depend *mainly* on the power of agents (2), (3), and (4) for their violence.

178. *Storms in the interior attracted by the Gulf Stream.* These agents, singly and together, produce rarefaction, diminish pressure, and call for an inward rush of air from either side. Mr. Espy asserts, and quotes actual observation to sustain the assertion, that the storms of the United States, even those which arise in the Mississippi Valley, travel east, and often march out to sea, where they join the Gulf Stream in its course.[5] That those which have their origin at sea, on the other side of the Gulf Stream, do (§ 174) often make right for it, is a fact well known to most observant sailors. Hence the interest that is attached to a proper series of observations on the meteorology of the Gulf Stream.

[5] [James P. Espy, *Fourth Meteorological Report* (34th Cong., 3rd sess., *Sen. Ex. Doc.* no. 65), 1857, p. 106.]

179. *Storms of — dreaded by seamen.* Sailors dread its storms more than they do the storms in any other part of the ocean. It is not the fury of the storm alone that they dread, but it is the "ugly sea" which these storms raise. The current of the stream running in one direction, and the wind blowing in another, create a sea that is often frightful.

180. *Routes formerly governed by the Gulf Stream. The influence of the Gulf Stream upon commerce and navigation.* Formerly the Gulf Stream controlled commerce across the Atlantic by governing vessels in their routes through this ocean to a greater extent than it does now, and simply for the reason that ships are faster, nautical instruments better, and navigators are more skillful now than formerly they were.

181. *Difficulties with early navigators.* Up to the close of the last century, the navigator *guessed* as much as he *calculated* the place of his ship; vessels from Europe to Boston frequently made New York, and thought the landfall by no means bad. Chronometers, now so accurate, were then an experiment. The Nautical Ephemeris itself was faulty, and gave tables which involved errors of thirty miles in the longitude. The instruments of navigation erred by *degrees* quite as much as they now do by *minutes*; for the rude "cross staff" and "back staff," the "sea-ring" and "mariner's bow," had not yet given place to the nicer sextant and circle of reflection of the present day. Instances are numerous of vessels navigating the Atlantic in those times being 6°, 8°, and even 10° of longitude out of their reckoning in as many days from port.

182. *Finding longitude by the Gulf Stream.* Though navigators had been in the habit of crossing and recrossing the Gulf Stream almost daily for three centuries, it never occurred to them to make use of it as a means of giving them their longitude, and of warning them of their approach to the shores of this continent. Dr. Franklin was the first to suggest this use of it. The contrast afforded by the temperature of its waters and that of the sea between the Stream and the shores of America was striking. The dividing line between the warm and the cool waters was sharp (§ 70); and this dividing line, especially that on the western side of the stream, seldom changed its position as much in longitude as mariners often erred in their reckoning.

183. *Folger's Chart.* When he was in London, in 1770, he happened to be consulted as to a memorial which the Board of Customs at Boston sent to the Lords of the Treasury, stating that the Falmouth packets were generally a fortnight longer to Boston than common traders were from London to Providence, Rhode Island. They therefore asked that the Falmouth packets might be sent to Providence instead of to Boston. This appeared strange to the doctor, for London was much farther than Falmouth, and from Falmouth the routes were the same, and the difference should have been the other way. He, however, consulted Captain Folger, a Nantucket whaler, who chanced to be in London also; the fisherman

explained to the philosopher that the difference arose from the circumstance that the Rhode Island captains were acquainted with the Gulf Stream, while those of the English packets were not. The latter kept in it, and were set back sixty or seventy miles a day, while the former avoided it altogether. He had been made acquainted with it by the whales which were found on either side of it, but never in it (§ 158). At the request of the doctor, he then traced on a chart the course of this stream from the Straits of Florida. The doctor had it engraved at Tower Hill, and sent copies of it to the Falmouth captains, who paid no attention to it. The course of the Gulf Stream, as laid down by that fisherman from his general recollection of it, has been retained and quoted on the charts for navigation, we may say, until the present day. But the investigations of which we are treating are beginning to throw more light upon this subject; they are giving us more correct knowledge in every respect with regard to it, and to many other new and striking features in the physical geography of the sea.

184. *Using the Gulf Stream in winter.* No part of the world affords a more difficult or dangerous navigation than the approaches of our northern coast in winter. Before the warmth of the Gulf Stream was known, a voyage at this season from Europe to New England, New York, and even to the Capes of the Delaware or Chesapeake, was many times more trying, difficult, and dangerous than it now is. In making this part of the coast, vessels are frequently met by snow-storms and gales which mock the seaman's strength and set at naught his skill. In a little while his bark becomes a mass of ice; with her crew frosted and helpless, she remains obedient only to her helm, and is kept away for the Gulf Stream. After a few hours' run, she reaches its edge, and almost at the next bound passes from the midst of winter into a sea at summer heat. Now the ice disappears from her apparel; the sailor bathes his stiffened limbs in tepid waters; feeling himself invigorated and refreshed with the genial warmth about him, he realizes, out there at sea, the fable of Antæus and his mother Earth. He rises up, and attempts to make his port again, and is again, perhaps, as rudely met and beat back from the northwest; but each time that he is driven off from the contest, he comes forth from this stream, like the ancient son of Neptune, stronger and stronger, until, after many days, his freshened strength prevails, and he at last triumphs, and enters his haven in safety, though in this contest he sometimes falls to rise no more, for it is terrible. Many ships annually founder in these gales; and I might name instances, for they are not uncommon, in which vessels bound to Norfolk or Baltimore, with their crews enervated in tropical climates, have encountered, as far down as the Capes of Virginia, snow-storms that have driven them back into the Gulf Stream time and again, and have kept them out for forty, fifty, and even for sixty days, trying to make an anchorage.

185. *Running south to spend the winter.* Nevertheless, the presence of the warm waters of the Gulf Stream, with their summer heat in mid-winter, off the shores of New England, is a great boon to navigation. At this season of the year especially, the number of wrecks and the loss of life along the Atlantic sea-front are frightful. The month's average of wrecks has been as high as three a day. How many escape by seeking refuge from the cold in the warm waters of the Gulf Stream is matter of conjecture. Suffice it to say, that before their temperature was known, vessels thus distressed knew of no place of refuge short of the West Indies; and the newspapers of that day — Franklin's Pennsylvania Gazette among them — inform us that it was no uncommon occurrence for vessels bound for the Capes of the Delaware in the winter to be blown off and to go to the West Indies, and there wait for the return of spring before they would attempt another approach to this part of the coast.

186. *Thermal navigation.* Accordingly, Dr. Franklin's discovery with regard to the Gulf Stream temperature was looked upon as one of great importance, not only on account of its affording to the frosted mariner in winter a convenient refuge from the snow-storm, but because of its serving the navigator with an excellent landmark or beacon for our coast in all weathers. And so viewing it, the doctor, through political considerations, concealed his discovery for a while. The prize of £20,000, which had been offered, and partly paid, by the British government, to Harrison, the chronometer maker, for improving the means of finding longitude at sea, was fresh in the minds of navigators. And here it was thought a solution of the grand problem — for longitude at sea was a grand problem — had been stumbled upon by chance; for, on approaching the coast, the current of warm water in the Gulf Stream, and of cold water on this side of it, if tried with the thermometer, would enable the mariner to judge with great certainty, and in the worst of weather, as to his position. Jonathan Williams afterward, in speaking of the importance which the thermal use of these warm and cold currents would prove to navigation, pertinently asked the question, "If these stripes of water had been distinguished by the colors of red, white, and blue, could they be more distinctly discovered than they are by the constant use of the thermometer?" And he might have added, could they have marked the position of the ship more clearly?

187. *Commodore Truxtun.* When his work on Thermometrical Navigation appeared, Commodore Truxtun wrote to him: "Your publication will be of use to navigation by rendering sea voyages secure far beyond what even you yourself will immediately calculate, for I have proved the utility of the thermometer very often since we sailed together. It will be found a most valuable instrument in the hands of mariners, and particularly as to those who are unacquainted with astronomical observations; * * * * these particularly stand in need of a simple method of

ascertaining their approach to or distance from the coast, especially in the winter season; for it is then that passages are often prolonged, and ships blown off the coast by hard westerly winds, and vessels get into the Gulf Stream without its being known; on which account they are often hove to by the captains' supposing themselves near the coast when they are very far off (having been drifted by the currents). On the other hand, ships are often cast on the coast by sailing in the eddy of the Stream, which causes them to outrun their common reckoning. Every year produces new proofs of these facts, and of the calamities incident thereto." [6]

188. *The discovery of the high temperature of the Gulf Stream followed by a decline in Southern commerce.* Though Dr. Franklin's discovery was made in 1775, yet, for political reasons, it was not generally made known till 1790. Its immediate effect in navigation was to make the ports of the North as accessible in winter as in summer. What agency this circumstance had in the decline of the direct trade of the South, which followed this discovery, would be, at least to the political economist, a subject for much curious and interesting speculation. I have referred to the commercial tables of the time, and have compared the trade of Charleston with that of the northern cities for several years, both before and after the discovery of Dr. Franklin became generally known to navigators. The comparison shows an immediate decline in the Southern trade and a wonderful increase in that of the North. But whether this discovery in navigation and this revolution in trade stand in the relation of cause and effect, or be merely a coincidence, let others judge.

189. *Statistics.* In 1769 the commerce of the two Carolinas equaled that of all the New England States together; it was more than double that of New York, and exceeded that of Pennsylvania by one third.[7] In 1792, the exports from New York amounted in value to two millions and a half; from Pennsylvania, to $3,820,000; and from Charleston alone, to $3,834,-

---

[6] [Jonathan Williams, *Thermometrical Navigation* (Philadelphia, 1799), "Postscript."]

[7] Exports and Imports in 1769, valued in Sterling Money. From [David] Macpherson, *Annals of Commerce* [*Manufactures, Fisheries and Navigation* . . . III (London, 1805), 571–572].

Exports

	To Gr. Britain			Sou. of Europe			West Indies			Africa			Total		
	£	s.	d.	£	s.	d.	£	s.	d.	£	s.	d.	£	s.	d.
New England	142,775	12	9	81,173	16	2	308,427	9	6	17,713	0	9	550,089	19	2
New York	113,382	8	8	50,885	13	0	66,324	17	5	1,313	2	6	231,906	1	7
Pennsylvania	28,112	6	9	203,762	11	11	178,331	7	8	560	9	9	410,756	16	1
N. & S. Carolina	405,014	13	1	76,119	12	10	87,758	19	3	691	12	1	569,584	17	3

Imports

New England	223,695	11	6	25,408	17	9	314,749	14	5	180	0	0	564,034	3	8
New York	75,930	19	7	14,927	7		897,420	4	0	697	10	0	188,976	1	3
Pennsylvania	204,979	17	4	14,249	8	4	180,591	12	4				399,830	18	0
N. & S. Carolina	327,084	8	6	7,099	5	10	76,269	17	11	137,620	10	9	535,714	2	3

000. But in 1795 — by which time the Gulf Stream began to be as well understood by navigators as it now is, and the average passages from Europe to the North were shortened nearly one half, while those to the South remained about the same — the customs at Philadelphia alone amounted to $2,941,000,[8] or more than one half of those collected in all the states together.

190. *The shortening of voyages.* Nor did the effect of the doctor's discovery end here. Before it was made, the Gulf Stream was altogether insidious in its effects. By it, vessels were often drifted many miles out of their course without knowing it; and in bad and cloudy weather, when many days would intervene from one observation to another, the set of the current, though really felt but for a few hours during the interval, could only be proportioned out equally among the whole number of days. Therefore navigators could have only very vague ideas either as to the strength or the actual limits of the Gulf Stream, until they were marked out to the Nantucket fishermen by the whales, or made known by Captain Folger to Dr. Franklin. The discovery, therefore, of its high temperature assured the navigator of the presence of a current of surprising velocity, and which, now turned to certain account, would hasten, as it had retarded his voyage in a wonderful degree. Such, at the present day, is the degree of perfection to which nautical tables and instruments have been brought, that the navigator may now detect, and with great certainty, every current that thwarts his way. He makes great use of them. Colonel Sabine, in his passage, a few years ago, from Sierra Leone to New York, was drifted one thousand six hundred miles of his way by the force of currents alone; and, since the application of the thermometer to the Gulf Stream, the average passage from England has been reduced from upward of eight weeks to a little more than four. Some political economists of America have ascribed the great decline of Southern commerce which followed the adoption of the Constitution of the United States to

[8] [From "Value of Exports of Domestic and Foreign Produce and Manufacture, Annually, from 1791 to 1837,"] 25th Cong., 2nd sess., H. R. Doc. no. 330 [pp. 4, 18]. Some of its statements do not agree with those taken from Macpherson and quoted [in n. 7].

Value of Exports in Dollars

	1791	1792	1793	1794	1795	1796
Massachusetts	2,519,651	2,888,104	3,755,347	5,292,441	7,117,907	9,949,345
New York	2,505,465	2,535,790	2,932,370	5,442,000	10,304,000	12,208,027
Pennsylvania	3,436,000	3,820,000	6,958,000	6,643,000	11,518,000	17,513,866
South Carolina	2,693,000	2,428,000	3,191,000	3,868,000	5,998,009	7,620,000

Duties on Imports in Dollars

	1791	1792	1793	1794	1795	1796	1833
Massachusetts	1,006,000	723,000	1,044,000	1,121,000	1,520,000	1,460,000	3,055,000
New York	1,334,000	1,173,000	1,204,000	1,878,000	2,028,000	2,187,000	10,713,000
Pennsylvania	1,466,000	1,100,000	1,823,000	1,498,000	2,300,000	2,050,000	2,207,000
South Carolina	523,000	359,000	360,000	661,000	722,000	66,000	389,000

the protection given by legislation to Northern interests. But I think these statements and figures show that this decline was in no small degree owing to the Gulf Stream, the water-thermometer, and the improvements in navigation; for they changed the relations of Charleston — the great Southern emporium of the times — removing it from its position as a half-way house, and placing it in the category of an outside station.

191. *The scope of these researches.* The plan of our work takes us necessarily into the air, for the sea derives from the winds some of the most striking features in its physical geography. Without a knowledge of the winds, we can neither understand the navigation of the ocean, nor make ourselves intelligently acquainted with the great highways across it. As with the land, so with the sea; some parts of it are as untraveled and as unknown as the great Amazonian wilderness of Brazil, or the inland basins of Central Africa. To the south of a line extending from Cape Horn to the Cape of Good Hope (Plate V) is an immense waste of waters. None of the commercial thoroughfares of the ocean lead through it; only the adventurous whaleman finds his way there now and then in pursuit of his game; but for all the purposes of science and navigation, it is a vast unknown region. Now, were the prevailing winds of the South Atlantic northerly or southerly, instead of easterly or westerly, this unplowed sea would be an oft-used thoroughfare. Nay, more, the sea supplies the winds with food for the rain which these busy messengers convey away from the ocean to "the springs in the valleys which run among the hills." To the philosopher, the places which supply the vapors are as suggestive and as interesting for the instruction they afford, as the places are upon which the vapors are showered down. Therefore, as he who studies the physical geography of the land is expected to make himself acquainted with the regions of precipitation, so he who looks into the physical geography of the sea should search for the regions of evaporation, and for those springs in the ocean which supply the reservoirs among the mountains with water to feed the rivers; and, in order to conduct this search properly, he must consult the winds, and make himself acquainted with their "circuits." Hence, in a work on the Physical Geography of the Sea and its Meteorology, we treat also of the ATMOSPHERE.

CHAPTER IV

§ 200–261. THE ATMOSPHERE

200. *Likened to a machine.* There is no employment more ennobling to man and his intellect than to trace the evidences of design and purpose, which are visible in many parts of the creation. Hence, to the right-minded mariner, and to him who studies the physical relations of earth, sea, and air, the atmosphere is something more than a shoreless ocean, at the bottom of which he creeps along. It is an envelope or covering for the dispersion of light and heat over the surface of the earth; it is a sewer into which, with every breath we draw, we cast vast quantities of dead animal matter; it is a laboratory for purification, in which that matter is recompounded, and wrought again into wholesome and health-ful shapes; it is a machine (§ 191) for pumping up all the rivers from the sea, and for conveying the water from the ocean to their sources in the mountains; it is an inexhaustible magazine, marvelously stored. Upon the proper working of this machine depends the well-being of every plant and animal that inhabits the earth. How interesting, then, ought not the study of it to be! An examination of the uses which plants and animals make of the air is sufficient to satisfy any reasoning mind in the conviction that when they were created, the necessity of this adaptation was taken into account. The connection between any two parts of an artificial machine that work into each other, does not render design in its construction more patent than is the fact that the great atmospherical machine of our planet was constructed by an Architect who designed it for certain purposes; therefore the management of it, its movements, and the performance of its offices, can not be left to chance. They are, we may rely upon it, guided by laws that make all parts, functions, and movements of this machinery as obedient to order and as harmonious as are the planets in their orbits.

201. *The air and the ocean governed by stable laws.* An examination into the economy of the universe will be sufficient to satisfy the well-balanced minds of observant men that the laws which govern the atmos-phere and the laws which govern the ocean (§ 164) are laws which were put in force by the Creator when the foundations of the earth were laid, and that therefore they are laws of order; else, why should the Gulf Stream, for instance, be always where it is, and running from the Gulf of Mexico, and not somewhere else, and sometimes running into it? Why should there be a perpetual drought in one part of the world, and con-

tinual showers in another? Or why should the conscious winds ever heed the voice of rebuke, or the "waves of the sea clap their hands with joy?"

202. *Importance of observing the works of nature.* To one who looks abroad to contemplate the agents of nature, as he sees them at work upon our planet, no expression uttered or act performed by them is without meaning. By such an one, the wind and rain, the vapor and the cloud, the tide, the current, the saltness, and depth, and warmth, and color of the sea, the shade of the sky, the temperature of the air, the tint and shape of the clouds, the height of the tree on the shore, the size of its leaves, the brilliancy of its flowers — each and all may be regarded as the exponent of certain physical combinations, and therefore as the expression in which Nature chooses to announce her own doings, or, if we please, as the language in which she writes down or elects to make known her own laws. To understand that language and to interpret aright those laws is the object of the undertaking which we now have in hand. No fact gathered from such a volume as the one before us can therefore come amiss to those who tread the walks of inductive philosophy; for, in the hand-book of nature, every such fact is a syllable; and it is by patiently collecting fact after fact, and by joining together syllable after syllable, that we may finally seek to read aright from the great volume which the mariner at sea as well as the philosopher on the mountain each sees spread out before him.

203. *Materials for this chapter.* There have been examined at the Observatory more than a million of observations on the force and direction of the winds at sea.[1] The discussion of such a mass of material has thrown much light upon the circulation of the atmosphere; for, as in the ocean (§ 201), so in the air, there is a regular system of circulation.

204. *Different belts of winds.* Before we proceed to describe this system, let us point out the principal belts or bands of wind that actual observation has shown to exist at sea, and which, with more or less distinctness of outline, extend to the land also, and thus encircle the earth. If we imagine a ship to take her departure from Greenland for the South Shetland Islands, she will, between the parallels of 60° north and south, cross these several bands or belts of winds and calms nearly at right angles, and in the following order: (1.) At setting out she will find herself in the region of southwest winds, or counter trades of the north — called *counter* because they blow in the direction whence come the trade-winds of their hemisphere. (2.) After crossing 50°, and until reaching the parallel of 35° N., she finds herself in the belt of westerly winds, a region in which winds from the southwest and winds from the northwest contend for the mastery, and with nearly equal persistency. (3.) Between 35° and 30°, she finds herself in a region of variable winds and calms; the winds blowing

---

[1] ["The Winds at Sea,"] *Nautical Monographs,* No. 1.

all around the compass, and averaging about three months from each quarter during the year. Our fancied ship is now in the "horse latitudes." Hitherto winds with *westing* in them have been most prevalent; but, crossing the calm belt of Cancer, she reaches latitudes where winds with *easting* become more prevalent. (4.) Crossing into these, she enters the region of northeast trades, which now become the prevailing winds, until she reaches the parallel of 10° N., and enters the equatorial calm belt, which, like all the other wind-bands, does not hold stationary limits. (5.) Crossing the parallel of 5° N., she enters where the southeast trades are the prevailing winds, and so continue until the parallel of 30° S. is reached. (6.) Here is the calm belt of Capricorn, where, as in that of Cancer (3), she again finds herself in a region of shifting winds, light airs, and calms, and where the winds with westing in them become the prevailing winds. (7.) Between the parallels of 35° and 40° S., the northwest and southwest winds contend with equal powers for the mastery. (8.) Crossing 40°, the counter-trades (1) — the northwest winds of the southern hemisphere — become the prevailing winds, and so remain, as far toward the south pole as our observations at sea extend.

Such are the most striking movements of the winds at the surface of the sea. But, in order to treat of the general system of atmospherical circulation, we should consider where those agents reside which impart to that system its dynamical force. They evidently reside near the equator on one side, and about the poles on the other. Therefore, if, instead of confining our attention to the winds at the surface, and their relative prevalence from each one of the four quarters, we direct our attention to the upper and lower currents, and to the *general* movements *back* and *forth* between the equator and the poles, we shall be enabled the better to understand the general movements of this grand machine.

205. *The trade-wind belts.* Thus treating the subject, observations show that from the parallel of about 30° or 35° north and south to the equator, we have, extending entirely around the earth, two zones of perpetual winds, viz., the zone of northeast trades on this side, and of southeast on that. With slight interruptions, these winds blow perpetually, and are as steady and as constant as the currents of the Mississippi River, always moving in the same direction (Plate I) except when they are turned aside by a desert here and there to blow as monsoons, or as land and sea breezes. As these two main currents of air are constantly flowing from the poles toward the equator, we are safe in assuming that the air which they keep in motion must return by *some* channel to the place toward the poles whence it came in order to supply the trades. If this were not so, these winds would soon exhaust the polar regions of atmosphere, and pile it up about the equator, and then cease to blow for the want of air to make more wind of.

206. *The return current.* This return current, therefore, *must* be in the upper regions of the atmosphere, at least until it passes over those parallels between which the trade-winds are usually blowing on the surface. The return current must also move in the direction opposite to that wind the place of which it is intended to supply. These direct and counter currents are also made to move in a sort of spiral or loxodromic curve, turning *to* the west as they go *from* the poles to the equator, and in the opposite direction as they move from the equator toward the poles. This turning is caused by the rotation of the earth on its axis.

207. *Effect of diurnal rotation on the course of the trade-winds.* The earth, we know, moves from west to east. Now if we imagine a particle of atmosphere at the north pole, where it is at rest, to be put in motion in a straight line toward the equator, we can easily see how this particle of air, coming from the very axis of diurnal rotation, where it did not partake of the diurnal motion of the earth, would, in consequence of its *vis inertiæ*, find, as it travels south, the earth slipping from under it, as it were, and thus it would appear to be coming from the northeast and going toward the southwest; in other words, it would be a northeast wind. The better to explain, let us take a common terrestrial globe for the illustration. Bring the island of Madeira, or any other place about the same parallel, under the brazen meridian; put a finger of the left hand on the place; then, moving the finger down along the meridian to the south, to represent the particle of air, turn the globe on its axis from west to east, to represent the diurnal rotation of the earth, and when the finger reaches the equator, stop. It will now be seen that the place on the globe under the finger is to the southward and westward of the place from which the finger started; in other words, the track of the finger over the surface of the globe, like the track of the particle of air upon the earth, has been *from* the northward and eastward. On the other hand, we can perceive how a like particle of atmosphere that starts from the equator, to take the place of the other at the pole, would, as it travels north, in consequence of its *vis inertiæ*, be going toward the east faster than the earth. It would therefore appear to be blowing *from* the southwest, and going toward the northeast, and exactly in the opposite direction to the other. Writing south for north, the same takes place between the south pole and the equator.

208. *Two grand systems of currents.* Such is the process which is actually going on in nature; and if we take the motions of these two particles as the type of the motion of all, we shall have an illustration of the great currents in the air (§ 204), the equator being near one of the nodes, and there being at least two systems of currents, an upper and an under, between it and each pole. Halley, in his theory of the trade-winds, pointed out the key to the explanation, so far, of the atmospherical circu-

lation; but, were the explanation to rest here, a northeast trade-wind extending from the pole to the equator would satisfy it; and were this so, we should have, on the surface, no winds but the northeast trade-winds on this side, and none but southeast trade-winds on the other side, of the equator.

209. *From the Pole to 35°–30°.* Let us return now to our northern particle (§ 207), and follow it in a round from the north pole across the equator to the south pole, and back again. Setting off from the polar regions, this particle of air, for some reason which does not appear to have been very satisfactorily explained by philosophers, instead of traveling (§ 208) on the surface all the way from the pole to the equator, travels in the upper regions of the atmosphere until it gets near the parallel of 30°–35°. Here it meets, also in the clouds, the hypothetical particle that is coming from the south, and going north to take its place.

210. *The "horse latitudes."* About this parallel of 30°–35° north, then, these two particles press against each other with the whole amount of their motive power, and produce a calm and an accumulation of atmosphere: this accumulation is sufficient to balance the pressure of the two winds from the north and south. From under this bank of calms, which seamen call the "horse latitudes," two surface currents of wind are ejected or drawn out; one toward the equator, as the northeast trades, the other toward the pole, as the southwest "passage-winds," or counter-trades. These winds come out at the lower surface of the calm region, and consequently the place of the air borne away in this manner must be supplied, we may infer, by downward currents from the superincumbent air of the calm region. Like the case of a vessel of water which has two streams from opposite directions running in at the top, and two of equal capacity discharging in opposite directions at the bottom, the motion of the water would be downward; so is the motion of the air in this calm zone.

211. *The barometer there.* The barometer, in this calm region, stands higher than it does either to the north or to the south of it; and this is another proof as to the accumulation of the atmosphere here, and pressure from its downward motion. And because the pressure under this calm belt is greater than it is on either side of it, the tendency of the air will be to flow out on either side; therefore, supposing we were untaught by observation, reason would teach us to look for the prevailing winds on each side of this calm belt to be *from* it.

212. *The equatorial calm belt.* Following our imaginary particle of air, however, from the north across this calm belt of Cancer, we now feel it moving on the surface of the earth as the northeast trade-wind; and as such it continues till it arrives near the equator, where it meets a like hypothetical particle, which, starting from the south at the same time

the other started from the north pole, has blown as the southeast trade-wind. Here, at this equatorial place of meeting, there is another conflict of winds and another calm region, for a northeast and southeast wind can not blow at the same time in the same place. The two particles have been put in motion by the same power; they meet with equal force; and, therefore, at their place of meeting, they are arrested in their course. Here, therefore, there is a calm belt, as well as at Capricorn and Cancer. Warmed now by the heat of the sun, and pressed on each side by the whole force of the northeast and southeast trades, these two hypothetical particles, taken as the type of the whole, cease to move onward and ascend. This operation is the reverse of that which took place at the meeting (§ 210) near the parallel of 30°–35°.

213. *The calm belt of Capricorn.* This imaginary particle then, having ascended to the upper regions of the atmosphere again, travels there counter to the southeast trades, until it meets, near the calm belt of Capricorn, another particle from the south pole; here there is a descent as before (§ 210); it then (§ 211) flows on toward the south pole as a surface wind from the northwest.

214. *The polar calms and the return current.* Entering the polar regions obliquely, it is pressed upon by similar particles flowing in oblique currents across every meridian; and here again is a calm place or node; for, as our imaginary particle approaches the parallels near the polar calms more and more obliquely, it, with all the rest, is whirled about the pole in a continued circular gale; finally, reaching the vortex or the calm place, it is carried upward to the regions of atmosphere above, whence it commences again its reflow to the north as an upper current, as far as the calm belt of Capricorn; here it encounters (§ 213) its fellow from the north (§ 207); they stop, descend, and flow out as surface currents (§ 210), the one with which the imagination is traveling, to the equatorial calm as the southeast trade-wind; here (§ 212) it ascends, traveling thence to the calm belt of Cancer as an upper current counter to the northeast trades. Here (§ 210 and 209) it ceases to be an upper current, but, descending (§ 210), travels on with the southwest passage-winds toward the pole.

215. *Diagram of the winds — Plate I.* Now the course we have imagined an atom of air to take, as illustrated by the "diagram of the winds," is this: an ascent in a place of calms about the north pole, as at V, P; an efflux thence as an upper current, A, B, C, until it meets R, S (also an upper current) over the calms of Cancer. Here there is supposed to be a descent, as shown by the arrows C D, S T. This current, A, B, C, D, from the pole, now becomes the northeast trade-wind, D, E, on the surface, until it meets the southeast trades, O, Q, in the equatorial calms, where it ascends as E, F, and travels as F, G with the upper current to the calms of

Capricorn, thence as H, J, K, with the prevailing northwest surface current to the south pole, thence up with the arrow P, and around with the hands of a watch, and back, as indicated by the arrows along L, M, N, O, Q, R, S, T, U, V.

216. *As our knowledge of the laws of nature has increased, so have our readings of the Bible improved.* The Bible frequently makes allusion to the laws of nature, their operation and effects. But such allusions are often so wrapped in the folds of the peculiar and graceful drapery with which its language is occasionally clothed, that the meaning, though peeping out from its thin covering all the while, yet lies in some sense concealed, until the lights and revelations of science are thrown upon it; then it bursts out and strikes us with exquisite force and beauty. As our knowledge of nature and her laws has increased, so has our understanding of many passages in the Bible been improved. The Psalmist called the earth "the round world;" yet for ages it was the most damnable heresy for Christian men to say the world is round; and, finally, sailors circumnavigated the globe, proved the Bible to be right, and saved Christian men of science from the stake. "Canst thou bind the sweet influences of Pleiades?" Astronomers of the present day, if they have not answered this question, have thrown so much light upon it as to show that, if ever it be answered by man, he must consult the science of astronomy. It has been recently all but proved, that the earth and sun, with their splendid retinue of comets, satellites, and planets, are all in motion around some point or centre of attraction inconceivably remote, and that that point is in the direction of the star Alcyon, one of the Pleiades! Who but the astronomer, then, could tell their "sweet influences?" And as for the general system of atmospherical circulation which I have been so long endeavoring to describe, the Bible tells it all in a single sentence: "The wind goeth toward the south, and turneth about unto the north; it whirleth about continually, and the wind returneth again according to his circuits." — Eccl., i., 6.

217. *Sloughing off from the counter trades.* Of course, as the surface winds, H, J, K and T, U, V, approach the poles, there must be a sloughing off, if I may be allowed the expression, of air from them, in consequence of their approaching the poles. For as they near the poles, the parallels become smaller and smaller, and the surface current must either extend much higher up, and blow with greater rapidity, or else a part of it must be sloughed off above, and so turn back before reaching the calms about the poles. The latter is probably the case. Such was the conjecture. Subsequent investigations [2] have, by *proving the converse,* established its correctness, and in this way: they show that the southeast trade-winds, as in the Atlantic, blow, on the average, during the year,

[2] "The Winds at Sea."

124 days between the parallels of 25° and 30° S., and that as you approach the equator their average annual duration increases until you reach 5° S. Here between 5° and 10° S. they blow on the average for 329 out of the 365 days.

218. *The air which supplies the southeast trade-wind in the band 5° does not cross the band 25°.* Now the question may be asked, Where do the supplies which furnish air for these winds for 329 days come from? The "trades" could not convey this fresh supply of air across the parallel of 25° S. during the time annually allotted for them to blow in that latitude. They can not for these reasons: (1.) Because the trade-winds in lat. 5° are stronger than they are in lat. 25°, and therefore, in equal times, they waft larger volumes of air across 5° than they do across 25°. (2.) Because the girdle of the earth near the equator is larger than it is farther off, as at 25°; therefore, admitting equal heights and velocities for the wind at the two parallels, it would, in equal times, bear more air across the one of larger circumference. Much less, therefore, can the air which crosses the parallel of 25° S. annually in the 124 trade-wind days of that latitude be sufficient to supply the trade-winds with air for their 329 days in lat. 5°. Whence comes the extra supply for them in this latitude? (3.) Of all parts of the ocean the trade-winds obtain their best development between 5° and 10° S. in the Atlantic Ocean, for it is there only that they attain the unequaled annual average duration of 329 days. But referring now to the average annual duration of the southeast trade-wind in all seas, we may, for the sake of illustration, liken this belt of winds which encircles the earth, say between the parallels of 5° and 25° S., to the frustum of a hollow cone, with its base toward the equator.

219. *Winds with northing and winds with southing in them contrasted.* Now, dividing the winds into only two classes, as winds with *northing* and winds with *southing* in them, actual observations show, taking the world around, that winds having southing in them blow *into* the southern or smaller end of this cone for 209 days annually, and out of the northern and larger end for 286 days.[3] They appear (§ 221) to come out of the larger end with *greater* velocity than they enter the smaller end. But we *assume* the velocity at going in and at coming out to be the same, merely for illustration. During the rest of the year, either winds with *northing* in them are blowing *in* at the big end, or *out* at the little end of the imaginary cone, or no wind is blowing at all: that is, it is calm. Now, if we suppose, merely for the sake of assisting farther in the illustration, that these winds with *northing* and these winds with *southing* move equal volumes of air in equal times, we may subtract the days of the one from the days of the other, and thus ascertain how much more air comes out at one end than goes in at the other of our frustum. Winds with northing in

[3] [*Ibid.*]

them blow in at the big end for 72 days, and out at the little end for 146 days annually. Now, if we subtract the whole number of winds (146) with northing in them that blow *out* at the south or small end, from the whole number (209) with southing in them that blow *in*, we shall have for the quantity that is to pass through, or go from the parallel of 25° to 5°, the value expressed by the transporting power of the southeast trade-winds at latitude 25° for 63 days (209 − 146 = 63). In like manner we obtain, in similar terms, an expression for the volume which these winds bring out at the large or equatorial end, and find it to be as much air as the southeast trade-winds can transport across the parallel of 5° S. in 214 days (286 − 72 = 214). Again,

220. *Southeast trade-winds stronger near the equatorial limits.* The southeast trade-winds, as they cross the parallel of 5° and come out of this belt, appear to be stronger [4] than they are when they enter it. But assuming the velocity at each parallel to be the same, we have (§ 219) just three times as much air with *southing* in it coming out of this belt on the equatorial side as with *southing* in it we find entering (§ 218) on the polar side. From this it is made plain that if all the air, whether from the southward and eastward, or from the southward and westward, which *enters* the southeast trade-wind belt near its polar borders, were to come out at its equatorial edge as southeast trade-winds, there would not be enough, by the volume required to supply them for 151 days (214 − 63 = 151), to keep up such a constant current of air as we find the southeast trade-winds actually conveying across this parallel of 5° S.

221. *Speed of vessels through the trade-winds.* The average speed which vessels make in sailing through the trade-winds in different parts of the world has been laboriously investigated at the National Observatory.[5] By this it appears that their average speed through the southeast trade-winds of the Atlantic is, between the parallels of 5° and 10°, 6.1 knots an hour, and 5.7 between 25° and 30°.

222. *The question, Whence are the southeast trade-winds supplied with air? answered.* All these facts being weighed, they indicate that the volume of air which investigations show that the southeast trade-winds of the world annually waft across the parallels of 10°–5° S. in 285 days [6] — for that is their average duration − is at least twice as great as the volume which they annually sweep across the parallel of 25° on 139 days, which is their average here. Hence the answer to the question (§ 218), Whence comes the excess? is, it can only come from above, and in this way, viz.: the southeast trade-winds, as they rush from 25° S. toward the equator,

[4] The force of the trade-winds, as determined by the average speed of 2235 vessels sailing through them, is greater between 5° and 10° S. than it is between 25° and 30° S. — *Sailing Directions* [8th ed., II, 860].

[5] "Average Force of the Trade-winds," *ibid.*, p. 857.

[6] "The Winds at Sea," pl. I.

act upon the upper air like an under-tow. Crossing, as they approach the equator, parallels of larger and larger circumference, they draw down and turn back from the counter current above air enough to supply *pabulum* to larger and larger, and to stronger and stronger currents of wind.

223. *Whither it goes.* The air which the trade-winds pour into the equatorial calm belt (§ 213) rises up, and *has* to flow off as an upper current, to make room for that which the trade-winds are continually pouring in below. They bring it from toward the poles — back, therefore, toward the poles the upper currents must carry it. On their journey they cross parallel after parallel, each smaller than the other in circumference. There is, therefore, a constant tendency with the air that these upper currents carry polarward to be *crowded* out, so to speak — to slough off and turn back. Thus the upper current is ever ready to supply the trade-winds, as they approach the equator, with air exactly at the right place, and in quantities just sufficient to satisfy the demand.

224. *How it is drawn down from above.* This upper air, having supplied the equatorial cloud-ring (§ 514) with vapor for its clouds, and with moisture for its rains, flows off polarward as comparatively dry air. The dryest air is the heaviest. This dry and heavy air is therefore the air most likely to be turned back with the trade-winds, imparting to them that elasticity, freshness, and vigor for which they are so famous, and which help to make them so grateful to man and beast in tropical climates. The curved arrows, *f g* and *f′ g′*, *r s* and *r′ s′*, are intended to represent, in the "diagram of the winds" (Plate I), this sloughing off and turning back of air from the upper currents to the trade-winds below.

225. *Velocity of southeast shown to be greater than northeast trade-winds.* According to investigations which are stated at length in Maury's Sailing Directions, on his Wind and Current Charts, and in the Monographs of the Observatory, the average strength and annual duration of the southeast trade-winds of the Atlantic may be thus stated for every band of 5° of latitude in breadth, from 30° to the equator. For the band between

	Ann. duration	Force	No. of obs.
30° and 25° S	124 days	5.6 miles [7]	19,817
25° and 20°	157 "	5.7 "	20,762
20° and 15°	244 "	5.9 "	17,844
15° and 10°	295 "	6.3 "	14,422
10° and 5°	329 "	6.1 "	13,714
5° and 0°	314 "	6.0 "	15,463

It thus appears that the southeast trade-winds of the Atlantic blow with most regularity between 10° and 5°, and with most force between 10° and 15°.

[7] Distance per hour that vessels average while sailing through it.

226. *The air sloughed off from the counter trades, moist air.* On the polar side of 35°–40°, and in the counter trades (§ 204 [7]), a different process of sloughing off and turning back is going on. Here the winds are blowing *toward* the poles; they are going from parallels of large to parallels of smaller circumference, while the upper return current is doing the reverse; it is widening out with the increasing circumference of parallels, and creating room for more air, while the narrowing current below is crowding out and sloughing off air for its winds.

227. *The air sloughed off from the upper trade current dry.* In the other case (§ 224), it was the heavy dry air that was sloughed off to join the winds below. In this case it is the moist and lightest air that is crowded out to join the current above.

228. *The meteorological influences of ascending columns of moist air.* This is particularly the case in the southern hemisphere, where, entirely around between the parallels of 40° and 60° or 65°, all, or nearly all, is water. In this great austral band the winds are in contact with an evaporating surface all the time. Aqueous vapor is very much lighter than atmospheric air; as this vapor rises, it becomes entangled with the particles of air, some of which it carries up with it, thus producing, through the *horizontal* flow of air with the winds, numerous little *ascending* columns. As these columns of air and vapor go up, the superincumbent pressure decreases, the air expands and cools, causing precipitation or condensation of the vapor. The heat that is set free during this process expands the air still further, thus causing here and there in those regions, and wherever it may chance to be raining, intumescences, so to speak, from the wind stratum below; the upper current, sweeping over these protuberances, bears them off in its course toward the equator, and thus we have another turning back — a constant mingling. The curved arrows, *h j k* and *h′ j′ k′*, are intended, on the "diagram of the winds" (Plate I), to represent this rising up from the counter trades and turning back with the upper current.

229. *Supposing the air visible, the spectacle that would be presented between the upper and lower currents.* Let us imagine the air to be visible, that we could see these different strata of winds, and the air as it is sloughed off from one stratum to join the other. We can only liken the spectacle that would be presented between the upper and the lower stratum of these winds to the combing of a succession of long waves as they come rolling in from sea, and, one after another, breaking upon the beach. They curl over and are caught up, leaving foam from their white caps behind, but nevertheless stirring up the sea and mixing up its waters so as to keep them all alike.

230. *The importance of atmospherical circulation.* If the ordinances of nature require a constant circulation and continual mixing up of the

water in the sea, that it become not stagnant, and that it may be kept in a wholesome state for its inhabitants, and subserve properly the various offices required of it in the terrestrial economy, how much more imperative must they not be with the air? It is more liable to corruption than water; stagnation is ruinous to it. It is both the sewer and the laboratory for the whole animal and vegetable kingdoms. Ceaseless motion has been given to it; perpetual circulation and intermingling of its ingredients are required of it. Personal experience teaches us this, as is manifest in the recognized necessity of ventilation in our buildings — the wholesome influences of fresh air, and the noxious qualities of "an atmosphere that has no circulation." Hence, continual mixing up of particles in the atmosphere being required of the winds in their circuits, is it possible for the human mind to conceive of the appointment of "circuits" for them (§ 216) which are so admirably designed and exquisitely adapted for the purpose as are those which this view suggests?

231. *Its vertical movements — how produced.* As a physical necessity, the vertical circulation of the air seems to be no less important than its horizontal movements, which we call wind. One begets the other. The wind, when it blows across parallels of latitude — as it always must, except at the equator, for it blows in arcs of great circles, and not in small ones [8] — creates a vertical circulation either by dragging down air from the upper regions (§ 224), or by sloughing it off and forcing it up from the lower (§ 228), according as the wind is approaching the pole or the equator.

232. *Vertical and horizontal movements in the air consequents of, and dependent upon each other.* Indeed, the point may be well made whether the horizontal circulation of the air be not dependent upon and a consequent of its vertical circulation — so nearly allied are the two motions in their relations as cause and effect. Upward and downward movements in fluids are consequent upon each other, and they involve lateral movements. The sea, with its vapor, is the great engine which gives upward motion in the air (§ 227). As soon as aqueous vapor is formed it rises; the air resists its ascent; but it is lighter than the air, therefore it forces the re-

[8] The tendency of all bodies, when put in motion on the surface of the earth, is, whether fluid, solid, or gaseous, to go from the point of departure to the point of destination by the shortest line possible; and this, when the motion is horizontal, is an arc of a great circle. If we imagine a partial vacuum to be formed at the north pole, we can readily enough perceive that the wind for 5°, 10°, 20° of polar distance, all around, would tend to rush north and strive to get there along the meridians — arcs of great circles. This would be the case whether the earth be supposed to be with or without diurnal rotation, or motion of any sort. Now suppose the place of rarefaction to be any where away from the poles, then draw great circles to a point in the middle of it, and the air all around would, in rushing into the vacuum, seek to reach it by these great circles. *Force* may turn it aside, but such is the tendency (§ 120).

sisting particles of air up along with it, and so produces ascending columns in the atmosphere. The juxta air comes in to occupy the space which that carried up by the vapor left behind it, and so there is a wind produced. The wind arising from this source alone is so slight generally, as scarcely to be perceived. But when the ascending vapor is condensed, and its latent heat liberated and set free in the upper air, we often have the most terrific storms.

233. *Cold belts.* Now suppose the surface from which this vapor rises, or on which it is condensed, be sufficiently large to produce a rush of wind from afar; suppose it, moreover, to be an oblong lying east and west somewhere, for example, in the temperate zone of the northern hemisphere. The wind that comes rushing in from the south side will be in the category of the counter trades of the southern hemisphere (§ 228), going from larger to smaller parallels, and giving rise to ascending columns; while that from the northern side, moving in the opposite direction, is, like the trade-winds (§ 223), bringing down air from above.

234. *The upper currents — their numbers and offices.* By the motion of the clouds upper currents of wind are discerned in the sky. They are arranged in layers or strata one above the other. The clouds of each stratum are carried by its winds in a direction and with a velocity peculiar to their stratum. How many of these superimposed currents of wind there may be between the top and bottom of the atmosphere we know not. As high up as the cloud-region several are often seen at the same time. They are pinions and rachets in the atmospherical machinery. We have seen (§ 230) some of their uses: let us examine them more in detail. Now, as the tendency of air in motion is (§ 120) to move in arcs of great circles, and as all great circles that can be drawn about the earth must cross each other in two points, it is evident that the particles of the atmosphere which are borne along as wind must have their paths all in *diverging* or *converging* lines, and that consequently each wind must either be, like the trade-winds (§ 222), drawing down and sucking in air from above, or, like the counter trades (§ 226), crowding out and forcing it off into the upper currents.

235. *Tendency of air when put in motion to move in the plane of a great circle.* This tendency to move in great circles is checked by the forces of diurnal rotation, or by the pressure of the wind when it blows toward a common centre, as in a cyclone. In no case is it entirely overcome in its tendency, but in all, it is diverted from the great circle path and forced to take up its line of march either in spirals about a point on the surface of the earth, or in loxodromics about its axis. In either case the pushing up or pulling down of the combing, curdling air from layer to layer is going on.

236. *The results upon its circulation of this tendency.* Thus the laws

of motion, the force of gravity, and the figure of the earth all unite in requiring every wind that blows either to force air up from the surface into
the regions above, or to draw it down to the earth from the crystal vaults
of the upper sky. Add to these the storm-king — traversing the air, he
thrusts in the whirlwind or sends forth the cyclone, the tornado, and the
hurricane to stir up and agitate, to mix and mingle the whole in one
homogeneous mass. By this perpetual stirring up, this continual agitation,
motion, mixing, and circulation, the airy covering of the globe is kept in
that state which the well-being of the organic world requires. Every
breath we draw, every fire we kindle, every blade of grass that grows or
decays, every blaze that shines and burns adds something that is noxious,
or takes something that is healthful away from the surrounding air. Diligent, therefore, in their offices must the agents be which have been appointed to maintain the chemical status of the atmosphere, to preserve its
proportions, to adjust its ingredients, and to keep them in that state of
admixture best calculated to fit it for its purposes.

237. *Experiments by the French Academy.* Several years ago the
French Academy sent out bottles and caused specimens of air from
various parts of the world to be collected and brought home to be analyzed. The nicest tests which the most skillful chemists could apply were
incapable of detecting any, the slightest, difference as to ingredients in
the specimens from either side of the equator; so thorough in the performance of their office are these agents. Nevertheless, there are a great
many more demands on the atmosphere by the organic world for *pabulum* in one hemisphere than in the other; and consequently a great many
more inequalities for these agents to restore in one than in the other. Of
the two, the land of our hemisphere most teems with life, and here the
atmosphere is most taxed. Here the hearth-stone of the human family has
been laid. Here, with our fires in winter and our crops in summer, with
our work-shops, steam-engines, and fiery furnaces going night and day —
with the ceaseless and almost limitless demands which the animal and
vegetable kingdoms are making upon the air overhead, we can not detect
the slightest difference between atmospherical ingredients in different
hemispheres; and yet, notwithstanding the compensations and adjustments between the two kingdoms of the organic world, there are almost
in every neighborhood causes at work which would produce a difference
were it not for these ascending and descending columns of air; for the
obedient winds; for this benign system of circulation; these little cogs and
rachets which have been provided for its perfect working. The study of
its mechanism is good and wholesome in its influences, and the contemplation of it well calculated to excite in the bosom of right-minded philosophers the deepest and the best of emotions.

238. *How supplies of fresh air are brought down from the upper sky.*

Upon the proper adjustments of the dynamical forces which keep up these ceaseless movements the life of organic nature depends. If the air that is breathed were not taken away and renewed, warm-blooded life would cease; if carbon, and oxygen, and hydrogen, and water were not in due quantities dispensed by the restless air to the flora of the earth, all vegetation would perish for lack of food. That our planet may be liable to no such calamity, power has been given to the wayward wind, as it "bloweth where it listeth," to bring down from the pure blue sky fresh supplies of life-giving air wherever it is wanted, and to catch up from the earth wherever it may be found, that which has become stale; to force it up, there to be deflagrated among the clouds, purified and renovated by processes known only to Him whose ministers they are. The slightest change in the purity of the atmosphere, though it may be too slight for recognition by chemical analysis in the laboratory, is sure to be detected by its effects upon the nicer chemistry of the human system, for it is known to be productive of disease and death. No chemical tests are sensitive enough to tell us what those changes are, but experience has taught us the necessity of ventilation in our buildings, of circulation through our groves. The cry in cities for fresh air from the mountains or the sea, reminds use continually of the life-giving virtues of circulation. Experience teaches that all air when pent up and deprived of circulation becomes impure and poisonous.

239. *Beautiful and benign arrangements.* How minute, then, pervading, and general, benignant, sure, and perfect must be that system of circulation which invests the atmosphere and makes the "whole world kin!" In the system of vertical circulation which I have been endeavoring to describe, we see, as in a figure, the lither sky filled with crystal vessels continually ascending and descending between the bottom and the top of the atmospherical ocean full of life-giving air; these buckets are let down by invisible hands from above, and, as they are taken up again, they carry off from the surface, to be purified in the laboratory of the skies, phials of mephitic vapors and noxious gases, with the dank and deadly air of marshes, ponds, and rivers.

240. *Their influences upon the mind.* Whenever, by study and research, we succeed in gaining an insight, though never so dim, into any one of the offices for which any particular part of the physical machinery of our planet was designed by the great Architect, the mind is enriched with the conviction that it has comprehended a thought that was entertained at the creation. For this reason the beautiful compensations which philosophers have discovered in terrestrial arrangements are sources of never-failing wonder and delight. How often have we been called on to admire the beautiful provision by which fresh water is so constituted that it expands from a certain temperature down to freezing! We recognize in

the formation of ice on the top instead of at the bottom of freezing water, an arrangement which subserves, in manifold ways, wise and beneficent purposes. So, too, when we discern in the upper sky (§ 234) currents of wind arranged in strata one above the other, and running hither and thither in different directions, may we not say that we can here recognize also at least one of the foreordained offices of these upper winds? That by sending down fresh air and taking up foul, they assist in maintaining the world in that state in which it was made and for which it is designed — "a habitation fit for man?"

241. *The effect of downward currents in producing cold.* The phenomenon of cold belts or bands is often observed in the United States, and I suppose in other parts of the world also; and here in these downward currents we have the explanation and the cause of sudden and severe local spells of cold. These belts generally lie east and west rather than north and south, and we have often much colder weather in them than we have even several degrees to the north of them. The conditions required for one of these "cold snaps" appear to be a north or northwest wind of considerable breadth from west to east. As it goes to the south, its tendency is, if it reach high enough, to bring down cold air from above in the manner of the trade-winds (§ 238); and when the air thus brought down chances to be, as it often is, dry and cold, we have the phenomenon of a cold belt, with warmer weather both to the north and the south of it. While I write the thermometer is −4° in Mississippi, lat. 32°, and they are having colder weather there than we have either in Washington or Cincinnati, 7° farther to the north.

242. *The winter northers of Texas.* The winter "northers" of Texas sometimes bring down the cold air there with terrific effect. These bitter cold winds are very severe at Nueces, in the coast country, or the southwest corner of Texas bordering the Gulf of Mexico, lat. 27°.5. They are often felt to the west in Mexico, but rarely in eastern or northern Texas. The fact that they are not known in northern Texas goes to show that the cold they bring is not translated by the surface winds from the north.

243. *Their severe cold.* A correspondent in Nueces, lat. 27° 36′ N., long. 97° 27′ W., has described these winds there during the winter of 1859–60. They prevail from November to March, and commence with the thermometer about 80° or 85°. A calm ensues on the coast; black clouds roll up from the north; the wind is heard several minutes before it is felt; the thermometer begins to fall; the cold norther bursts upon the people, bringing the temperature down to 28°, and sometimes even to 25°, before the inhabitants have time to change clothing and make fires. So severe is the cold, so dry the air, that men and cattle have been known to perish in them.[9] These are the winds which, entering the Gulf and

[9] "Two men," says Mr. M. A. Taylor, in a letter dated January 11th, 1860, at

sucking up heat and moisture therefrom, still retain enough of strength to make themselves terrible to mariners — they are the far-famed northers of Vera Cruz.

244. *The wind in his circuits.* We now see the general course of the "wind in his circuits," as we see the general course of the water in a river. There are many abrading surfaces, irregularities, &c., which produce a thousand eddies in the main stream; yet, nevertheless, the general direction of the whole is not disturbed nor affected by those counter currents; so with the atmosphere and the variable winds which we find here in this latitude. Have I not, therefore, very good grounds for the opinion (§ 200) that the "wind in his circuits," though apparently to us never so wayward, is as obedient to law and as subservient to order as were the morning stars when first they "sang together?"

245. *Forces which propel the wind.* There are at least two forces concerned in driving the wind through its circuits. We have seen (§ 207) whence that force is derived which gives easting to the winds as they approach the equator, and westing as they approach the poles; and allusion, without explanation, has been made (§ 212) to the source whence they derive their northing and their southing. Philosophers formerly held that the trade-winds are drawn toward the equator by the influence of the direct rays of the sun upon the atmosphere there. They heated it, expanded it, and produced rarefaction, thereby causing a rush of the wind both from the north and the south; and as they played with greatest effect at the equator, there the ascent of the air and the meeting of the two winds would naturally be. Such was the doctrine.

246. *Effect of the direct heat of the sun upon the trade-winds.* But the direct rays of the sun, instead of being most powerful upon the air at the equator, is most powerful where the sun is vertical; and if the trade-winds were produced by direct heat alone from the sun, the place of meeting would follow the sun in declination much more regularly than it does. But, instead of so following the sun, the usual place of meeting between the trade-winds is neither at the equator nor where the sun is vertical. It is at a mean between the parallels of 5° and 10° or 12° N. It is in the northern hemisphere, notwithstanding the fact that in the southern summer, when the sun is on the other side of the line, the earth is in perihelion, and the amount of heat received from the vertical ray in a day there is very much greater ($\frac{1}{15}$) than it is when she is in aphelion, as in our midsummer. For this reason the southern summer is really hotter than

---

Nueces, Texas, "were actually frozen to death within a few miles of this place this winter in a norther. Animals seem to tell by instinct when the norther is coming, and make their way from the open prairies to timber and other shelter, starting often on a run when the heat is not oppressive. This is when the change is to be sudden and violent. Many cattle, horses, and sheep are frozen to death at such times."

the northern; yet, notwithstanding this, the southeast trade-winds actually blow the air away from under this hot southern sun, and bring it over into the northern hemisphere. They cross over into the northern hemisphere annually, and blow between 0° and 5° N. for 193 days,[10] whereas the northeast trades have rarely the force to reach the south side of the equator at all.

247. *The two systems of trade-winds unequal both in force, duration, and stability.* By examining the log-books of vessels while sailing through the northeast and southeast trade-wind belts, and comparing their rate of sailing, it has been ascertained that ships sail faster with the southeast than they do with the northeast trade-winds, and that the southeast blow more days during the year than do the northeast trades.[11] The logs of vessels that spent no less than 166,000 days in sailing through these two belts of wind show that the average sailing speed through the southeast trade-wind belt, which lies between the equator and 30° S., is about eight miles an hour, and the average number of uninterrupted southeast trade-wind days in the year is 227. For the northeast it is 183 days, with strength enough to give ships an average speed of only 5.6 miles an hour. Hence it appears that the two systems of trade-winds are very unequal both as to force and stability, the southeast surpassing in both.

248. *Effects of heat and vapor.* Moreover, the hottest place within the trade-wind regions is not at the equator, it is where these two winds meet (§ 246). Lieutenant Warley has collated from the abstract logs the observations on the temperature of the air made by 100 vessels, indiscriminately taken, during their passage across the trade-wind and equatorial calm belts of the Atlantic. The observations were noted at each edge of the calm belt, in the middle of it, and 5° from each edge in the trade-winds, with the following averages: In the northeast trades, 5° north of the north edge of the equatorial calm belt, say in latitude 14° N., air 78°.69. North edge calm belt, say 9° N., air 80°.90. Middle of calm belt, say 4½ N., air 82°. South edge, say 0°, air 82°.30; and 5° S. (in southeast trades), air 81°.14. These thermometers had not all been compared with standards, but their *differences* are probably correct, notwithstanding the means themselves may not be so. Hence we infer the south edge of the calm belt is 1°.4 warmer than the north. The extreme difference between the annual isotherms that lie between the parallels of 30° N. and 30° S. — between which the trade-wind belts are included — does not probably exceed 12°. According to the experiments of Gay-Lussac and Dalton, the

---

[10] "The Winds at Sea."
[11] *Sailing Directions*, 8th ed., II[858-860; Maury, *Pilot Chart of the North Atlantic*, Sheet no. 2 (Washington, 1853); *Pilot Chart of the South Atlantic*, ser. C, Sheets nos. 1 & 2 (Washington, 1853)].

dilatation of atmospheric air due a change of 12° in temperature is 2½ per cent.; that is, a column of atmosphere 100 feet high will, after its temperature has been raised 12°, be 102½ feet high. However, only about one third of the direct heat of the sun is absorbed in its passage down through the atmosphere. The other two thirds are employed in lifting vapor up from the sea, or in warming the crust of the earth, thence to be radiated off again, or to raise the temperature of sea and air by conduction. The air at the surface of the earth receives most heat directly from the sun; as you ascend, it receives less and less, and the consequent temperature becomes more and more uniform; so that the height within the tropics to which the direct rays of the sun [heat the air] is not, as reason suggests, and as the snow-lines of Chimborazo and other mountains show, very great or very variable.

249. *Hurricanes not due to direct heat of the sun.* Moreover, daily observations show most conclusively that the strong winds and the great winds, the hurricanes and tornadoes, do not arise from the direct heat of the sun, for they do not come in the hottest weather or in the clearest skies. On the contrary, winter is the stormy period in the extra-tropical regions of the north;[12] and in the south, rains and gales — not gales and sunshine [13] — accompany each other. The land and sea breezes express more than double the amount of wind force which the direct heat of the sun is capable of exerting upon the trade-winds. I say more than double, because in the land and sea breezes the wind-producing power acts alternately on the land and on the sea — in opposite scales of the balance; whereas in the trade-winds it acts all the time in one scale — in the sea scale; and the thermal impression which the solar ray makes through the land upon the air is much greater than that which it makes by playing upon the water.

250. *The influence of other agents required.* From these facts it is made obvious that other agents besides the direct and reflected heat of the sun are concerned in producing the trade-winds. Let us inquire into the nature of these agents.

251. *Where found.* They are to be found in the unequal distribution of land and sea, and rains, as between the two hemispheres. They derive their power from heat, it is true, but it is chiefly from the latent heat of vapor which is set free during the processes of precipitation. The vapor itself, as it rises from the sea, is (§ 232) no feeble agent in the production of wind, nor is it inconsiderable in its influence upon the trade-winds.

252. *Vapor as one of the causes of the trade-winds.* Let us consider

[12] [Maury] *Gales in the Atlantic* [*Wind and Current Charts* (Washington, 1857); also as 24 unnumbered plates, *Sailing Directions*, 8th ed., I].
[13] "Storm and Rain Chart [of the North Atlantic" (Washington, 1853)]; "Storm and Rain Chart [of the South Atlantic" (Washington, 1854). (*Wind and Current Charts*, ser. E)].

this influence. A cubic foot of water, being converted into vapor, occupies the space of 1800 cubic feet.[14] This vapor is also lighter than the 1800 cubic feet of air which it displaces. Thus, if the displaced air weigh 1000 ounces, the vapor will weigh 623; consequently, when air is surcharged with vapor, the atmosphere is bulged out above, and the barometric pressure is diminished in proportion to the volume which flows off above in consequence of this bulging out. Thus, if we imagine the air over the Atlantic Ocean to be all in a state of rest, and that suddenly during this calm columns of vapor were to commence rising from the middle of this ocean, we can understand how the wind would commence to flow into this central space from all around. Now, if we imagine no other disturbing cause to arise, but suppose the evaporation from this central area to go on with ceaseless activity, we can see that there would be a system of winds in the Atlantic as steady, but perhaps not so strong as the trades, yet owing their existence, nevertheless, merely to the formation of aqueous vapor. But this is not all.

253. *Black's law.* "During the conversion of solids into liquids, or of liquids into vapors, heat is absorbed, which is again given out on their recondensation." [15] In the process of converting one measure of water into vapor, heat enough is absorbed — *i. e.*, rendered latent, without raising the temperature of the vapor in the least — to raise the temperature of 1000 such measures of water 1°; when this vapor is condensed again into water, wherever the place of recondensation may be, this heat is set free again. If it be still further condensed, as into hail or snow, the latent heat rendered sensible during the process of congelation would be sufficient to raise the temperature of 140 additional measures of water 1°.

254. *The latent heat transported in vapor.* In this heat rendered latent by the processes of evaporation, and transported hither and thither by the winds, resides the chief source of the dynamical power which gives them motion. In some aspects vapor is to the winds what fuel is to the steam-engine: they carry it to the equatorial calm belt; there it rises, entangling the air, and carrying it up along with it as it goes. As it ascends it expands, as it expands it grows cool; and as it does this its vapor is condensed, the latent heat of which is thus liberated; this raises the temperature of the upper air, causing it to be rarefied and to ascend still higher. This increased rarefaction calls for increased velocity on the part of the inpouring trade-winds below.

255. *The effect of the deserts upon the trade-winds.* Thus the vapors uniting with the direct solar ray would, were there no counteracting influences, cause the northeast and southeast trade-winds to rush in with equal force. But there is on the polar side of the northeast trade-winds an

[14] Black and Watt's Experiments on Heat.
[15] Black's law. It is an important one, and should be remembered.

immense area of arid plains for the heat of the solar ray to beat down upon, also an area of immense precipitation. These two sources of heat hold back the northeast trade-winds, as it were, and, when the two are united, as they are in India, they are sufficient not only to hold back the northeast trade-wind, but to reverse it, causing the southwest monsoon to blow for half the year instead of the northeast trade.

256. *Indications of a crossing at the calm belts.* We have, in this difference as to strength and stability (§ 247) between the northeast and southeast trade-winds, another link in the chain of facts tending to show that there is a crossing of the winds at the calm belts. The greatest amount of evaporation takes place in the southern hemisphere, which is known by the simple circumstance that there is so much more sea-surface there. The greatest quantity of rain falls in the northern hemisphere, as both the rain-gauge and the rivers show. So likewise does the thermometer; for the vapor which affords this excess of precipitation brings the heat — the dynamical power — from the southern hemisphere; this vapor transports the heat in the upper regions from the equatorial cloud-ring to the calms of Cancer, on the polar side of which it is liberated as the vapor is precipitated, thus assisting to make the northern warmer than the southern hemisphere. In those northern latitudes where the precipitation of vapor and liberation of heat take place, aerial rarefaction is produced, and the air in the calm belt of Cancer, which is about to blow northeast trade, is turned back and called in to supply the indraught toward the north. Thus, the northeast trade-winds being checked, the southeast are called on to supply the largest portion of the air that is required to feed the ascending columns in the equatorial calm belt.

257. *The counter trades — they approach the pole in spirals.* On the north side of the trade-wind belt in the northern, and on the south side in the southern hemisphere, the prevailing direction of the winds is not toward the equator, but exactly in the opposite direction. In the extra-tropical region of each hemisphere the prevailing winds blow from the equator toward the poles. These are the counter-trades (§ 204). The precipitation and congelation that go on about the poles produce in the amount of heat set free, according to Black's law (§ 253), a rarefaction in the upper regions, and an ascent of air about the poles similar to that about the equator, with this difference, however: the place of ascent over the equator is a line, or band, or belt; about the poles it is a disc. The air rushing in from all sides will give rise to a wind, which, being operated upon by the forces of diurnal rotation as it flows north, for example, will approach the north pole by a series of spirals from the southwest.

258. *They turn with the hands of a watch about the south pole, against them about the north.* If we draw a circle about this pole on a common terrestrial globe, and intersect it by spirals to represent the direction of

the wind, we shall see that the wind enters all parts of this circle from the southwest, and that, consequently, there should be about the poles a disc or circular space of calms, in which the air ceases to move forward as wind, and ascends as in a calm; about this calm disc, therefore, there should be a whirl, in which the ascending column of air revolves from right to left, or *against* the hands of a watch. At the south pole the winds come from the northwest (§ 213), and consequently there they revolve about it *with* the hands of a watch. That this should be so will be obvious to any one who will look at the arrows on the polar sides of the calms of Cancer and Capricorn (Plate I, § 215). These arrows are intended to represent the prevailing direction of the wind at the surface of the earth on the polar side of these calms.

259. *The arrows in the diagram of the winds.* The arrows that are drawn about the axis of this diagram are intended to represent, by their flight, the mean direction of the wind, and by their length and their feathers the mean annual duration from each quadrant. Only the arrows nearest to the axis in each belt of 5° of latitude are drawn with such nicety. The largest arrow indicates that the wind in that belt blows annually, on the average, for ten months as the arrow flies. The arrow from the next most prevalent quarter is half-feathered, provided the average annual duration of the wind represented is not less than four months. The unfeathered arrows represent winds having an average duration of less than three months. The arrows are on the decimal scale; the longest arrow — which is that representing the southeast trade-winds between 5° and 10° S., where their average duration is ten months — being half an inch. Winds that blow five months are represented by an arrow half this length, and so on. The half-bearded arrows are on a scale of two for one. It appears, at first, as a singular coincidence that the wind should whirl in these discs about the poles as it does in cyclones, viz., against the hands of a watch in the northern, with them in the southern hemisphere.

260. *The offices of sea and air in the physical economy.* To act and react upon each other, to distribute moisture over the surface of the earth, and to temper the climate of different latitudes, it would seem, are two of the many offices assigned by their Creator to the ocean and the air. When the northeast and southeast trades meet and produce the equatorial calms (§ 212), the air, by the time it reaches this calm belt, is heavily laden with moisture, for in each hemisphere it has traveled obliquely over a large space of the ocean. It has no room for escape but in the upward direction (§ 223). It expands as it ascends, and becomes cooler; a portion of its vapor is thus condensed, and comes down in the shape of rain. Therefore it is that, under these calms, we have a region of constant precipitation. Old sailors tell us of such dead calms of long continuance here, of such heavy and constant rains, that they have

scooped up fresh water from the sea to drink. The conditions to which this air is exposed here under the equator are probably not such as to cause it to precipitate all the moisture that it has taken up in its long sweep across the waters. Let us see what becomes of the rest; for Nature, in her economy, permits nothing to be taken away from the earth which is not to be restored to it again in some form, and at some time or other. Consider the great rivers — the Amazon and the Mississippi, for example. We see them day after day, and year after year, discharging immense volumes of water into the ocean. "All the rivers run into the sea, yet the sea is not full." — Eccl., i., 7. Where do the waters so discharged go, and where do they come from? They come from their sources, you will say. But whence are their sources supplied? for, unless what the fountain sends forth be returned to it again, it will fail and be dry. We see simply, in the waters that are discharged by these rivers, the amount by which the precipitation exceeds the evaporation throughout the whole extent of valley drained by them; and by precipitation I mean the total amount of water that falls from, or is deposited by the atmosphere, whether as dew, rain, hail, or snow. The springs of these rivers (§ 191) are supplied from the rains of heaven, and these rains are formed of vapors which are taken up from the sea, that "it be not full," and carried up to the mountains through the air. "Note the place whence the rivers come, thither they return again." Behold now the waters of the Amazon, of the Mississippi, the St. Lawrence, and all the great rivers of America, Europe, and Asia, lifted up by the atmosphere, and flowing in invisible streams back through the air to their sources among the hills (§ 191), and that through channels so regular, certain, and well defined, that the quantity thus conveyed one year with the other is nearly the same: for that is the quantity which we see running down to the ocean through these rivers; and the quantity discharged annually by each river is, as far we can judge, nearly a constant.

261. *Powerful machinery.* We now begin to conceive what a powerful machine the atmosphere must be; and, though it is apparently so capricious and wayward in its movements, here is evidence of order and arrangement which we must admit, and proof which we can not deny, that it performs this mighty office with regularity and certainty, and is therefore as obedient to law as is the steam-engine to the will of its builder. It, too, is an engine. The South Seas themselves, in all their vast intertropical extent, are the boiler for it, and the northern hemisphere is its condenser (§ 24). The mechanical power exerted by the air and the sun in lifting water from the earth, in transporting it from one place to another, and in letting it down again, is inconceivably great. The utilitarian who compares the water-power that the Falls of Niagara would afford if applied to machinery, is astonished at the number of figures

which are required to express its equivalent in horse-power. Yet what is the horse-power of the Niagara, falling a few steps, in comparison with the horse-power that is required to lift up as high as the clouds and let down again all the water that is discharged into the sea, not only by this river, but by all the other rivers in the world. The calculation has been made by engineers, and, according to it, the force for making and lifting vapor from each area of one acre that is included on the surface of the earth is equal to the power of 30 horses.

CHAPTER V

§ 270–303. RAINS AND RIVERS

270. *Rivers considered as rain-gauges — the ten largest.* Rivers are the rain-gauges of nature. The volume of water annually discharged by any river into the sea expresses the total amount by which the precipitation upon the valley drained by such river exceeds the evaporation from the same valley during the year. There are but ten rivers that we shall treat as rain-gauges; and there are only ten in the world whose valleys include an area of more than 500,000 square miles. They are:

	Square miles
The Amazon, including the Tocantines and Orinoco	2,048,000
Mississippi	982,000
La Plata	886,000
Yenisei	785,000
Obi	725,000
Lena	594,000
Amoor	583,000
Yang-tse-kiang	548,000
Hoang-ho	537,000
Nile	520,000

These areas are stated in round numbers, and according to the best authorities. The basin of the Amazon is usually computed at 1,512,000 square miles; but such computation excludes the Tocantines, 204,000 square miles, which joins the Amazon near its mouth, and the Orinoco, with a hydrographic area of 252,000 square miles, which, by means of the Casiquiare, is connected also with the Amazon. We think that these three rivers should all be regarded as belonging to one hydrographic basin, for a canoe may pass inland from any one to either of the others without portage. Of these hydrographic basins, three, including an area of 3,916,000 square miles, are American; six, which contain an area of 3,772,000 square miles, belong to Asia, one to Africa, and none to Europe. The three largest rivers of Asia, the Yenisei, Obi, and Lena (2,104,000 square miles), discharge their waters into the Arctic Ocean; their outlets are beyond the reach of the commercial world; consequently they do not possess the interest which, in the minds of men generally, is attached to the rest. The three others of Asia drain 1,668,000 square miles, and run into the Pacific; while the whole American system feed with their waters and their commerce the Atlantic Ocean. These rivers, with their springs, give drink to man and beast, and nourish with their waters plants and reptiles, with fish and fowl not a few. The capacity of their basins for production and wealth is without limits. These streams are the great arteries of inland commerce. Were they to dry up, political communities would be torn asunder, the harmonies of the earth would be destroyed, and that beautiful adaptation of physical forces to terrestrial machinery, by which climates are regulated, would lose its adjustment and run wild, like a watch without a balance.

271. *Heat required to lift vapor for these rivers.* We see these majestic streams pouring their waters into the sea, and from the sea we know those waters must come again, else the sea would be full. We know, also, that the sunbeam and the sea-breeze suck them up again; and it is curious to fancy such volumes of water as this mighty company of ten great rivers is continually discharging into the sea, taken up by the winds and the sun, and borne away through the invisible channels of the air to the springs among the hills that are the source of all rivers. This operation is perpetually going on, yet we perceive it not. It is the work of that invisible, imponderable, omnipresent, and wonderful agent called heat. This is the agent which controls both sea and air in their movements and in many of their offices. The average amount of heat daily dispensed to our planet from the source of light in the heavens is enough to melt a coating of ice completely encasing the earth with a film 1½ in. in thickness.[1] Heat is the agent that distills for us fresh water from the sea. It pumps up out of the ocean all the water for our lakes and rivers, and gives power to the winds to transport it as vapor thence to the mountains. And though this is but a part of the work which in the terrestrial economy has been assigned to this mighty agent, we may acquire much profitable knowledge by examining its operations here in various aspects. To assist in this undertaking I have appealed to the ten greatest rivers for terms and measures in which some definite idea may be conveyed as to the magnitude of the work and the immense physico-mechanical power of this imponderable and invisible agent called heat. Calculations have been made which show that the great American lakes contain 11,000 cubic miles of water. This, according to the best computation, is twice as much as is contained in all the other fresh water lakes, and rivers, and cisterns of the world. The Mississippi River does not, during a hundred years, discharge into the sea so large a volume of water as is at this moment contained in the great northern lakes of this continent; and yet this agent, whose works we are about to study, operating through the winds, has power annually to lift up from the sea and pour down upon the earth in grateful showers fresh water enough to fill the lakes at least twenty times over.

272. *Rain-fall in the Mississippi Valley.* That we may be enabled the

---

[1] *Deduced* from the experiments of Pouillet. [M. Pouillet, Éléments de physique experimentale et de météorologie, (7th ed; Paris, 1856), II, 714–715].

better to appreciate the power and the majesty of the thermal forces of the sun, and comprehend in detail the magnitude and grandeur of their operations, let us inquire how much rain falls annually upon the watersheds of one of these streams, as of the Mississippi; how much is carried off by the river; how much is taken up by evaporation; and how much heat is evolved in hoisting up and letting down all this water. In another chapter we shall inquire for the springs in the sea that feed the clouds with rain for these rivers. If we had a pool of water one mile square and six inches deep to be evaporated by artificial heat, and if we wished to find out how much would be required for the purpose, we should learn from Mr. Joule's experiments that it would require about as much as is evolved in the combustion of 30,000 tons of coal. Thus we obtain (§ 271) our unit of measure to help us in the calculation; for if the number of square miles contained in the Mississippi Valley, and the number of inches of rain that fall upon it annually be given, then it will be easy to tell how many of such huge measures of heat are set free during the annual operation of condensing the rain for our hydrographic basin. And then, if we could tell how many inches of this rain-water are again taken up by evaporation, we should have the data for determining the number of these monstrous measures of heat that are employed for that operation also.

273. *Its area, and the latent heat liberated during the processes of condensation there.* The area of the Mississippi Valley is said by physical geographers to embrace 982,000 square miles; and upon every square mile there is an annual average rain-fall of 40 inches. Now if we multiply 982,000 by the number of times 6 will go into 40, we shall have the number of our units of heat that are annually set free among the clouds that give rain to the Mississippi Valley. Thus the imagination is startled, and the mind overwhelmed with the announcement that the quantity of heat evolved from the vapors as they are condensed to supply the Mississippi Valley with water is as much as would be set free by the combustion of 30,000 tons of coal multiplied 6,540,000 times. Mr. Joule, of Manchester, is our authority for the heating power of one pound of coal; the Army Meteorological Register, compiled by Lorin Blodget, and published by the Surgeon General's Office in 1855, is the authority on which we base our estimate as to the average annual fall of rain; and the annals of the National Observatory show, according to the observations made by Lieutenant Marr at Memphis, the annual fall of rain there to be 49 inches, the annual evaporation 43, and the quantity of water that annually passes by in the Mississippi to be 93 cubic miles. The water required to cover to the depth of 40 inches an area of 982,000 square miles would, if collected together in one place, make a sea one mile deep, with a superficial area of 620 square miles.

274. *Annual discharge of the Mississippi River.* It is estimated that the tributaries which the Mississippi River receives below Memphis increase the volume of its waters about one eighth, so that its annual average discharge into the sea may be estimated to be about 107 cubic miles, or about one sixth of all the rain that falls upon its water-shed. This would leave 513 cubic miles of water to be evaporated from this river-basin annually. All the coal that the present mining force of the country could raise from its coal measures in a thousand years would not, during its combustion, give out as much heat as is rendered latent annually in evaporating this water. Utterly insignificant are the sources of man's mechanical powers when compared with those employed by nature in moving machinery which brings the seasons around and preserves the harmonies of creation!

275. *Physical adaptations.* The amount of heat required to reconvert these 513 cubic miles of rain-water into vapor and bear it away, had accumulated in the Mississippi Valley faster than the earth could throw it off by radiation. Its continuance there would have been inconsistent with the terrestrial economy. From this standpoint we see how the rain-drop is made to preserve the harmonies of nature, and how water from the sea is made to carry off by re-evaporation from the plains and valleys of the earth their surplusage of heat, which could not otherwise be got rid of without first disturbing the terrestrial arrangements, and producing on the land desolation and a desert. Behold now the offices of clouds and vapor — the adaptations of heat. Clouds and vapor do something more than brew storms, fetch rain, and send down thunder-bolts. The benignant vapors cool our climates in summer by rendering latent the excessive heat of the noonday sun; and they temper them in winter, by rendering sensible and restoring again to the air the equivalent of that selfsame heat.

276. *Whence come the rains for the Mississippi.* Whence came, and by what channels did they come, these cubic miles of water which the Mississippi River pours annually into the sea? The wisest of men has told us they come from the sea. Let us explore the sea for their place and the air for their channel. The Gulf of Mexico can not furnish rain for all the Mississippi Valley. The Gulf lies within the region of the northeast trades, and these winds carry its vapors off to the westward, and deliver them in rain to the hills, and the valleys, and the rivers of Mexico and Central America. The winds that bring the rains for the upper Mississippi Valley come not from the south; they come from the direction of the Rocky Mountains, the Sierra Nevada, and the great chain that skirts the Pacific coast. It is, therefore, needless to search in the Gulf, for the rain that comes from it upon that valley is by no means sufficient to feed one half of its springs. Let us next examine the Atlantic Ocean, and include its slopes also in the investigation.

277. *The northeast trades of the Atlantic supply rains only for the rivers of Central and South America.* The northeast trade-wind region of this ocean extends (§ 210) from the parallel of 30° to the equator. They carry their vapor before them, and, meeting the southeast trade-wind, the two form clouds which give rain not only to Central America, but they drop down, also, water in abundance for the Atrato, the Magdalena, the Orinoco, the Amazon, and all the great rivers of intertropical America; also for the Senegal, the Niger, and the Congo of Africa. So completely is the rain wrung out of these winds for these American rivers by the Andes, that they become dry and rainless after passing this barrier, and as such reach the western shores of the continent, producing there, as in Peru, a rainless region. The place in the sea whence our rivers come, and whence Europe is supplied with rains, is clearly not to be found in this part of the ocean.

278. *The calm belt of Cancer furnishes little or no rain.* Between the parallels of 30° and 35° N. lies the calm belt of Cancer, a region where there is no *prevailing* wind (see Diagram of the Winds, Plate I). It is a belt of light airs and calms — of airs so baffling that they are often insufficient to carry off the "loom," or that stratum of air which, being charged with vapor, covers calm seas as with a film, and as if to prevent farther evaporation. This band of the ocean can scarcely be said to furnish any vapor to the land, for a rainless country, both in Africa, and Asia, and America, lies within it.

279. *The North Atlantic insufficient to supply rain for so large a portion of the earth as one sixth of all the land.* All Europe is on the north side of this calm belt. Let us extend our search, then, to that part of the Atlantic which lies between the parallels of 35° and 60° N., to see if we have water surface enough there to supply rains for the 8½ millions of square miles that are embraced by the water-sheds under consideration. The area of this part of the Atlantic is not quite 5 millions of square miles, and it does not include more than one thirtieth of the entire sea surface of our planet, while the water-sheds under consideration contain one sixth part of its entire land surface. The natural proportion of land and water surface is nearly as 1 to 3. According to this ratio, the extent of sea surface required to give rain for these 8½ millions of square miles would be a little over 25, instead of a little less than 5 millions of square miles.

280. *Daily rate of evaporation at sea less than on land — observations wanted.* The state of our knowledge concerning the actual amount of evaporation that is daily going on at sea has, notwithstanding the activity in the fields of physical research, been but little improved. Records as to the amount of water daily evaporated from a plate or dish on shore affords us no means of judging as to what is going on even in the same latitude at sea. Sea-water is salt, and does not throw off its vapor as freely as fresh-

water. Moreover, the wind that blows over the evaporating dish on shore is often dry and fresh. It comes from the mountains, or over the plains where it found little or no water to drink up; therefore it reaches the observer's dish as thirsty wind, and drinks up vapor from it greedily. Now had the same dish been placed on the sea, the air would come to it over the water, drinking as it comes, and arriving already quite or nearly saturated with moisture; consequently, the observations of the amount of evaporation on shore give no idea of it at sea.

281. *Rivers are gauges for the amount of effective evaporation.* There is no physical question of the day which is more worthy of attention than the amount of effective evaporation that is daily going on in the sea. By *effective* I mean the amount of water that, in the shape of vapor, is daily transferred from the sea to the land. The volume discharged by the rivers into the sea expresses (§ 270) that quantity; and it may be ascertained with considerable accuracy by gauging the other great rivers as I procured the Mississippi to be gauged at Memphis in 1849.

282. *Importance of rain and river gauges.* The monsoons supply rains to feed the rivers of India, as the northeast and southeast trade-winds of the Atlantic supply rains to feed the rivers of Central and South America. Now rain-gauges which will give us the mean annual rain-fall on these water-sheds, and river-gauges which would give us the mean annual discharge of the principal watercourses, would afford data for an excellent determination as to the amount of evaporation from some parts of the ocean at least, especially for the trade-wind belts of the Atlantic and the monsoon region of the Indian Ocean. All the rain which the monsoons of India deliver to the land the rivers of India return to the sea. And if, in measuring this for the whole of India, our gauges should lead us into a probable error, amounting in volume to half the discharge of the Mississippi River, it would not make a difference in the computed rate of the *effective* daily evaporation from the North Indian Ocean exceeding the one two-thousandth part of an inch (0.002 in.).

283. *Hypsometry in the North Atlantic peculiar.* That part of the extra-tropical North Atlantic under consideration is peculiar as to its hypsometry. It is traversed by large icebergs, which are more favorable to the recondensation of its vapors than so many islets would be. Warm waters are in the middle of it, and both the east and the west winds, which waft its vapors to the land, have, before reaching the shores, to cross currents of cool waters, as the inshore current counter to the Gulf Stream on the western side, and the cool drift from the north on the east side. In illustration of this view, and of the influence of the icebergs and cold currents of the Atlantic upon the hypsometry of that ocean, it is only necessary to refer to the North Pacific, where there are no icebergs nor marked contrasts between the temperature of its currents. Ireland and the

Aleutian Islands are situated between the same parallels. On the Pacific islands there is an uninterrupted rain-fall during the entire winter. At other seasons of the year sailors describe the weather, in their log-books, there as "raining pretty much all the time." This is far from being the case even on the western coasts of Ireland, where there is a rain-fall of only 47 inches [2] — probably not more than a third of what Oonalaska receives. And simply for this reason: the winds reach Ireland after they have been robbed (partially) of the vapors by the cool temperatures of the icebergs and cold currents which lie in their way; whereas, such being absent from the North Pacific, they arrive at the islands there literally reeking with moisture. Oregon in America, and France on the Bay of Biscay, are between the same parallels of latitude; their situation with regard both to wind and sea is the same, for each has an ocean to windward. Yet their annual rain-fall is, for Oregon,[3] 65 inches, for France, 30. None of the islands which curtain the shores of Europe are visited as abundantly by rains as are those in the same latitudes which curtain our northwest coast. The American water-shed receives about twice as much rain as the European. How shall we account for this difference except upon the supposition that the winds from the Pacific carry more rain than the winds from the Atlantic? Why should they do this, except for the icebergs and cool streaks already alluded to? [4]

284. *Limited capacity of winds to take up and transport, for the rivers of Europe and America, vapor from the North Atlantic.* It may well be doubted whether the southwesterly winds — which are the prevailing winds in this part of the Atlantic — carry into the interior of Europe much more moisture than they bring with them into the Atlantic. They enter it with a mean annual temperature of 60°, with an average dew point of about 55°. They leave it at a mean temperature varying from 60° to 40°, according to the latitude in which they reach the shore, and consequently with an average dew-point *not higher* than the mean temperature. Classifying the winds of this part of the ocean according to the halves of the horizon as east and west, the mean of 44,999 observations in the log-books of the Observatory shows that, on the average, the west winds blow annually 230 and the east winds 122 days.

285. *The vapor-springs for all these rivers not in the Atlantic Ocean.* Taking all these facts and circumstances into consideration, and without pretending to determine how much of the water which the rivers of America and Europe carry into this part of the ocean comes from it again, we may with confidence assume that the winds do not get vapor enough

---

[2] [Alexander] Keith Johnston [*The Physical Atlas of Natural Phenomena*, a new and enlarged edition (Edinburgh and London, 1856), pl. 20].

[3] *Army Meteorological Register* [*for twelve Years, from 1843 to 1854, Inclusive* (Washington, D.C.], 1855) [p. 677 and map opp. p. 734].

[4] [Johnston, *Physical Atlas*, pp. 47–48.]

from this part of the ocean to give rain to Europe, to the Mississippi Valley, to our Atlantic slopes, and the western half of Asiatic Russia. We are authorized in this conclusion, just as we have authority to say that the evaporation from the Mediterranean is greater in amount than the volume of water discharged into it again by the rivers and the rains; only in this case the reverse takes place, for the rivers empty more water into the Atlantic than the winds carry from it. This fact also is confirmed by the hydrometer, for it shows that the water of the North Atlantic is, parallel for parallel, lighter than water in the Southern Ocean.

286. *The places in the sea whence come the rivers of the north, discovered — proves the crossing at the calm belts.* The inference, then, from all this is, that the place in the sea (§ 276) whence come the waters of the Mississippi and other great rivers of the northern hemisphere is to be found in these southern oceans, and the channels by which they come are to be searched out aloft, in the upper currents of the air. Thus we bring evidence and facts which seem to call for a crossing of air at the calm belts, as represented by the diagram of the winds, Plate I. It remains for those who deny that there is any such crossing — who also deny that extra-tropical rivers of the northern are fed by rains condensed from vapors taken up in the southern hemisphere — to show whence come the *hundreds* of cubic miles of water which these rivers annually pour into the Atlantic and the Arctic Oceans. In finding the "place" of all this water, it is incumbent upon them to show us the winds which bring it also, and point out its channels.

287. *Spirit in which the search for truth should be conducted.* "In the greater number of physical investigations some hypothesis is requisite, in the first instance, to aid the imperfection of our senses; and when the phenomena of nature accord with the assumption, we are justified in believing it to be a general law." [5]

288. *The number of known facts that are reconciled by the theory of a crossing at the calm belts.* In this spirit this hypothesis has been made. Without any evidence bearing upon the subject, it would be as philosophical to maintain that there is no crossing at the calm belts as it would be to hold that there is; but nature suggests in several instances that there must be a crossing. (1.) In the homogeneousness of the atmosphere (§ 237). The vegetable kingdom takes from it the impurities with which respiration and combustion are continually loading it; and in the winter, when the vegetable energies of the northern hemisphere are asleep, they are in full play in the southern hemisphere. And is it consistent with the spirit of true philosophy to deny the existence, because we may not comprehend the nature of a contrivance in the machinery of the universe

---

[5] [Mary] Somerville [*On the Connexion of the Physical Sciences* (9th ed; London, 1858), p. 104].

which guides the impure air that proceeds from our chimneys and the nostrils of all air-breathing creatures in our winter over into the other hemisphere for re-elaboration, and which conducts across the calm places and over into this that which has been replenished from the plains and sylvas of the south? (2.) Most rain, notwithstanding there is most water in the southern hemisphere, falls in this. How can vapor thence come to us except the winds bring it, and how can the winds fetch it except by crossing the calm places? (3.) The "sea-dust" of the southern hemisphere, as Ehrenberg calls the red fogs of the Atlantic, has its *locus* on the other side of the equator, but it is found in the winds of the North Atlantic Ocean. If this be so, it must cross one or more of the calm belts.[6] (4.) Parallel for parallel, from the equator to 40° or 45° S., the southern hemisphere is the cooler. This fact is consistent with the supposition that the heat which is rendered latent and abstracted from that hemisphere by its vapors is set free by their condensation in this. Upon no other hypothesis than by these supposed crossings can this fact be reconciled, for the amount of heat annually received from the sun by the two hemispheres is, as astronomers have shown, precisely the same.[7] (5.) Well-conducted

[6] After this had been written, I received from my colleague, Lieut. Andrau, an account of the following little tell-tale upon this subject:

"I found a confirmation of your theory in a piece of vegetable substance caught in a small sack (hoisted up above the tops) between 22°–25° lat. N., and 38°–39½° long. W. This piece is of the following dimensions: 14 millim. long, 1 to 1½ mm. large, 1/5 mm. thick, and weighing 1½ milligrams. Our famous microscopist and naturalist, Professor P. Harting, at Utrecht, told me, after an exact inquiry, 'that this vegetable fragment issued from a leaf of the family Monocotyledon, probably not from a palm-tree, but from a Pardanaceæ or Scitamineæ' — consequently, from trees belonging to the tropical regions. Now I am sure it comes from the tropics. I am greatly surprised to perceive that a piece of leaf of this dimension could run off a distance of more than 1200 geographical miles in the upper regions of the atmosphere; for the nearest coast-lines of the two continents, America and Africa, lay at the said distance from the place where this vegetable fragment was caught, by the carefulness of Capt. S. Stapert, one of our most zealous co-operators. There can be no doubt that it comes from South America, because the direction of the trade-winds on the west coast of Africa is too northerly to bring this fragment to the finding-place in 25° N. and 38° W." — *Letter from Lieut. Andrau*, of the Dutch Navy, dated Utrecht, Jan. 2, 1860.

[7] The amount of solar heat annually impressed upon the two hemispheres is identically the same; yet within certain latitudes the southern hemisphere is, parallel for parallel, the cooler. How does it become so? If it be the cooler by radiation, then it must be made so by radiating more heat than it receives; such a process would be cumulative in its effects, and were it so, the southern hemisphere would be gradually growing cooler. There is no evidence that it is so growing, and the inference that it is seems inadmissible. In fact, the southern hemisphere radiates less heat than the northern, though it receives as much from the sun. And it radiates less for this reason: there is more land in the northern — land is a better radiator than water — therefore the northern radiates more heat than the southern hemisphere; the southern has more water and more clouds — clouds prevent radiation — therefore the southern hemisphere radiates less heat than the northern; still it is the cooler. How is this paradox to be reconciled but upon the supposition that the southern surplusage is

observations made with the hydrometer [8] (§ 285) for every parallel of latitude in the Atlantic Ocean from 40° S. to 40° N., show that, parallel for parallel, and notwithstanding the difference of temperature, the specific gravity of sea-water is greater in the southern than it is in the northern hemisphere. This difference as to the average condition of the sea on different sides of the line is reconciled by the hypothesis which requires a crossing at the calm belts. The vapor which conveys fresh water and caloric from the southern hemisphere to the northern will in part account for this difference both of specific gravity and temperature, and no other hypothesis will. This hydrometric difference indicates the amount of fresh water which, as vapor in the air, as streams on the land, and as currents in the sea,[9] is constantly in transitu between the two hemispheres. All these facts are inconsistent with the supposition that there is no crossing at the calm belts, and consistent with the hypothesis that there is. It is no argument against the hypothesis that assumes a crossing, to urge our ignorance of any agent with power to conduct the air across the calm belts. It would be as reasonable to deny the red to the rose or the blush to the peach, because we do not comprehend the processes by which the coloring matter is collected and given to the fruit or flower, instead of the wood or leaves of the plant. To assume that the direction of air is, after it enters the calm belts, left to chance, would be inconsistent with our notions of the attributes of the great Architect. The planets have their orbits, the stars their courses, and the wind "his circuits." And in the construction of our hypotheses, it is pleasant to build them up on the premiss that He can and has contrived all the machinery necessary for guiding every atom of air in the atmosphere through its channels and according to its circuits, as truly and as surely as He has contrived it for holding comets to their courses and binding the stars in their places. These circumstances, and others favoring this hypothesis as to these air-crossings, will be presented more in detail.

289. *The atmosphere to be studied like any other machinery, by its operations.* In observing the workings and studying the offices of the various parts of the physical machinery which keeps the world in order, we should ever remember that it is all made for its purposes, that it was planned according to design, and arranged so as to make the world as we behold it — a place for the habitation of man. Upon no other hypothesis can the student expect to gain profitable knowledge concerning the physics of sea, earth, or air. Regarding these elements of the old philosophers as parts only of the same piece of machinery, we are struck with the fact,

---

stowed away in vapors, transported thence across the calm belts by the winds, and liberated by precipitation on our side of the equator?

[8] [Commodore John] Rodgers, in the *Vincennes. Sailing Directions,* 8th ed., I, 235.

[9] The water which the rivers empty into the North Atlantic has to find its way south with the currents of the sea.

and disposed to inquire why is it that the proportion of land and water in the northern hemisphere is very different from the proportion that obtains between them in the southern? In the northern hemisphere, the land and water are nearly equally divided. In the southern, there is several times more water than land. Is there no connection between the machinery of the two hemispheres? Are they not adapted to each other? Or, in studying the physical geography of our planet, shall we regard the two hemispheres as separated from each other? Rather let us regard them as made for each other, as adapted to each other, and the one as an essential to the other. So regarding them, we observe that all the great rivers in the world are in the northern hemisphere, where there is less ocean to supply them. Whence, then, are their sources replenished? Those of the Amazon are, as we have seen (§ 277), supplied with rain from the equatorial calms and trade-winds of the Atlantic. That river runs east, its branches come from the north and south; it is always the rainy season on one side or the other of it; consequently, it is a river without periodic stages of a very marked character. It is always near its high-water mark. For one half of the year its northern tributaries are flooded, and its southern for the other half. It discharges under the line, and as its tributaries come from both hemispheres, it can not be said to belong exclusively to either. It is supplied with water made of vapor that is taken up from the Atlantic Ocean. Taking the Amazon, therefore, out of the count, the Rio de la Plata is the only great river of the southern hemisphere. There is no large river in New Holland. The South Sea Islands give rise to none, nor is there one in South Africa entitled to be called great that we know of.

290. *Arguments furnished by the rivers.* The great rivers of North America and North Africa, and all the rivers of Europe and Asia, lie wholly within the northern hemisphere. How is it, then, considering that the evaporating surface lies mainly in the southern hemisphere — how is it, I say, that we should have the evaporation to take place in one hemisphere and the condensation in the other? The total amount of rain which falls in the northern hemisphere is much greater, meteorologists tell us, than that which falls in the southern. The annual amount of rain in the north temperate zone is half as much again as that of the south temperate. How is it, then, that this vapor gets, as stated, from the southern into the northern hemisphere, and comes with such regularity that our rivers never go dry and our springs fail not? It is because of these air-crossings — these beautiful operations, and the exquisite *compensation* of this grand machine, the atmosphere. It is exquisitely and wonderfully counterpoised. Late in the autumn of the north, throughout its winter, and in early spring, the sun is pouring his rays with the greatest intensity down upon the seas of the southern hemisphere, and this powerful engine which we are contemplating is pumping up the water there (§ 261) with

the greatest activity, and sending it over here for our rivers. The heat which this heavy evaporation absorbs becomes latent, and, with the moisture, is carried through the upper regions of the atmosphere until it reaches our climates. Here the vapor is formed into clouds, condensed, and precipitated. The heat which held this water in the state of vapor is set free, it becomes sensible heat, and it is that [(4), § 288] which contributes so much to temper our winter climate. It clouds up in winter, turns warm, and we say we are going to have falling weather. That is because the process of condensation has already commenced, though no rain or snow may have fallen: thus we feel this southern heat, that has been collected from the rays of the sun by the sea, been bottled away by the winds in the clouds of a southern summer, and set free in the process of condensation in our northern winter. If Plate I fairly represent the course of the winds, the southeast trade-winds would enter the northern hemisphere, and, as an upper current, bear into it all their moisture, except that which is precipitated in the region of equatorial calms, and in the crossing of high mountain ranges, as the Cordilleras of South America.

291. *More rain in the northern than in the southern hemisphere.* The South Seas, then, according to § 290, should supply mainly the water for this engine, while the northern hemisphere condenses it; we should, therefore, have more rain in the northern hemisphere. The rivers tell us that we have — the rain-gauge also. The yearly average of rain in the north temperate zone is, according to Johnston, thirty-seven inches. He gives but twenty-six in the south temperate. The observations of mariners are also corroborative of the same. Log-books, containing altogether the records for upward of 260,000 days in the Atlantic Ocean north and south (Plate IX), have been carefully examined for the purpose of ascertaining, for comparison, the number of calms, rains, and gales therein recorded for each hemisphere. Proportionally the number of each is given as decidedly greater for the north than it is for the south. The result of this examination is very instructive, for it shows the status of the atmosphere to be much more unstable in the northern hemisphere, with its excess of land, than in the southern, with its excess of water. Rains, and fogs, and thunder, and calms, and storms, all occur much more frequently, and are more irregular also as to the time and place of their occurrence on this side, than they are on the other side of the equator. Moisture is never extracted from the air by subjecting it from a low to a higher temperature, but the reverse. Thus all the air which comes loaded with moisture from the other hemisphere, and is borne into this with the southeast trade-winds, travels in the upper regions of the atmosphere (§ 213) until it reaches the calms of Cancer; here it becomes the surface wind that prevails from the southward and westward. As it goes north it grows cooler, and the process of condensation commences. We may now liken it to the

wet sponge, and the decrease of temperature to the hand that squeezes that sponge. Finally reaching the cold latitudes, all the moisture that a dew-point of zero, and even far below, can extract, is wrung from it; and this air then commences "to return according to his circuits" as dry atmosphere. And here we can quote Scripture again: "The north wind driveth away rain." This is a meteorological fact of high authority, and one of great significance too.

292. *The trade-winds the evaporating winds.* By reasoning in this manner and from such facts, we are led to the conclusion that our rivers are supplied with their waters principally from the trade-wind regions — the extra-tropical northern rivers from the southern trades, and the extra-tropical southern rivers from the northern trade-winds, for the trade-winds are the evaporating winds.

293. *The saltest part of the sea.* Taking for our guide such faint glimmerings of light as we can catch from these facts, and supposing these views to be correct, then the saltest portion of the sea should be in the trade-wind regions, where the water for all the rivers is evaporated; and there the saltest portions are found. There, too, the rains fall less frequently (Plate IX). Dr. Ruschenberger, of the Navy, on his last voyage to India, was kind enough to conduct a series of observations on the specific gravity of sea water. In about the parallel of 17° north and south — midway of the trade-wind regions — he found the heaviest water. Though so warm, the water there was heavier than the cold water to the south of the Cape of Good Hope. Lieutenant D. D. Porter, in the steam-ship Golden Age, found the heaviest water about the parallels of 20° north and 17° south. Captain Rodgers, in the United States ship Vincennes, found the heaviest water in 17° N., and between 20° and 25° S.

294. *Seeing that the southern hemisphere affords the largest evaporating surface, how, unless there be a crossing, could we have most rain and the great rivers in the northern?* In summing up the evidence in favor of this view of the general system of atmospherical circulation, it remains to be shown how it is, if the view be correct, there should be smaller rivers and less rain in the southern hemisphere. The winds that are to blow as the northeast trade-winds, returning from the polar regions, where the moisture (§ 292) has been compressed out of them, remain, as we have seen, dry winds until they cross the calm zone of Cancer, and are felt on the surface as the northeast trades. About two thirds of them only can then blow over the ocean; the rest blow over the land, over Asia, Africa, and North America, where there is comparatively but a small portion of evaporating surface exposed to their action. The zone of the northeast trades extends, on an average, from about 29° north to 7° north. Now, if we examine the globe, to see how much of this zone is land and how much water, we shall find, commencing with China and coming over

Asia, the broad part of Africa, and so on, across the continent of America to the Pacific, land enough to fill up, as nearly as may be, just one third of it. This land, if thrown into one body between these parallels, would make a belt equal to 120° of longitude by 22° of latitude, and comprise an area of about twelve and a half millions of square miles, thus leaving an evaporating surface of about twenty-five millions of square miles in the northern against about seventy-five millions in the southern hemisphere. According to the hypothesis, illustrated by Plate I, as to the circulation of the atmosphere, it is these northeast trade-winds that take up and carry over, after they rise up in the belt of equatorial calms, the vapors which make the rains that feed the rivers in the extra-tropical regions of the southern hemisphere. Upon this supposition, then, two thirds only of the northern trade-winds are fully charged with moisture, and only two thirds of the amount of rain that falls in the northern hemisphere should fall in the southern; and this is just about the proportion (§ 292) that observation gives. In like manner, the southeast trade-winds take up the vapors which make our rivers, and as they prevail to a much greater extent at sea, and have exposed to their action about twice as much ocean as the northeast trade-winds have, we might expect, according to this hypothesis, more rains in the northern — and, consequently, more and larger rivers — than in the southern hemisphere. A glance at Plate V will show how very much larger that part of the ocean over which the southeast trades prevail is than that where the northeast trade-winds blow. This estimate as to the quantity of rain in the two hemispheres is one which is not capable of verification by any more than the rudest approximations; for the greater extent of southeast trades on one side, and of high mountains on the other, must each of necessity, and independent of the other agents, have their effects. Nevertheless, this estimate gives as close an approximation as we can make out from our data.

295. *The rainy seasons, how caused.* The calm and trade-wind regions or belts move up and down the earth, annually, in latitude nearly a thousand miles. In July and August the zone of equatorial calms is found between 7° north and 12° north; sometimes higher; in March and April, between latitude 5° south and 2° north.[10] With this fact and these points of view before us, it is easy to perceive why it is that we have a rainy season in Oregon, a rainy and dry season in California, another at Panama, two at Bogotá, none in Peru, and one in Chili. In Oregon it rains every month, but about five times more in the winter than in the summer months. The winter there is the summer of the southern hemisphere, when this steam-engine (§ 24) is working with the greatest pressure. The vapor that is taken up by the southeast trades is borne along over the region of

---

[10] [Maury] "Trade-Wind Chart [of the Atlantic Ocean" (*Wind and Current Charts*, ser. B), 1851].

northeast trades to latitude 35° or 40° north, where it descends and appears on the surface with the southwest winds of those latitudes. Driving upon the highlands of the continent, this vapor is condensed and precipitated, during this part of the year, almost in constant showers, and to the depth of about thirty inches in three months.

296. *The rainy seasons of California and Panama.* In the winter the calm belt of Cancer approaches the equator. This whole system of zones, viz., of trades, calms, and westerly winds, follows the sun; and they of our hemisphere are nearer the equator in the winter and spring months than at any other season. The southwest winds commence at this season to prevail as far down as the lower part of California. In winter and spring the land in California is cooler than the sea air, and is quite cold enough to extract moisture from it. But in summer and autumn the land is the warmer, and can not condense the vapors of water held by the air. So the same cause which made it rain in Oregon now makes it rain in California. As the sun returns to the north, he brings the calm belt of Cancer and the northeast trades along with him; and now, at places where, six months before, the southwest winds were the prevailing winds, the northeast trades are found to blow. This is the case in the latitude of California. The prevailing winds, then, instead of going from a warmer to a cooler climate, as before, are going the opposite way. Consequently, if, under these circumstances, they have the moisture in them to make rains of, they can not precipitate it. Proof, if proof were wanting that the prevailing winds in the latitude of California are from the westward, is obvious to all who cross the Rocky Mountains or ascend the Sierra Madre. In the pass south of the Great Salt Lake basin those west winds have worn away the hills and polished the rock by their ceaseless abrasion and the scouring effects of the driving sand. Those who have crossed this pass are astonished at the force of the wind and the marks there exhibited of its GEOLOGICAL AGENCIES. Panama is in the region of equatorial calms. This belt of calms travels during the year, back and forth, over about 17° of latitude, coming farther north in the summer, where it tarries for several months, and then returning so as to reach its extreme southern latitude some time in March or April. Where these calms are it is always raining, and the chart [11] shows that they hang over the latitude of Panama from June to November; consequently, from June to November is the rainy season at Panama. The rest of the year that place is in the region of the northeast trades, which, before they arrive there, have to cross the mountains of the isthmus, on the cool tops of which they deposit their moisture, and leave Panama rainless and pleasant until the sun returns north with the belt of equatorial calms after him. They then push the belt of northeast trades farther to the north, occupy a

[11] [*Ibid.*]

part of the winter zone, and refresh that part of the earth with summer rains. This belt of calms moves over more than double of its breadth, and nearly the entire motion from south to north is accomplished generally in two months, May and June. Take the parallel of 4° north as an illustration: during these two months the entire belt of calms crosses this parallel, and then leaves it in the region of the southeast trades. During these two months it was pouring down rain on that parallel. After the calm belt passes it the rains cease, and the people in that latitude have no more wet weather till the fall, when the belt of calms recrosses this parallel on its way to the south. By examining the "Trade-wind Chart," it may be seen what the latitudes are that have two rainy seasons, and that Bogotá is within the bi-rainy latitudes.

297. *The rainless regions.* The coast of Peru is within the region of perpetual southeast trade-winds. Though the Peruvian shores are on the verge of the great South Sea boiler, yet it never rains there. The reason is plain. The southeast trade-winds in the Atlantic Ocean first strike the water on the coast of Africa. Traveling to the northwest, they blow obliquely across the ocean till they reach the coast of Brazil. By this time they are heavily laden with vapor, which they continue to bear along across the continent, depositing it as they go, and supplying with it the sources of the Rio de la Plata and the southern tributaries of the Amazon. Finally they reach the snow-capped Andes, and here is wrung from them the last particle of moisture that that very low temperature can extract. Reaching the summit of that range, they now tumble down as cool and dry winds on the Pacific slopes beyond. Meeting with no evaporating surface, and with no temperature *colder* than that to which they were subjected on the mountain-tops, they reach the ocean before they again become charged with fresh vapor, and before, therefore, they have any which the Peruvian climate can extract. The last they had to spare was deposited as snow on the tops of the Cordilleras, to feed mountain streams under the heat of the sun, and irrigate the valleys on the western slopes. Thus we see how the top of the Andes becomes the reservoir from which are supplied the rivers of Chili and Peru. The other rainless or almost rainless regions are the western coasts of Mexico, the deserts of Africa, Asia, North America, and Australia. Now study the geographical features of the country surrounding those regions; see how the mountain ranges run; then turn to Plate V to see how the winds blow, and where the sources are (§ 276) which supply them with vapors. This plate shows the prevailing direction of the wind only at sea; but, knowing it there, we may infer what it is on the land. Supposing it to prevail on the land as it generally does in corresponding latitudes at sea, then the Plate will suggest readily enough how the winds that blow over these deserts came to be robbed of their moisture, or, rather, to have so much of it taken from

them as to reduce their dew-point below the Desert temperature; for *the air can never deposit its moisture when its temperature is higher than its dew-point.* We have a rainless region about the Red Sea, because the Red Sea, for the most part, lies within the northeast trade-wind region, and these winds, when they reach that region, are dry winds, for they have as yet, in their course, crossed no wide sheets of water from which they could take up a supply of vapor. Most of New Holland lies within the southeast trade-wind region; so does most of intertropical South America. But intertropical South America is the land of showers. The largest rivers and most copiously watered country in the world are to be found there, whereas almost exactly the reverse is the case in Australia. Whence this difference? Examine the direction of the winds with regard to the shore-line of these two regions, and the explanation will at once be suggested. In Australia — east coast — the shore-line is stretched out in the direction of the trades; in South America — east coast — it is perpendicular to their direction. In Australia they fringe this shore only with their vapor; thus that thirsty land is so stinted with showers that the trees can not afford to spread their leaves out to the sun, for it evaporates all the moisture from them; their vegetable instincts teach them to turn their edges to his rays. In intertropical South America the trade-winds blow perpendicularly upon the shore, penetrating the very heart of the country with their moisture. Here the leaves, measuring many feet square — as the plantain, &c. — turn their broad sides up to the sun, and court his rays.

298. *The rainy side of mountains. Why there is more rain on one side of a mountain than on the other.* We may now, from what has been said, see why the Andes and all other mountains which lie athwart the course of the winds have a dry and a rainy side, and how the prevailing winds of the latitude determine which is the rainy and which the dry side. Thus, let us take the southern coast of Chili for illustration. In our summer time, when the sun comes north, and drags after him the belts of perpetual winds and calms, that coast is left within the regions of the northwest winds — the winds that are counter to the southeast trades — which, cooled by the winter temperature of the highlands of Chili, deposit their moisture copiously. During the rest of the year, the most of Chili is in the region of the southeast trades, and the same causes which operate in California to prevent rain there, operate in Chili; only the dry season in one place is the rainy season of the other. Hence we see that the weather side of all such mountains as the Andes is the wet side, and the lee side the dry. The same phenomenon, from a like cause, is repeated in intertropical India, only in that country each side of the mountain is made alternately the wet and the dry side by a change in the prevailing direction of the wind. Plate V shows India to be in one of the monsoon regions: it is

the most famous of them all. From October to April the northeast trades prevail. They evaporate from the Bay of Bengal water enough to feed with rains, during this season, the western shores of this bay and the Ghauts range of mountains. This range holds the relation to these winds that the Andes of Peru (§ 297) hold to the southeast trades; it first cools and then relieves them of their moisture, and they tumble down on the western slopes of the Ghauts, Peruvian-like, cool, rainless, and dry; wherefore that narrow strip of country between the Ghauts and the Arabian Sea would, like that in Peru between the Andes and the Pacific, remain without rain forever, were it not for other agents which are at work about India and not about Peru. The work of the agents to which I allude is felt in the monsoons, and these prevail in India and not in Peru. After the northeast trades have blown out their season, which in India ends in April, the great arid plains of Central Asia, of Tartary, Thibet, and Mongolia, become heated up; they rarefy the air of the northeast trades, and cause it to ascend. This rarefaction and ascent, by their demand for an indraught, are felt by the air which the southeast trade-winds bring to the equatorial Doldrums of the Indian Ocean: it rushes over into the northern hemisphere to supply the upward draught from the heated plains as the southwest monsoons. The forces of diurnal rotation assist (§ 113) to give these winds their westing. Thus the southeast trades, in certain parts of the Indian Ocean, are converted, during the summer and early autumn, into southwest monsoons. These then come from the Indian Ocean and Sea of Arabia loaded with moisture, and, striking with it perpendicularly upon the Ghauts, precipitate upon that narrow strip of land between this range and the Arabian Sea an amount of water that is truly astonishing. Here, then, are not only the conditions for causing more rain, now on the west, now on the east side of this mountain range, but the conditions also for the most copious precipitation. Accordingly, when we come to consult rain gauges, and to ask meteorological observers in India about the fall of rain, they tell us that on the western slopes of the Ghauts it sometimes reaches the enormous depth of twelve or fifteen inches in one day.[12] Were the Andes stretched along the eastern instead of the western coast of America, we should have an amount of precipitation on their eastern slopes that would be truly astonishing; for the water which the Amazon and the other majestic streams of South America return to the ocean would still be precipitated between the sea-shore and the crest of these mountains. These winds of India then continue their course to the Himalaya range as dry winds. In crossing this range, they are subjected to a lower temperature than that to which they were exposed in crossing the Ghauts. Here they drop more of their moisture in the shape of snow and rain, and then pass over into the

[12] [Johnston, *Physical Atlas*, p. 66.]

thirsty lands beyond with scarcely enough vapor in them to make even a cloud. Thence they ascend into the upper air, there to become counter-currents in the general system of atmospherical circulation. By studying Plate V, where the rainless regions and inland basins, as well as the course of the prevailing winds, are shown, these facts will become obvious.

299. *The regions of greatest precipitation.* We shall now be enabled to determine, if the views which I have been endeavoring to present be correct, what parts of the earth are subject to the greatest fall of rain. They should be on the slopes of those mountains which the trade-winds or monsoons first strike after having blown across the greatest tract of ocean. The more abrupt the elevation, and the shorter the distance be-tween the mountain top and the ocean (§ 298), the greater the amount of precipitation. If, therefore, we commence at the parallel of about 30° north in the Pacific, where the northeast trade-winds first strike that ocean, and trace them through their circuits till they first meet high land, we ought to find such a place of heavy rains. Commencing at this parallel of 30°, therefore, in the North Pacific, and tracing thence the course of the northeast trade-winds, we shall find that they blow thence, and reach the region of equatorial calms near the Caroline Islands. Here they rise up; but, instead of pursuing the same course in the upper stratum of winds through the southern hemisphere, they, in consequence of the rotation of the earth (§ 207), are made to take a southeast course. They keep in this upper stratum until they reach the calms of Capricorn, between the parallels of 30° and 40°, after which they become the pre-vailing northwest winds of the southern hemisphere, which correspond to the southwest of the northern. Continuing on to the southeast, they are now the surface winds; they are going from warmer to cooler lati-tudes; they become as the wet sponge (§ 292), and are abruptly inter-cepted by the Andes of Patagonia, whose cold summit compresses them, and with its low dew-point squeezes the water out of them. Captain King found the astonishing fall of water here of nearly thirteen feet (one hun-dred and fifty-one inches) in forty-one days; and Mr. Darwin reports that the sea water along this part of the South American coast is some-times quite fresh, from the vast quantity of rain that falls. A similar rain-fall occurs on the sides of Cherraponjie, a mountain in India. Colonel Sykes reports a fall there during the southwest monsoons of 605¼ inches. This is at the *rate* of 86 *feet* during the year; but King's Patagonia rain-fall is at the rate of 114 feet during the year. Cherraponjie is not so near the coast as the Patagonia range, and the monsoons lose moisture before they reach it. We ought to expect a corresponding rainy region to be found to the north of Oregon; but there the mountains are not so high, the obstruction to the southwest winds is not so abrupt, the highlands are farther from the coast, and the air which these winds carry in their circu-

lation to that part of the coast, though it be as heavily charged with moisture as at Patagonia, has a greater extent of country over which to deposit its rain, and, consequently, the fall to the square inch will not be as great. In like manner, we should be enabled to say in what part of the world the most equable climates are to be found. They are to be found in the equatorial calms, where the northeast and southeast trades meet fresh from the ocean, and keep the temperature uniform under a canopy of perpetual clouds.

300. *Amount of evaporation.* The mean annual fall of rain on the entire surface of the earth is estimated at about five feet. To evaporate water enough annually from the ocean to cover the earth, on the average, five feet deep with rain; to transport it from one zone to another; and to precipitate it in the right places, at suitable times, and in the proportions due, is one of the offices of the grand atmospherical machine. This water is evaporated principally from the torrid zone. Supposing it all to come thence, we shall have, encircling the earth, a belt of ocean three thousand miles in breadth, from which this atmosphere evaporates a layer of water annually sixteen feet in depth. And to hoist up as high as the clouds, and lower down again all the water in a lake sixteen feet deep, and three thousand miles broad, and twenty-four thousand long, is the yearly business of this invisible machinery. What a powerful engine is the atmosphere! and how nicely adjusted must be all the cogs, and wheels, and springs, and *compensations* of this exquisite piece of machinery, that it never wears out nor breaks down, nor fails to do its work at the right time and in the right way! The abstract logs at the Observatory in Washington show that the water of the Indian Ocean is warmer than that of any other sea; therefore it may be inferred that the evaporation from it is also greater. The North Indian Ocean contains about 4,500,000 square miles, while its Asiatic water-shed contains an area of 2,500,000. Supposing all the rivers of this water-shed to discharge annually into the sea four times as much water as the Mississippi (§ 274) discharges into the Gulf, we shall have an average annual evaporation (§ 282) from the North Indian Ocean of 6.0 inches, or 0.0165 per day.

301. *The rivers of India, and the measure of the* effective *evaporation from that ocean.* The rivers of India are fed by the monsoons, which have to do their work of distributing their moisture in about three months. Thus we obtain 0.065 inch as the average daily rate of *effective* (§ 282) evaporation from the warm waters of this ocean. If it were all rained down upon India, it would give it a drainage which would require rivers having sixteen times the capacity of the Mississippi to discharge. Nevertheless, the evaporation from the North Indian Ocean required for such a flood is only one sixteenth of an inch daily throughout the year.[13] Availing

[13] In his annual report to the Society (*Trans. Bombay Geogr. Soc.* from May, 1849,

myself of the best lights — dim at best — as to the total amount of evaporation that annually takes place in the trade-wind region generally at sea, I estimate that it does not exceed *four feet*.

302. *Physical adjustments.* We see the light beginning to break upon us, for we now begin to perceive why it is that the proportions between the land and water were made as we find them in nature. If there had been more water and less land, we should have had more rain, and *vice versa;* and then climates would have been different from what they are now, and the inhabitants, animal or vegetable, would not have been as they are. And as they are, that wise Being who, in his kind providence, so watches over and regards the things of this world that he takes notice of the sparrow's fall, and numbers the very hairs of our head, doubtless designed them to be. The mind is delighted, and the imagination charmed, by contemplating the physical arrangements of the earth from such points of view as this is which we now have before us; from it the sea, and the air, and the land, appear each as a part of that grand machinery upon which the well-being of all the inhabitants of earth, sea, and air depends; and which, in the beautiful adaptations that we are pointing out, affords new and striking evidence that they all have their origin in ONE omniscient idea, just as the different parts of a watch may be considered to have been constructed and arranged according to *one* human design. In some parts of the earth the precipitation is greater than the evaporation; thus the amount of water borne down by every river that runs into the sea (§ 270) may be considered as the excess of the precipitation over the evaporation that takes place in the valley drained by that river. In other parts of the earth the evaporation and precipitation are exactly equal, as in those inland basins such as that in which the city of Mexico, Lake Titicaca, the Caspian Sea, etc., etc., are situated, which basins have no ocean drainage. If more rain fell in the valley of the Caspian Sea than is evaporated from it, that sea would finally get full and overflow the whole of that great basin. If less fell than is evaporated from it again, then that sea, in the course of time, would dry up, and plants and animals there would all perish for the want of water. In the sheets of water which we find distributed over that and every other inhabitable inland basin, we see reservoirs or evaporating surfaces just

---

to August, 1850, IX[(1850), cv], Dr. Buist, the secretary, states, on the authority of Mr. Laidlay, the evaporation at Calcutta to be "about fifteen feet annually; that between the Cape and Calcutta it averages, in October and November, nearly three fourths of an inch daily; between 10° and 20° in the Bay of Bengal, it was found to exceed an inch daily. Supposing this to be double the average throughout the year, we should," continues the doctor, "have eighteen feet of evaporation annually." All the heat received by the intertropical seas from the sun annually would not be sufficient to convert into vapor a layer of water from them sixteen feet deep. It is these observations as to the *rate* of evaporation on shore that have led (§ 280) to such extravagant estimates as to the rate at sea.

sufficient for the supply of that degree of moisture which is best adapted to the well-being of the plants and animals that people such basins. In other parts of the earth still, we find places, as the Desert of Sahara, in which neither evaporation nor precipitation takes place, and in which we find neither plant nor animal to fit the land for man's use.

303. *Adaptations: their beauties and sublimity.* In contemplating the system of terrestrial adaptations, these researches teach one to regard the mountain ranges and the great deserts of the earth as the astronomer does the counterpoises to his telescope — though they be mere dead weights, they are, nevertheless, necessary to make the balance complete, the adjustment of his machine perfect. These counterpoises give ease to the motions, stability to the performance, and accuracy to the workings of the instrument. They are "*compensations.*" Whenever I turn to contemplate the works of nature, I am struck with the admirable system of compensation, with the beauty and nicety with which every department is adjusted, adapted, and regulated according to the others; things and principles are meted out in directions apparently the most opposite, but in proportions so exactly balanced that results the most harmonious are produced. It is by the action of opposite and compensating forces that the earth is kept in its orbit, and the stars are held suspended in the azure vault of heaven; and these forces are so exquisitely adjusted, that, at the end of a thousand years, the earth, the sun, and moon, and every star in the firmament, is found to come and twinkle in its proper place at the proper moment. Nay, philosophy teaches us that when the little snow-drop, which in our garden-walks we see raising its head at "the singing of birds," to remind us that "the winter is passed and gone," was created, the whole mass of the earth, from pole to pole, and from circumference to centre, must have been taken into account and weighed, in order that the proper degree of strength might be given to its tiny fibres. Botanists tell us that the constitution of this plant is such as to require that, at a certain stage of its growth, the stalk should bend, and the flower should bow its head, that an operation may take place which is necessary in order that the herb should produce seed after its kind; and that, after this fecundation, its vegetable health requires that it should lift its head again and stand erect. Now, if the mass of the earth had been greater or less, the force of gravity would have been different; in that case, the strength of fibre in the snow-drop, as it is, would have been too much or too little; the plant could not bow or raise its head at the right time, fecundation could not take place, and its family would have become extinct with the first individual that was planted, because its "seed" would not have been "in itself," and therefore could not have reproduced itself, and its creation would have been a failure. Now, if we see such a perfect adaptation, such exquisite adjustment in the case of

one of the smallest flowers of the field, how much more may we not expect "compensation" in the atmosphere and the ocean, upon the right adjustment and due performance of which depends not only the life of that plant, but the well-being of every individual that is found in the entire vegetable and animal kingdoms of the world? When the east winds blow along the Atlantic coast for a little while, they bring us air saturated with moisture from the Gulf Stream, and we complain of the sultry, oppressive, heavy atmosphere; the invalid grows worse, and the well man feels ill, because, when he takes this atmosphere into his lungs, it is already so charged with moisture that it can not take up and carry off that which encumbers his lungs, and which nature has caused his blood to bring and leave there, that respiration may take up and carry off. At other times the air is dry and hot; he feels that it is conveying off matter from the lungs too fast; he realizes the idea that it is consuming him, and he calls the sensation burning. Therefore, in considering the general laws which govern the physical agents of the universe, and regulate them in the due performance of their offices, I have felt myself constrained to set out with the assumption that, if the atmosphere had had a greater or less capacity for moisture, or if the proportion of land and water had been different — if the earth, air, and water had not been in exact counterpoise — the whole arrangement of the animal and vegetable kingdoms would have varied from their present state. But God, for reasons which man may never know, chose to make those kingdoms what they are; for this purpose it was necessary, in his judgment, to establish the proportions between the land and water, and the desert, just as they are, and to make the capacity of the air to circulate heat and moisture just what it is, and to have it to do all its work in obedience to law and in subservience to order. If it were not so, why was power given to the winds to lift up and transport moisture, and to feed the plants with nourishment? or why was the property given to the sea by which its waters may become first vapor, and then fruitful showers or gentle dews? If the proportions and properties of land, sea, and air were not adjusted according to the reciprocal capacities of all to perform the functions required by each, why should we be told that HE "measured the waters in the hollow of his hand, and comprehended the dust in a measure, and weighed the mountains in scales, and the hills in a balance?" Why did he span the heavens but that he might mete out the atmosphere in exact proportion to all the rest, and impart to it those properties and powers which it was necessary for it to have, in order that it might perform all those offices and duties for which he designed it? Harmonious in their action, the air and sea are obedient to law and subject to order in all their movements; when we consult them in the performance of their manifold and marvelous offices, they teach us lessons concerning the wonders of the deep, the mysteries of the sky, the greatness, and the wisdom, and goodness of the Creator, which make us wiser and better men. The investigations into the broad-spreading circle of phenomena connected with the winds of heaven and the waves of the sea are second to none for the good which they do and the lessons which they teach. The astronomer is said to see the hand of God in the sky; but does not the right-minded mariner, who looks aloft as he ponders over these things, hear his voice in every wave of the sea that "claps its hands," and feel his presence in every breeze that blows?

CHAPTER VI

§ 311–332. RED FOGS AND SEA BREEZES

311. *The alternations of land and sea breezes.* The inhabitants of the sea-shore in tropical countries wait every morning with impatience the coming of the sea breeze. It usually sets in about ten o'clock. Then the sultry heat of the oppressive morning is dissipated, and there is a delightful freshness in the air which seems to give new life to all for their daily labors. About sunset there is again another calm. The sea breeze is now done, and in a short time the land breeze sets in. This alternation of the land and sea breeze — a wind from the sea by day and from the land by night — is so regular in intertropical countries, that they are looked for by the people with as much confidence as the rising and setting of the sun.

312. *The sea breeze at Valparaiso.* In extra-tropical countries, especially those on the polar side of the trade-winds, this phenomenon is presented only in summer and fall, when the heat of the sun is sufficiently intense to produce the requisite degree of atmospherical rarefaction over the land. This depends in a measure, also, upon the character of the land upon which the sea breeze blows; for when the surface is arid and the soil barren, the heating power of the sun is exerted with most effect. In such cases the sea breeze amounts to a gale of wind. In the summer of the southern hemisphere the sea breeze is more powerfully developed at Valparaiso than at any other place to which my services afloat have led me. Here regularly in the afternoon, at this season, the sea breeze blows furiously; pebbles are torn up from the walks and whirled about the streets; people seek shelter; the Almendral is deserted, business interrupted, and all communication from the shipping to the shore is cut off. Suddenly the winds and the sea, as if they had again heard the voice of rebuke, are hushed, and there is a great calm.

313. *The contrast.* The lull that follows is delightful. The sky is without a cloud; the atmosphere is transparency itself; the Andes seem to draw near; the climate, always mild and soft, becomes now doubly sweet by the contrast. The evening invites abroad, and the population sally forth — the ladies in ball costume, for now there is not wind enough to disarrange the lightest curl. In the southern summer this change takes place day after day with the utmost regularity, and yet the calm always seems to surprise, and to come before one has time to realize that the

furious sea-wind could so soon be hushed. Presently the stars begin to peep out, timidly at first, as if to see whether the elements here below had ceased their strife, and if the scene on earth be such as they, from bright spheres aloft, may shed their sweet influences upon. Sirius, or that blazing world η Argus, may be the first watcher to send down a feeble ray; then follow another and another, all smiling meekly; but presently, in the short twilight of the latitude, the bright leaders of the starry host blaze forth in all their glory, and the sky is decked and spangled with superb brilliants. In the twinkling of an eye, and faster than the admiring gazer can tell, the stars seem to leap out from their hiding-places. By invisible hands, and in quick succession, the constellations are hung out; but first of all, and with dazzling glory, in the azure depths of space appears the Great Southern Cross. That shining symbol lends a holy grandeur to the scene, making it still more impressive. Alone in the night-watch, after the sea breeze has sunk to rest, I have stood on the deck under those beautiful skies gazing, admiring, rapt. I have seen there, above the horizon at once, and shining with a splendor unknown to these latitudes, every star of the first magnitude — save only six — that is contained in the catalogue of the 100 principal fixed stars of astronomers. There lies the city on the sea-shore, wrapped in sleep. The sky looks solid, like a vault of steel set with diamonds. The stillness below is in harmony with the silence above, and one almost fears to speak, lest the harsh sound of the human voice, reverberating through those vaulted "chambers of the south," should wake up echo, and drown the music that fills the soul. On looking aloft, the first emotion gives birth to a homeward thought: bright and lovely as they are, those, to northern sons, are not the stars nor the skies of fatherland. Alpha Lyræ, with his pure white light, has gone from the zenith, and only appears for one short hour above the top of the northern hills. Polaris and the Great Bear have ceased to watch from their posts; they are away down below the horizon. But, glancing the eye above and around, you are dazzled with the splendors of the firmament. The moon and the planets stand out from it; they do not seem to touch the blue vault in which the stars are set. The Southern Cross is just about to culminate. Climbing up in the east are the Centaurs, Spica, Bootes, and Antares, with his lovely little companion, which only the best telescopes have power to unveil. These are all bright particular stars, differing from one another in color as they do in glory. At the same time, the western sky is glorious with its brilliants too. Orion is there, just about to march down into the sea; but Canopus and Sirius, with Castor and his twin brother, and Procyon, η Argus, and Regulus — these are high up in their course; they look down with great splendor, smiling peacefully as they precede the Southern Cross on its western way. And yonder, farther still, away to the south, float the Magellanic clouds, and the "Coal Sacks"

— those mysterious, dark spots in the sky, which seem as though it had been rent, and these were holes in the "azure robe of night," looking out into the starless, empty, black abyss beyond. One who has never watched the southern sky in the stillness of the night, after the sea breeze with its turmoil is done, can have no idea of its grandeur, beauty, and loveliness.

314. *Land and sea breezes along the shores of intertropical countries.* Within the tropics, however, the land and sea breezes are more gentle, and, though the night-scenes there are not so suggestive as those just described, yet they are exceedingly delightful and altogether lovely. The oppressive heat of the sun and the climate of the sea-shore is mitigated and made both refreshing and healthful by the alternation of those winds which invariably come from the coolest place — the sea, which is the cooler by day, and the land, which is the cooler by night. About ten in the morning the heat of the sun has played upon the land with sufficient intensity to raise its temperature above that of the water. A portion of this heat, being imparted to the superincumbent air, causes it to rise, when the air, first from the beach, then from the sea, to the distance of several miles, begins to flow in with a most delightful and invigorating freshness.

315. *Cause of land and sea breezes.* When a fire is kindled on the hearth, we may, if we will observe the moats floating in the room, see that those nearest to the chimney are the first to feel the draught and to obey it — they are drawn into the blaze. The circle of inflowing air is gradually enlarged, until it is scarcely perceived in the remote parts of the room. Now the land is the hearth, the rays of the sun the fire, and the sea, with its cool and calm air, the room; and thus we have at our firesides the sea-breeze in miniature. When the sun goes down the fire ceases; then the dry land commences to give off its surplus heat by radiation, so that by dew-fall it and the air above it are cooled below the sea temperature. The atmosphere on the land thus becomes heavier than on the sea, and, consequently, there is a wind seaward which we call the land-breeze.

316. *Lieut. Jansen on the land and sea breezes in the Indian Archipelago.* "A long residence in the Indian Archipelago, and, consequently, in that part of the world where the investigations of the Observatory at Washington have not extended, has given me," says Jansen,[1] in his

---

[1] I have been assisted in my investigations into these phenomena of the sea by many thinking minds; among those whose debtor I am stands first and foremost the clear head and warm heart of a foreign officer, Lieutenant Marin Jansen, of the Dutch Navy, whom I am proud to call my friend. He has served many years in the East Indies, and has enriched my humble contributions to the "Physical Geography of the Sea" with contributions from the store-house of his knowledge, set off and presented in many fine pictures, and has appended them to a translation of the second edition of this work in the Dutch language. He has added a chapter on the land and sea breezes; another on the changing of the monsoons in the East Indian Archipelago: he has also extended his remarks to the northwest monsoon, to hurricanes, the south-

Appendix to the Physical Geography of the Sea, "the opportunity of studying the phenomena which there occur in the atmosphere, and to these phenomena my attention was, in the first place, directed. I was involuntarily led from one research to another, and it is the result of these investigations to which I would modestly give a place at the conclusion of Maury's Physical Geography of the Sea, with the hope that these first-fruits of the log-books of the Netherlands may be speedily followed by more and better. Upon the northern coast of Java, the phenomenon of daily land and sea breezes is finely developed. There, as the gorgeous 'eye of day' rises almost perpendicularly from the sea with fiery ardor, in a cloudless sky, it is greeted by the volcanoes with a column of white smoke, which, ascending from the conical summits high in the firmament above, forms a crown, or assumes the shape of an immense bouquet,[2] that they seem to offer to the dawn; then the joyful land-breeze plays over the flood, which, in the torrid zone, furnishes, with its fresh breath, so much enjoyment to the inhabitants of that sultry belt of the earth, for, by means of it, every thing is refreshed and beautified. Then, under the influence of the glorious accompaniments of the break of day, the silence of the night is awakened, and we hear commencing every where the morning hymn of mute nature, whose gesticulation is so expressive and sublime. All that lives feels the necessity of pouring forth, each in its way, and in various tones and accents, from the depths of inspiration, a song of praise. The air, still filled with the freshness of the evening dew, bears aloft the enraptured song, as, mingled with the jubilee tones which the contemplation of nature every where forces from the soul, it gushes forth in deep earnestness to convey the daily thank-offering over the sea, over hill and dale.[3] As the sun ascends the sky, the azure vault is bathed in dazzling light; now the land-breeze, wearied with play, goes to rest. Here and there it still plays over the water, as if it could not sleep; but finally becoming exhausted, it sinks to repose in the stillness of the calm. But not so with the atmosphere: it sparkles, and glitters, and twinkles, becoming clear under the increasing heat, while the gentle swelling of the now polished waves reflects, like a thousand

---

east trades of the South Atlantic, and to winds and currents generally. [M. F. Maury, *Natuurkundige beschrijving der zeeën*, vertaald door M. H. Jansen (Dordrecht, 1855). Jansen's appendix to the translation (*Bijdrage*) occupies pp. 265–303. In identification of Maury's numerous quotations it will be cited "*Bijdrage*." The long quotation introduced here appears on pp. 265–269. Maury ignores Jansen's division of the quoted passages into paragraphs.]

[2] Upon the coast of Java I saw daily, during the east monsoon, such a column of smoke ascending at sunrise from Bromo, Lamongan, and Smiro. Probably there is then no wind above. — [*Bijdrage*, p. 266, n. 1.]

[3] In the very fine mist of the morning, a noise — for example, the firing of cannon — at a short distance is scarcely heard, while at midday, with the sea-breeze, it penetrates for miles with great distinctness. — [*Ibid.*, n. 2.]

mirrors, the rays of light which dance and leap to the tremulous but vertical movements of the atmosphere. Like pleasant visions of the night, that pass before the mind in sleep, so do sweet phantoms hover about the land-breeze as it slumbers upon the sea. The shore seems to approach and to display all its charms to the mariner in the offing. All objects become distinct and more clearly delineated,[4] while, upon the sea, small fishing-boats loom up like large vessels. The seaman, drifting along the coast, and misled by the increasing clearness and mirage, believes that he has been driven by a current toward the land; he casts the lead, and looks anxiously out for the sea-breeze, in order to escape from what he believes to be threatening danger.[5] The planks burn under his feet; in vain he spreads the awning to shelter himself from the broiling sun. Its rays are oppressive; repose does not refresh; motion is not agreeable. The inhabitants of the deep, awakened by the clear light of day, prepare themselves for labor. Corals, and thousands of crustacea, await, perhaps impatiently, the coming of the sea-breeze, which shall cause evaporation to take place more rapidly, and thus provide them with a bountiful store of building material for their picturesque and artfully constructed dwellings: these they know how to paint and to polish in the depths of the sea more beautifully than can be accomplished by any human art. Like them, also, the plants of the sea are dependent upon the winds, upon the clouds, and upon the sunshine; for upon these depend the vapor and the rains which feed the streams that bring nourishment for them into the sea.[6] When the sun reaches the zenith, and his stern eye, with burning glare, is turned more and more upon the Java Sea, the air seems to fall into a magnetic sleep; yet, even as the magnetizer exercises his will upon his subject, and the latter, with uncertain and changeable gestures, gradually puts himself in motion, and sleeping obeys that will, so also we see the slow efforts of the sea-breeze to repress the vertical movements of the air, and to obey the will which calls it to the land. This vertical movement appears to be not easily overcome by the horizontal which we call wind. Yonder, far out upon the sea, arises and disappears alternately a darker tint upon the otherwise shining sea-carpet; finally, that tint remains and approaches; that is the long-wished-for sea-breeze: and yet it is sometimes one, yes,

---

[4] The transparency of the atmosphere is so great that we can sometimes discover Venus in the sky in the middle of the day. — [*Bijdrage*, p. 267, n. 1.]

[5] Especially in the rainy season the land looms very greatly; then we see mountains which are from 5000 to 6000 feet high at a distance of 80 or 100 English miles. [*Ibid.*, n. 2.]

[6] The archipelago of coral islands on the north side of the Straits of Sunda is remarkable. Before the salt water flowed from the Straits it was deprived of the solid matter of which the *Thousand Islands* are constructed. A similar group of islands is found between the Straits of Macassar and Balie. — [*Ibid.*, n. 3. In Jansen's original all verbs in the second sentence of the note are in the present tense.]

even two hours before the darker tint is permanent, before the sea-breeze has regularly set in. Now small white clouds begin to rise above the horizon; to the experienced seaman they are a prelude to a fresh sea-breeze. We welcome the first breath from the sea; it is cooling, but it soon ceases; presently it is succeeded by other grateful puffs of air, which continue longer; presently they settle down into the regular sea-breeze, with its cooling and refreshing breath. The sun declines, and the sea-wind — that is, the common trade-wind or monsoon which is drawn toward the land — is awakened. It blows right earnestly, as if it would perform its daily task with the greatest possible ado. The air, itself refreshed upon the deep, becomes gray from the vapor which envelops the promontories in mist, and curtains the inland with dark clouds. The land is discernible only by the darker tint which it gives to the mist; but the distance can not be estimated. The sailor thinks himself farther from shore than he really is, and steers on his course carelessly, while the capricious wind lashes the waters, and makes a short and broken sea, from the white caps of which light curls are torn, with sportive hand, to float away like party-colored streamers in the sunbeam. In the mean while clouds appear now and then high in the air, yet it is too misty to see far. The sun approaches the horizon. Far over the land the clouds continue to heap up; already the thunder is heard among the distant hills; the thunder-bolts reverberate from hill-side to hill-side, while through the mist the sheets of lightning are seen.[7] Finally, the 'king of day' sinks to rest; now the mist gradually disappears, and as soon as the wind has laid down the lash, the sea, which, chafing and fretting, had with curled mane resisted its violence, begins to go down also. Presently both winds and waves are hushed, and all is again still. Above the sea, the air is clearer or slightly clouded; above the land, it is thick, dark, and swollen. To the feelings, this stillness is pleasant. The sea-breeze, the driving brine, that has made a salt-pan of the face, the short, restless sea, the dampness — all have grown wearisome, and welcome is the calm. There is, however, a somewhat of dimness in the air, an uncertain but threatening appearance. Presently, from the dark mass of clouds, which hastens the change of day into night, the thunder-storm peals forth. The rain falls in torrents in the mountains, and the clouds gradually overspread the whole sky. But for the wind, which again springs up, it would be alarming to the sailor, who is helpless in a calm. What change will take place in the air? The experienced seaman, who has to work against the trade-wind or against the monsoon, is off the coast, in order to take advantage of the land-breeze (the destroyer of the trade) so soon as it shall come. He rejoices when

[7] At Buitenzorg, near Batavia, 40 English miles from the shore, five hundred feet above the sea, with high hills around, these thunder-storms occur between 4 P.M. and 8 P.M. — [*Bijdrage*, p. 268, n. 1.]

the air is released from the land and the breeze comes, at first feebly, but afterward growing stronger, as usual, during the whole night. If the land-breeze meets with a squall, then it is brief, and becomes feeble and uncertain. We sometimes find then the permanent sea-breeze close to the coast, which otherwise remains twenty or more English miles from it. One is not always certain to get the land-breeze at the fixed time. It sometimes suffers itself to be waited for; sometimes it tarries the whole night long. During the greatest part of the rainy season, the land-breeze in the Java Sea can not be depended upon. This is readily explained according to the theory which ascribes the origin of the sea and land breezes to the heating of the soil by day, and the cooling by means of radiation by night; for, during the rainy season, the clouds extend over land and sea, interrupting the sun's rays by day, and the radiation of heat by night, thus preventing the variations of temperature; and from these variations, according to this theory, the land and sea breezes arise. Yet there are other tropical regions where the land and sea breezes, even in the rainy season, regularly succeed each other."

317. *Sanitary influences of land and sea breezes.* One of the causes which make the west coast of Africa so very unhealthy when compared with places in corresponding latitudes on the opposite side of the Atlantic, as in Brazil, is no doubt owing to the difference in the land and sea breezes on the two sides. On the coast of Africa the land-breeze is "universally scorching hot." [8] There the land-breeze is the trade-wind. It has traversed the continent, sucking up by the way disease and pestilence from the dank places of the interior. Reeking with miasm, it reaches the coast. Peru is also within the trade-wind region, and the winds reach the west coast of South America, as they do the west coast of Africa, by an overland path; but, in the former case, instead of sweeping over dank places, they come cool and fresh from the pure snows of the Andes. Between this range and the coast, instead of marshes and a jungle, there is a desert — a rainless country, upon which the rays of the sun play with sufficient force not only to counteract the trade-wind power and produce a calm, but to turn the scale, and draw the air back from the sea, and so cause the sea-breeze to blow regularly.

318. *Influences which regulate their strength.* On the coast of Africa, on the contrary, a rank vegetable growth screens the soil from the scorching rays of the sun, and the rarefaction is not every day sufficient to do more than counteract the trade-wind force and produce a calm. The same intensity of ray, however, playing upon the intertropical vegetation of a lee shore, is so much force added to the sea-breeze; and hence, in Brazil, the sea-breeze is fresh, and strong, and healthful; the land breeze feeble,

[8] [*Bijdrage*, p. 270.]

and therefore not so sickly. Thus we perceive that the strength as well as regularity of the land and sea breezes not only depend upon the topography of a place, but also upon its situation with regard to the prevailing winds; and also that a given difference of temperature between land and water, though it may be sufficient to produce the phenomena of land and sea breezes at one place, will not be adequate to the same effect at another; and the reason is perfectly philosophical.

319. *Land breezes from the west coast of Africa scorching hot.* It is easier to obstruct and turn back the current in a sluggish than in a rapid stream. So, also, in turning a current of air first upon the land, then upon the sea — very slight alternations of temperature would suffice for this on those coasts where calms would prevail were it not for the land and sea breezes, as, for instance, in and about the region of equatorial calms; there the air is in a state of rest, and will obey the slightest call in any direction; not so in regions where the trades blow over the land, and are strong. It requires, under such circumstances, a considerable degree of rarefaction to check them and produce a calm, and a still farther rarefaction to turn them back, and convert them into a regular sea-breeze. Hence the scorching land-breeze (§ 317) on the west coast of Africa: the heat there may not have been intense enough to produce the degree of rarefaction required to check and turn back the southeast trades. In that part of the world, their natural course is from the land to the sea, and therefore, if this view be correct, the sea-breeze should be more feeble than the land-breeze, neither should it last so long.

320. *Land-breeze in Brazil and Cuba.* But on the opposite side — on the coast of Brazil, as at Pernambuco, for instance — where the trade-wind comes from the sea, we should have this condition of things reversed, and the sea-breeze will prevail for most of the time — then it is the land-breeze which is feeble and of short duration: it is rarely felt. Again, the land and sea breezes in Cuba, and along the Gulf shores of the United States, will be more regular in their alternations than they are along the shores of Brazil or South Africa, and for the simple reason that the Gulf shores lie nearly parallel with the prevailing direction of the winds. In Rio de Janeiro, the sea-breeze is the regular trade-wind made fresher by the daily action of the sun on the land. It is worthy of remark, also, that, for the reason stated by Jansen, the land and sea breezes in the winter time are almost unknown in countries of severe cold, though in the summer the alternation of wind from land to sea, and sea to land, may be well marked.

321. *Night scenes when sailing with the land-breeze.* "Happy he," remarks Jansen, "who, in the Java Sea at evening, seeking the land-breeze off the coast, finds it there, after the salt-bearing, roaring sea-wind, and can, in the magnificent nights of the tropics, breathe the refreshing land-

breeze, ofttimes laden with delicious odors.[9] The veil of clouds, either after a squall, with or without rain, or after the coming of the land-breeze, is speedily withdrawn, and leaves the sky clearer during the night, only now and then flecked with dark clouds floating over from the land. Without these floating clouds the land-breeze is feeble. When the clouds float away from the sea, the land-breeze does not go far out from the coast, or is wholly replaced by the sea-breeze, or, rather, by the trade-wind. If the land-breeze continues, then the stars loom forth, as if to free themselves from the dark vault of the heavens, but their light does not wholly vanquish its deep blue, which causes the 'Coal Sacks' to come out more distinctly near the Southern Cross, as it smiles consolingly upon us, while Scorpio, the emblem of the tropical climate, stands like a warning in the heavens. The starlight, which is reflected by the mirrored waters, causes the nights to vie in clearness with the early twilight in high latitudes. Numerous shooting stars weary the eye, although they break the monotony of the sparkling firmament. Their unceasing motion in the unfathomable ocean affords a great contrast to the seeming quiet of the gently-flowing, aerial current of the land-breeze. But at times, when, 30° or 40° above the horizon, a fire-ball arises which suddenly illumines the whole horizon, appearing to the eye the size of the fist, and fading away as suddenly as it appeared, falling into fiery nodules, then we perceive that, in the apparent calm of nature, various forces are constantly active, in order to cause, even in the invisible air, such combinations and combustions, the appearance of which amazes the crews of ships. When the slender keel glides quickly over the mirrored waters upon the wings of the wind, it cuts for itself a sparkling way, and disturbs in their sleep the monsters of the deep, which whirl and dart quicker than an eight-knot ship; sweeping and turning around their disturber, they suddenly clothe the dark surface of the water in brilliancy. Again, when we go beyond the limits of the land-breeze, and come into the continuous trade-wind, we occasionally see from the low-moving, round black clouds (unless it thunders), light blue sparks collected upon the extreme points of the iron belaying-pins, etc.;[10] then the crew appear to fear a new danger, against which courage is unavailing, and which the mind can find no power to endure. The fervent, fiery nature inspires the traveler with deep awe. They who, under the beating of the storm and terrible violence of the ocean, look danger courageously in the face, feel, in the

[9] In the Roads of Batavia, however, they are not very agreeable. — [*Ibid.*, n. 3.]

[10] I have seen this in a remarkable degree upon the south coast of Java; these sparks were then seen six feet above the deck, upon the frames of timber (*koussen der blokken*), in the implements, etc. — [*Bijdrage,* p. 271, n. 2. My translation of Jansen's note: I once saw this phenomenon in an extreme form on the south coast of Java, when these sparks were to be seen within three ells of the deck on the iron fastenings of the rigging. — J. L.]

presence of these phenomena, insignificant, feeble, anxious. Then they perceive the mighty power of the Creator over the works of his creation. And how can the uncertain, the undetermined sensations arise which are produced by the clear yet sad light of the moon? she who has always great tears in her eyes, while the stars look sweetly at her, as if they loved to trust her and to share her affliction.[11] In the latter part of the night the land-breeze sinks to sleep, for it seldom continues to blow with strength, but is always fickle and capricious. With the break of day it again awakes, to sport a while, and then gradually dies away as the sun rises. The time at which it becomes calm after the land and sea breezes is indefinite, and the calms are of unequal duration. Generally, those which precede the sea-breeze are rather longer than those which precede the land-breeze. The temperature [12] of the land, the direction of the coast-line with respect to the prevailing direction of the trade-wind in which the land is situated, the clearness of the atmosphere, the position of the sun, perhaps also that of the moon, the surface over which the sea-breeze blows, possibly also the degree of moisture and the electrical state of the air, the heights of the mountains, their extent, and their distance from the coast, all have influence thereon. Local observations in regard to these can afford much light, as well as determine the distance at which the land-breeze blows from the coast, and beyond which the regular trade-wind or monsoon continues uninterruptedly to blow. The direction of land and sea winds must also be determined by local observations, for the idea is incorrect that they should always blow perpendicularly to the coast-line. Scarcely has one left the Java Sea — which is, as it were, an inland sea between Sumatra, Borneo, Java, and the archipelago of small islands between both of the last named — than, in the blue waters of the easterly part of the East Indian Archipelago, nature assumes a bolder aspect, more in harmony with the great depth of the ocean. The beauty of the Java Sea, and the delightful phenomena which air and ocean display, have here ceased. The scene becomes more earnest. The coasts of the eastern islands rise boldly out of the water, far in whose

[11] Some one has ventured the remark that at full moon, near the equator, more and darker [*Bijdrage: zwaarder;* i. e., heavier] dew falls than at new moon, and to this are ascribed the moonheads (*maanhoofden*), which I have seen, however, but once during all the years which I have spent between the tropics. — [*Bijdrage,* p. 272, n. 1. The word "moonheads" means nothing in English. As for the Dutch word so translated, *maanhoofden,* J. A. van Duijnen Montijn, Director of the Division of Oceanography and Maritime Meteorology of the Royal Netherlands Meteorological Institute, De Bilt, writes me under date of January 3, 1962: "It is not a present-day Dutch word and I have also been unable to find it in any other Dutch writings on meteorology and optical phenomena." Presumably Jansen used *maanhoofd* to designate the nimbus or glory seen about the shadow of the head of an observer on a grassy surface bearing a heavy deposit of dew in bright moonlight, in German called *Heiligenschein.* — J. L.]

[12] [*Bijdrage,* p. 272: *gesteldheid;* i. e., nature or character.]

depths they have planted their feet. The southeast wind, which blows upon the southern coasts of the chain of islands, is sometimes violent, always strong through the straits which separate them from each other, and this appears to be more and more the case as we go eastward. Here, also, upon the northern coast, we find land-breezes, yet the trade-wind often blows so violently that they have not sufficient power to force it beyond the coast. Owing to the obstruction which the chain of islands presents to the southeast trade-wind, it happens that it blows with violence away over the mountains, apparently as the land-breeze does upon the north coast; [13] yet this wind, which only rises when it blows hard from the southeast upon the south coast, is easily distinguished from the gentle land-breeze. The regularity of the land and sea breezes in the Java Sea and upon the coasts of the northern range of islands, Banca, Borneo, Celebes, etc., during the east monsoon, must in part be ascribed to the hinderances which the southeast trade-wind meets in the islands which lie directly in its way — in part to the inclination toward the east monsoon [14] which the trade-wind undergoes after it has come within the archipelago — and, finally, to its abatement as it approaches the equator. The causes which produce the land-breezes thus appear collectively not sufficiently powerful to be able to turn back a strong trade-wind in the ocean." [15]

322. *Red fogs in the Mediterranean.* Seamen tell us of "red fogs" which they sometimes encounter, especially in the vicinity of the Cape de Verd Islands. In other parts of the sea also they meet showers of dust. What these showers precipitate in the Mediterranean is called "sirocco dust," and in other parts "African dust," [16] because the winds which accompany them are supposed to come from the Sirocco desert, or some other parched land of the continent of Africa. It is of a brick-red or cinnamon color, and it sometimes comes down in such quantities as to obscure the sun, darken the horizon, and cover the sails and rigging with a thick coating of dust, though the vessel may be hundreds of miles from the land.

323. *Red fogs near the equator.* Dr. Clymer, Fleet-surgeon of the African squadron, reports a red fog which was encountered in February, 1856, by the U.S. ship Jamestown. "We were," says he, "immersed in the dust-fog six days, entering it abruptly on the night of the 9th of February, in lat. 7° 30′ N., and long. 15° W., and emerging from it (and at the

---

[13] Such is the case, among others, in the Strait of Madura, upon the heights of Bezoekie. — [*Bijdrage*, p. 273, n. 1: *op de hoogte van Bezoukie; i. e.*, off Bezoukie.]

[14] [*Bijdrage*, p. 273: *buiging tot oost-mousson; i. e.*, deflection, implying transformation, of the trade wind into the east monsoon.]

[15] [*Ibid.: in zee; i. e.*, to sea. The entire quotation appears on pp. 270–273.]

[16] Prof. Ehrenberg calls it "Sea-dust." [I find no such expression in Ehrenberg. — J. L.]

same time from the zone of the equatorial calms into the northeast trades)
on the 15th instant, in lat. 9° N., and long. 19° W. With these winds we
beat to Porto Praya (in lat. 14° 54′ N., and long. 23° 30′ W.), crossing a
southwest current of nearly a mile an hour, arriving at Porto Praya on the
22d of February. The red dust settled thickly on the sails, rigging, spars,
and decks, from which it was easily collected. It was an impalpable
powder, of a brick-dust or cinnamon color. The atmosphere was so dusky
that we could not have seen a ship at midday beyond a quarter of a
mile." [17]

324. *Putting tallies on the wind.* Now the patient reader, who has
had the heart to follow me in a preceding chapter (IV) around with "the
wind in his circuits," will perceive that evidence in detail is yet wanting
to establish it as a fact that the northeast and southeast trades, after
meeting and rising up in the equatorial calms, do cross over and take the
paths represented by R S and F G, Plate I. Statements, and reasons, and
arguments enough have already been made and adduced (§ 288) to
make it highly probable, according to human reasoning, that such is the
case; and though the theoretical deductions showing such to be the case
be never so plausible, positive proof that they are true can not fail to be
received with delight and satisfaction. Were it possible to take a portion
of this air, which should represent, as it travels along with the southeast
trades, the general course of atmospherical circulation, and to put a tally
on it by which we could follow it in its circuits and always recognize it,
then we might hope actually to prove, by evidence the most positive, the
channels through which the air of the trade-winds, after ascending at the
equator, returns whence it came. But the air is invisible; and it is not
easily perceived how either marks or tallies may be put upon it, that it
may be traced in its paths through the clouds. The skeptic, therefore, who
finds it hard to believe that the general circulation is such as Plate I
represents it to be, might consider himself safe in his unbelief were he to
declare his willingness to give it up the moment any one should put
tallies on the wings of the wind, which would enable him to recognize
that air and those tallies again, when found at other parts of the earth's
surface. As difficult as this seems to be, it has actually been done. Ehrenberg, with his microscope, has established, almost beyond a doubt, that
the air which the southeast trade-winds bring to the equator does rise up
there and pass over into the northern hemisphere. The Sirocco or African
dust, which he has been observing so closely, has turned out to be tallies
put upon the wind in the other hemisphere; and this beautiful instrument of his enables us to detect the mark on these little tallies as plainly
as though those marks had been written upon labels of wood and tied
to the wings of the wind.

[17] *Sailing Directions*, 8th ed., II, 377.

325. *They tell of a crossing at the calm belts.* This dust, when sub-jected to microscopic examination, is found to consist of infusoria and organisms whose *habitat* is not Africa, but South America, and in the southeast trade-wind region of South America. Professor Ehrenberg has examined specimens of sea-dust from the Cape de Verds and the regions thereabout — from Malta, Genoa, Lyons, and the Tyrol — and he has found a similarity among them as striking as it would have been had these specimens been all taken from the same pile. South American forms he recognizes in all of them; indeed, they are the prevailing forms in every specimen he has examined. It may, I think, be now regarded as an established fact that there is a perpetual upper current of air from South America to North Africa; and that the volume of air which flows to the northward in these upper currents is nearly equal to the volume which flows to the southward with the northeast trade-winds, there can be no doubt. The "rain dust" has been observed most frequently to fall in spring and autumn; that is, the fall has occurred after the equinoxes, but at intervals from them varying from thirty to sixty days, more or less. To account for this sort of periodical occurrence of the falls of this dust, Ehrenberg thinks it "necessary to suppose *a dust-cloud to be constantly swimming in the atmosphere by continuous currents of air, and lying in the region of the trade-winds, but suffering partial and periodical devia-tions.*" [18] It has already been shown (§ 295) that the rain or calm belt between the trades travels up and down the earth from north to south, making the rainy season wherever it goes. The reason of this will be explained in another place. This dust is probably taken up in the dry, and not in the wet season; instead, therefore, of its being "held in clouds suffering partial and periodical deviations," as Ehrenberg suggests, it more probably comes from one place about the vernal, and from another about the autumnal equinox; for places which have their rainy season at one equinox have their dry season at the other. At the time of the vernal equinox, the valley of the Lower Orinoco is then in its dry season — every thing is parched up with the drought; the pools are dry, and the marshes and plains become arid wastes. All vegetation has ceased; the great serpents and reptiles have buried themselves for hibernation; [19] the hum of insect life is hushed, and the stillness of death reigns through the valley. Under these circumstances, the light breeze, raising dust from lakes that are dried up, and lifting motes from the brown savannas, will bear them away like clouds in the air. This is the period of the year when the surface of the earth in this region, strewed with impalpable and feather-light remains of animal and vegetable organisms, is swept over by

[18] [Christian Gottfried] Ehrenberg [*Passat-Staub und Blut-Regen . . . Mehrere Vorträge* (Berlin, 1849), p. 39.]
[19] [See reference in next section.]

whirlwinds, gales, and tornadoes of terrific force; this is the period for the general atmospheric disturbances which have made characteristic the equinoxes. Do not these conditions appear sufficient to afford the "rain dust" for the spring showers? At the period of the autumnal equinox, another portion of the Amazonian basin is parched with drought, and liable to winds that fill the air with dust, and with the remains of dead animal and vegetable matter; these impalpable organisms, which each rainy season calls into being, to perish the succeeding season of drought, are perhaps distended and made even lighter by the gases of decomposition which has been going on in the period of drought. May not, therefore, the whirlwinds which accompany the vernal equinox, and sweep over the lifeless plains of the Lower Orinoco, take up the "rain dust" which descends in the northern hemisphere in April and May? and may it not be the atmospherical disturbances which accompany the autumnal equinox that take up the microscopic organisms from the Upper Orinoco and the great Amazonian basin for the showers of October?

326. *Humboldt's description of the dust-whirlwinds of the Orinoco.* The Baron von Humboldt, in his *Aspects of Nature*, thus contrasts the wet and the dry seasons there: "When, under the vertical rays of the never-clouded sun, the carbonized turfy covering falls into dust, the indurated soil cracks asunder as if from the shock of an earthquake. If at such times two opposing currents of air, whose conflict produces a rotary motion, come in contact with the soil, the plain assumes a strange and singular aspect. Like conical-shaped clouds, the points of which descend to the earth, the sand rises through the rarefied air on the electrically-charged centre of the whirling current, resembling the loud water-spout, dreaded by the experienced mariner. The lowering sky sheds a dim, almost straw-colored light on the desolate plain. The horizon draws suddenly nearer, the steppe seems to contract, and with it the heart of the wanderer. The hot, dusty particles which fill the air increase its suffocating heat, and the east wind, blowing over the long-heated soil, brings with it no refreshment, but rather a still more burning glow. The pools which the yellow, fading branches of the fan-palm had protected from evaporation, now gradually disappear. As in the icy north the animals become torpid with cold, so here, under the influence of the parching drought, the crocodile and the boa become motionless and fall asleep, deeply buried in the dry mud. . . . The distant palm-bush, apparently raised by the influence of the contact of unequally heated and therefore unequally dense strata of air, hovers above the ground, from which it is separated by a narrow intervening margin. Half-concealed by the dense clouds of dust, restless with the pain of thirst and hunger, the horses and cattle roam around, the cattle lowing dismally, and the horses stretching out their long necks and snuffing the wind, if haply a

moister current may betray the neighborhood of a not wholly dried-up pool. . . . At length, after the long drought, the welcome season of the rain arrives; and then how suddenly is the scene changed! . . . Hardly has the surface of the earth received the refreshing moisture, when the previously barren steppe begins to exhale sweet odors, and to clothe itself with killingias, the many panicles of the paspulum, and a variety of grasses. The herbaceous mimosas, with renewed sensibility to the influence of light, unfold their drooping, slumbering leaves to greet the rising sun; and the early song of birds and the opening blossoms of the water plants join to salute the morning." [20]

327. *Are the great deserts centres of circulation?* The arid plains and deserts, as well as high mountain ranges, have, it may well be supposed, an influence upon the movements of the great aerial ocean, as shoals and other obstructions have upon the channels of circulation in the sea. The deserts of Asia, for instance, produce (§ 299) a disturbance upon the grand system of atmospherical circulation, which, in summer and autumn, is felt in Europe, in Liberia, and away out upon the Indian Ocean, as far to the south as the equinoctial line. There is an indraught from all these regions toward these deserts. These indraughts are known as monsoons at sea; on the land, as the prevailing winds of the season. Imagine the area within which this indraught is felt, and let us ask a question or two, hoping for answers. The air which the indraught brings into the desert places, and which, being heated, rises up there, whither does it go? It rises up in a column a few miles high and many in circumference, we know, and we can imagine that it is like a shaft many times thicker than it is tall; but how is it crowned? Is it crowned like the stem of a mushroom, with an efflorescence or ebullition of heated air flaring over and spreading out in all directions, and then gradually thinning out as an upper current, extending even unto the verge of the area whence the indraught is drawn? If so, does it then descend and return to the desert plains as an indraught again? Then these desert places would constitute centres of circulation for the monsoon period; and if they were such centres, whence would these winds get the vapor for their rains in Europe and Asia? Or, instead of the mushroom shape, and the flare at the top in all directions from centre to circumference, does the uprising column, like one of those submarine fountains which are said to be in the Gulf Stream off the coast of Florida, bubble up and join in with the flow of the upper current? The right answers and explanations to these questions would add greatly to our knowledge concerning the general circulation of the atmosphere. It may be in the power of observation and the microscope to give light here. Let us hope.

[20] [Alexander von Humboldt, *Aspects of Nature in Different Lands and Different Climates*, transl. by Mrs. Sabine (Philadelphia, 1849), pp. 36–38.]

328. *The color of "sea-dust."* The color of the "rain dust," when collected in parcels and sent to Ehrenberg, is "brick-red," or "yellow ochre;"
when seen by Humboldt in the air, it was less deeply shaded, and is
described *by him* as imparting a "straw color" to the atmosphere. In the
search of spider-lines for the diaphragm of my telescopes, I procured the
finest and best threads from a cocoon of a dirty-red color; but the threads
of this cocoon, as seen singly in the diaphragm, were of a golden color;
there would seem, therefore, no difficulty in reconciling the difference
between the colors of the rain dust when viewed in little piles by the
microscopist, and when seen attenuated and floating in the wind by the
great traveler.

329. *A clew leading into the chambers of the south.* It appears, therefore, that we here have placed in our hands a clew, which, attenuated
and gossamer-like though it at first appears, is nevertheless palpable and
strong enough to guide us along through the "circuits of the wind" even
unto "the chambers of the south." The frequency of the fall of "rain dust"
between the parallels of 17° and 25° north, and in the vicinity of the Cape
Verd Islands, is remarked upon with emphasis by the microscopist. It is
worthy of remark, because, in connection with the investigations at the
Observatory, it is significant. The latitudinal limits of the northern edge
of the northeast trade-winds are variable. In the spring they are nearest
to the equator, extending sometimes at this season not farther from the
equator than the parallel of 15° north. The breadth of the calms of Cancer
is also variable; so also are their limits. The extreme vibration of this
zone is between the parallels of 17° and 38° north, according to the
season of the year.

330. *Red fogs do not always occur at the same place, but they occur
on a northeast and southwest range.* According to the hypothesis (§ 210)
suggested by my researches, this is the region in which the upper currents
of atmosphere that ascended in the equatorial calms, and flowed off to
the northward and eastward, are supposed to descend. This, therefore,
is the region in which the atmosphere that bears the "rain dust," or
"African sand," descends to the surface; and this, therefore, is the region,
it might be supposed, which would be the most liable to showers of this
"dust." This is the region in which the Cape Verd Islands are situated;
they are in the direction which theory gives to the upper current of air
from the Orinoco and Amazon with its "rain dust," and they are in the
region of the most frequent showers of "rain dust:" all of which, though
they do not absolutely prove, are nevertheless strikingly in conformity
with this theory as to the circulation of the atmosphere.

331. *Conditions requisite to the production of a sea fog.* It is true that,
in the present state of our information, we can not tell why this "rain
dust" should not be gradually precipitated from this upper current, and

descend into the stratum of trade-winds, as it passes from the equator to higher northern latitudes; neither can we tell why the vapor which the same winds carry along should not, in like manner, be precipitated on the way; nor why we should have a thunder-storm, a gale of wind, or the display of any other atmospherical phenomenon to-morrow, and not to-day: all that we can say is, that the conditions of to-day are not such as the phenomenon requires for its own development. Therefore, though we can not tell why the "sea-dust" should not always fall in the same place, we may nevertheless suppose that it is not always in the atmosphere, for the storms that take it up occur only occasionally, and that when up, and in passing the same parallels, it does not, any more than the vapor from a given part of the sea, always meet with the conditions — electrical and others — favorable to its descent, and that these conditions, as with the vapor, may occur now in this place, now in that. But that the fall does occur always in the same atmospherical vein or general direction, my investigations would suggest, and Ehrenberg's researches prove. Judging by the fall of sea or rain dust, we may suppose that the currents in the upper regions of the atmosphere are remarkable for their general regularity, as well as for their general direction and sharpness of limits, so to speak. We may imagine that certain electrical conditions are necessary to a shower of "sea-dust" as well as to a thunderstorm; and that the interval between the time of the equinoctial disturbances in the atmosphere and the occurrence of these showers, though it does not enable us to determine the true rate of motion in the general system of atmospherical circulation, yet assures us that it is not less on the average than a certain rate. We can not pretend to prescribe the conditions requisite for bringing the dust-cloud down to the earth. The radiation from the smoke-dust — as the particles of visible smoke may be called — has the effect of loading each little atom of smoke with dew, causing it to descend in the black fogs of London. Any circumstances, therefore, which may cause the dust that ascends as a straw-colored cloud from the Orinoco to radiate its caloric and collect moisture in the sky, may cause it to descend as a red fog in the Atlantic or Mediterranean.

332. *What is the agent that guides the air across the calm belts?* I do not offer these remarks as an explanation with which we ought to rest satisfied, provided other proof can be obtained; I rather offer them in the true philosophical spirit of the distinguished microscopist himself, simply as affording, as far as they are entitled to be called an explanation, that explanation which is most in conformity with the facts before us, and which is suggested by the results of a novel and beautiful system of philosophical research. It is not, however, my province, or that of any other philosopher, to dictate belief. Any one may found hypotheses if he

will state his facts and the reasoning by which he derives the conclusions which constitute the hypothesis. Having done this, he should patiently wait for time, farther research, and the judgment of his peers, to expand, confirm, or reject the doctrine which he may have conceived it his duty to proclaim. Thus, though we have tallied the air, and put labels on the wind, to "tell whence it cometh and whither it goeth," yet there evidently is an agent concerned in the circulation of the atmosphere whose functions are manifest, but whose presence has never yet been clearly recognized. When the air which the northeast trade-winds bring down meets in the equatorial calms that which the southeast trade-winds convey, and the two rise up together, what is it that makes them cross? where is the power that guides that from the north over to the south, and that from the south up to the north? The conjectures in the next chapter as to "the relation between magnetism and the circulation of the atmosphere" may perhaps throw some light upon the answer to this question.

CHAPTER VII

§ 341–369. THE EASTING OF THE TRADE-WINDS, THE CROSSING AT THE CALM BELTS, AND THE MAGNETISM OF THE ATMOSPHERE

341. *Halley's theory not fully confirmed by observations.* Halley's theory of the trade-winds, especially so much of it as ascribes their easterly direction to the effect of the diurnal rotation of the earth, seems to have been generally received as entirely correct. But it is only now, since all the maritime nations of the world have united in a common system of research concerning the physics of the sea, and occupied it with observers that we have been enabled to apply the *experimentum crucis* to this part of that famous theory. The abstract logs, as the observing-books are called, have placed within my reach no less than 632,460 observations — each one itself being the mean of many separate ones — upon the force and direction of the trade-winds. It appears from these that diurnal rotation being regarded as the *sole cause does not entirely* account for the *easting* of these winds.

342. *Observed course of the trade-winds.* From these observations Table 2 has been compiled. It shows the mean annual direction of the trade-winds in each of the six belts, north and south, between the parallels of 30° and the equator, together with the number of observations from which the mean for the belt is derived. Between the equator and 5° north, the annual average duration of the trades is 67 days for the northeast, and 199 for the southeast, with a mean direction for the latter — which are the prevailing winds between those parallels — of S. 47° 30′ E. According to the Halleyan theory they should be southwest winds.

Table 2

[Mean annual direction of the trade-winds]

Between	N. E. Trades		S. E. Trades	
	Course	No. of Obs.	Course	No. of Obs.
30° and 25°	N. 51° E.	68,777	S. 46° E.	66,635
25° and 20°	51° 30′	44,527	49° 20′	66,395
20° and 15°	53° 30′	33,103	52°	46,604
15° and 10°	52° 30′	30,339	49° 40′	43,817
10° and 5°	53° 30′	36,841	51° 40′	54,648
5° and 0°	54° 30′	67,829	48° 40′	72,945
Mean . . .	N. 52° 45′ E.		S. 49° 33′ E.	

343. *Velocities of the trade-winds.* In the Atlantic the average velocity of the southeast is greater than the average velocity of the northeast trades.[1] I estimate one to be from 14 to 18, the other from about 25 to 30 miles an hour. Assuming their velocity to be 14 and 25, the *departures* [presented in Table 3] show the miles of *easting* which the trade-winds average per hour through each of the above-named belts.

Table 3

Hourly Rate of Departure across the Trade-wind Belts

Between	N.E. Trades	S.E. Trades
	Easting per Hour	Easting per Hour
30° and 25°	10.9 miles	18. miles
25° and 20°	11. "	19. "
20° and 15°	11.2 "	19.7 "
15° and 10°	11. "	19.1 "
10° and 5°	11.2 "	19.6 "
5° and 0°	11.4 "	18.8 "

344. *Difference between observation and theory.* That diurnal rotation does impart easting to these winds there is no doubt; but the path suggested by Table 3 does not conform to that which, according to any reasonable hypothesis, the trade-winds would follow if left to obey the forces of diurnal rotation alone, as they would do were diurnal rotation the *sole* cause of their easting. As these winds approach the equator, the effect of diurnal rotation becomes more and more feeble. But Table 3 shows no such diminution of effect. They have as much easting between 5° and 0° as they have between 30° and 25°. Nay, the southeast trades between the equator and 5° N. — where, by the Halleyan theory, they *should* have *westing* — have as much easting (§ 342) as they have between 30° and 25° S. We can not tell how much the air is checked in its easterly tendency by resisting agents, by friction, etc., but we know that tendency is ten times stronger between 30° and 25° than it is between 5° and 0°, and yet actual observations show no difference in their course. This table reminds us that diurnal rotation should not, until more numerous and accurate observations shall better satisfy the theory than those half a million and more now do, be regarded as the *sole* cause of the easterly direction of the trade-winds. It suggests either that other agents are concerned in giving the trade-winds their easting, or that the effect of the upper and counter current, when drawn down and turned back (§ 223), is such as to counteract their unequal turning in obedience to the varying forces of diurnal rotation. No apology is needed for applying the tests of actual observation to this part of the Halleyan theory, not-

[1] "Average Force of the Trade-Winds," *Sailing Directions* [8th ed.], II, 857.

withstanding the general concurrence of opinion as to its sufficiency. With equal favor that feature of it also was received which ascribes the rising up in the belt of equatorial calms to the direct influence of the solar ray. But the advancement which has been made in our knowledge of physical laws since Halley expounded his trade-wind theory suggested a review of that feature, and it was found that, though the direct heat of the sun is one of the agents which assists the air to rise there, it is not the *sole* agent; the latent heat which is set free by condensing vapor for the equatorial cloud-ring and its rains is now also (§ 245) recognized as an agent of no feeble power in this calm belt.

345. *Faraday's discovery of magnetism in the air.* Where shall those who are disposed to search look for this other agent that is supposed to be concerned with the trade-winds in their easting? I can not say where it is to be found, but, considering the recent discoveries in terrestrial magnetism — considering the close relations between many of its phenomena and those both of heat and electricity, the question may be asked whether some power capable of guiding "the wind in his circuits" may not lurk there. Oxygen comprises more than one fifth part (two ninths) of the atmosphere, and Faraday has discovered that oxygen is para-magnetic. If a bar of iron be suspended between the poles of a magnet, it will arrange itself axially, and point toward them; but if, instead of iron, a bar of bismuth be used, it will arrange itself equatorially, and point in a direction perpendicular to that of the iron. To distinguish these two kinds of forces, Dr. Faraday has said iron is para-magnetic, bismuth diamagnetic. Oxygen and iron belong to the same class, and all substances in nature belong to one or the other of the two classes of which iron and bismuth are the types.

346. *Lines of magnetic force.* This eminent philosopher has also shown that if you place a magnetized bar of iron on a smooth surface, and sift fine iron filings down upon it, these filings will arrange themselves in curved lines, as in Figure 1, *a*; or, if the bar be broken, they will arrange themselves as in Figure 1, *b*. The earth itself, or the atmospheric envelope by which it is surrounded, is a most powerful magnet, and the lines of force which proceed whether from its interior, its solid shell, or vaporous covering, are held to be just such lines as those are which surround artificial magnets; proceed whence they may, they are supposed to extend through the atmosphere, and to reach even to the planetary spaces. Many eminent men and profound thinkers, Sir David Brewster among them, suspect that the atmosphere itself is the seat of terrestrial magnetism. All admit that many of those agents, both thermal and electrical, which play highly important parts in the meteorology of our planet, exercise a marked influence upon the magnetic condition of the atmosphere also.

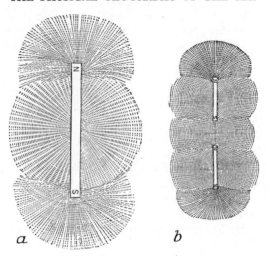

Figure 1
[Lines of magnetic force about bar magnets.]

347. *The magnetic influences of the oxygen of the air and of the spots on the sun.* Now, when, referring to Dr. Faraday's discovery (§ 345), and the magnetic lines of force as shown by the iron filings (§ 346), we compare the particles of oxygen gas to these minute bits of ferruginous dust that arrange themselves in lines and curves about magnets — when we reflect that this great magnet, the earth, is surrounded by a paramagnetic gas, to the molecules of which the finest atom from the file is in comparison gross and ponderous matter — that the entire mass of this air is equivalent to a sea of mercury covering the earth around and over to the depth of 30 inches, and that this very subtile mass is in a state of unstable equilibrium, and in perpetual commotion by reason of various and incessant disturbing causes — when we reflect farther upon the recent discoveries of Schwabe and of Sabine concerning the spots on the sun and the magnetic elements of the earth, which show that if the sun or its spots be not the great fountain of magnetism, there is at least reason to suspect a close alliance between solar and terrestrial magnetism; that certain well-known meteorological phenomena, as the aurora, come also within the category of magnetic phenomena; that the magnetic poles of the earth and the poles of maximum cold are at the same spot; that the thermal equator is not parallel to or coincident with either the terrestrial or with that which the direct solar ray would indicate, but that it follows, and in its double curvatures conforms to the magnetic equator; moreover, when we reflect upon Barlow's theory and Fox's observations, which go to show that the direction of metallic veins of the northern hemisphere,

which generally lie northeast and southwestwardly, must have been influenced by the direction of the magnetic meridians of the earth or air — when, I say, we reflect upon magnetism in all its aspects, we may well inquire whether such a mass of highly magnetic gas as that which surrounds our planet does not intervene, by reason of its magnetism, in influencing the circulation of the atmosphere and the course of the winds.

348. *The needle in its diurnal variations, the barometer in its readings, and the atmosphere in its electrical tension, all have the same hours for their maxima and minima.* This magnetic sea, as the atmosphere may be called, is continually agitated; it is disturbed in its movements by various influences which prevent it from adjusting itself to any permanent magnetic or other dynamical status; and its para-magnetic properties are known to vary with every change of pressure or of temperature. The experiments of Faraday show that the magnetic force of the air changes with temperature; that it is least near the equator, and greatest at the poles of maximum cold; that it varies with the seasons, and changes night and day; nay, the atmosphere has regular variations in its electrical conditions expressed daily at stated hours of maximum and minimum tension. Coincident with this, and in all parts of the world, but especially in sub-tropical latitudes, the barometer also has its maxima and minima readings for the day. So also, and at the same hours, the needle attains the maxima and minima of its diurnal variations. Without other time-piece, the hour of the day may be told by these maxima and minima, each group of which occurs twice a day and at six-hour intervals. These invisible ebbings and flowings — the diurnal change in the electrical tension — the diurnal variation of the needle, and the diurnal rising and falling of the barometer, follow each other as closely and as surely, if not quite as regularly, as night the day. Any cause which produces changes in atmospheric pressure invariably puts it in motion, giving rise to gentle airs or furious gales, according to degree; and here, at least, we have a relation between the movements in the air and the movements of the needle so close that it is difficult to say which is cause, which effect, or whether the two be not the effects of a common cause.

349. *The question raised by modern researches.* Indeed, such is the nature of this imponderable called magnetism, and such the suggestions made by Faraday's discoveries, that the question has been raised in the minds of the most profound philosophers of the age whether the various forces of light, heat, and gravitation, of chemical affinity, electricity, and magnetism, may not yet be all traced to one common source. Surely, then, it can not be considered as unphilosophical to inquire of magnetism for some of the anomalous movements that are observed in the atmosphere. These anomalies are many; they are not confined to the easting of the trade-winds; they are to be found in the counter-trades and the calm belts

also. There is reason to believe, as has already been stated (§ 288), that there is a crossing of the winds at the calm belts (§ 212), and it was promised to go more into detail concerning the circumstances which seem to favor this belief. Our researches have enabled us, for instance, to trace from the belt of calms, near the tropic of Cancer, which extends entirely across the seas, an efflux of air both to the north and to the south. From the south side of this belt the air flows in a steady breeze, called the northeast trade-winds, toward the equator (Plate I); on the north side of it, the prevailing winds come from it also, but they go toward the northeast. They are the well-known westerly winds which prevail along the route from this country to England in the ratio of two to one. But why should we suppose a crossing to take place here? We suppose so from these facts: because throughout Europe, the land upon which these westerly winds blow, precipitation is in excess of evaporation, and because at sea they are going from a warmer to a colder climate, and therefore it may be inferred that nature exacts from them what we know she exacts from the air under similiar circumstances, but on a smaller scale, before our eyes, viz., more precipitation at sea than evaporation. In other words, they probably leave in the Atlantic as much vapor as they take up from the Atlantic. Then where, it may be asked, does the vapor which these winds carry along, for the replenishing of the whole extra-tropical regions of the north, come from? They did not get it as they came along in the upper regions, a counter-current to the northeast trades, unless they evaporated the trade-wind clouds, and so robbed those winds of their vapor. They certainly did not get it from the surface of the sea in the calm belt of Cancer, for they did not tarry long enough there to become saturated with moisture. Thus circumstances again pointed to the southeast trade-wind regions as the place of supply. This question has been fully discussed in Chapter V, where it has been shown they did not get it from the Atlantic. Moreover, these researches afforded grounds for the supposition that the air of which the northeast trade-winds are composed, and which comes out of the same zone of calms as do these southwesterly winds, so far from being saturated with vapor at its exodus, is dry; for near their polar edge, the northeast trade-winds are, for the most part, dry winds.

350. *Wet and dry air of the calm belts.* Facts seem to confirm this, and the calm belts of Cancer and Capricorn both throw a flood of light upon the subject. These are two bands of light airs, calms, and baffling winds, which extend entirely around the earth. The air flows out north and south from these belts. That which comes out on the equatorial side goes to feed the trades, and makes a dry wind; that which flows out on the polar side goes to feed the counter-trades (§ 349), and is a rain wind. How is it that we can have from the same trough or receiver, as these

calm belts may be called, an efflux of dry air on one side and of moist on the other? Answer: upon the supposition that the air without rain comes from one quarter, that with rain from another — that, coming from opposite directions to this place of meeting, where there is a crossing, they pass each other in their circuits. They both meet here as upper currents, and how could there be a crossing without an agent or influence to guide them? and why should we not look to magnetism for this agent as well as to any other of the influences which are concerned in giving to the winds their force and direction?

351. *Principles according to which the physical machinery of our planet should be studied.* HE that established the earth "created it not in vain; He formed it to be inhabited." And it is presumptuous, arrogant, and impious to attempt the study of its machinery upon any other theory: *it was made to be inhabited.* How could it be inhabitable but for the sending of the early and the latter rain? How can the rain be sent except by the winds? and how can the fickle winds do their errands unless they have a guide? Suppose a new piece of human mechanism were shown to one of us, and we were told the object of it was to measure time; now, if we should seek to examine it with the view to understand its construction, would we not set out upon the principle — the theory — that it was made to measure time? By proceeding on any other supposition or theory we should be infallibly led into error. And so it is with the physical machinery of the world. The theory upon which this work is conducted is that *the earth was made for man*; and I submit that no part of the machinery by which it is maintained in a condition fit for him is left to chance, any more than the bit of mechanism by which man measures time is left to go by chance.

352. *Division into wind-bands.* That I might study to better advantage the workings of the atmospherical machinery in certain aspects, I divided the sea into bands 5° of latitude in breadth, and stretching east and west entirely around the earth, but skipping over the land. There are twelve of these bands on each side of the equator that are traversed more or less frequently by our fleet of observers; they extend to the parallel of 60° in each hemisphere. To determine the force and direction of the wind for each one of these bands, the abstract logs were examined until all the data afforded by 1,159,533 observations were obtained; and the mean direction of the wind for each of the four quarters in every band was ascertained. Considering difference of temperature between these various bands to be one of the chief causes of movement in the atmosphere — that the extremes on one hand are near the equator, and on the other about the poles — considering that the tendency of every wind (§ 234) is to blow along the arc of a great circle, and therefore that every wind that was observed in any one of these bands must have moved in a

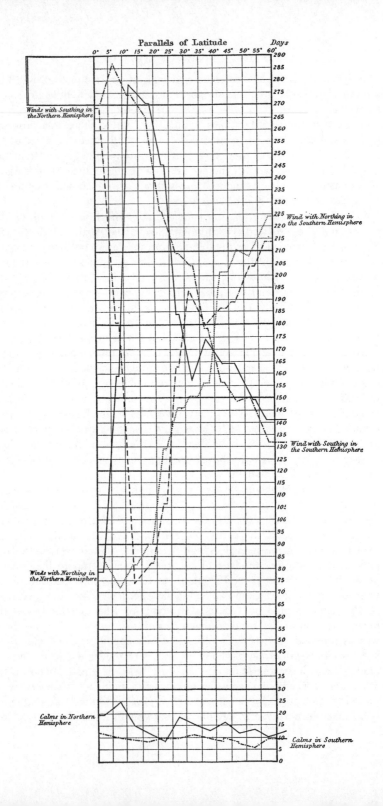

Parallels of Latitude

Days

Winds with Southing in
the Northern Hemisphere

Wind with Northing in
the Southern Hemisphere

Wind with Southing in
the Southern Hemisphere

Winds with Northing in
the Northern Hemisphere

Calms in Northern
Hemisphere

Calms in Southern
Hemisphere

path crossing these bands more or less obliquely, and that therefore the general movements in the atmosphere might be classed accordingly, as winds either with northing or with southing in them. We have so classed them; and we have so classed them that we might study to more advantage the general movements of the great atmospherical machinery. See Figure 2.

353. *The medial bands.* Thus, when, after so classing them, we come to examine those movements in the band between 5° and 10° south, and to contrast them with the movements in the band between 55° and 60° south, for example, we find the general movements to be exactly in opposite directions. Observations show that during the year the winds in the former blow toward the equator 283, and from it 73 days; and in the latter they blow toward the pole for 224, and from it 132 days. These facts show that there must be a place of rarefaction — of low barometer, an indraught toward the poles as well as the equator, and that consequently, also, there must be a medial line or band somewhere between the parallels of 10° and 55° south, on one side of which the prevailing direction of the wind is toward the equator, on the other toward the pole. So, in the northern hemisphere, the same series of observations point this medial band out to us. They show that one is near the calm belt of Capricorn, the other near the calm belt of Cancer, and that they both probably lie between the parallels of 35° and 40°, where the winds north and south are equal, as per Table 4. The wind curves (Figure 2 and Table 4) afford a very striking view of these medial bands, as the parallels in either hemisphere between which the winds with northing and the winds with southing are on the yearly average exactly equal. In the northern hemisphere the debatable ground appears by the table to extend pretty nearly from 25° to 50° N. By the figure the two winds first become equal between 25° and 30°; the two curves then recede and approach very closely again, but without crossing, between 35° and 40°. In the southern hemisphere, the conflict between the polar and equatorial indraught, as expressed by winds with southing and winds with northing, is more decided. There the two curves march, one up, the other down, and cross between the parallels of 35° and 40° S., thus confirming what from other data we had already learned, viz., that the condition of the atmosphere is more unstable in the northern than it is in the southern hemisphere.

354. *The rainless regions and the calm belts.* Such, for the winds at sea, is their distribution between the two halves of the horizon in the several bands and in each hemisphere. Supposing a like distribution to

←   Figure 2

[Average number of days annually, and for every degree of latitude between the parallels of 60° north and 60° south, that the winds have northing in them on the one hand, and southing in them on the other; also the proportion of calm days.]

Table 4

Winds with Northing and Winds with Southing in each Hemisphere, expressed by the Average Number of Days for which they blow annually

Bands	Northern Hemisphere			Southern Hemisphere		
	Northing	Southing	No. of Obs.	Northing	Southing	No. of Obs.
Between	Days	Days		Days	Days	
0° and 5°	78	268	67,829	84	269	72,945
5° and 10°	158	182	36,841	73	283	54,648
10° and 15°	278	73	27,339	82	275	43,817
15° and 20°	272	81	33,103	91	266	46,604
20° and 25°	246	101	44,527	128	227	66,395
25° and 30°	185	162	68,777	146	208	66,635
30° and 35°	155	195	62,514	150	204	76,254
35° and 40°	173	178	41,233	178	177	107,231
40° and 45°	163	186	33,252	202	155	63,669
45° and 50°	164	188	29,461	209	148	29,132
50° and 55°	147	204	41,570	208	151	14,286
55° and 60°	141	213	17,874	224	132	13,617
			504,320			655,233

1,159,553

obtain on shore, we shall find it suggestive to trace the calm belts of the tropics across the continents (Plate V), and to examine, in connection with them, the rainless regions of the earth, and those districts of country which, though not rainless, are nevertheless considered as "*dry* countries," by reason of the small amount of precipitation upon them. So, tracing the calm belt of Cancer, which at sea lies between the parallels of 28° and 37° (Plate V), but which, according to Sir John Herschel,[2] reaches higher latitudes on shore, it will be perceived that the winds that flow out on the north side blow over countries abounding in rivers, which countries are therefore abundantly supplied with rains. Hence we infer (§ 350) that those winds are *rain* winds. On the other hand, the winds that flow out on the equatorial side blow either over deserts, rainless regions, or dry countries. Hence we infer that these winds are *dry* winds. These "dry" winds traverse a country abounding in springs and rivers in India, but it is the monsoons there which bring the water for them. The winds which come out of this calm belt on its equatorial side give out no moisture, except as dew, until they reach the sea, and are replenished with vapor thence in sufficient quantities to make rain of; whereas the winds which come out on the polar side leave moisture enough as they come for such rivers as the Obi, the Yenisei, the Lena, and the Amoor, in Asia; the Missouri, the Sascatchawan, the Red River of the North, and others, in America. Between this calm belt and the head waters of these rivers there are no seas or other evaporating surfaces, neither are they so

[2] "Physical Geography," *Encyclop. Britan.*, XVII, 614.

situated with regard to the sea-coast that they may be, as the shores of Eastern China and the Atlantic slopes of the United States are, supplied with vapor by the winds from the sea-board. When we consider the table (§ 353), the situation of the rainless regions and dry countries with regard to the calm belt of Cancer, we are compelled to admit that, come whence it may and by what channels it may, there are flowing out of this calm belt two kinds of air, one well charged with moisture, the other dry and thirsty to a degree.

355. *The theory of the crossings restated, and the facts reconciled by it.* The supposition that the dry air came from the north and the moist from the south, and both as an upper current, is the only hypothesis that is consistent with all the known facts of the case. The dry air gave up all its moisture when, as a surface wind, it played upon the frozen summits of the northern hills; the wet obtained its moisture when, as the southeast trade-winds, it swept across the bosom of intertropical seas of the southern hemisphere. Rising up at the equator, it did not leave all its moisture with the cloud-ring, but, retaining a part, conveyed it through the cloud region, above the northeast trades, to this calm belt, where there was a descent and a crossing. The fact that these dry places are all within or on the equatorial side of this calm belt, while countries abounding with rains and well watered with running streams are to be found all along its polar side, is clearly indicative of a crossing. Upon no other supposition can we account for the barrenness on one side, the fertility on the other. The following are also links in the chain of facts and circumstances which give strength to the supposition that the rains for the Lena and the Missouri are brought across the calm belt of Cancer by those currents of air which flow thence toward the pole as the prevailing counter-trades or southwesterly winds of the extra-tropical north. We have already seen (§ 353) that, on the north side of this calm zone of Cancer, the prevailing winds on the surface are from this zone toward the pole, and (Plate I, § 215) that these winds return as A, B, C through the upper regions from the pole; that, arriving at the calms of Cancer, this upper current, A, B, C, meets another upper current, S, R, from the equator, where they neutralize each other, produce a calm, descend, and come out as surface winds, D, E, or the trade-winds; and as T, U, or the counter-trades. Now observations have shown that the winds represented by T, U are rain winds; those represented by D, E, dry winds; and it is evident that A, B, C could not bring any vapors to these calms to serve for T, U to make rains of; for the winds represented by A, B, C have already performed the circuit of surface winds as far as the pole, during which journey they parted with all their moisture, and, returning through the upper regions of the air to the calm belt of Cancer, they arrived there as dry winds. The winds represented by D, E are dry winds; therefore it

was supposed that these are, for the most part, but a continuation of the winds A, B, C. On the other hand, if the winds A, B, C, after descending, do turn about and become the surface winds T, U, they would first have to remain a long time in contact with the sea, in order to be supplied with vapor enough to feed the great rivers, and supply the rains for the whole earth between us and the north pole. In this case, we should have an evaporating region at sea and a rainless region ashore on the north as well as on the south side of this zone of Cancer; but investigation shows no such region. Hence it was inferred that B, C and R, S do come out on the surface as represented by Plate I. But what is the agent that should lead them out by such opposite paths? According to this mode of reasoning, the vapors which supply the rains for T, U would be taken up in the southeast trade-wind region by O, Q, and conveyed thence along the route Q, R, S to T. And if this mode of reasoning be admitted as plausible — if it be true that R, S carry the vapor which, by condensation, is to water with showers the extra-tropical regions of the northern hemisphere, Nature, we may be sure, has provided a guide for conducting S, T across this belt of calms, and for sending it on in the right way. Here it was, then, at this crossing of the winds, that I thought I first saw the footprints of an agent whose character I could not comprehend. Can it be the magnetism that resides in the oxygen of the air? Heat and cold, the early and the latter rain, clouds and sunshine, are not, we may rely upon it, distributed over the earth by chance; they are distributed in obedience to laws that are as certain and as sure in their operations as the seasons in their rounds. If it depended upon chance whether the dry air should come out on this side or on that of this calm belt, or whether the moist air should return or not whence it came — if such were the case in nature, we perceive that, so far from any regularity as to seasons, we should have, or might have, years of drought the most excessive, and then again seasons of rains the most destructive; but, so far from this, we find for each place a mean annual proportion of both, and that so regulated withal, that year after year the quantity is preserved with remarkable regularity. Having thus shown that there is no reason for supposing that the upper currents of air, when they meet over the calms of Cancer and Capricorn, are turned back to the equator, but having shown that there is reason for supposing that the air of each current, after descending, continues on in the direction toward which it was traveling before it descended, we may go farther, and, by a similar train of circumstantial evidence, afforded by these researches and other sources of information, show that the air, kept in motion on the surface by the two systems of trade-winds, when it arrives at the belt of equatorial calms and ascends, continues on thence, each current toward the pole which it was approaching while on the surface. In a problem like this, demonstration in the positive way is difficult, if not impossible. We must rely for our proof upon philosophical deduction, guided by the lights of reason; and in all cases in which positive proof can not be adduced, it is permitted to bring in circumstantial evidence; and the circumstantial evidence afforded by my investigations goes to show that the winds represented by O, Q, § 215, *do* become those represented by R, S, T, U, V, A, and A, B, C, D, E, F respectively. In the first place, O, Q represents the southeast trade-winds — *i. e.*, all the winds of the southern hemisphere as they approach the equator; and is there any reason for supposing that the atmosphere does not pass freely from one hemisphere to another? On the contrary, many reasons present themself for supposing that it does. If it did not, the proportion of land and water, and consequently of plants and warm-blooded animals, being so different in the two hemispheres, we might imagine that the constituents of the atmosphere in them would, in the course of ages, probably become different also, and that consequently, in such a case, man could not safely pass from one hemisphere to the other. Consider the manifold beauties in the whole system of terrestrial adaptations; remember what a perfect and wonderful machine (§ 261) is this atmosphere; how exquisitely balanced and beautifully compensated it is in all its parts. We know that it is perfect; that in the performance of its various offices it is never left to the guidance of chance — no, not for a moment. Wherefore I was led to ask myself why the air of the southeast trades, when arrived at the zone of equatorial calms, should not, after ascending, rather return to the south than go on to the north? Where and what is the agency by which its course is decided? Here I found circumstances which again induced me to suppose it probable that it neither turned back to the south nor mingled with the air which came from the regions of the northeast trades, ascended, and then flowed indiscriminately to the north or the south. But I saw reasons for supposing that what came to the equatorial calms as the southeast trade-winds continued to the north as an upper current, and that what had come to the same zone as northeast trade-winds ascended and continued over into the southern hemisphere as an upper current, bound for the calm zone of Capricorn. And these are the principal reasons and conjectures upon which these suppositions were based: At the seasons of the year when the area covered by the southeast trade-winds is large, and when they are evaporating most rapidly in the southern hemisphere, even up to the equator, the most rain is falling in the northern. Therefore it is fair to suppose that much of the vapor which is taken up on that side of the equator is precipitated on this. The evaporating surface in the southern hemisphere is greater, much greater, than it is in the northern; still, all the great rivers are in the northern hemisphere, the Amazon being regarded as common to both; and this fact, as far as it goes, tends to corroborate the suggestion as to the crossing of the trade-winds at the equatorial calms. Taking the laws and rates of evaporation into consideration, I could find (Chapter V) no part of the ocean of the northern hemisphere from which the sources of the Mississippi, the St. Lawrence, and the other great rivers of our hemisphere could be supplied. A regular series of meteorological observations has been carried on at the military posts of the United States since 1819. Rain maps of the whole country [3] have been prepared from these observations by Mr. Lorin Blodget at the surgeon general's office, and under the direction of Dr. Cooledge, U. S. A. These maps, as far as they go, sustain these views in a remarkable manner, for they bring out facts in a most striking way to show that the dry season in California and Oregon is the wet season in the Mississippi Valley. The winds coming from the southwest, and striking upon the coasts of California and Oregon in winter, precipitate there copiously. They then pass over the mountains robbed in part of their moisture. Of course, after watering the Pacific shores, they have not as much vapor to make rains of, especially for the upper Mississippi Valley, as they had in the summer time, when they dispensed their moisture, in the shape of rains, most sparingly upon the Pacific coasts. According to these views, the dry season on the Pacific slopes should be the wet, especially in the upper Mississippi Valley, and *vice versa*. Blodget's maps show that such is actually the case. Meteorological observations in the "Red River country" and other parts of British America would throw farther light and give farther confirmation, I doubt not, both to these views and to this interesting question. These army observations, as expressed in Blodget's maps, reveal other interesting features, also, touching the physical geography of the country. I allude to the two isothermal lines 45° and 65° (Plate V), which include between them all places that have a mean annual temperature between 45° and 65°. I have drawn, for the sake of comparison, similar lines on the authority of Dove and Johnston (A. K., of Edinburgh), across Europe and Asia. The isotherm of 65° skirts the northern limits of the sugarcane, and separates the intertropical from the extra-tropical plants and productions. I have drawn these two lines across America in order to give a practical exemplification of the nature of the advantages which the industrial pursuits and the political economy of the country would derive by the systematic extension of our meteorological observations from the sea to the land. These lines show how much we err when we reckon climates according to parallels of latitude. The space that these two isotherms of 45° and 65° comprehend between the Mississippi and the Rocky Mountains, owing to the singular effect of those mountains upon the climate, is larger than the space they comprehend between the Mississippi and the Atlantic. Hyetographically it is also different, being dryer, and possessing a purer atmosphere. In this grand range of climate be-

[3] *Army Meteorological* [*Register,* maps opp. p. 734.]

tween the meridians of 100° and 110° W., the amount of precipitation is just about one half of what it is between those two isotherms east of the Mississippi. In this new country west of it, winter is the dry, and spring the rainy season. It includes the climates of the Caspian Sea, which Humboldt regards as the most salubrious in the world, and where he found the most delicious fruits that he saw during his travels. Such was the purity of the air there, that polished steel would not tarnish even by night exposure. These two isotherms, with the remarkable loop which they make to the northwest, beyond the Mississippi, embrace the most choice climates for the olive, the vine, and the poppy; for the melon, the peach, and almond. The finest of wool may be grown there; and the potato, with hemp, tobacco, maize, and all the cereals, may be cultivated there in great perfection. No climate of the temperate zone will be found to surpass in salubrity that of this Piedmont trans-Mississippi country. The calm zone of Capricorn is the duplicate of that of Cancer, and the winds flow from it as they do from that, both north and south, but with this difference: that on the polar side of the Capricorn belt they prevail from the northwest instead of the southwest, and on the equatorial side from the southeast instead of the northeast. Now if it be true that the vapor of the northeast trade-winds is condensed in the extra-tropical regions of the southern hemisphere, the following path, on account of the effect of diurnal rotation of the earth upon the course of the winds, would represent the mean circuit of a portion of the atmosphere moving according to the general system of its circulation over the Pacific Ocean, viz.: coming down from the north as an upper current, and appearing on the surface of the earth in about longitude 120° west, and near the tropic of Cancer, it would here commence to blow the northeast trade-winds of that region. To make this clear, see Plate IV, on which I have marked the course of such vapor-bearing winds; A being the breadth or *swath* of winds in the northeast trades; B, the same wind as the upper and counter-current to the southeast trades; and C, the same wind after it has descended in the calm belt of Capricorn, and come out on the polar side thereof, as the rain winds and prevailing northwest winds of the extra-tropical regions of the southern hemisphere. This, as the northeast trades, is the evaporating wind. As the northeast trade-wind, it sweeps over a great waste of waters lying between the tropic of Cancer and the equator. Meeting no land in this long oblique track over the tepid waters of a tropical sea, it would, if such were its route, arrive, somewhere about the meridian of 140° or 150° west, at the belt of equatorial calms, which always divides the northeast from the southeast trade-winds. Here, depositing a portion of its vapor as it ascends, it would, with the residuum, take, on account of diurnal rotation, a course in the upper region of the atmosphere to the southeast, as far as the calms of Capricorn. Here it descends and continues on toward the coast of South America, in the same direction, appearing now as the prevailing northwest wind of the extra-tropical regions of the southern hemisphere. Traveling on the surface from warmer to colder regions, it must, in this part of its circuit, precipitate more than it evaporates. Now it is a coincidence, at *least*, that this is the route by which, on account of the land in the northern hemisphere, the northeast trade-winds have the fairest sweep over that ocean. This is the route by which they are longest in contact with an evaporating surface; the route by which all circumstances are most favorable to complete saturation; and this is the route by which they can pass over into the southern hemisphere most heavily laden with vapors for the extra-tropical regions of that half of the globe; and this is the supposed route which the northeast trade-winds of the Pacific take to reach the equator and to pass from it. Accordingly, if this process of reasoning be good, that portion of South America between the calms of Capricorn and Cape Horn, upon the mountain ranges of which this part of the atmosphere, whose circuit I am considering as a type, first impinges, ought to be a region of copious precipitation. Now let us turn to the works on Physical Geography, and see what we can find upon this subject. In Berghaus and Johnston — department Hyetography — it is stated, on the authority of Captain King, R. N., that upward of twelve feet (one hundred and fifty-three inches) of rain fell in forty-one days on that part of the coast of Patagonia which lies within the sweep of the winds just described.[4] So much rain falls there, navigators say, that they sometimes find the water on the top of the sea fresh and sweet. After impinging upon the cold hill-tops of the Patagonian coast, and passing the snow-clad summits of the Andes, this same wind tumbles down upon the eastern slopes of the range as a dry wind; as such, it traverses the almost rainless and barren regions of cis-Andean Patagonia and South Buenos Ayres, Plate V. These conditions, the direction of the prevailing winds, and the amount of precipitation, may be regarded as evidence afforded by nature, if not in favor of, certainly not against, the conjecture that such may have been the voyage of this vapor through the air. At any rate, here is proof of the immense quantity of vapor which these winds of the extra-tropical regions carry along with them toward the poles; and I can imagine no other place than that suggested, whence these winds could get so much vapor.

356. *The question, How can two currents of air cross? answered.* Notwithstanding the amount of circumstantial evidence that has already been brought to show that the air which the northeast and the southeast trade-winds discharge into the belts of equatorial calms, does, in ascending, cross — that from the southern passing over into the northern, and that from the northern passing over into the southern hemisphere (see

_____

[4] [*Physical Atlas*, p. 65. But the passage contains no reference to Captain King.]

O, Q, R, S, and D, E, F, G, § 215) — yet some have implied doubt by asking the question, "How are two such currents of air to pass each other?" And, for the want of light upon this point, the correctness of my reasoning, facts, inferences, and deductions have been questioned. In the first place, it may be said in reply, the belt of equatorial calms is often several hundred miles across, seldom less than sixty; whereas the depth of the volume of air that the trade-winds pour into it is only about three miles, for that is supposed to be about the height to which the trade-winds extend. Thus we have the air passing into these calms by an opening on the north side for the northeast trades, and another on the south for the southeast trades, having a cross section of three miles vertically to each opening. It then escapes by an opening upward, the cross section of which is sixty or one hundred, or even three hundred miles. A very slow motion upward there will carry off the air in that direction as fast as the two systems of trade-winds, with their motion of twenty miles an hour, can pour it in; and that *curds* or *flakes* of air can readily cross each other and pass in different directions without interfering the one with the other, or at least without interfering to that degree which prevents, we all know. The brown fields in the summer afford evidence in a striking manner of the fact that, in nature, flakes, or streamlets, or curdles of air do really move among each other without obstruction. That tremulous motion which we so often observe above stubble-fields, barren wastes, or above any heated surface, is caused by the ascent and descent, at one and the same time, of flakes of air at different temperatures, the cool coming down, the warm going up. They do not readily commingle, for the astronomer, long after nightfall, when he turns his telescope upon the heavens, perceives and laments the unsteadiness they produce in the sky. If the air brought to the calm belt by the northeast trade-winds differ in temperature (and why not?) from that brought by the southeast trades, we have the authority of nature for saying that the two currents would not readily commingle (§ 98). Proof is daily afforded that they would not, and there is reason to believe that the air of each current, in streaks, or patches, or *flakes*, does thread its way through the air of the other without difficulty. Therefore we may assume it as a postulate which nature concedes, that there is no difficulty as to the two currents of air, which come into those calm belts from different directions, crossing over, each in its proper direction, without mingling.

357. *The rain winds in the Mississippi Valley.* The same process of reasoning which conducted us (§ 355) into the trade-wind region of the northern hemisphere for the sources of the Patagonian rains, now invites us into the trade-wind regions of the South Pacific Ocean to look for the vapor springs of the Mississippi. If the rain winds of the Mississippi Valley come from the east, then we should have reason to suppose that

their vapors were taken up from the Atlantic Ocean and Gulf Stream; if the rain winds come from the south, then the vapor springs might, perhaps, be in the Gulf of Mexico; if the rain winds come from the north, then the great lakes might be supposed to feed the air with moisture for the fountains of that river; but if the rains come from the west, where, short of the great Pacific Ocean, should we look for the place of evaporation? Wondering where, I addressed a circular letter to farmers and planters of the Mississippi Valley, requesting to be informed as to the direction of their rain winds. I received replies from Virginia, Mississippi, Tennessee, Missouri, Indiana, and Ohio; and, subsequently, from Colonel W. A. Bird, Buffalo, New York, who says, "The southwest winds are our fair-weather winds; we seldom have rain from the southwest." Buffalo may get much of its rain from the Gulf Stream with easterly winds. But I speak of the Mississippi Valley; all the respondents there, with the exception of one in Missouri, said, "The southwest winds bring us our rains." These winds certainly can not get their vapors from the Rocky Mountains, nor from the Salt Lake, for they rain quite as much upon that basin as they evaporate from it again; if they did not, they would, in the process of time, have evaporated all the water there, and the lake would now be dry. These winds, that feed the sources of the Mississippi with rain, like those between the same parallels upon the ocean, are going from a higher to a lower temperature; and the winds in the Mississippi Valley, not being in contact with the ocean, or with any other evaporating surface to supply them with moisture, must bring with them from some sea or another that which they deposit. Therefore, though it may be urged, inasmuch as the winds which brought the rains to Patagonia (§ 355) came direct from the sea, that they therefore took up their vapors as they came along, yet it can not be so urged in this case; and if these winds could pass with their vapors from the equatorial calms through the upper regions of the atmosphere to the calms of Cancer, and then as surface winds into the Mississippi Valley, it was not perceived why the Patagonian rain winds should not bring their moisture by a similar route. These last are from the northwest, from warmer to colder latitudes; therefore, being once charged with vapors, they must precipitate as they go, and take up less moisture than they deposit. The circumstance that the rainy season in the Mississippi Valley (§ 355) alternates with the dry season on the coast of California and Oregon, indicates that the two regions derive vapor for their rains from the same fountains.

358. *Ehrenberg and his microscope.* During the discussion of this subject, my friend Baron von Gerolt, the Prussian minister, had the kindness to place in my hand Ehrenberg's work, "Passat-Staub und Blut-Regen." Here I found another clew leading across the calm places. That celebrated microscopist reports that he found South American infusoria

in the blood-rains and sea-dust of the Cape Verd Islands, Lyons, Genoa, and other places (§ 325); thus confirming, as far as such evidence can, the indications of our observations, and increasing the probability that the general course of atmospherical circulation is in conformity with the suggestions of the facts gathered from the sea as I had interpreted them, viz., that the trade-winds of the southern hemisphere, after arriving at the belt of equatorial calms, ascend and continue in their course toward the calms of Cancer as an upper current from the southwest, and that, after passing this zone of calms, they are felt on the surface as the prevailing southwest winds of the extra-tropical parts of our hemisphere; and that, for the most part, they bring their moisture with them from the trade-wind regions of the opposite hemisphere. I have marked on Plate IV the supposed track of the "Passat-Staub," showing where it was taken up in South America, as at P, P, and where it was found, as at S, S; the part of the line in dots denoting where it was in the upper current, and the unbroken line where it was wafted by a surface current; also on the same plate is designated the part of the South Pacific in which the vapor-springs for the Mississippi rains are supposed to be. The hands (☞) point out the direction of the wind. Where the shading is light, the vapor is supposed to be carried by an upper current. Such is the character of the circumstantial evidence which induced me to suspect that some agent, whose office in the grand system of atmospherical circulation is neither understood nor recognized, was at work in these calm belts and other places. It may be electrical, or it may be magnetic, or both conjoined.

359. *Quetelet's observations.* The more we study the workings of the atmospherical machinery of our planet, the more are we impressed with the conviction that we as yet know very little concerning its secret springs, and the little "governors" here and there which regulate its movements. My excellent friend M. Quetelet, the astronomer royal at Brussels, has instituted a most excellent series of observations upon atmospherical electricity. He has shown that there is in the upper regions of air a great reservoir of positive electricity, which increases as the temperature diminishes. So, too, with the magnetism of the oxygen in the upper regions.

360. *At sea in the southern hemisphere we have the rule, on land in the northern the exceptions, as to the general circulation of the atmosphere.* In the southern hemisphere, we may, by reason of its great aqueous area, suppose the general law of atmospherical movements to be better developed than it is in the northern hemisphere. We accordingly see by the table (§ 353) that the movements north and south between 45° and 50° correspond with the movements south and north between 25° and 30°; that as you go from the latter band toward the equator the winds with southing in them increase, while the winds with northing in them increase as you go from the former toward the pole.

361. *The magnetic poles, the poles of the wind and of cold coincident.*
This is the law in both hemispheres, thus indicating that there must be in
the polar regions, as in the equatorial, a calm place, where these polar-
bound winds cease to go forward, rise up, and commence their return
(§ 214) as an upper current. So we have theoretically a calm disc, a poly-
gon — not a belt — about each pole. The magnetic poles and the poles of
maximum cold (§ 347) are coincident. Do not those calm discs, or "poles
of the wind," and the magnetic poles, cover the same spot, the two
standing in the relation of cause and effect? This question was first asked
several years ago,[5] and I was then moved to propound it by the inductions
of theoretical reasoning. Observers, perhaps, may never reach those inhospitable regions with their instruments to shed more light upon this subject; but Parry and Barrow have found reasons to believe in the existence
of a perpetual calm about the north pole, and, later, Bellot has reported
the existence of a calm region within the frigid zone. Professor J. H.
Coffin, in an elaborate and valuable paper [6] on the "Winds of the Northern
Hemisphere," arrives by deduction at a like conclusion. In that paper he
has discussed the records at no less than five hundred and seventy-nine
meteorological stations, embracing a totality of observations for two
thousand eight hundred and twenty-nine years. He places his "meteorological pole" — pole of the winds — near latitude 84° north, longitude
105° west. The pole of maximum cold, by another school of philosophers,
Sir David Brewster among them, has been placed in latitude 80° north,
longitude 100° west; and the magnetic pole, by still another school,[7] in
latitude 73° 35′ north, longitude 95° 39′ west. Neither of these poles is a
point susceptible of definite and exact position. The polar calms are no
more a point than the equatorial calms are a line; and, considering that
these poles are areas or discs, not points, it is a little curious that philosophers in different parts of the world, using different data, and following
up investigation each through a separate and independent system of
research, and each aiming at the solution of different problems, should
nevertheless agree in assigning very nearly the same position to them all.
Are these three poles grouped together by chance or by some physical
cause? By the latter, undoubtedly. Here, then, we have another of those
gossamer-like clews, that sometimes seem almost palpable enough for
the mind, in its happiest mood, to lay hold of, and follow up to the very
portals of knowledge, where we pause and linger, fondly hoping that the
chambers of hidden things may be thrown open, and that we may be
permitted to behold and contemplate the mysteries of the winds, the

---

[5] *Sailing Directions* [(1851), p. 171].

[6] *Smithsonian Contributions to Knowledge,* VI [no. 6] (1854), [p. 132 and pl. I].

[7] [Carl Friedrich] Gauss, ["Allgemeine Theorie des Erdmagnetismus," *Resultate
aus den Beobachtungen des magnetischen Vereins im Jahre 1838* (Leipzig, 1839),
pp. 1–57, esp., p. 44].

frost, and the trembling needle. In the polar calms there is (§ 215) an ascent of air; if an ascent, a diminution of pressure and an expansion; and if expansion, a decrease of temperature. Therefore we have palpably enough a connecting link here between the polar calms and the polar place of maximum cold. Thus we establish a relation between the pole of the winds and the pole of cold, with evident indications that there is also a physical connection between these and the magnetic pole. Here the outcroppings of a relation between magnetism and the circulation of the atmosphere again appear.

362. *The barometer in the wind bands.* Thousands of observations, made by mariners and recorded in their abstract logs, have enabled us to determine approximately the mean height of the barometer for the various bands (§ 352) at sea. Between the parallels of 36° S. and 50° N., Lieut. Andrau, of the Dutch Navy, has collected from the abstract logs at the Meteorological Institute of Utrecht no less than 83,334 observations on the height of the barometer in the bands [listed in Table 5].

Table 5

Number of Observations and Mean Height of the Barometer (temperature 32°) between the Parallels of 36° S. and 50° N.

North	Barometer	No.	South	Barometer	No.
0° and 5° [a]	29.915	5114	0° and 5°	29.940	3692
5° and 10°	29.922	5343	5° and 10°	29.981	3924
10° and 15°	29.964	4496	10° and 15°	30.028	4156
15° and 20°	30.018	3592	15° and 20°	30.060	4248
20° and 25°	30.081	3816	20° and 25°	30.102	4536
25° and 30°	30.149	4392	25° and 30°	30.095	4780
30° and 35°	30.210	4989	30° and 36° [a]	30.052	6970
35° and 40°	30.124	5103	42° 53′	29.90 [e]	
40° and 45°	30.077	5899	45° 0′	29.66 [f]	
45° and 50°	30.060	8282	49° 08′	29.47	
51° 29′	29.99 [b]		51° 33′	29.50	
59° 51′	29.88 [c]		54° 26′	29.35	
78° 37′	29.759 [d]		55° 52′	29.36	
			60° 0′	29.11	
			66° 0′	29.08	
			74° 0′	28.93	

(a) From 50° N. to 36° S. the observations are the mean of 83,334 taken from "Maandelijksche Zeilaanwijzingen van Java naar het Kanaal" (Koninklijk Nederlandsch Meteorologisch Instituut, 1859).

(b) Greenwich; mean of 4 years' observations.

(c) St. Petersburg; mean of 10 years' observations.

(d) Dr. Kane; 12,000 observations (mean of 17 months' observations).

(e) Hobart Town; mean of 10 years' observations.

(f) Sir J. C. Ross [A *Voyage of Discovery and Research* . . . . *during the Years 1839–43*, II (London, 1847), p. 385.]

363. *More atmosphere in the northern than in the southern hemisphere.* The diagram of the winds (Plate I) has been constructed so as to show by its shaded border this unequal distribution of the atmosphere between

the two hemispheres. Have we not here proof that the southern hemisphere (§ 261) is indeed the boiler to this mighty atmospherical engine? The aqueous vapor rising from its waste of waters drives the air away from the austral regions, just as the vapor that is formed in the real steam-boiler expels the air from it. This difference of atmosphere over the two halves of the globe, as indicated by the barometer, is very suggestive.

364. *A standard of comparison for the barometer at sea.* Admiral Fitzroy has also reduced from the abstract logs in the Meteorological Department of the Board of Trade in London a great number of barometrical observations. He claims to have discovered that about 5° N. in the Atlantic Ocean the pressure of the atmosphere is so uniform as to afford navigators a natural standard by which, out there at sea, they may, as they pass to and fro, compare their barometers. This pressure is said to be so uniform, that after allowing for the six-hourly fluctuations, the mariner may detect any error in his barometer amounting to the two or three thousandth part of an inch.

365. *Southeast trade-winds having no moisture traced over into rainless regions of the northern hemisphere.* According to the views presented in § 358 and Plate IV, the southeast trade-winds, which reach the shores of Brazil near the parallel of Rio, and which blow thence for the most part over the land, should be the winds which, in the general course of circulation, would be carried, after crossing the Andes and rising up in the belt of equatorial calms, toward Northern Africa, Spain, and the South of Europe. They might carry with them the infusoria of Ehrenberg (§ 358), but, according to this theory, they would be wanting in moisture. Now, are not those portions of the Old World, for the most part dry countries, receiving but a small amount of precipitation? Hence the general rule: those countries to the north of the calms of Cancer, which have large bodies of land situated to the southward and westward of them, in the southeast trade-wind region of the earth, should have a scanty supply of rain, and *vice versa.* Let us try this rule: The extratropical part of New Holland comprises a portion of land thus situated in the southern hemisphere. Tropical India is to the northward and westward of it; and tropical India is in the northeast trade-wind region, and should give extra-tropical New Holland a slender supply of rain. But what modifications the monsoons of the Indian Ocean may make to this rule, or what effect they may have upon the rains in New Holland, my investigations in that part of the ocean have not been carried far enough for final decision; though New Holland is a dry country.

366. *Each hemisphere receives from the sun the same amount of heat.* The earth is nearer to the sun in the summer of the southern hemisphere than it is in the summer of the northern; consequently, it has been held that one hemisphere annually receives more heat than the other. But the

northern summer is 7.7 days longer than the southern; and Sir John Herschel has shown, and any one who will take the trouble may demonstrate, that the total amount of direct solar heat received annually by each hemisphere is identically the same, and therefore the northern hemisphere in its longer summer makes up with heat for the greater intensity but shorter duration of the southern summer. But though the amount of heat annually impressed by the sun upon each hemisphere be identically the same, it by no means follows that the amount radiated off into space by each hemisphere again is also identically the same. There is no reason to believe that the earth is growing warmer or cooler, and therefore we infer that the total amount of heat received annually by the *whole* earth is again annually radiated from the whole earth. Nevertheless, the two hemispheres may radiate very unequally.

367. *The northern radiates most.* Direct observations concerning the amount of radiation from different parts of the surface of our planet are meagre, and the results as to quantity by no means conclusive; but we have in the land and sea breezes a natural index to the actinometry of sea and land, which shows that the radiating forces of the two are very different. Notwithstanding the temperature of the land is raised so much above that of the waters during the day, its powers of radiation are so much greater than those of water that its temperature falls during the night below that of the sea, and so low as to produce the land-breeze. From this fact it may be inferred that the hemisphere that has most land dispenses most heat by radiation.

368. *Another proof of the crossings at the calm belts.* The question now may be well put: Since the two hemispheres receive annually the same amount of heat from the sun, and since the northern hemisphere, with its greater area of land, radiates most, whence does it derive the surplus? The theory of the crossing at the calm belts indicates both the way and the means, and suggests the answer; for it points to the latent heat of vapor that is taken up in the southern hemisphere, transported by the winds across the calm belts, and liberated, as the clouds drop down their fatness upon northern fields. It is not only the difference of radiating power between land and water that makes the northern continents the *chimneys of the earth*, but the difference of cloud in a continental and an oceanic sky must also greatly quicken the radiating powers of the northern hemisphere. Radiation goes on from the upper surface of the clouds and from the atmosphere itself, but we know that clouds in a great measure obstruct radiation from the *surface* of the earth; and as the surface of the earth receives more of the direct heat of the sun than the atmosphere, the point under discussion relates to the mode in which the *surface* of the earth gets rid of that heat. It gets rid of it chiefly in three ways: some is carried off by *convection* in the air; some by evaporation;

and some by radiation; and such is the interference of clouds with this last-named process, that we are told that during the rainy season in intertropical countries, as on the coast of Africa, there is often not radiation enough to produce the phenomena of land and sea breezes. The absence of dew in cloudy nights is a familiar instance of the anti-radiating influence of clouds. The southern hemisphere, being so much more aqueous, is no doubt much more enveloped with clouds where its oceans lie, than is the northern where its continents repose, and therefore it is that one hemisphere radiates more than the other.

369. *Facts and pearls.* Thus, by observing and discussing, by resorting to the force of reason and to the processes of induction, we have gathered for the theory that favors the air-crossings at the calm belts fact upon fact, which, like pearls for the necklace, only require a string to hang them together.

CHAPTER VIII

§ 370–409. CURRENTS OF THE SEA

§ 370. *Obedient to order.* We here set out with the postulate that the sea, as well as the air, has its system of circulation, and that this system, whatever it be, and wherever its channels lie, whether in the waters at or below the surface, is in obedience to physical laws. The sea, by the circulation of its waters, doubtless has its offices to perform in the terrestrial economy; and when we see the currents in the ocean running hither and thither, we feel that they were not put in motion without a cause. On the contrary, we know they move in obedience to some law of Nature, be it recorded down in the depths below, never so far beyond the reach of human ken; and being a law of Nature, we know who gave it, and that neither chance nor accident had any thing to do with its enactment. Nature grants us all that this postulate demands, repeating it to us in many forms of expression; she utters it in the blade of green grass which she causes to grow in climates and soils made kind and genial by warmth and moisture that some current of the sea or air has conveyed far away from under a tropical sun. She murmurs it out in the cooling current of the north; the whales of the sea tell of it (§ 158); and all its inhabitants proclaim it.

371. *The fauna and flora of the sea.* The fauna and the flora of the sea are as much the creatures of climate (§ 164), and are as dependent for their well-being upon temperature as are the fauna and the flora of the dry land. Were it not so, we should find the fish and the algæ, the marine insect and the coral, distributed equally and alike in all parts of the ocean. The arctic whale would delight in the torrid zone, and the habitat of the pearl oyster would be also under the iceberg, or in the frigid waters of polar seas.

372. *Those of southern unlike those of northern seas.* Nevertheless, though the constituents of sea water be the same in kind, we must not infer that they are the same in degree for all parts of the ocean, for there is a peculiarity, perhaps of temperature, perhaps of transparency, which marks the inhabitants of trans-equatorial seas. MM. Peron and Le Sueur, who have turned their attention to the subject, assert that out of many thousand cases they did not find a single one in which the inhabitants of trans-equatorial were not distinguishable from those of their species in cis-equatorial seas.[1]

[1] [F. Péron and Louis Freycinet (eds.), *Voyage de découvertes aux terres australes . . . pendant les années 1800- . . . 1804,* II(Paris, 1816), 348.]

373. *The capacity of water to convey heat.* Water, while its capacities for heat are scarcely exceeded by those of any other substance, is one of the most complete of non-conductors. Heat does not permeate water as it does iron, for instance, or other good conductors. Heat the top of an iron plate, and the bottom becomes warm; but heat the top of a sheet of water, as in a pool or basin, and that at the bottom remains cool. The heat passes through iron by conduction, but to get through water it requires to be conveyed by a motion, which in fluids we call currents. Therefore the study of the climates of the sea involves a knowledge of its currents, both cold and warm. They are the channels through which the waters circulate, and by means of which the harmonies of old ocean are preserved.

374. *Currents of the sea to be considered in pairs.* Hence, in studying the system of oceanic circulation, we set out with the very simple assumption, viz., that from whatever part of the ocean a current is found to run, to the same part a current of equal volume is bound to return; for upon this principle is based the whole system of currents and counter-currents of the air as well as of the water. Hence the advantage of considering them as the anatomist does the nerves of the human system — in pairs. Currents of water, like currents of air, meeting from various directions, create gyrations, which in some parts of the sea, as on the coast of Norway, assume the appearance of whirlpools, as though the water were drawn into a chasm below. The celebrated Maelstrom is caused by such a conflict of tidal or other streams. The late Admiral Beechey, R.N.,[2] gave diagrams illustrative of many "rotatory streams in the English Channel, a number of which occur between the outer extremities of the channel tide and the stream of the oceanic or parent wave." "They are clearly to be accounted for," says he, "by the streams acting obliquely upon each other."

375. *Marine currents do not, like those on land, run of necessity from higher to lower levels.* It is not necessary to associate with oceanic currents the idea that they must, of necessity, as on land, run from a higher to a lower level. So far from this being the case, some currents of the sea actually run up hill, while others run on a level. The Gulf Stream is of the first class (§ 83).

376. *The Red Sea current.* The currents which run from the Atlantic into the Mediterranean, and from the Indian Ocean into the Red Sea, are the reverse of this. Here the bottom of the current is probably a water-level, and the top an inclined plane, running *down hill.* Take the Red Sea current as an illustration. That sea lies, for the most part, within a rainless and riverless district. It may be compared to a long and

---

² [F. W.] Beechey, "Tidal Streams of the North Sea and English Channel," *Philos. Trans.*, Pt. 2(1851), pp. 703–718 [esp., p. 709].

narrow trough. Being in a rainless district, the evaporation from it is immense; none of the water thus taken up is returned to it either by rivers or rains. It is about one thousand miles long; it lies nearly north and south, and extends from latitude 13° to the parallel of 30° north. From May to October, the water in the upper part of this sea is said to be two feet lower than it is near the mouth.[3] This change or difference of level is ascribed to the effect of the wind, which, prevailing from the north at that season, is supposed to blow the water out. But from May to October is also the hot season; it is the season when evaporation is going on most rapidly; and when we consider how dry and how hot the winds are which blow upon this sea at this season of the year; that it is a narrow sea; that they blow across it, and are not saturated, we may suppose the daily evaporation to be immense. The evaporation from this sea and the Persian Gulf is probably greater than it is from any other arms of the ocean. We know that the waste from canals by evaporation, in the summer time, is an element which the engineer, when taking the capacity of his feeders into calculation, has to consider. With him it is an important element; how much more so must the waste by evaporation from this sea be when we consider the physical conditions under which it is placed. Its feeder, the Arabian Sea, is a thousand miles from its head; its shores are burning sands; the evaporation is *ceaseless*; it is a natural evaporating dish (§ 525) on a grand scale; none of the vapors which the scorching winds that blow over it carry away are returned to it again in the shape of rains. The Red Sea vapors are carried off and precipitated elsewhere. The depression in the level of its head waters in the summer time, therefore, it appears, is owing to the effect of evaporation, as well as to that of the wind blowing the waters back. The evaporation in certain parts of the Indian Ocean is supposed to be (§ 103) from three fourths of an inch to an inch daily. Whatever it be, it is doubtless greater in the Red Sea. Let us assume it, then, in the summer time to average only half an inch a day. Now, if we suppose the velocity of the current which runs into that sea to average, from mouth to head, twenty miles a day, it would take the water fifty days to reach the head of it. If it lose half an inch from its surface by evaporation daily, it would, by the time it reaches the Isthmus of Suez, have lost twenty-five inches from its surface. Thus the waters of the Red Sea ought to be lower at the Isthmus of Suez than they are at the Straits of Babelmandeb. Independently of the forcing out by the wind, the waters there ought to be lower from two other causes, viz., evaporation and temperature; for the temperature of that sea is necessarily lower at Suez, in latitude 30°, than it is at Babelmandeb, in latitude 13°. To make it quite clear that the surface of the Red Sea is not a sea level, but is an inclined plane, suppose the channel of the Red Sea to have a

[3] *Physical Atlas* [p. 50].

perfectly smooth and level floor, with no water in it, and a wave ten feet high to enter the Straits of Babelmandeb, and to flow up the channel, like the present surface current, at the rate of twenty miles a day for fifty days, losing daily, by evaporation, half an inch; it is easy to perceive that, at the end of the fiftieth day, this wave would not be so high by two feet (twenty-five inches) as it was the first day it commenced to flow. The top of that sea, therefore, may be regarded as an inclined plane, made so by evaporation.

377. *Upper and under currents through straits explained.* But the salt water, which has lost so much of its freshness by evaporation, becomes salter, and therefore heavier. The lighter water at the Straits can not balance the heavier water at the Isthmus, and the colder and salter, and therefore heavier water, must either run out as an under current, or it must deposit its surplus salt in the shape of crystals, and thus gradually make the bottom of the Red Sea a salt-bed, or it must abstract all the salt from the ocean to make the Red Sea brine — and we know that neither the one process nor the other is going on. Hence we infer that there is from the Red Sea an under and outer current, as there is from the Mediterranean through the Straits of Gibraltar, and that the surface waters near Suez are salter than those near the mouth of the Red Sea. And, to show why there should be an outer and under current from each of these two seas, let us suppose the case of a vat of oil and a vat of wine connected by means of a narrow trough — the trough being taken to represent the straits connecting seas the waters of which differ as to specific gravity. Suppose the trough to have a flood-gate, which is closed until we are ready for the experiment. Now let the two vats be filled, one with wine the other with oil, up to the same level. The oil is introduced to represent the lighter water as it enters either of these seas from the ocean, and the wine the same water after it has lost some of its freshness by evaporation, and therefore has become salter and heavier. Now suppose the flood-gate to be raised, what would take place? Why, the oil would run in as an upper current, overflowing the wine, and the wine would run out as an under current.

378. *The Mediterranean current.* The rivers which discharge their waters into the Mediterranean are not sufficient to supply the waste of evaporation, and it is by a process similar to this that the salt which is carried in from the ocean is returned to the ocean again; were it not so, the bed of that sea would be a mass of solid salt. The equilibrium of the seas is a physical necessity. Were it to be lost, the consequences would be as disastrous as would be any other derangement in the forces of attraction. Without doubt, the equilibrium of the sea is preserved by a system of compensation as exquisitely adjusted as are those by which the "music of the spheres" is maintained. It is difficult to form an adequate conception

of the immense quantities of solid matter which the current from the Atlantic, holding in solution, carries into the Mediterranean. In his abstract log for March 8th, 1855, Lieutenant William Grenville Temple, of the United States ship Levant, homeward bound, has described the indraught there: "Weather fine; made 1¼ pt. lee-way. At noon, stood in to Almiria Bay, and anchored off the village of Roguetas. Found a great number of vessels waiting for a chance to get to the westward, and learned from them that at least a thousand sail are weather-bound between this and Gibraltar. Some of them have been so for six weeks, and have even got as far as Malaga, only to be swept back by the current. Indeed, no vessel had been able to get out into the Atlantic for three months past." Now, suppose this current, which baffled and beat back this fleet for so many days, ran no faster than two knots the hour. Assuming its depth to be 400 feet only, and its width seven miles, and that it carried in with it the average proportion of solid matter — say one thirtieth — contained in sea water; and admitting these postulates into calculation as the basis of the computation, it appears that salts enough to make no less than 88 cubic miles of solid matter, of the density of water, were carried into the Mediterranean during these 90 days. Now, unless there were some escape for all this solid matter, which has been running into that sea, not for 90 days merely, but for ages, it is very clear that the Mediterranean would, ere this, have been a vat of very strong brine, or a bed of cubic crystals.

379. *The Suez Canal.* We have in this fact, viz., the difficulty of egress from the Mediterranean, and the tedious character of the navigation, under canvas, within it, the true secret of the indifference which, in commercial circles in England and the Atlantic states of Europe, is manifested toward the projected Suez Canal. But to France and Spain on the Mediterranean, to the Italian States, Greece, and Austria, it would be the greatest commercial boon of the age. The Mediterranean is a great gulf running from west to east, penetrating the Old World almost to its very centre, and separating its most civilized from its most savage parts. Its southern shores are inhabited, for the most part, by an anti-commercial and thriftless people. On the northern shores the climates of each nation are nearly duplicates of the climates of her neighbors to the east and the west; consequently, these nations all cultivate the same staples, and their wants are similar: for a commerce among themselves, therefore, they lack the main elements, viz., difference of production, and the diversity of wants which are the consequence of variety of climates. To reach these, the Mediterranean people have had to encounter the tedious navigation and the sometimes difficult egress — just described — from their sea. Clearing the Straits of Gibraltar, their vessels do not even then find themselves in a position so favorable for reaching the markets of the world as they would be were they in Liverpool or off the Lizard. Such is the ob-

struction which the winds and the current from the Atlantic offer to the navigation there, that vessels bound to India from the United States, England, or Holland, may often double the Cape of Good Hope before one sailing with a like destination from a Mediterranean port would find herself clear of the Straits of Gibraltar. It is therefore not surprising that none of the great commercial marts of the present day are found on the shores of this classic sea. The people who inhabit the hydrographic basin of the Mediterranean — which includes the finest parts of Europe — have, ever since the discovery of the passage around the Cape of Good Hope, been commercially pent up. A ship-canal across the Isthmus of Suez will let them out into the commercial world, and place them within a few days of all the climates, wants, supplies, and productions of India. It will add largely to their wealth and prosperity. As these are increased, trading intercourse is enhanced, and so by virtue of this canal they will become better customers for England and Holland, and all other trading nations whose ports are havens of the Atlantic. Occupying this stand-point in their system of commercial economy, the people of the United States await with a lively interest the completion of the Suez Canal.

380. *Hydrometrical observations at sea wanted.* Of all parts of the ocean, the warmest water, the saltest and the heaviest too, is said to be found in the seas of the Indian Ocean. A good series of observations there with the hydrometer, at the different seasons of the year, is a desideratum. Taking, however, such as we have upon the density of the water in the Red Sea and the Mediterranean, and upon the under currents that run out from these seas, let us examine results.

381. *Specific gravity of Red Sea water.* Several years ago, Mr. Morris, chief engineer of the Oriental Company's steam-ship Ajdaha, collected specimens of Red Sea water all the way from Suez to the Straits of Babelmandeb, which were afterward examined by Dr. Giraud, who reported the results [summarized in Table 6].[4] These observations agree with the

Table 6

[*Specific gravity and saline content of Red Sea water*]

	Latitude	Longitude	Spec. Grav.	Saline Cont.
	°	°		1000 parts
No. 1. Sea at Suez	—	—	1027	41.0
No. 2. Gulf of Suez	27.49	33.44	1026	40.0
No. 3. Red Sea	24.29	36.	1024	39.2
No. 4. do.	20.55	38.18	1026	40.5
No. 5. do.	20.43	40.03	1024	39.8
No. 6. do.	14.35	42.43	1024	39.9
No. 7. do.	12.39	44.45	1023	39.2

[4] [G. Buist, "On the Saltness of the Red Sea"], *Trans. Bombay Geogr. Soc.*, IX(1850) [38–43, esp., p. 38].

theoretical deductions just announced, and show that the surface waters at the head are heavier and salter than the surface waters at the mouth of the Red Sea.

382. *Evaporation from.* In the same paper, the temperature of the air between Suez and Aden often rises, it is said, to 90°, "and probably averages little less than 75° day and night all the year round. The surface of this sea varies in heat from 65° to 85°, and the difference between the wet and dry bulb thermometers often amounts to 25° — in the kamsin, or desert winds, to from 30° to 40°; the average evaporation at Aden is about eight feet for the year." "Now assuming," says Dr. Buist, "the evaporation of the Red Sea to be no greater than that of Aden, a sheet of water eight feet thick, equal in area to the whole expanse of that sea, will be carried off annually in vapor; or, assuming the Red Sea to be eight hundred feet in depth at an average — and this, most assuredly, is more than double the fact — the whole of it would be dried up, were no water to enter from the ocean, in one hundred years. The waters of the Red Sea, throughout, contain some four per cent. of salt by weight — or, as salt is a half heavier than water, some 2.7 per cent. in bulk — or, in round numbers, say three per cent. In the course of three thousand years, on the assumptions just made, the Red Sea ought to have been one mass of solid salt, if there were no current running out." Now we know the Red Sea is more than three thousand years old, and that it is not filled with salt; and the reason is, that as fast as the upper currents bring the salt in at the top, the under currents carry it out at the bottom.

383. *Mediterranean Currents.* With regard to an under current from the Mediterranean, we may begin by remarking that we know that there is a current always setting in at the surface from the Atlantic, and that this is a salt-water current, which carries an immense amount of salt into that sea. We know, moreover, that that sea is not salting up; and therefore, independently of the postulate (§ 374) and of observations, we might infer the existence of an under current, through which this salt finds its way out into the broad ocean again.[5]

[5] Dr. Smith appears to have been the first to *conjecture* this explanation, which he did in 1684 [Tho. Smith, "A Conjecture about an Under-current at the Streights-Mouth,"] *Philos. Trans.* [XIV] (1684) [564–566, esp., p. 565]. This continual indraught into the Mediterranean appears to have been a vexed question among the navigators and philosophers even of those times. Dr. Smith alludes to several hypotheses which had been invented to solve these phenomena, such as subterraneous vents, cavities, exhalation by the sun's beams, etc., and then offers his *conjecture*, which, in his own words, is, "that there is an under current, by which as great a quantity of water is carried out as comes flowing in. To confirm which, besides what I have said above about the difference of tides in the offing and at the shore in the Downs, which necessarily supposes an under current, I shall present you with an instance of the like nature in the Baltic Sound, as I received it from an able seaman, who was at the making of the trial. He told me that, being there in one of the king's frigates, they went with their pinnace into the mid stream, and were carried violently by the

384. *The drift of the Phœnix.* With regard to this outer and under current, we have observations telling of its existence as long ago as 1712. "In the year 1712," says Dr. Hudson, in a paper communicated to the Philosophical Society in 1724, "Monsieur du L'Aigle, that fortunate and generous commander of the privateer called the Phœnix, of Marseilles, giving chase near Ceuta Point to a Dutch ship bound to Holland, came up with her in the middle of the Gut between Tariffa and Tangier, and there gave her one broadside, which directly sunk her, all her men being saved by Monsieur du L'Aigle; and a few days after, the Dutch ship, with her cargo of brandy and oil, arose on the shore near Tangier, which is at least four leagues to the westward of the place where she sunk, and directly against the strength of the current, which has persuaded many men that there is a recurrency in the deep water in the middle of the Gut that sets outward to the grand ocean, which this accident very much demonstrates; and, possibly, a great part of the water which runs into the Straits returns that way, and along the two coasts before mentioned; otherwise this ship must, of course, have been driven toward Ceuta, and so upward. The water in the Gut must be very deep; several of the commanders of our ships of war having attempted to sound it with the longest lines they could contrive, but could never find any bottom." [6]

385. *Saltness of the Mediterranean.* In 1828, Dr. Wollaston, in a paper before the Philosophical Society, stated that he found the specific gravity of a specimen of sea water, from a depth of six hundred and seventy fathoms, fifty miles within the Straits, to have a "density exceeding that of distilled water by more than four times the usual excess, and accordingly leaves, upon evaporation, more than four times the usual quantity of saline residuum. Hence it is clear that an under current outward of such denser water, if of equal breadth and depth with the current inward near the surface, would carry out as much salt below as is brought in above, although it moved with less than one fourth part of the velocity, and would thus prevent a perpetual increase of saltness in the Mediterranean Sea beyond that existing in the Atlantic." [7] The doctor obtained this specimen of sea water from Captain, now Admiral Smyth, of the English Navy, who had collected it for Dr. Marcet. Dr. Marcet died before receiving it, and it had remained in the admiral's hands some time before it came into those of Wollaston. It may, therefore, have lost some-

---

current; that, soon after this, they sunk a bucket with a heavy cannon ball to a certain depth of water, which gave a check to the boat's motion; and, sinking it still lower and lower, the boat was driven ahead to the windward against the upper current: the current aloft, as he added, not being over four or five fathoms deep, and that the lower the bucket was let fall, they found the under current the stronger."

    [6] ["Of the Currents at the Streights Mouth. By Capt. ———, Communicated by Dr. Hudson," *Philos. Trans.*, XXX(1728), 191–192.]

    [7] [William Hyde Wollaston, "On the Water of the Mediterranean," *Philos. Trans.*, CXIX(1829), 29–31, esp., p. 30.

thing by evaporation; for it is difficult to conceive that all the river water, and three fourths of the sea water which runs into the Mediterranean, is evaporated from it, leaving a brine for the under current having *four times* as much salt as the water at the surface of the sea usually contains. Very recently, M. Coupvent des Bois is said to have shown, by actual observation, the existence of an outer and under current from the Mediterranean.

386. *The escape of salt and heavy water by under currents.* However that may be, these facts, and the statements of the Secretary of the Geographical Society of Bombay (§ 382), seem to leave no room to doubt as to the existence of an under current both from the Red Sea and Mediterranean, and as to the cause of the surface current which flows into them. I think it a matter of demonstration. It is accounted for (§ 377) by the salts of the sea. Writers whose opinions are entitled to great respect differ with me as to the conclusiveness of this demonstration. Among those writers are Admiral Smyth, of the British Navy, and Sir Charles Lyell, who also differ with each other. In 1820, Dr. Marcet, being then engaged in studying the chemical composition of sea water, the admiral, with his usual alacrity for doing "a kind turn," undertook to collect for the doctor specimens of Mediterranean water from various depths, especially in and about the Straits of Gibraltar. Among these was the one (§ 385) taken fifty miles within the Straits from the depth of six hundred and seventy fathoms (four thousand and twenty feet), which, being four times salter than common sea water, left, as we have just seen, no doubt in the mind of Dr. Wollaston as to the existence of this under current of brine. But the indefatigable admiral, in the course of his celebrated survey of the Mediterranean, discovered that, while inside of the Straits the depth was upward of nine hundred fathoms, yet in the Straits themselves the depth across the shoalest section is not more than one hundred and sixty [8] fathoms. "Such being the case, we can now prove," exclaims Sir Charles Lyell, "that the vast amount of salt brought into the Mediterranean *does not* pass out again by the Straits; for it appears by Captain Smyth's soundings, which Dr. Wollaston had not seen, that between the Capes of Trafalgar and Spartel, which are twenty-two miles apart, and where the Straits are shallowest, the deepest part, which is on the side of Cape Spartel, is only two hundred and twenty fathoms.[9] It is therefore evident, that if water sinks in certain parts of the Mediterranean, in consequence of the increase of its specific gravity, to greater depths than two hundred and twenty fathoms, it can never flow out again into the Atlantic, since it must be stopped by the submarine barrier which crosses the shallowest part of the Straits of Gibraltar." [10]

[8] Smyth, *The Mediterranean* [p. 159].
[9] One hundred and sixty, Smyth [*ibid.*].
[10] [Charles] Lyell, *Principles of Geology*, 9th ed.(1853), pp. 334–335.

387. *Vertical circulation in the sea a physical necessity.* According to this reasoning, all the cavities, the hollows, and the valleys at the bottom of the sea, especially in the trade-wind region, where evaporation is so constant and great, ought to be salting up or filling up with brine. Is it probable that such a process is actually going on? No. According to this reasoning, the water at the bottom of the great American lakes ought to remain there forever, for the bottom of Erie is far below the barrier which separates this lake from the Falls of Niagara, and so is the bottom of every one of the lakes below the shallows in the straits or rivers that connect them as a chain. We may presume that the water at the bottom of every extensive and quiet sheet of water, whether salt or fresh, is at the bottom by reason of specific gravity; but that it does not remain there forever we have abundant proof. If so, the Niagara River would be fed by Lake Erie only from that layer of water which is above the level of the top of the rock at the Falls. Consequently, wherever the breadth of that river is no greater than it is at the Falls, we should have a current as rapid as it is at the moment of passing the top of the rock to make the leap. To see that such is not the way of Nature, we have but to look at any common mill-pond when the water is running over the dam. The current in the pond that feeds the overflow is scarcely perceptible, for "still water runs deep." Moreover, we know it is not such a skimming current as the geologist would make, which runs from one lake to another; for wherever above the Niagara Falls the water is deep, there we are sure to find the current sluggish, in comparison with the rate it assumes as it approaches the Falls; and it is sluggish in deep places, rapid in shallow ones, because it is fed from below. The common "wastes" in our canals teach us this fact.

388. *The bars at the mouths of the Mississippi an illustration.* The reasoning of this celebrated geologist appears to be founded upon the assumption that when water, in consequence of its specific gravity, once sinks below the bottom of a current where it is shallowest, there is no force of *traction*, so to speak, in fluids, nor any other power, which can draw this heavy water up again. If such were the case, we could not have deep water immediately inside of the bars which obstruct the passage of the great rivers into the sea. Thus the bar at the mouth of the Mississippi, with only fifteen feet of water on it, is estimated to travel out to sea at rates varying from one hundred to twenty yards a year. In the place where that bar was when it was one thousand yards nearer to New Orleans than it now is, whether it were fifteen years ago or a century ago, with only fifteen or sixteen feet of water on it, we have now four or five times that depth. As new bars were successively formed seaward from the old, what dug up the sediment which formed the old, and lifted it up from where specific gravity had placed it, and carried it out to sea over a bar-

rier not more than a few feet from the surface? Indeed, Sir Charles himself makes this majestic stream to tear up its own bottom to depths far
below the top of the bar at its mouth. He describes the Mississippi as a
river having nearly a uniform breadth to the distance of two thousand
miles from the sea.[11] He makes it cut a bed for itself out of the soil,
which is heavier than Admiral Smyth's deep sea water, to the depth of
more than two hundred feet [12] below the top of the bar which obstructs
its entrance into the sea. Could not the same power which scoops out
this solid matter for the Mississippi draw the brine up from the pool in
the Mediterranean, and pass it out across the barrier in the Straits? The
currents which run over the bars and shoals in our rivers are fed from
the pools above with water which we know comes from depths far below
the top of such bars. The breadth of the river where the bar is may be
the same as its breadth where the deep pool is, yet the current in the pool
may be so sluggish as scarcely to be perceptible, while it may dash over
the bar or down the rapids with mill-tail velocity. Were the brine not
drawn out again from the hollow places in the sea, it would be easy to
prove that this indraught into the Mediterranean has taken, even during
the period assigned by Sir Charles to the formation of the Delta of the
Mississippi — one of the newest formations — salt enough to fill up the
whole basin of the Mediterranean with crystals. Admiral Smyth brought
up bottom with his briny sample of deep-sea water (six hundred and
seventy fathoms), but no salt crystals.

389. *Views of Admiral Smyth and Sir C. Lyell.* The gallant admiral
— appearing to withhold his assent both from Dr. Wollaston in his conclusions as to this under current, and from the geologist in his inferences
as to the effect of the barrier in the Straits — suggests the probability
that, in sounding for the heavy specimen of sea water, he struck a brine
spring. But the specimen, according to analysis, was of sea water, and it
is not necessary to call in the supposition of a brine spring to account for
this heavy specimen. If we admit the principle assumed by Sir Charles
Lyell, that water from the great pools and basins of the sea can never
ascend to cross the ridges which form these pools and basins, then the
harmonies of the sea are gone, and we are forced to conclude they never
existed. Every particle of water that sinks below a submarine ridge is,
ipso facto, by his reasoning, stricken from the channels of circulation, to
become thenceforward forever motionless matter. The consequence
would be "cold obstruction" in the depths of the sea, and a system of circulation between different seas of the waters only that float above the

---

[11] "From near its mouth at the Balize, a steam-boat may ascend for 2000 miles
with scarcely any perceptible difference in the width of the river." — *Ibid.*, p. 264.
[12] "The Mississippi is continually shifting its course in the great alluvial plain,
cutting frequently to the depth of one hundred, and even sometimes to the depth of
two hundred and fifty feet." — *Ibid.*, p. 273.

shoalest reefs and barriers. If the water in the depths of the sea were to be confined there, doomed to everlasting repose, then why was it made fluid, or why was the sea made any deeper than just to give room for its surface currents to skim along? If water once below them must remain below them — and they are shallow — why were the depths of the ocean filled with fluid instead of solid matter? Doubtless, when the seas were measured and the mountains stood in the balance, the solid and fluid matter of the earth were adjusted in exact proportions to insure perfection in the terrestrial machinery. I do not believe in the existence of any such imperfect mechanism, or in any such failure of design as the imparting of useless properties to matter, such as fluidity to that which is doomed to be stationary would imply. To my mind, the proofs — the theoretical proofs — the proofs derived exclusively from reason and analogy — are as clear in favor of this under current from the Mediterranean as they were in favor of the existence of Leverrier's planet before it was seen through the telescope at Berlin. Now suppose, as Sir Charles Lyell maintains, that none of these vast quantities of salt which this surface current takes into the Mediterranean find their way out again. It would not be difficult to show, even to the satisfaction of that eminent geologist, that this indraught conveys salt away from the Atlantic faster than all the *fresh*-water streams empty fresh supplies of salt into the ocean. Now, besides this drain, vast quantities of salts are extracted from sea water for coral reefs, shell banks, and marl beds; and by such reasoning as this, which is perfectly sound and good, we establish the existence of this under current, or else we are forced to the very unphilosophical conclusion that the sea must be losing its salts, and becoming less and less briny.

390. *The Currents of the Indian Ocean.* By carefully examining the physical features of this sea (Plates V and VI), and studying its conditions, we are led to look for warm currents that have their genesis in this ocean, and that carry from it volumes of overheated water, probably exceeding in quantity many times that which is discharged by the Gulf Stream from its fountains (Plate III). The Atlantic Ocean is open at the north, but tropical countries bound the Indian Ocean in that direction. The waters of this ocean are hotter than those of the Caribbean Sea, and the evaporating force there (§ 300) is much greater. That it is greater we might, without observation, infer from the fact of a higher temperature and a greater amount of precipitation on the neighboring shores (§ 298). These two facts, taken together, tend, it would seem, to show that large currents of warm water have their genesis in the Indian Ocean. One of them is the well-known Mozambique current, called at the Cape of Good Hope the Lagulhas current. Another of these warm currents from the Indian Ocean makes its escape through the Straits of Malacca, and, being joined by other warm streams from the Java and

China Seas, flows out into the Pacific, like another Gulf Stream, between the Philippines and the shores of Asia. Thence it attempts the great circle route (§ 118) for the Aleutian Islands, tempering climates, and losing itself in the sea as its waters grow cool on its route toward the northwest coast of America.

391. *The Black Stream of the Pacific contrasted with the Gulf Stream of the Atlantic.* Between the physical features of this, the "Black Stream" of the Pacific, and the Gulf Stream of the Atlantic, there are several points of resemblance. Sumatra and Malacca correspond to Florida and Cuba; Borneo to the Bahamas, with the Old Providence Channel to the south, and the Florida Pass to the west. The coasts of China answer to those of the United States, the Philippines to the Bermudas, the Japan Islands to Newfoundland. As with the Gulf Stream, so also here with this China current, there is a counter current of cold water between it and the shore. The climates of the Asiatic coast correspond with those of America along the Atlantic, and those of Columbia, Washington, and Vancouver resemble those of Western Europe and the British Islands; the climate of California (State) resembles that of Spain; the sandy plains and rainless regions of Lower California reminding one of Africa, with its deserts between the same parallels, etc. Moreover, the North Pacific, like the North Atlantic, is enveloped, where these warm waters go, with mists and fogs, and streaked with *lightning*. The Aleutian Islands are almost as renowned for fogs and mists as are the Grand Banks of Newfoundland. A surface current flows north from Behring's Strait into the Arctic Sea; but in the Atlantic the current is from, not into the Arctic Sea: it flows south on the surface, north below; Behring's Strait being too shallow to admit of mighty under currents, or to permit the introduction from the polar basin of any large icebergs into the Pacific. Behring's Strait, in geographical position, answers to Davis' Strait in the Atlantic; and Alaska, with its Aleutian chain of islands, to Greenland. But instead of there being to the east of Alaska, as there is to the east of Greenland, an escape into the polar basin for these warm waters of the Pacific, a shore-line intervenes; being cooled here, and having their specific gravity changed, they are turned down through a sort of North Sea along the western coast of the continent toward Mexico. They appear here as a cold current. The effect of this body of cool water upon the littoral climate of California is very marked. Being cool, it gives freshness and strength to the sea-breeze of that coast in summer time, when the "cooling sea-breeze" is most grateful. These contrasts show the principal points of resemblance and of contrast between the currents and aqueous circulation in the two oceans. The ice-bearing currents of the North Atlantic are not repeated as to volume in the North Pacific, for there is no nursery for icebergs like the frozen ocean and its Atlantean arms. The

seas of Okotsk and Kamtschatka alone, and not the frozen seas of the Arctic, cradle the icebergs for the North Pacific.

392. *The Lagulhas Current and the storms of the Cape.* The Lagulhas current, as the Mozambique is sometimes called, skirts the coast of Natal as our Gulf Stream does the coast of Georgia, where it gives rise to the most grand and terrible displays of thunder and lightning that are any where else to be witnessed. Missionaries thence report to me the occurrence there of thunder-storms in which for hours consecutively they have seen an uninterrupted blaze of lightning, and heard a continuous peal of thunder. Reaching the Lagulhas banks, the current spreads itself out there in the midst of cooler waters, and becomes the centre of one of the most remarkable storm-regions in the world. My friend and fellow-laborer, Lieut. Andrau, of the Dutch Navy, has made the storms upon these banks a specialty for study. He has pointed out from the abstract logs at Utrecht the existence there of some curious and interesting atmospherical phenomena to which this body of warm water gives rise. The storms that it calls up come rushing from the westward — sweeping along parallel with the coast of Africa, they curve along it. Though so near the land, they seldom reach it. They march into these warm waters with furious speed; reaching them with a low barometer, they pause and die out. That officer has conferred a boon upon the Indiamen of all flags, for he has taught them how to avoid these dreadful winter storms of the Cape.

393. *The currents and drift of the Indian Ocean.* There is sometimes, if not always, another exit of warm water from the Indian Ocean. It seems to be an overflow of the great intertropical caldron of India — seeking to escape thence, it works its way polarward more as a drift than as a current. It is to the Mozambique current what the northern flow of warm waters in the Atlantic (§ 141) is to the Gulf Stream. This Indian overflow is very large. The best indication of it is afforded by the sperm whale curve (Plate VI). This overflow finds its way south midway between Africa and Australia, and appears to lose itself in passing around a sort of Sargasso Sea, thinly strewed with patches of weed. Nor need we be surprised at such a vast flow of warm water as these three currents indicate from the Indian Ocean, when we recollect that this ocean (§ 392) is land-locked on the north, and that the temperature of its waters is frequently as high as 90° Fahr. There must, therefore, be immense volumes of water flowing into the Indian Ocean to supply the waste created by these warm currents.

394. *The ice-bearing currents from the Antarctic regions.* On either side of this warm current that escapes from the intertropical parts of the Indian Ocean, but especially on the Australian side, an ice-bearing current (Plate VI) is found wending its way from the Antarctic regions with

supplies of cold water to modify climates and restore the aqueous equilibrium in that part of the world. There is a general drift up into the South Atlantic of ice-bearing waters from Antarctic seas. The icebergs brought thence, being often very large and high, are drifted to the eastward by the "brave west winds" of those regions. Hence the icebergs that are so often seen to the south of the Cape of Good Hope. They set off for the Atlantic, but were driven to the eastward by the west winds of these latitudes. The Gulf Stream seldom permits icebergs from Arctic waters to reach the parallel of 40° in the North Atlantic, but I have known the ice-bearing current which passes east of Cape Horn into the South Atlantic to convey its bergs as far as the parallel of 37° south latitude. This is the nearest approach of icebergs to the equator. These currents which run out from the intertropical basin of that immense sea — Indian Ocean — convey along immense volumes of water containing vast quantities of salt, and we know that sea water enough to convey back equal quantities of salt, and salt to keep up supplies for the outgoing currents, must flow into the intertropical regions of the same sea; therefore, if observations were silent upon the subject, reason would teach us to look for currents here that keep in motion immense volumes of water.

395. *The Currents of the Pacific.* The contrast has been drawn (§ 391) between the Japan or "Black Stream" of the North Pacific, and the Gulf Stream of the North Atlantic. The course of the former has never been satisfactorily traced out. There is (Plate VI), along the coast of California and Mexico, a southwardly movement of waters, as there is along the west coast of Africa toward the Cape de Verd Islands. In the open space west of this southwardly set along the African coast there is the famous Sargasso Sea (Plate VI), which is the general receptacle of the drift-wood and sea-weed of the Atlantic. So, in like manner, to the west from California of this other southwardly set, lies the pool into which the drift-wood and sea-weed of the North Pacific are generally gathered, but in small quantities. I have received from Johnston's Islands (17° N., 169° 30′ W), which are near the edge of this pool, specimens of drift-wood from the Columbia, and also of the red cedar of California. The immense trees from Oregon and California that have been cast up on these guano islands were probably drifted down with the cool California current into the northeast trades, and by them wafted along to the west, thus showing that the currents of the North Pacific flow in a sort of circle, on the outer edge of which lie the Japanese and Aleutian Islands, and the northwest coast of America.

396. *The Black Current of the Pacific, like the Gulf Stream, salter than the adjacent waters.* The natives of the Aleutian Islands, where no trees grow, depend upon the drift-wood cast ashore there for all the timber used in the construction of their boats, fishing-tackle, and house-

hold gear. Among this timber, the camphor-tree, and other woods of China and Japan, are said to be often recognized. In this fact we have additional evidence touching this China Stream, as to which (§ 395) but little, at best, is known. "The Japanese," says Lieutenant Bent,[13] in a paper read before the American Geographical Society, January, 1856, "are well aware of its existence, and have given it the name of 'Kuro-Siwo,' or Black Stream, which is undoubtedly derived from the deep blue color of its water, when compared with that of the adjacent ocean." From this we may infer (§ 71) that the blue waters of this China Stream also contain more salt than the neighboring waters of the sea.

397. *The Cold Current of Okotsk.* Inshore of, but counter to the "Black Stream," along the eastern shores of Asia, is found (§ 391) a streak, or layer, or current of cold water answering to that between the Gulf Stream and the American coast. This current, like its fellow in the Atlantic, is not strong enough at all times sensibly to affect the course of navigation; but, like that in the Atlantic, it is the nursery (§ 158) of most valuable fisheries. The fisheries of Japan are quite as extensive as those of Newfoundland, and the people of each country are indebted for their valuable supplies of excellent fish to the cold waters which the currents of the sea bring down to their shores.

398. *Humboldt's Current.* The currents of the Pacific are but little understood. Among those about which most is thought to be known is the Humboldt Current of Peru, which the great and good man whose name it bears was the first to discover. It has been traced on Plate VI according to the best information — defective at best — upon the sub-ject. This current is felt as far as the equator, mitigating the rainless climate of Peru as it goes, and making it delightful. The Andes, with their snowcaps, on one side of the narrow Pacific slopes of this inter-tropical republic, and the current from the Antarctic regions on the other, make its climate one of the most remarkable in the world; for, though torrid as to latitude, it is such as to temperature that cloth clothes are seldom felt as oppressive during any time of the year, especially after nightfall.

399. *The "desolate" region.* Between Humboldt's Current and the great equatorial flow there is an area marked as the "desolate region," Plate VI. It was observed that this part of the ocean was rarely visited by the whale, either sperm or right; why, it did not appear; but observations asserted the fact. Formerly, this part of the ocean was seldom whitened by the sails of a ship, or enlivened by the presence of man. Neither the

---

[13] Lieut. [Silas?] Bent ["The Japanese Gulf Stream," *Bull. Amer. Geogr. and Stat. Soc.*, II(1857), 203–213, esp. p. 205]. Lieutenant Bent was in the Japan Expedition with Commodore Perry and used the opportunities thus afforded to study the phenomena of the stream.

industrial pursuits of the sea nor the highways of commerce called him into it. Now and then a roving cruiser or an enterprising whaleman passed that way; but to all else it was an unfrequented part of the ocean, and so remained until the gold-fields of Australia and the guano islands of Peru made it a thoroughfare. All vessels bound from Australia to South America now pass through it, and in the journals of some of them it is described as a region almost void of the signs of life in both sea and air. In the South Pacific Ocean especially, where there is such a wide expanse of water, sea-birds often exhibit a companionship with a vessel, and will follow and keep company with it through storm and calm for weeks together. Even those kinds, as the albatross and Cape pigeon, that delight in the stormy regions of Cape Horn and the inhospitable climates of the Antarctic regions, not unfrequently accompany vessels into the perpetual summer of the tropics. The sea-birds that join the ship as she clears Australia will, it is said, follow her to this region, and then disappear. Even the chirp of the stormy-petrel ceases to be heard here, and the sea itself is said to be singularly barren of life.

400. *Polynesian drift.* In the intertropical regions of the Pacific, and among the heated waters of Polynesia, a warm current or drift of immense volume has its genesis. It rather drifts than flows to the south, laving, as it goes, the eastern shore of Australia and both shores of New Zealand. These are the waters in which the little corallines delight to build their atolls and their reefs. The intertropical seas of the Pacific afford an immense surface for evaporation. No rivers empty there; the annual fall of rain there, except in the "Equatorial Doldrums," is small, and the evaporation is all that both the northeast and the southeast trade-winds can take up and carry off. I have marked on Plate VI the direction of the supposed warm-water current which conducts these overheated and briny waters from the tropics in mid ocean to the extratropical regions where precipitation is in excess. Here, being cooled, and agitated, and mixed up with waters that are less salt, these overheated and over-salted waters from the tropics are replenished and restored to their rounds in the wonderful system of oceanic navigation.

401. *Equatorial currents.* There are also about the equator in this ocean some curious currents, which I have called the "Doldrum Currents" of the Pacific, but which I do not understand, and as to which observations are not sufficient yet to afford the proper explanation or description. There are many of them, some of which, at times, run with great force. On a voyage from the Society to the Sandwich Islands I encountered one running at the rate of ninety-six miles a day. These currents are generally found setting to the west. They are often, but not always, encountered in the equatorial Doldrums on the voyage between the Society and the Sandwich Islands. In Captain Pichon's abstract log of the French corvette

"L'Eurydice," from Honolulu to Tahiti, in August, 1857, a "doldrum" current is recorded at 79 miles a day, west by north. He encountered it between 1° N. and 4° S., where it was 300 miles broad. On the voyage to Honolulu in July of the same year, he experienced no such current; but in 6° N. he encountered one of 36 miles, setting southeast, or nearly in the opposite direction. This current does not appear to have been more than 60 miles broad. What else should we expect in this ocean but a system of currents and counter-currents apparently the most uncertain and complicated? The Pacific Ocean and the Indian Ocean may, in the view we are about to take, be considered as one sheet of water. This sheet of water covers an area quite equal in extent to one half of that embraced by the whole surface of the earth; and, according to Professor Alexander Keith Johnston, who so states it in the new edition of his splendid Physical Atlas, the total annual fall of rain on the earth's surface is one hundred and eighty-six thousand, two hundred and forty cubic imperial miles. Not less than three fourths of the vapor which makes this rain comes from this waste of waters; but supposing that only half of this quantity, *i.e.*, ninety-three thousand, one hundred and twenty cubic miles of rain falls upon this sea, and that that much, at least, is taken up from it again as vapor, this would give two hundred and fifty-five cubic miles as the quantity of water which is daily lifted up and poured back again into this expanse. It is taken up at one place and rained down at another, and in this process, therefore, we have agencies for multitudes of partial and conflicting currents – all, in their set and strength, apparently as uncertain as the winds.

402. *The influence of rains and evaporation upon currents.* The better to appreciate the operation of such agencies in producing currents in the sea, now here, now there, first this way, and then that, let us, by way of illustration, imagine a district of two hundred and fifty-five square miles in extent to be set apart, in the midst of the Pacific Ocean, as the scene of operations for one day. We must now conceive a machine capable of pumping up, in the twenty-four hours, all the water to the depth of one mile in this district. The machine must not only pump up and bear off this immense quantity of water, but it must discharge it again into the sea on the same day, but at some other place. Now here is a force for creating currents that is equivalent in its results to the effects that would be produced by bailing up, in twenty-four hours, two hundred and fifty-five cubic miles of water from one part of the Pacific Ocean, and emptying it out again upon another part. The currents that would be created by such an operation would overwhelm navigation and desolate the sea; and, happily for the human race, the great atmospherical machine which actually does perform every day, on the average, all this lifting up, transporting, and letting down of water upon the face

of the grand ocean, does not confine itself to an area of two hundred and fifty-five square miles, but to an area three hundred thousand times as great; yet, nevertheless, the same quantity of water is kept in motion, and the currents, in the aggregate, transport as much water to restore the equilibrium as they would have to do were all the disturbance to take place upon our hypothetical area of one mile deep over the space of two hundred and fifty-five square miles. Now when we come to recollect that evaporation is lifting up, that the winds are transporting, and that the clouds are letting down every day actually such a body of water, we are reminded that it is done by little and little at a place, and by hair's breadths at a time, not by parallelopipedons one mile thick, and that the evaporation is most rapid and the rains most copious, not always at the same place, but now here, now there. We thus see actually existing in nature a force perhaps quite sufficient to give rise to just such a system of currents as that which mariners find in the Pacific (§ 401) — currents which appear to rise in mid ocean, run at unequal rates, sometimes east, sometimes west, but which always lose themselves where they rise, viz., in mid ocean.

403. *Under Currents — Parker's deep-sea sounding.* Lieutenant J. C. Walsh, in the U.S. schooner "Taney," and Lieutenant S. P. Lee, in the U.S. brig "Dolphin," both, while they were carrying on a system of observations in connection with the *Wind and Current Charts,* had their attention directed to the subject of submarine currents. They made some interesting experiments upon the subject. A block of wood was loaded to sinking, and, by means of a fishing-line or a bit of twine, let down to the depth of one hundred or five hundred fathoms, at the will of the experimenter. A small barrel as a float, just sufficient to keep the block from sinking farther, was then tied to the line, and the whole let go from the boat. To use their own expressions, "It was wonderful, indeed, to see this *barrega* move off, against wind, and sea, and surface current, at the rate of over one knot an hour, as was generally the case, and on one occasion as much as 1¾ knots. The men in the boat could not repress exclamations of surprise, for it really appeared as if some monster of the deep had hold of the weight below, and was walking off with it." [14] Both officers and men were amazed at the sight. The experiments in deep-sea soundings have also thrown much light upon the subject of under currents. There is reason to believe that they exist in all, or almost all parts of the deep sea, for never in any instance yet has the deep-sea line ceased to run out, even after the plummet had reached the bottom. If the line be held fast in the boat, it invariably parts, showing, when two or three miles of it are out, that the under currents are sweeping against the bight of it with

[14] "Lieutenant [J. C.] Walsh [to Lieutenant Maury," *Sailing Directions* (1851), pp. 62–69, esp. p. 64].

what seamen call a *swigging force*, that no sounding twine has yet proved strong enough to withstand. Lieutenant J. P. Parker, of the United States frigate Congress, attempted, in 1852, a deep-sea sounding off the coast of South America. He was engaged with the experiment eight or nine hours, during which time a line nearly ten miles long was paid out. Night coming on, he had to part the line (which he did simply by attempting to haul it in) and return on board. Examination proved that the ocean there, instead of being over ten miles in depth, was not over three, and that the line was swept out by the force of one or more under currents. But in what direction these currents were running is not known.

404. *The compressibility of water — effect of in the oceanic circulation.* Vertical circulation is as important in the sea as it is in the air (§ 231). In striving to understand the physical machinery of our planet and to comprehend its workings, we *must*, if we would learn, proceed upon the principle (§ 351) that at creation the waters were measured, the hills weighed, and the atmosphere meted out, and that each was endowed with its peculiar properties so proportioned and so adjusted as exactly to answer its purposes in the grand design. And, consequently, we are entitled to infer that fluidity instead of solidity was imparted to a certain quantity of matter which we call water, to enable it to perform the offices to be required of fluid matter, and which solid matter was not adapted to perform in the terrestrial economy. By this mode of reasoning we are taught to regard the fluidity of all the water in the sea as a physical necessity — and by this mode of reasoning we are required to reject as insufficient any hypothesis touching the system of aqueous circulation on our planet which ignores, even in the profoundest depths of the ocean, an interchange of its particles between the bottom and the top. Were such interchange not to take place — were the water in the sea which once sinks below the level of its horizontal circulation doomed to remain there forever, it would not be difficult to show that the sea would lose its balance and its counterpoises; that, not being able to preserve its status, the water at the bottom would have grown heavier and heavier, while that at the top would have become lighter and lighter, until the one became saturated with salt, the other entirely fresh. To prevent this state of things, we recognize the influences of the winds and tides, as well as the necessity of vertical movements in the sea. Whence, therefore, let us inquire, when a given quantity of water once finds its way to the bottom of the sea, whence — since it goes there by virtue of its own specific gravity — whence is power to be derived for bringing it up again? for sooner or later, according to this view, up it must come. We thus arrive precisely at one of those points (§ 287) at which hypothesis becomes absolutely necessary if we would make further progress. Here, therefore, let us pause to search among the physics of the sea for such a power and the foundation for hypothesis. Leslie has pointed out exactly such a

power for the atmospheric ocean — a power which, after the heaviest air has settled at the bottom of its subtile sea — after the lightest has come to rest at the top, and the whole arranged itself according to specific gravity — can haul that which is below to the top, and send that which is on the top down into the recesses and cavities below. Suppose the entire atmosphere to be, from the bottom to the top, nearly of the same temperature, and in a perfect state of quiescent equilibrium, and that from some cause a certain volume of air above has its specific gravity so changed that it commences to descend. As it descends the pressure upon it increases — and air, being compressed, contracts and gives out heat. A like volume ascends to take its place, and in ascending it expands and grows cool. Thus the total mass, and the total pressure, and the total amount of caloric remain the same; but there is a transfer of heat from the top to the bottom, by which the equilibrium of the mass is destroyed, and a force established at the bottom of the atmospherical ocean which, with the assistance of an agent at the top to alter specific gravity, is capable of sending up the heavy air from the bottom, of drawing down the light from the top, and of turning, in course of time, the whole atmosphere upside down. All philosophers acknowledge the power of this omnipresent agent in the air, and that, by alternately assuming the latent and the sensible form, it, to say the least, assists to give to the atmosphere the dynamical force required for its system of vertical circulation as well as its horizontal. So with water and the *salt* sea where we do have an agent that is continually altering specific gravity at the surface. Notwithstanding the Florentine experiment upon water in the gold ball, it has since been abundantly proved that water is compressible — so much so, that at the depth of ninety-three miles its density would be doubled. Consequently, a given quantity of water — such, for instance, as a cubic foot measured at the surface — would not, if sunk to the depth of four miles, measure a cubic foot by seventy-two cubic inches. As a rule, the compressibility of water in the depths of the sea is one per cent. for every 1000 fathoms. Here, then, in the latent heat which is liberated in the processes of descent, have we not a power which is capable of sending up to the top water from the uttermost depths of the sea? Suppose that this cubic measure of water, by supplying vapor to the winds at the surface, to have its saltness so increased as to alter its specific gravity to sinking: Like the air, it is compressed, and contracts in its descent, giving out heat, raising the temperature, and changing the specific gravity of like quantities in the various thermal strata through which it has to pass. Thus heat is conveyed from the top to the bottom of the sea, there to be liberated and impart to its waters dynamical force for their upward movement. This is the power we paused to search for: whatever be its amount it is in the nature of a *vera causa*, and we must therefore recognize it, if not as the sole agent, nevertheless as one of the principal agents which Nature employs in the system of vertical circulation that has been ordained for the waters of the sea.

405. *Assisted by its salts.* Now, but for the salts of the sea, this process could not go on so long as the laws of thermal dilatation remain as they are for sea water. Unlike fresh water, which expands as it is cooled below 39°.5, sea water contracts until it has passed its freezing point and attained the temperature of 25°.6.[15] Were it not for its salts, sea water once on the surface within the tropics would, by reason of its warmth and thermal dilatation, remain on the surface. Vertical circulation would be confined to polar seas, and many of the living creatures that inhabit its waters would perish for the lack of currents to convey them their food.

406. *The origin of currents.* If we except the tides, and the partial currents of the sea, such as those that may be created by the wind, we may lay it down as a rule (§ 103) that all the currents of the ocean owe their origin to difference of specific gravity between sea water at one place and sea water at another; for wherever there is such a difference, whether it be owing to difference of temperature or to difference of saltness, etc., it is a difference that disturbs equilibrium, and currents are the consequence. The heavier water goes toward the lighter, and the lighter whence the heavier comes; for two fluids differing in specific gravity (§ 106), and standing at the same level, can no more balance each other than unequal weights in opposite scales of a true balance. It is immaterial, as before stated, whether this difference of specific gravity be caused by temperature, by the matter held in solution, or by any other thing; the effect is the same, namely, a current. That the sea, in all parts, holds in solution the same kind of solid matter; that its waters in this place, where it never rains, are not salter than the strongest brine; and that in another place, where the rain is incessant, they are not entirely without salt, may be taken as evidence in proof of a system of currents or of circulation in the sea, by which its waters are shaken up and kept mixed together as though they were in a phial.

407. *Currents of the Atlantic.* The principal currents of the Atlantic have been described in the chapter on the Gulf Stream. Besides this, its eddies and its offsets, are the equatorial current (Plate III), and the St. Roque or Brazil Current. Their fountain-head is the same: it is in the warm waters about the equator, between Africa and America. The former, receiving the Amazon and the Orinoco as tributaries by the way, flows into the Caribbean Sea, and becomes, with the waters (§ 103) in which the vapors of the trade-winds leave their salts, the feeder of the Gulf Stream. The Brazil Current, coming from the same fountain, is supposed to be divided by Cape St. Roque, one branch going to the south under this name (Plate VI), the other to the westward. This last has been a great bugbear to navigators, principally on account of the

[15] See Prof. Hubbard's experiments, *Sailing Directions* [8th ed.], I [239–241].

difficulties which a few dull vessels falling to leeward of St. Roque have found in beating up against it. It was said to have caused the loss of some English transports in the last century, which fell to leeward of the Cape on a voyage to the other hemisphere; and navigators, accordingly, were advised to shun it as a danger.

408. *The St. Roque current.* This current has been an object of special investigation during my researches connected with the Wind and Current Charts, and the result has satisfied me that it is neither a dangerous nor a constant current, notwithstanding older writers. Horsburgh,[16] in his East India Directory, cautions navigators against it; and Keith Johnston, in his grand Physical Atlas, published in 1848, thus speaks of it: "This current greatly impedes the progress of those vessels which cross the equator west of 23° west longitude, impelling them beyond Cape St. Roque, when they are drawn toward the northern coast of Brazil, and can not regain their course till after weeks or months of delay and exertion." [17] So far from this being the case, my researches abundantly prove that vessels which cross the equator five hundred miles to the west of longitude 23° have no difficulty on account of this current in clearing that cape. I receive almost daily the abstract logs of vessels that cross the equator west of 30° west, and in three days from that crossing they are generally clear of that cape. A few of them report the current in their favor; most of them experience no current at all; but, now and then, some do find a current setting to the northward and westward, and operating against them at the rate of twenty miles a day. The intertropical regions of the Atlantic, like those of the other oceans (§ 401), abound with conflicting currents, which no researches yet have enabled the mariner to unravel so that he may at all times know where they are and tell how they run, in order that he may be certain of their help when favorable, or sure of avoiding them if adverse.

409. *The Greenland Current.* There are other currents, such as the Greenland Current, the cold current from Davis' Strait, the ice-bearing current from the Antarctic regions, all setting into the Atlantic and the Gulf Stream, one branch of which finds its way into the Arctic Sea; the other (§ 89) finds its way back to the south partly as Rennell's current, all of which have been well treated of in Chap. II, or are delineated on Plates III and VI. Judging by these, there would seem to be a larger flow of polar waters into the Atlantic than of other waters from it, and I can not account for the preservation of the equilibrium of this ocean by any other hypothesis than that which calls in the aid of under currents. They, I have no doubt, bear an important part in the system of oceanic circulation.

[16] [James Horsburgh, *The India Directory* . . ., I(London, 1852), 52.]
[17] [*Physical Atlas*, "Hydrology," text to Chart no. 1, p. 4.]

CHAPTER IX

§ 420–460. THE SPECIFIC GRAVITY OF THE SEA, AND THE OPEN WATER IN THE ARCTIC OCEAN

420. *Interesting physical inquiries.* The crust of the planet upon which we live, with the forces that have been and are at work upon it, is the most interesting subject of physical inquiry and study that can claim the attention of diligent students. Precisely as the progress of man has been upward and onward, precisely has he looked more earnestly and with deeper longings toward the mysteries that encircle this crust. It is but a shell, and at most we can reach only a little way either above or below its very surface, and yet upon the tablets of this thin shell are the records of all that he may ever know concerning this his cosmical hearthstone.

421. *Voyages of discovery to the North Pole.* Researches have been carried on from the bottom of the deepest pit to the top of the highest mountain, but these have not satisfied. Voyages of discovery, with their fascinations and their charms, have led many a noble champion both into the torrid and frigid zones; and notwithstanding the hardships, sufferings, and disasters to which many northern parties have found themselves exposed, seafaring men, as science has advanced, have looked with deeper and deeper longings toward the mystic circles of the polar regions. There icebergs are framed and glaciers launched. There the tides have their cradle, the whales their nursery. There the winds complete their circuits, and the currents of the sea their round in the wonderful system of oceanic circulation. There the Aurora Borealis is lighted up and the trembling needle brought to rest; and there too, in the mazes of that mystic circle, terrestrial forces of occult power and of vast influence upon the well-being of man are continually at play. Within the arctic circle is the pole of the winds and the poles of the cold, the pole of the earth and of the magnet. It is a circle of mysteries; and the desire to enter it, to explore its untrodden wastes and secret chambers, and to study its physical aspects, has grown into a longing. Noble daring has made Arctic ice and waters classic ground. It is no feverish excitement nor vain ambition that leads man there. It is a higher feeling, a holier motive — a desire to look into the works of creation, to comprehend the economy of our planet, and to grow wiser and better by the knowledge. Soon after the discovery of America, John Cabot and his sons, with five ships, sailed upon the first Arctic expedition. Between that year and the

present no less than 155 vessels, besides boat and land parties, have at various periods, and with divers objects in view, been sent from Europe and America up into those inhospitable regions. Whatever may have been the immediate object of these various expeditions, whether to enlarge the fields of commerce, to carry the Bible, to spread civilization, to push conquests, or to bring back contributions to science, it has never lost sight of the promise made by Columbus of a western route to India.

422. *The first suggestions of an open sea in the Arctic Ocean.* Like the air, like the body, the ocean *must* have a system of circulation for its waters. No other hypothesis will explain the fact which observations reveal concerning the saltness of the sea, the constituents of sea water, and many other phenomena. An attentive study of the currents of the sea, and a close examination of the laws which govern the movements of the waters in their channels of circulation through the ocean, will lead any one irresistibly to the conclusion that always, in summer and winter, there must be, somewhere within the arctic circle, a large body of open water. This open water must impress a curious feature upon the physical aspects of those regions. The whales had taught us to suspect the existence of open water in the arctic basin, and in their mute way told of a passage there, at least sometimes. It is the custom among whalers to have their harpoons marked with date and the name of the ship; and Dr. Scoresby, in his work on arctic voyages, mentions several instances of whales that have been taken near the Behring's Strait side with harpoons in them bearing the stamp of ships that were known to cruise on the Baffin's Bay Side of the American continent; and as, in one or two instances, a very short time had elapsed between the date of capture in the Pacific and the date when the fish must have been struck on the Atlantic side, it was argued therefore that there was a northwest passage by which the whales passed from one side to the other, since the stricken animal could not have had the harpoon in him long enough to admit of a passage around either Cape Horn or the Cape of Good Hope.

423. *Harpoons — habits of the whales.* The whale-fishing is, among the industrial pursuits of the sea, one of no little importance; and when the system of investigation out of which the "Wind and Current Charts" have grown was commenced, the haunts of this animal did not escape attentive examination. The log-books of whalers were collected in great numbers, and patiently examined, co-ordinated, and discussed, in order to find out what parts of the ocean are frequented by this kind of whale, what parts by that, and what parts by neither. (See Plate VI.) Log-books containing the records by different ships for hundreds of thousands of days were examined, and the observations in them co-ordinated for this chart. And this investigation, as Plate VI shows, led to the discovery that the tropical regions of the ocean are to the right whale as a sea of fire,

through which he can not pass, and into which he never enters. The fact was also brought out that the same kind of whale that is found off the shores of Greenland, in Baffin's Bay, etc., is found also in the North Pacific, and about Behring's Strait, and that the right whale of the northern hemisphere is a different animal from that of the southern. Thus the fact was established that the harpooned whales did not pass around Cape Horn or the Cape of Good Hope, for they were of the class that could not cross the equator. In this way we were furnished with circumstantial evidence affording the most irrefragable proof that there is, at times at least, open water communication through the Arctic Sea from one side of the continent to the other, for it is known that the whales can not travel under the ice for such a great distance as is that from one side of this continent to the other. But this did not prove the existence of an open sea there; it only established the existence — the occasional existence, if you please — of a channel through which whales had passed. Therefore we felt bound to introduce other evidence before we could expect the reader to admit our proof, and to believe with us in the existence of an open sea in the Arctic Ocean.

424. *The under current into the Arctic Ocean — its influences.* There is an under current setting from the Atlantic through Davis' Strait into the Arctic Ocean, and there is a surface current setting out. Observations have pointed out the existence of this under current there, for navigators tell of immense icebergs which they have seen drifting rapidly to the north, and against a strong surface current. These icebergs were high above the water, and their depth below, supposing them to be parallelopipeds, was at least seven times greater than their height above. No doubt they were drifted by a powerful under current. Now this under current comes from the south, where it is warm, and the temperature of its waters is perhaps not below 30°; at any rate, they are comparatively warm. There must be a place somewhere in the Arctic seas where this under current ceases to flow north, and begins to flow south as a surface current; for the surface current, though its waters are mixed with the fresh waters of the rivers and of precipitation in the polar basin, nevertheless bears out vast quantities of salt, which is furnished neither by the rivers nor the rains. These salts are supplied by the under current; for as much salt as one current brings in, other currents must take out, else the polar basin would become a basin of salt; and where the under current transfers its waters to the surface, there is, it is supposed, a basin in which the waters, as they rise to the surface, are at 30°, or whatever be the temperature of the under current, which we know must be above the freezing point, for the current is of water in a fluid, not in a solid state. An arrangement in nature, by which a basin of considerable area in the frozen ocean could be supplied by water coming in at the bottom and

rising up at the top, with a temperature not below 30°, or even 27°.2 — the freezing point of sea water — would go far to mitigate the climate in the regions round about.

425. *Indications of a milder climate.* And that there is a warmer climate somewhere in the inhospitable sea, the observations of many of the explorers who have visited it indicate. Its existence may be inferred also from the well-known fact that the birds and animals are found at certain seasons migrating to the north, evidently in search of milder climates. The instincts of these dumb creatures are unerring, and we can imagine no mitigation of the climate in that direction, unless it arise from the proximity, or the presence there of a large body of open water. It is another furnace (§ 151) in the beautiful economy of Nature for tempering climates there.

426. *How the littoral waters, by being diluted from the rivers and the rains, serve as a mantle for the salter and warmer sea water below.* The hydrographic basin of the Arctic Ocean is large, and it delivers into that sea annually a very copious drainage. Such an immense volume of fresh water discharged into so small a sea as the Arctic Ocean is must go far toward diluting its brine. Figure 3, *b* (§ 433), shows the extent to which the brine of our littoral seas is diluted by the drainage from the Atlantic slopes of the United States. It will be observed by that figure that suddenly after crossing the parallel of 34° N. the water begins to grow *cooler* and *lighter*. The observations for these two curves are a part of the celebrated series made by Captain Rodgers in the U.S. ship Vincennes all the way from Behring's Straits by the way of Cape Horn to New York. He cleared the inner edge of the Gulf Stream in 34°, where the waters began to grow cooler and lighter, and so continued to do as he approached the shore. The remarkable and sudden approach of the thermal and specific gravity curves after crossing 34° N. can be explained by no other hypothesis than this, viz.: the surface water of the sea was so diluted with the fresh water from the Chesapeake, the Delaware, and New York Bays, that, notwithstanding the temperature decreased as Rodgers approached the shore, yet the specific gravity *decreased* also, because the saltness decreased by reason of the increasing proportion of river water as he neared the shore. And thus we have in our own waters an illustration and an example of how cool and light — because not so salt — water may be made to cover and protect as with a mantle a sheet of warmer, but salter and heavier water below.

427. *An under current of warm but salt and heavy water.* The mean specific gravity of the Arctic Ocean water as observed by Rodgers, and reduced to the freezing point (27°.2) of sea water, was 1.0263. The specific gravity of the Gulf Stream water, as observed by him, and reduced to the same temperature (27°.2), was 1.0303. If these be taken

as fair specimens of the water of the torrid and frigid zones, it would appear that the waters of intertropical seas have 15 per cent. more salt in them than the surface water of the Arctic Ocean has. It is to be regretted that the hydrometer has not been more frequently used in the Arctic Ocean, for a careful series of observations upon the specific gravity of the water there at the surface and at various depths would indicate to us not only the extent to which the water there is diluted by the rivers and the rains, but it would yield other highly interesting results. Now this salt and heavy water, whose specific gravity at 27°.2 would have been 1.0303, is the very water which Rodgers observed in the Gulf Stream on its way to the arctic regions. This is the water which, after passing the Grand Banks, and meeting the diluted water as an ice-bearing current from the north, dips down, but continues its course as an under current. It is protected from farther loss of heat, after the manner of our own littoral waters, by the colder but lighter current from the north, until it enters the Arctic Ocean, there to rise up like a boiling spring in the centre of an open sea.

428. *De Haven's water sky.* Relying upon a process of reasoning like this, and the deductions flowing therefrom, Lieutenant De Haven, when he went in command of the American expedition in search of Sir John Franklin and his companions, was told, in his letter of instructions, to look, when he should get well up into Wellington Channel, for an open sea to the northward and westward. He looked, and saw in that direction a "water sky." Captain Penny afterward went there, found open water, and sailed upon it. The open sea in the Arctic Ocean is probably not always in the same place, as the Gulf Stream (§ 126) is not always in one place. It probably is always where the waters of the under currents are brought to the surface; and this, we may imagine, would depend upon the freedom of ingress for the under current. Its course may perhaps be modified more or less by the ice on the surface, by changes, from whatever cause, in the course or velocity of the surface current, for obviously the under current could not bring more water into the frozen ocean than the surface current would carry out again, either as ice or water. Every winter, an example of how very close warm water in the sea and a very severe climate on the land or the ice may be to each other is afforded to us in the case of the Gulf Stream and the Labrador-like climate of New England, Nova Scotia, and Newfoundland. In these countries, in winter, the thermometer frequently sinks far below zero, notwithstanding that the tepid waters of the Gulf Stream may be found with their summer temperature within one day's sail of these very, very cold places.

429. *Dr. Kane.* Dr. Kane reports an open sea north of the parallel of 82°. To reach it, his party crossed a barrier of ice 80 or 100 miles broad.

Before gaining this open water, he found the thermometer to show the extreme temperature of −60°. Passing this ice-bound region by traveling north, he stood on the shores of an iceless sea, extending in an unbroken sheet of water as far as the eye could reach toward the pole. Its waves were dashing on the beach with the swell of a boundless ocean. The tides ebbed and flowed in it, and I apprehend that the tidal wave from the Atlantic can no more pass under this icy barrier to be propagated in seas beyond, than the vibrations of a musical string can pass with its notes a fret upon which the musician has placed his finger. The swell of the sea can not pass wide fields or extensive barriers of ice; for De Haven, during his long imprisonment and drift (§ 474), found the ice so firm that he observed regularly from an artificial horizon placed upon it, and found the mercury always "perfectly steady." These tides, therefore, must have been born in that cold sea, having their cradle about the North Pole. If these statements and deductions be correct, then we infer that most, if not all the unexplored regions about the pole are covered with deep water; for, were this unexplored area mostly land or shallow water, it could not give birth to regular tides. Indeed, the existence of these tides, with the immense flow and drift which annually take place from the polar seas into the Atlantic, suggests many conjectures concerning the condition of these unexplored regions. Whalemen have always been puzzled as to the place of breeding for the right whale. It is a cold-water animal, and, following up this train of thought, the question is prompted, Is not the nursery for the great whale in this polar sea, which has been so set about and hemmed in with a hedge of ice that man may not trespass there? This providential economy is still farther suggestive, prompting us to ask, Whence comes the food for the young whales there? Do the teeming waters of the Gulf Stream (§ 160) convey it there also, and in channels so far down in the depths of the sea that no enemy may waylay and spoil it on the long journey? Seals were sporting and water-fowl feeding in this open sea of Dr. Kane's. Its waves came rolling in at his feet, and dashing with measured tread, like the majestic billows of old ocean, against the shore. Solitude, the cold and boundless expanse, and the mysterious heavings of its green waters, lent their charm to the scene. They suggested fancied myths, and kindled in the ardent imagination of the daring mariners many longings. The temperature of its waters was only 36°! Such warm water could get there from the south only as a current far down in the depths below. The bottom of the ice of this eighty miles of barrier was no doubt many — perhaps hundreds of — feet below the surface level. Under this ice there was also doubtless water above the freezing point.

430. *Under currents change temperature slowly.* Nor need the presence of warm water within the arctic circle excite surprise, when we

recollect that the cold waters of the frigid zone are transferred to the torrid without changing their temperature perhaps more than 7° or 8° by the way. This transfer of cold waters for a part of the way may take place on the surface, and until the polar flow (§ 89) dips down and becomes submarine. At any rate, officers on the Coast Survey have found water at the bottom of the Gulf Stream, in latitude 25° 30′ N., as low in temperature as 35°. Now, if water flowing out of the polar basin at the temperature of 28° may, by passing along the secret paths of the sea, reach the Gulf of Mexico in summer at a temperature of only 3° above the freezing point of fresh water, why may not water, leaving the torrid zone at a temperature of 82°, and traveling by the same hidden ways, reach the frigid zone without losing more than the cold currents gained in temperature, viz., 7°? In 1840, Sir James C. Ross, being in the antarctic regions with the surface water at 32°, found the temperature in depth to be 38°.8 at 400 fathoms, and 39°.8 at 600. At a greater depth there is a greater pressure; and there ought to be (§ 404) a stated temperature, that after passing a certain depth in the deep sea grows higher and higher as the depth increases. The thermal laws of "deep-sea" temperatures for fresh and for salt water are very different. In September, when the surface water of Loch Lomond and Loch Katrine — Scottish lakes — which are between 500 and 600 feet deep, is 58°, that at the bottom is uniformly 41°, which is very near the point of maximum density for fresh water. Saussure has shown the same for the Italian lakes; only, at the depth of 1000 feet in the Lake of Geneva, it was a little warmer, probably on account of pressure (§ 404), than it was at less depth in Lakes Lucerne and Thun. In these it was 41°, or 1° colder than the bottom of Geneva, their surface water being about 60°. In Lago Sabatino, near Rome, with the surface water at 77°, Barlocci reports 44° at the depth of 490 feet. The winter in Rome is not severe enough to cool such a mass of water below 44°. But with the exception of the Lake of Geneva, which is deep enough to have the temperature of its water somewhat influenced by pressure (§ 404), the law is uniform; as you descend in fresh-water lakes, the temperature decreases to that of maximum density. Saussure extended his experiments to the Gulfs of Nice and Genoa — salt-water bays in the neighborhood of his fresh-water lakes. Here, with the surface temperature of 69°, he found even at the depth of 1720 feet, the water no cooler than 55°.8. This salt water might have been cooled 30° lower before it would have reached the maximum density (25°.6) of average sea water. We see that the severest winters are not sufficient to bridge our deep fresh-water lakes over with ice, though their waters, being cooled below 39°.5, grow light, and remain on the surface to be frozen. On the contrary, sea water contracts, grows heavy, and sinks, until the whole basin, from the bottom to the top, be reduced to 27°.2. Yet many confess no surprise at the open water in fresh-water lakes that are comparatively shallow, while they can conceive of no such thing in the Arctic Ocean, though it be very much deeper than the deepest fresh-water lakes!

431. *Solid matter annually drifted out of the polar basin.* At the very time that the doctor was gazing with longing eyes upon these strange green waters (§ 429), there is known to have been a powerful drift setting out from another part of this Polar Sea, and carrying with it from its mooring the English exploring ship Resolute, which her officers and men had abandoned fast bound in the ice several winters before. This drift carried a field of ice that covered an area not less than 300,000 square miles, through a distance of a thousand miles to the south. The drift of this ship was a repetition of De Haven's celebrated drift (§ 474); for in each case the ice in which the vessel was fastened floated out and carried the vessel along with it; by which I mean to be understood as wishing to convey the idea that the vessel was not drifted through a line or an opening in the ice, but, remaining fast in the ice, she was carried along with the whole icy field or waste. This field of ice averaged a thickness of not less than seven feet; at least that was the case with De Haven. A field of ice covering to the depth of seven feet an area of 300,000 square miles, would weigh not less than 18,000,000,000 tons. This, then, is the quantity of *solid* matter that is drifted out of the polar seas through one opening — Davis' Straits — alone, and during a part of the year only. The quantity of water which was required to float and drive this solid matter out was probably many times greater than this. A quantity of water equal in weight to these two masses had to go in. The basin to receive these inflowing waters, *i.e.*, the unexplored basin about the North Pole, includes an area of a million and a half square miles; and, as the outflowing ice and water are at the surface, the return current must be submarine. A part of the water that it bears probably flows in beneath Dr. Kane's barrier of ice (§ 429).

432. *Volume of water kept in motion by the arctic flow and reflow.* These two currents, therefore, it may be perceived, keep in motion between the temperate and polar regions of the earth a volume of water, in comparison with which the mighty Mississippi, in its greatest floods, sinks down to a mere rill. On the borders of this ice-bound sea Dr. Kane found subsistence for his party — another proof of the high temperature and comparative mildness of its climate.

433. *The hydrometer at sea.* The Brussels Conference recommended the systematic use of the hydrometer at sea. Captain Rodgers, Lieutenant Porter, and Dr. Ruschenberger, all of the United States Navy, with Dr. Raymond, in the American steamer Golden Age, and Captain Toynbee, of the English East Indiaman the Gloriana, have all returned to me valu-

able observations with this instrument. Rodgers, however, has afforded
the most extended series. It embraces 128° of latitude, extending from
71° in one hemisphere to 57° in the other. And here I beg to remark, that
those navigators who use the hydrometer systematically and carefully at
sea are quietly enlarging for us the bounds of knowledge and our field of
research. These observations have already led to the discovery of new and
beneficent relations in the workshops of the sea. In the physical machinery
of the universe there is no compensation to be found that is more exquisite or beautiful than that which, by means of this little instrument, has
been discovered in the sea between its salts, the air, and the sun. The
observations made with it by Captain Rodgers, on board the U.S. ship
Vincennes, have shown that the specific gravity of sea water varies but little in the trade-wind regions, notwithstanding the change of temperature.
The temperature was a little greater in the southeast trade-wind region
of the Pacific; less in the Atlantic. But, though the sea at the equatorial
borders of the trade-wind belt is some 20° or 25° warmer than it is on the
polar edge, yet the specific gravity of its waters at the two places in the
Atlantic differs but little. Though the temperature of the water was noted,
his observations on its specific gravity have not been corrected for temperature. The object which the Brussels Conference had in view when the
specific gravity column was introduced into the sea-journal was, that
hydrographers might find in it data for computing the dynamical force
which the sea derives for its currents from the difference in the specific
gravity of its waters in different climes. The Conference held, and rightly
held, that a given difference as to specific gravity between the water in
one part of the sea and the water in another would give rise to certain
currents, and that the set and strength of these currents would be the
same, whether such difference of specific gravity arose from difference of
temperature or difference of saltness, or both.

434. *Specific gravity of average sea water.* According to Rodgers'
observations, the average specific gravity of sea water, as it is taken from
the sea on the parallels of 34° north and south, at a mean temperature of
64°, is just what, according to thermal laws, it ought to be; but its specific gravity when taken from the equator, at a mean temperature of 81°,
is much greater than, according to the same laws, it ought to be. The observed difference of its specific gravity at 64° and 81° is .0015; whereas,
according to thermal laws, it ought to be .0025, or 67 per cent. greater
than it actually is. What makes this difference? Let us inquire:

435. *An Anomaly.* The anomaly is in the trade-wind region, and is best
developed (Figure 3, *b*) in the North Atlantic, between the parallel of
40° and the equator. Though it is sufficiently apparent both in the North
and South Pacific (Figure 3, *a*) — it is marked by the Gulf Stream in
the North Atlantic — commencing at the polar borders of these winds,

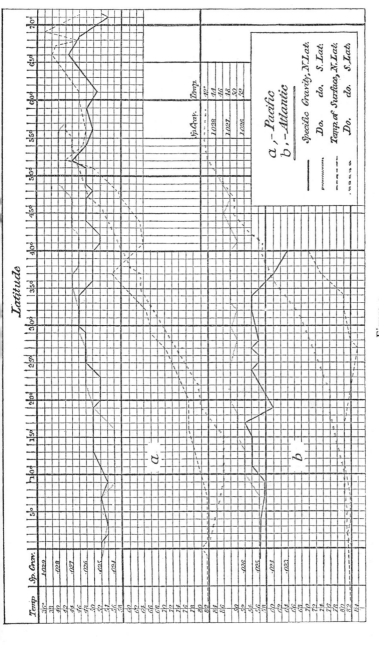

Figure 3

[Curves of specific gravity and temperature of the surface waters of the ocean, as observed by Captain John Rodgers in the U. S. ship *Vincennes*, on a voyage from Behring's Strait *via* California and Cape Horn to New York.]

the anomaly is developed as you approach the equator. The water grows warmer, but not proportionably lighter. This is the trade-wind region. These winds evaporate as they go; but can it be possible that they are so regulated and adjusted, counterpoised and balanced, that the salt which they, by evaporation, leave behind, is just sufficient to counterbalance the dilatation due the increasing warmth of the sea? It is even so.

436. *Influence of the trade-winds upon the specific gravity of sea water.* It is the trade-winds, then, which prevent the thermal and specific gravity curves from conforming with each other in intertropical seas. The water they suck up is fresh water, and the salt it contained, being left behind, is just sufficient to counterbalance, by its weight, the effect of thermal dilatation upon the specific gravity of sea water between the parallels of 34° north and south. As we go from 34° to the equator, the water grows warm and expands. It would become lighter, but the trade-winds, by taking up vapor without salt, make the water salter, and therefore heavier. The conclusion is, the proportion of salt in sea water, its expansibility between 62° and 82° (for its thermal *dilatability* varies with its temperature), and the thirst of the trade-winds for vapor are, where they blow, so balanced as to produce perfect compensation; and a more beautiful compensation can not, it appears to me, be found in the mechanism of the universe than that which we have here stumbled upon. It is a triple adjustment: the power of the sun to expand, the power of the winds to evaporate, and the quantity of salts in the sea — these are so proportioned and adjusted that when both the wind and the sun have each played with its forces upon the intertropical waters of the ocean, the residuum of heat and of salt should be just such as to balance each other in their effects, and to preserve the aqueous equilibrium of the torrid zone.

437. *Compensating influences.* Nor are these the only adjustments effected by this exquisite combination of compensations. If all the intertropical heat of the sun were to pass into the seas upon which it falls, simply raising the temperature of their waters, it would create a thermodynamical force in the ocean capable of transporting water scalding hot from the torrid zone, and spreading it, while still in the tepid state, around the poles. The annual evaporation from the trade-wind region of the ocean has been computed, according to the most reliable observations (§ 300), to be as much as 16 feet, which is at the rate of half an inch per day. The heat required for this evaporation would raise from the normal temperature of intertropical seas to the boiling point a layer of water covering the entire ocean to the depth of more than 100 feet. Such increase of temperature, by the consequent change which it would produce upon the specific gravity of the sea, would still further augment its dynamical power, until, even in the Atlantic, there would be force

enough to put in motion and feed with boiling-hot water many Gulf Streams. But the trade-winds and the seas are so adjusted that this heat, instead of penetrating into the depths of the ocean to raise the temperature of its waters, is sent off by radiation or taken up by the vapor, carried off by the winds, and dispensed by the clouds in the upper air of distant lands. Nor does this exquisite system of checks and balances, compensations and adjustments, end here. In equatorial seas the waters are dark blue, in extra-tropical they are green. This difference of color bears upon their heat-absorbing properties, and it comes in as a make-weight in the system of oceanic climatology, circulation, and stability. Now, suppose there were no trade-winds to evaporate and to counteract the dynamical force of the sun; this hot and light water, by becoming hotter and lighter, would flow off in currents, with almost mill-tail velocity, toward the poles, covering the intervening sea with a mantle of warmth as with a garment. The cool and heavy water of the polar basin, coming out in under currents, would flow equatorially with equal velocity. How much, if to any extent, the former warm climates of the British Islands and Northern Asia may be due to such a warm covering of the sea, may perhaps, at some future time, be considered worthy of special inquiry. We have already seen (§ 434) that there is something else besides temperature that is at work in effecting changes in the specific gravity of sea water. Whatever increases or diminishes its saltness, increases or diminishes its specific gravity; and the agents that are at work in the sea doing this are sea shells, the rivers, and the rains, as well as the winds. Between 35° or 40° and the equator, evaporation is in excess of precipitation; at any rate, there is but little precipitation except under the equatorial cloud-ring (see Storm and Rain Chart, Plate IX); and though, as we approach the equator on either side from these parallels, the solar ray warms and expands the surface water of the sea, the winds, by the vapor they carry off and the salt they leave behind, prevent it from making that water lighter.

438. *Nicely adjusted.* Thus two antagonistic forces are unmasked, and, being unmasked, we discover in them a most exquisite adjustment — a compensation — by which the dynamical forces that reside in the sunbeam and the trade-wind are made to counterbalance each other; by which the climates of intertropical seas are regulated; and by which the set, force, and volume of oceanic currents are measured. This compensation is most beautiful; it explains the paradox (§ 434), gives volume to the harmonies of the sea, and makes them louder in their songs of Almighty praise than the noise of many waters. Philosophers have admired the relations between the size of the earth, the force of gravity, and the strength of fibre in the flower-stalk of plants (§ 303), but how much more exquisite is the system of counterpoises and adjustments here presented between the sea and its salts, the winds and the heat of the sun! The

capacity of the sun to warm, of the sea water to expand, the quantity of salts these contain, and the power of the wind to suck up vapor, are all in such nice adjustment the one with the other, that there is the most perfect compensation. By it they make music in the sea, and the harmony that comes pealing thence, though not of so lofty a strain, is nevertheless, like the song of the stars, divine.

439. *A thermal tide.* Suppose there were no winds to suck up fresh water from the brine of the ocean; that its average depth were 3000 fathoms; that the solar ray were endowed with power to penetrate with its heat from the top to the bottom; and that, from bottom to top, the seas of each hemisphere, in thermal alternation with the seasons, were raised to summer heat and lowered to winter temperature: the change of sea level from summer to winter, and from winter to summer, in one hemisphere, would, from this cause alone, be upward of 125 feet; and in its rise and fall we should have, from pole to pole, the ebb and flow of a great thermal tide that would turn with the sun in the ecliptic, and tell the seasons by the march on the tide staff of its rising and falling waters. But difference of level would not be all that would give strength and volume to this tide; difference of specific gravity would lend its weight as so much dynamical force, which difference would create an upper and under annual tide from one hemisphere to the other. This double disturbance of equilibrium would not give rise to a tidal wave — mere motion without translation — but to a tidal flow and reflow of water from one hemisphere to the other in volumes of vast magnitude, power, and majesty. This is an exaggerated view of the dynamical force of the sunbeam; but it is presented to show the origin of the thermal tide shown in Figure 10, § 723. The difference between the actual and the supposed thermal tides is one of degree merely; for the sea water that is liable to any considerable change of temperature, instead of reaching from the bottom to the top, is scarcely more than a "pellicle" to the ocean. Nevertheless, it is a regular periodical flow and reflow between the poles and the equator. It is the annual ebb of this tide which fills the upper half of the North Atlantic with icebergs every spring and summer. The heated portion forms a stratum or layer which is thickest at the equator, and which comes to the surface near the polar edge of the temperate zones; it then dips again as it recedes toward the region of perpetual winter.

440. *The isothermal floor of the ocean.* The observations of Kotzebue, Admiral Beechey, and Sir James C. Ross first suggested the existence in the ocean of this isothermal floor. Its temperature, according to Kotzebue, is 36°. The depth of this bed of water of invariable and uniform temperature is computed to be 1200 fathoms at the equator. It gradually rises thence to the parallel of about 56° N. and S., where it crops out, and there the temperature of the sea, from top to bottom, is conjectured to

be permanently at 36°. The place of this outcrop, no doubt, shifts with the seasons, vibrating up and down, *i.e.*, north and south, after the manner of the calm belts. Proceeding, in our description, onward to the frigid zones, this aqueous stratum of an unchanging temperature dips again, and continues to incline till it reaches the poles at the depth of 740 fathoms. So that on the equatorial side of the outcrop the water above this floor is the warmer, but on the polar side the supernatant water is the colder. By this floor, with its waters of one uniform and permanent temperature, "the ocean," says Sir John Herschel, "is divided into three great regions — two polar basins in which the surface temperature is below, and one medial zone in which it is above 39°.5, being 80° at the equator;[1] and at the poles, of course, the freezing point of sea water. It will be very readily understood that in this statement there is nothing repugnant to hydrostatical laws, the compressibility of water insuring an increase of density in descending within much wider limits of temperature than here contemplated."

441. *Thermal dilatation of the water.* The temperature of 39°.5 was assigned to this floor probably under the supposition that sea water follows fresh in its laws of thermal dilatation. Not so: while fresh water attains its maximum density at 39°.5, average sea water does not arrive at its degree of maximum density until it passes its freezing point (27°.2) and reaches the temperature of 25°.6. In the winter of 1858 a very elaborate series of observations was conducted at the National Observatory, by Professor Hubbard, upon the thermal dilatation of sea water, and with the results [shown in Table 7], 60° being the standard temperature.

Table 7

Thermal Dilatation of Sea Water [2]

Temp.	Dilatation	Temp.	Dilatation	Temp.	Dilatation	Temp.	Dilatation
°		°		°		°	
22	0.99807	32	0.99795	50	0.99895	110	0950
23	801	33	797	55	943	120	1218
24	798	34	800	60	1.00000	130	1506
25	795	35	803	65	067	140	1804
26	793	36	806	70	142	150	2118
27	792	37	810	75	221	160	2460
28	791	38	814	80	309	170	2823
29	791	39	819	85	402	180	3192
30	792	40	.99823	90	503	190	3588
31	793	45	0.99856	100	0716	200	1.03993

[1] This remark was made by Sir John on the supposition, probably, that the maximum density of sea water was at the same temperature as that of fresh, but it is some 12° or 14° lower.

[2] This agrees more nearly with Despretz [cited in Pouillet, *Éléments de physique*, p. 251] than with Dr. Marcet ["On the Specific Gravity . . . of Sea Waters . . .," *Philos. Trans.*, CIX, 188]. The latter states that sea water decreases in weight to the

442. *Experiments on the freezing point.* The dilatation of the glass tube is included in this table. To determine the freezing point of average sea water, I filled a glass jar 18 inches high, and 3 inches in diameter, with specimens of average sea water obtained in mid-ocean and near the equator. On the 12th of February, 1858, the thermometer in the shade being 23°, I exposed this jar of water, with a standard thermometer immersed, to the out-door temperature. When the thermometer in the jar reached 27°, small crystals of ice, like macles of snow, were observed to form near the bottom, to rise, and to increase as they rose. In truth, the phenomenon presented most beautifully in miniature a snow-storm reversed, for the flakes appeared literally to "fall upward;" and while it was "snowing up" in the jar, covering the top with ice, the water in it rose in temperature from 27°.2 to 28°, thus showing the maximum density of the water to be not above 27°.2. As soon, and invariably as soon, as the first crystals of ice began to appear, the water immediately rose to 28°, and there remained as long as the process of congelation was going on. In some instances the water was brought down, as in a confined vessel, to 18° before freezing; but as soon as freezing commenced, the thermometer would mount up to 28°. The same water was used for the series of observations upon the thermal changes of the specific gravity of sea water [presented in Table 8], fresh water being the unit.

Table 8

[*Thermal changes of the specific gravity of sea water*]

Temperature	27°.1	Spec. grav.	1.0290	Temperature	38°.0	Spec. grav.	1.0287
"	28°.3	"	89	"	43°.5	"	86
"	28°.8	"	91	"	54°.7	"	775
"	29°.0	"	885	"	55°.5	"	77
"	29°.5	"	906	"	62°.5	"	69
"	30°.0	"	885	"	63°.5	"	675
"	32°.0	"	88	"	64°.5	"	665
"	34°.0	"	88	"	80°.5	"	43
"	34°.4	"	89	"	88°.3	"	30
"	35°.2	"	89	"	°93°.3	"	1.0221

° Specific gravity at 200° = 0.9908.

freezing point until actually congealed. In four experiments Dr. Marcet cooled sea water down to between 18° and 19° Fahr., and found that it decreased in bulk till it reached 22°, after which it expanded a little, and continued to do so till the fluid was reduced to between 19° and 18°, when it suddenly expanded, and became ice with a temperature of 28°. It should always be recollected that a saturated solution of common salt does not become solid, or converted into ice, at a less temperature than 4° Fahr.; and, therefore, if the sea should be, as is sometimes supposed, more saline at great depths, and as it appears to be in the Mediterranean from the experiments of Dr. Wollaston, ice could not be formed there at the same temperature as it could nearer the surface. — (*Vide* [Sir Henry Thomas] de la Beche [*A Geological*] *Manual* [Philadelphia, 1832], p. 22.)

443. *Sea water at summer more expansible than sea water at winter temperature.* All these experiments unite in showing that sea water at equatorial temperatures is many times more expansible than sea water at polar temperatures; that is, sea water, according to its rate of dilatation (§ 441), will expand about seventeen times as much for 5°, when its temperature is raised from 85°, as it will when raised from 28°; and yet, according to Figure 3, the curves of temperature and specific gravity are symmetrical in polar, non-symmetrical in equatorial seas. These experiments, and the compressibility of sea water (§ 404), show that we have not yet data sufficient to establish the depth, or even the existence of such an isothermal floor (§ 440) all the way from pole to pole.

444. *Data for Figure 3.* "The physical consequences of this great law, should it be found completely verified by farther research, are in the last degree important." The observations which furnished the data for Figure 3, *a*, were made in the North Pacific between the months of August, 1855, and April, 1856, and in the South Pacific during April and May; whereas for Figure 3, *b* the southern observations were made in May and June, the northern in June and July.

445. *A thermal tide: it ebbs and flows once a year.* It is well to bear this difference as to season north and south in mind, and to compare these curves with those of the thermal charts; for the two together indicate the existence in the ocean of a thermal tide, which, as before stated (§ 439), ebbs and flows but once a year. By this figure the South Atlantic appears to be cooler and heavier than the northern. The season of observation, however, is southern fall and winter *vice* northern summer. In January, February, and March, the waters of the southern ocean are decidedly warmer, as at the opposite six months they are decidedly cooler, parallel for parallel, than those of the northern oceans. Thus periodically differing in temperature, the surface waters of the two hemispheres vary also in specific gravity, and give rise to an annual ebb and flow — an upper and an under tide — not from one hemisphere to the other, but between each pole and equator. In contemplating the existence and studying the laws of this thermal tide we are struck with the compensations and adjustments that are allotted to it in the mechanism of the sea; for these feeble forces in the water remind one of the quantities of small value — residuals of compensation — with which the astronomer has to deal when he is working out the geometry of the heavens. He finds that it is these small quantities which make harmony in the celestial spaces; and so, too, it is the gentle forces like this in the waters which preserves the harmony of the seas. Equatorial and polar seas are of an invariable temperature, but in middle latitudes the sunbeam has power to wrinkle and crumple the surface of the sea by alternate expansion and contraction of its waters. In these middle latitudes is the cradle of the tiny

thermal tide here brought to light; feeble, indeed, and easily masked are its forces, but they surely *exist*. It may be that the thermometer and hydrometer are the only instruments which are nice enough to enable us to detect it. Its footprints, nevertheless, are well marked in our tables showing (§ 441) the thermal dilatation of sea-water. The movements of the isothermal lines, marching up and down the ocean, show by signs not to be mistaken its rate and velocity. These movements are well represented on the thermal charts. The tiny ripplings of this feeble tide have, we may be sure, their office to perform in the general system of aqueous circulation in the sea. Their influence may be feeble, like small perturbations in the orbits of planets; but the physicist is no more at liberty to despise the one than the astronomer is to neglect the other.

446. *Sea water of the southern cooler and heavier, parallel for parallel, than sea water of the northern hemisphere.* The problems that we now have in hand, and which is represented by the diagrams of Figure 3, is to put the seas in scales, the ocean in a balance, and to weigh in the specific gravity bottle the waters of the northern with the waters of the southern hemisphere. By *b* it would appear that both the water and the air of the south Atlantic are decidedly cooler and likewise heavier, parallel for parallel, than the waters of the north Atlantic; but this difference may be more apparent than real; for the observations were made in the northern summer on this side, and in the southern fall and winter on the other side of the equator. Had we a series of observations the converse of this, viz., winter in the north Atlantic, summer in the south, perhaps the latter would then appear to be specifically the lighter; at any rate, the mean summer temperature of each Atlantic, north and south, is higher than its mean winter temperature, and consequently the specific gravity of the waters of each must change with the seasons. A diagram — had we the data for such a one — to show these changes, would be very instructive; it would show beautifully, by its marks, the ebb and flow of this new-born tide of the ocean. By *a* the south Pacific also outweighs the north in specific gravity; but here again the true difference, whatever it be, is somewhat masked by the time of year when the observations were made. Those north were made during the fall, winter, and spring; those south, during the fall and first winter months of that hemisphere. Nevertheless, the weight of the observations presented in Figure 3 does, as far as they go, indicate that the seas of the southern do outweigh in specific gravity the seas of the northern hemisphere in the proportion of 1.0272 to 1.0262 of specific gravity.[3] Daubeny, Dové, *et al.*, have pointed out an excess of salt contained in sea water south of the equator, as compared with that contained in sea water north.

[3] According to Dr. Marcet, the southern ocean contains more salt than the northern in the proportion of 1.02919 to 1.02757. — ["On the Specific Gravity . . . of Sea Waters . . .," p. 173.]

447. *Testimony of the hydrometer in favor of the air crossings at the calm belts.* These indications, as far as they go, and this view of the subject, whatever future investigations may show to be its true worth, seem to lean in support of the idea advanced and maintained by facts and arguments in Chapter IV, viz., that the southern seas are the boiler and the northern hemisphere the condenser for the grand atmospherical engine, which sucks up vapor from the south to feed the northern hemisphere with rains. If it be true — and Dové also thinks it is — that the clouds which supply our fountains with rains for the great American lakes, and with rains for the majestic water-courses of Europe and Asia, Northern Africa and America, are replenished from seas beyond the equator, then the waters of the ocean south should be a little salter, and therefore specifically a little heavier, parallel for parallel, and temperature for temperature, than the waters of cis-equatorial seas. We begin to find that the hydrometer is bearing testimony in support of the evidence adduced in Chapters IV and VII, to show that when the trade-winds meet and rise up in the equatorial calm belt, the atmosphere which comes there as southeast trade-winds passes with its vapor over into the northern hemisphere. We had not anticipated that this little instrument could throw any light upon this subject; but if, as it indicates, the sea water of the other hemisphere be salter and heavier than the sea water of this, what makes it so but evaporation, and what prevents currents from restoring its equilibrium but the winds, which are continually sucking up vapor from the brine of trans-equatorial seas, and pouring it down as fresh water upon cis-equatorial seas and land? It is taking out of one scale of the balance and putting into the other; and the difference of specific gravity between the sea water of the opposite hemisphere may give us a measure for determining the amount of fresh water that is always in transitu. Certainly, if evaporation and rains were to cease, if the rivers were to dry up, and the sea-shells to perish, the waters of the ocean would, in the course of time, become all of the same saltness, and the only difference of specific gravity in the sea would be due to thermal agencies. After having thus ceased, if evaporation were then to commence only in the other hemisphere, and condensation take place only in this, half the difference, as to saltness of the sea water in opposite hemispheres, would express the ratio in volumes of fresh water, whether as vapor or liquid, that would then be kept in transitu between the two hemispheres. But it evaporates on both sides and precipitates on both; nevertheless, more on one side than on the other, and the difference of saltness will still indicate the proportion in transitu. If we follow the thermal and specific gravity curves from the parallels of $30° - 34°$ to the equator (Figure 3, *a*, *b*), we see, as I have said, that sea water in this part of the ocean does not grow lighter in proportion as it grows warmer. This is accounted for on the supposition (§ 435) that the effects of the thermal

dilatation on the specific gravity is counteracted by evaporation. Now, if we knew the thickness of the stratum which supplies the fresh water for this evaporation, we should not only have a measure for the amount of water which as vapor is sucked up and carried off from the trade-wind regions of the sea, to be deposited in showers on other parts of the earth, but we should be enabled to determine also the quantity which is evaporated in one hemisphere and transported by the clouds and the winds to be precipitated in the other. These are questions which are raised for contemplation merely; they can not be answered now; they grow out of some of the many grand and imposing thoughts suggested by the study of the revelations which the hydrometer is already beginning to make concerning the wonders of the sea. Returning from this excursion toward the fields of speculation, it will be perceived that these observations upon the temperature and density of sea water have for their object to weigh the seas, and to measure in the opposite scales of a balance the specific gravity of the waters of one hemisphere with the specific gravity of the waters of the other. This problem is quite within the compass of this exquisite system of research to solve. But, in order to weigh the seas in this manner, it is necessary that the little hydrometric balance by which it is to be done should be well and truly adjusted.

448. *Amount of salt in, and mean specific gravity of sea water.* From these premises it would not be difficult to show that the saltness of the sea is a physical necessity. In some of the aspects presented, the salts of the sea hold the relation in the terrestrial mechanism that the balance-wheel does to the machinery of a watch. Without them, the climates of the earth could not harmonize as they do; neither could the winds, by sucking up vapor, hold in check the expansive power (§ 437) of tropical heat upon the sea; nor counteract, by leaving the salts behind, the thermal influence of the sun in imparting dynamical force to marine currents; nor prevent the solar ray from unduly disturbing the aqueous equilibrium of our planet. As evaporation goes on from a sea of fresh water, the level only, and not the specific gravity, of the remaining water is changed. The waters of fresh intertropical seas would, instead of growing heavy by reason of evaporation between the tropics, becoming lighter and lighter by reason of the heat; while the water of fresh polar seas would grow heavier and heavier by reason of the cold — a condition which, by reason of evaporation and precipitation, is almost the very reverse of that which nature has ordained for the salt sea, and which, therefore, is the wisest and the best. The average amount of salts in sea water is not accurately known. From such data as I have, I estimate it to be about 4 per cent. (.039), and the mean specific gravity of sea water at 60° to be about 1.0272. Supposing these conditions to be accurate — and they are based on data which entitle them to be considered as not very wide of the

mark — the hydrometer and thermometer, with the aid of the table (§ 441), will give us a direct measure for the amount of salt in any specimen of sea water into which the navigator will take the trouble to dip these two instruments.

449. *Light cast by Figure 3 on the open sea in the Arctic Ocean.* These specific gravity and thermal curves, as they are presented in Figure 3, throw light also on the question of an open sea in the Arctic Ocean. That open sea is like a boiling spring (§ 427) in the midst of winter, which the severest cold can never seal up; only it is on a larger scale than any spring, or pool, or lake, and it is fed by the under currents with warm water from the south, which, by virtue of its saltness (see Figure 3, *b*), is heavier than the cool and upper current which runs out of the polar basin, and which is known as an ice-bearing current. It is the same which is felt by mariners as far down as the Grand Banks of Newfoundland, and recognized by philosophers off the coast of Florida. This upper current, though colder than its fellow below, is lighter, because it is not so salt. Figure 3, *a* reveals to us a portion of sea between the parallels of 34° and 40° north, exactly in such a physical category as that in which this theory presents the Arctic Ocean. Here, along our own shores, the thermal curve loses 12° of heat; and what does the specific gravity curve gain in the same interval? Instead of increasing up to 1.027, according to the thermal law, it decreases to 1.023 for the want of salt to sustain it. Now recollect that the great American chain of fresh water lakes never freezes over. Why? Because (§ 430) of their depth and their vertical circulation. The depths below are continually sending water above 32° to the surface, which, before it can be cooled down to the freezing point, sinks again. Now compare the shallow soundings in these lakes with the great depths of the Arctic Ocean; compute the vast extent of the hydrographic basin which holds this polar sea; gauge the rivers that discharge themselves into it; measure the rain, and hail, and snow that the clouds pour down upon it; and then contrast its area, and the fresh water drainage into it, with the like of Long Island Sound, Delaware Bay, and the Chesapeake; consider also the volume of diluted sea water between our shoreline and the Gulf Stream; strike the balance, and then see if the arctic supply of fresh water be not enough to reduce its salts as much as our own fresh water streams are diluting the brine of the sea under our own eyes. The very Gulf Stream water, which the observing vessel left as she crossed the parallel of 34° and entered into those light littoral waters, was bound northward. Suppose it to have flowed on as a surface current until it, with its salts, was reduced to the temperature of 40°. Its specific gravity at that temperature would have been 1.030, or specifically 30 per cent. heavier than the sea water of our own coasts. Could two such currents of water meet any where at sea, except as upper and under currents? If water that freezes

at 32°, that grows light and remains on the surface as you cool it below 39°, is prevented from freezing in our great fresh water lakes by vertical circulation, how much more would both vertical and horizontal circulation prevent congelation in the open polar sea, that is many times deeper and larger than the lakes, and the water of which contracts all the way down to its freezing point of 27°.2.

450. *The heaviest water.* The heaviest water in the sea, uncorrected for the temperature, as shown by the observations before us, is 1.028. This water was found (Figure 3, *a* and *b*) off Cape Horn. Let use examine a little more closely into the circumstances connected with the heaviest water on our side of the equator. It was a specimen of water from the Sea of Okotsk, which is a sea in a riverless region, and one where evaporation is probably in excess of precipitation — thus fulfilling the physical conditions for heavy water. The Red Sea is in a riverless and rainless region. Its waters ought to be heavier than those of any other mere arm of the ocean, and the dynamical force arising from the increase of specific gravity acquired by its waters after they enter it at Babelmandeb is sufficient to keep up a powerful inner and outer current through those straits. At the ordinary meeting of the Bombay Geographical Society for November, 1857, the learned secretary stated that recent observations then in his possession, and which were made by Mr. Ritchie and Dr. Giraud (§ 381), go to show that the saltest water in the Red Sea is where theory (§ 377) makes it, viz., in the Gulf of Suez; and that its waters become less and less salt thence to its mouth, and even beyond, till you approach the meridian of Socotra; after which the saltness again increases as you approach Bombay.

451. *Chapman's experiments.* Its waters, from the mouth of the straits for 300 or 400 miles up, have been found as high in temperature as 95° Fahrenheit — a sea at blood heat! The experiments of Professor Chapman, of Canada, which indicated as law — the salter the water the slower the evaporation, seem to suggest an explanation of this, at least in part. Evaporation ought to assist in keeping the surface of intertropical seas cool in the same way that it helps to cool other wet surfaces. And if the waters of the Red Sea become so salt that they can not make vapor enough to carry off the excessive heat of the solar ray, we may be sure that nature has provided means for carrying it off. But for the escape which these highly heated waters are, by means of their saltness, enabled to make from that sea, its climate, as well as the heat of its waters, would be more burning and blasting than the sands of Sahara. Even as it is, the waters of this sea are hotter than the air of the desert.

452. *The hydrometer indicates the rainy latitudes at sea.* There is another indication which this little instrument has afforded concerning the status of the sea, and which deserves notice. We are at first puzzled with

the remarkably light water between 9° and 16° S., Figure 3, *a* and in *b* between 7° and 9° N., as well as in 19° N. But, after a little examination, we are charmed with the discovery that the hydrometer points out the rainy regions at sea. Rodgers' observations on his homeward passage from San Francisco to Cape Horn furnish the data for the curves (*a*) between 37° N. and 57° S. Now Plate V shows that the equatorial calm belt lies south of the line where it is intersected by the homeward route from California. It also shows that when he crossed the "Doldrums" in the Atlantic, that belt was in north latitude about 7°–10°, and that when he was in 18°–20° N. (*b*) he was then passing through the offings of what are called the "Leeward Islands" of the West Indies, and that these are rainy latitudes at sea — the first two being under the cloud ring, the last being near the land in the trade-wind region, and confirming the remark so often made concerning the influence of islands at sea upon vapor, clouds, and precipitation.

453. *Astronomical view.* The most comprehensive view that we are permitted to take of cosmical or terrestrial arrangements and adaptations is at best narrow and contracted. Nevertheless, in studying the mechanism which Wisdom planned and the Great Architect of nature designed for the world, we sometimes fancy that we can discover a relation between the different parts of the wonderful machinery, and perceive some of the reasons and almost comprehend parts of the design which Omnipotent Intelligence had in view when those relations were established. Such fancies, rightly indulged, are always refreshing; and the developments of the hydrometer which we have been studying point us to one of them. This fancied discovery is, that a sea of fresh water instead of salt would not afford the compensations that are required in the terrestrial economy; we also fancy that we have almost discovered a relation between the orbit of the earth and the arrangement of land and water on its surface and their bearing upon climate. Our planet passes its perihelion during the southern summer, when it is nearer the centre and source of light and heat by more than three millions of miles than it is at its winter solstice, so that, on the 1st of January, the total amount of heat received by the earth is about $\frac{1}{15}$ more than it receives during a day in July, when it is in aphelion.[4] January is the midsummer month of the southern hemisphere, consequently that half of the globe receives more heat in a day of its summer than the other half receives in a day of the northern summer. But the northern summer is a week the longer, by reason of the ellipticity of the earth's orbit. What becomes of this diurnal excess of southern summer heat, be it in its aggregate never so small, and why does it not accumulate in trans-equatorial climes? So far from it, the southern hemisphere is the cooler.

[4] Herschel ["Meteorology," *Encyclop. Britan.*, XIV, 636–690, esp., p. 637].

454. *The latent heat of vapor.* In the southern hemisphere there is more sea and less land than in the northern. But the hydrometer indicates that the water in the seas of the former are salter and heavier than the waters of seas cis-equatorial; and man's reasoning faculties suggest, in explanation of this, that this difference of saltness or specific gravity is owing to the excess of evaporation in the southern half, excess of precipitation in the northern half of our planet. "When water passes, at 212° Fahrenheit, into steam, it absorbs 1000° of heat, which becomes insensible to the thermometer, or latent; and conversely, when steam is condensed into water, it gives out 1000° of latent heat, which thus becomes free, and affects both the thermometer and the senses. Hence steam of 212° Fahrenheit will, in condensing, heat five and a half times its own weight of water from the freezing to the boiling point." — *M'Culloch.* Now there is in the southern a very much larger water surface exposed to the sun than there is in the northern hemisphere, and this excess of heat is employed in lifting up vapor from that broad surface, in transporting it across the torrid zone and conveying it to extra-tropical northern latitudes, where the vapor is condensed to replenish our fountains, and where this southern heat is set free to mitigate the severity of northern climates.

455. *Its influence upon climates.* In order to trace a little farther, in our blind way, the evidences of wisdom and design, which we imagine we can detect in the terrestrial arrangement of land and water, let us fancy the southern hemisphere to have the land of the northern, and the northern to have the water of the southern, the earth's orbit remaining the same. Is it not obvious to our reason that by this change the whole system of climatology in both hemispheres would be changed? The climates of our planet are as obedient to law as the hosts of heaven. They are as they were designed to be; and all those agents which are concerned in regulating, controlling, and sustaining them are "ministers of His." Johnston, in the chapter to Plate XVIII of his great Physical Atlas, thus alludes to the seas, land, and climates of the two hemispheres: "The mild winter of the southern hemisphere, plus the contemporaneous hot summer of the northern hemisphere, necessarily gives a higher sum of temperature than the cool summer of the southern, plus the cold winter of the northern hemisphere. The above described relations appear to furnish the motive power in the machinery of the general atmosphere of the earth in the periodical conversion of the aqueous vapors into liquid form. In this manner the circuit of the fluid element, the essential support of all vegetable and animal life, no longer appears to depend on mere local coolings, or on the intermixture of atmospheric currents of different temperatures; but the unequal distribution of land and sea in the northern and southern hemispheres supplies an effectual provision, from whence it necessarily follows that the aqueous vapor,

which from the autumnal to the vernal equinox is developed to an immense extent over the southern hemisphere, returns to the earth, in the other half of the year, in the form of rain or snow. And thus the wonderful march of the most powerful steam-engine with which we are acquainted, the atmosphere, appears to be permanently regulated. The irregular distribution of physical qualities over the earth's surface is here seen to be a preserving principle for terrestrial life. Professor Dove considers the northern hemisphere as the condenser in this great steam-engine, and the southern hemisphere as its water reservoir; that the quantity of rain which falls in the northern hemisphere is, therefore, considerably greater than that which falls in the southern hemisphere; and that one reason of the higher temperature of the northern hemisphere is that the larger quantity of heat which becomes latent in the southern hemisphere in the formation of aqueous vapor is set free in the north in great falls of rain and snow." [5]

456. *The results of the marine hydrometer.* In this view of what our little hydrometer has developed or suggested, we trace the principles of compensation and adjustment, the marks of design, the evidence of adaptation between the orbit of the earth and the time from the vernal to the autumnal, and from the autumnal to the vernal equinox; between the arrangement of the land in one hemisphere and the arrangement of the water in the other; between the rains of the northern and the winds of the southern hemisphere; between the vapor in the air and the salts of the sea; and between climates on opposite sides of the equator. And all this is suggested by merely floating a glass bubble in sea water during a voyage in the Pacific! Thus even the little hydrometer, in its mute way, points the Christian philosopher to the evidences of design in creation. That the arrangements suggested above are adapted to each other, this instrument affords us evidence as clear as that which the telescope and the microscope bear in proof that the eye, in its structure, was adapted to the light of heaven. The universe is the expression of one thought, and that it is so every new fact developed in the progress of our researches is glorious proof.

457. *Barometer indications of an open sea.* In the course of our investigations into the physics of the sea, 100,000 observations of the barometer, and more than a million on the direction of the winds, have been discussed. They indicate an open water in the Arctic Ocean. They show that about the poles there is a high degree of aerial rarefaction — higher, indeed, than there is about the equator; for the barometer not only stands lower in this place of polar calms than it does in the equatorial calm belt, but the inrushing air comes from a greater distance to the cold than to the warm calms. [6]

[5] [Johnston, *Physical Atlas*, p. 60.]
[6] "The Winds at Sea," *Nautical Monographs*, No. 1, pl. IV.

458. *Polar rarefaction.* The question may be asked, Whence comes the heat that expands and rarefies the atmosphere in these polar places? The answer is, it comes from the condensation of vapor. The south pole is surrounded by water, the north pole by land. But the unexplored regions within the arctic basin are for the most part probably sea, within the antarctic, land. The rarefaction produced in the latter by the latent heat of vapor is such that the mean height of the barometer there is about 28 inches, while that in the arctic calm place is such as to reduce the barometer there to a mean not far from 29.5 inches. In the equatorial calm its mean height is about 29.9 inches. The hypothesis of an open sea in the Arctic Ocean becomes necessary to supply a source for this vapor; for the winds, entering the Arctic Ocean as they do after passing over the land and mountain heights of America, Europe, and Asia, must be robbed of much of their moisture ere they reach that ocean; it will require an abundant supply of vapor to create there by precipitation and the liberation of latent heat a degree of rarefaction sufficient to cause a general movement of the air polarward for the distance of 40° of latitude all around. That there is an immense volume of comparatively warm water going into the Arctic Ocean is abundantly shown by observation, and the rising up there of this water to the surface would afford heat and vapor enough for a vast degree of rarefaction.

459. *The middle ice.* The records of arctic explorations, together with the whalemen's accounts of "middle ice" in Baffin's Bay and Davis' Straits, go to confirm this view, which is further elaborated in the next chapter (§ 475). The facts there stated, and this "middle ice," go to show that every winter a drift takes place which brings out of the Frozen Ocean a tongue of ice a thousand miles or more in length: it is the compact and cold "middle ice." In our fresh-water streams it is the *middle* ice that first breaks up; that which is out of the way of the current remains longest. Not so in this bay and strait; there the littoral ice first gives way, leaving an open channel on either side in the spring and early summer, while the "middle ice" remains firm and impassable. The explanation is simple enough. The middle ice was formed in the severe cold of more northern latitudes, from which it has drifted down, while that on the sides was formed in the less severe climates of the bay and straits. This winter tongue of ice, which we know by actual observation is in motion from December till May, must, during that time, be detached from the main mass of ice in the Arctic Ocean, consequently there must be water between the ice that is in motion and the ice that is at rest. Not only so: in early summer the whalemen will run up to the north in the open water at the side of the "middle ice" in Davis' Strait and Baffin's Bay, even as far sometimes as Cape Alexander in 78°, to look for a crossing-place. Here, though so far north, they will find the "middle ice" gone, or so

broken up that they can cross over to the west side. They trace it up thus far, because at the south, and in spite of a higher thermometer, they find the "middle ice" compact and firm, so much so as to be impassable. In this fact we recognize another circumstance favoring the theory of an open sea at the north, and giving plausibility to the conjecture that this "middle ice" drifts out from the southern edge of the open sea as fast as it is formed during the winter. According to this conjecture, the thickest part of the "middle ice" should be that which has been exposed to the longest and severest cold, and this is probably that which began to be formed on the edge of the open sea in January. As it drifted to the south it continued to form and grow thick, and perhaps would be the last to melt; while that which began to be formed at the edge of the open sea in March or April would drift out, and not attain much thickness before it would cease to freeze and commence to thaw. It is this spring-made "middle ice" which, as it drifts to the south, would, being thin, be the first to break up; and experience has taught the whalemen to look north for the first breaking up and the earliest passage through the "middle ice."

460. *Position of the open sea.* The open sea, therefore, is, it may be inferred, at no great distance from the several straits which, leading in a northwardly direction, connect Baffin's Bay with the Arctic Ocean. It is through these straits that the winter drift takes place. The ice in which the Fox, the Resolute, the Advance, and the Rescue each drifted a thousand miles or more, came down through these straits. The fact of this annual winter drift from the Arctic Ocean is a most important one for future explorers. Had Captain Franklin known of it, he might have put his vessels in the line of it, and so escaped the rigors of that second winter. It would have brought him safely to the parallel of 65° or 60°, and set him free, as it did four other vessels, in the glad waters of the Atlantic Ocean.

CHAPTER X

§ 461–499. THE SALTS OF THE SEA

461. *The brine of the ocean.* The brine of the ocean is the ley of the earth (§ 43). From it the sea derives dynamical power, and its currents their main strength. Hence, to understand the dynamics of the ocean, it is necessary to study the effects of its salts upon the equilibrium of its waters; wherefore this chapter is added to assist in the elucidation of what has already been said concerning the currents and other phenomena of the sea. Why was the sea made salt? It is the salts of the sea that impart to its waters those curious anomalies in the laws of freezing and of thermal dilatation which have been described in a previous chapter (IX). It is the salts of the sea that assist the rays of heat to penetrate its bosom; [1] but for these, the solar ray, instead of heating large masses of water like the Gulf Stream, would play only at or near the surface, raising the temperature of the waters there, like the sand in the desert places, to an inordinate degree. The salts of the sea invest it with adaptations which it could not possess were its waters fresh. Were they fresh, they would attain their maximum density at 39°.5 instead of 25°.6, and the sea then would not have dynamical force enough to put the Gulf Stream in motion, nor could it regulate those climates we call marine.

462. *Were the sea of fresh water.* Were the sea fresh and not salt, Ireland would never have presented those ever-green shores which have won for her the name of "Emerald;" and the climate of England would have vied with Labrador for inhospitality. Had not the sea been salt, the torrid zone would have been hotter and the frigid colder for lack of aqueous circulation; had the sea not been salt, intertropical seas would have been at a constant temperature higher than blood heat, and the polar oceans would have been sealed up in everlasting fetters of ice, while certain parts of the earth would have been deluged with rain. Had the seas been of fresh water, the amount of evaporation, the quantity of rain, the volume and size of our rivers, would all have been different from what they are; the quantity of electricity in the air would have been

[1] Melloni has shown that the power of *salt* water to transmit heat is very much greater than that of fresh. [Macédoine (Macedonio) Melloni, "Mémoire sur la transmission libre de la chaleur rayonnante par differents corps solides et liquides," *Annales de Chimie et Physique*, LIII(1833), 5–73, esp., p. 55; *La thermochrôse ou la coloration calorifique* (Naples, 1850), p. 165. Melloni's experiments showed that a given thickness of a saturated solution of sodium chloride transmits 9% more infra-red radiation than the same thickness of distilled water.]

permanently changed from what it is, and its tension in the sky would have been exceedingly feeble. In the evaporation of fresh water at normal temperatures, but little of that fluid is evolved; while vapor from salt water carries off vitreous, and leaves behind resinous electricity in abundance. Hence, with seas of fresh water, our thunder-storms would be feeble contrivances, flashing only with such sparks as the vegetable kingdom might, when the juices of its plants were converted into vapor, lend to the clouds. It might seem strange, this idea that the thunderbolt of the sky, the sheet-lightning of the clouds, and the forked flashes of the storm, all have their genesis chiefly in the salts of the sea, and so it would be held were it not that Faraday has shown that a single grain of water and a little zinc can evolve electricity enough for a thunder-clap; therefore, were there no salts in the waters of the ocean, the sound of thunder would scarce be heard in the sky [2] — there would be no Gulf Stream, and no open sea in the Arctic Ocean.

463. *Uniform character of sea water.* As a general rule, the constituents of sea water are as constant in their proportions as are the components of the atmosphere. It is true that we sometimes come across arms of the sea, or places in the ocean, where we find the water more salt or less salt than sea water is generally; but this circumstance is due to local causes of easy explanation. For instance: when we come to an arm of the sea, as the Red Sea (§ 376), upon which it never rains, and from which the atmosphere is continually abstracting, by evaporation, fresh water from the salt, we may naturally expect to find a greater proportion of salt in the sea water that remains than we do near the mouth of some great river, as the Amazon, or in the regions of constant precipitation, or in other parts, as on the polar side of 40° in the North Atlantic, where it rains more than it evaporates. Yet in the case of the Red Sea, and all such natural salt-pans, as that and other rainless portions of the sea may be called, there is, on account of currents which are continually bearing away the water that has given off its vapors and bringing forward that which is less concentrated as to brine, a moderate degree of saltness which its waters can not exceed. We moreover find that, though the constituents of sea water, like those of the atmosphere, are not for every place invariably the same as to their proportions, yet they are the same, or nearly the same, as to their character. When, therefore, we take into considera-

---

[2] The great American lakes afford, it may be supposed, a considerable portion of the vapor which goes to make rain for the hydrographic basin in which they are. Visiting the Lake country in 1858, I was struck with the fact that so few trees bore the marks of lightning. The rule appeared to be, the nearer the lakes, the more rare was it for one of these ornaments of the forest to have been defaced by lightning; and, on inquiry from the Lake Board of Underwriters, I was informed that among the records of lake disasters there was not a single instance of a vessel having been struck by lightning on the great American lakes!

tion the fact that, as a general rule, sea water is, with the exception above stated, every where and always the same, and that it can only be made so by being well shaken together, we find grounds on which to base the conjecture that the ocean has its system of circulation, which is well calculated to excite our admiration, for it is as wonderful as the circulation of the blood.

464. *Hypotheses.* In order to investigate the effect of the salts of the sea upon its currents, and to catch a glimpse of the laws by which the circulation of its waters is governed, hypothesis, in the present meagre state of absolute knowledge with regard to the subject, seems to be as necessary to progress as is a cornerstone to a building. To make progress with such investigations, we want something to build upon. In the absence of facts, we are sometimes permitted to suppose them; only, in supposing them, we should take not only the possible, but the probable; and in making the selection of the various hypotheses which are suggested, we are bound to prefer that one by which the greatest number of phenomena can be reconciled. When we have found, tried, and offered such a one, we are entitled to claim for it a respectful consideration at least, until we discover it leading us into some palpable absurdity, or until some other hypothesis be suggested which will account equally as well, but for a greater number of phenomena. Then, as honest searchers after truth, we should be ready to give up the former, adopt the latter, and hold it until some other better than either of the two be offered. With this understanding, I venture to offer an hypothesis with regard to the agency of the salts or solid matter of the sea in imparting dynamical force to the waters of the ocean, and to suggest that one of the purposes which, in the grand design, it was probably intended to accomplish by having the sea salt, and not fresh, was to impart to its waters the forces and powers necessary to make their circulation complete. In the first place, we rely mainly upon hypothesis or conjecture for the assertion that there is a set of currents in the sea by which its waters are conveyed from place to place with regularity, certainty, and order. But this conjecture appears to be founded on reason, and we believe it to be true; for if we take a sample of water which shall fairly represent, in the proportion of its constituents, the average water of the Pacific Ocean, and analyze it, and if we do the same by a similar sample from the Atlantic, we shall find the analysis of the one to resemble that of the other as closely as though the two samples had been taken from the same bottle after having been well shaken. How, then, shall we account for this, unless upon the supposition that sea water from one part of the world is, in the process of time, brought in contact and mixed up with sea water from all other parts of the world? Agents, therefore, it would seem, are at work, which shake up the waters of the sea as though they were in a bottle, and which, in the

course of time, mingle those that are in one part of the ocean with those that are in another as thoroughly and completely as it is possible for a man to do in a vessel of his own construction. This fact as to uniformity of components appears to call for the hypothesis that sea water which today is in one part of the ocean, will, in the process of time, be found in another part the most remote. It must, therefore, be carried about by currents; and as these currents have their offices to perform in the terrestrial economy, they probably do not flow by chance, but in obedience to physical laws; they no doubt, therefore, assist to maintain the order and preserve the harmony which characterize every department of God's handy-work, and as such we treat them.

465. *Arguments afforded by corallines in favor of.* This hypothesis about currents is based upon our faith in the physical adaptations with which the sea is invested. Take, for example, the coral islands, reefs, beds, and atolls with which the Pacific Ocean is studded and garnished. They were built up of materials which a certain kind of insect quarried from the sea water. The currents of the sea ministered to this little insect — they were its *hod-carriers*. When fresh supplies of solid matter were wanted for the coral rock upon which the foundations of the Polynesian Islands were laid, these hod carriers brought them in unfailing streams of sea water, loaded with food and building materials for the coralline. The obedient currents, therefore, *must* thread the widest and the deepest seas, for they never fail to come at the right time, nor refuse to go after they have ministered to the hungry creature. Unless the currents of the sea were employed to carry off from this insect the waters that have been emptied by it of their lime, and to bring to it others charged with more, it is evident the little creature would have perished for want of food long before its task was half completed. But for currents, it would have been impaled in a nook of the very drop of water in which it was brought forth; for it would have soon secreted the lime contained in this drop, and then, without the ministering aid of currents to bring it more, it would have perished for the want of food for itself and materials for its edifice; and thus, but for the benign currents which took this exhausted water away, there we perceive this emptied drop would have remained, not only as the grave of the little architect, but as a monument in attestation of the shocking monstrosity that there had been a failure in the sublime system of terrestrial adaptations — that the sea had not been adapted by its Creator to the well-being of all its inhabitants. Now we do know that its adaptations are suited to all the wants of every one of its inhabitants — to the wants of the coral insect as well as to those of the whale. Thus our simple hypothesis acquires the majesty of truth, for we are now prepared boldly to assert *we know* that the sea has its system of circulation, because it transports materials for the coral rock from one

part of the world to another; because its currents receive them from the rivers, and hand them over to the little mason for the structure of the most stupendous works of solid masonry that man has ever seen — the coral islands of the sea. Thus, and, moreover, by a process of reasoning which is perfectly philosophical, we are irresistibly led to conjecture that there are regular and certain, if not appointed channels through which the water travels from one part of the ocean to another, and that those channels belong to an arrangement which may make, and which, for aught we know to the contrary, does make the system of oceanic circulation as complete, as perfect, and as harmonious as is that of the atmosphere or the blood. Every drop of water in the sea is as obedient to law and order as are the members of the heavenly host in the remotest regions of space; for when the morning stars sang together in the almighty anthem, we are told "the waves also lifted up their voice" in chorus; and doubtless, therefore, the harmony in the depths of the ocean is in tune with that which comes from the spheres above. We can not doubt it; for, were it not so, were there no channels of circulation from one ocean to another, and if, accordingly, the waters of the Atlantic were confined to the Atlantic, or if the waters of the arms and seas of the Atlantic were confined to those arms and seas, and had no channels of circulation by which they could pass out into the ocean, and traverse different latitudes and climates — if this were so, then the machinery of the ocean would be as incomplete as that of a watch without a balance-wheel.

466. *Ditto by the Red Sea.* For instance, take the Red Sea and the Mediterranean by way of illustration. Upon the Red Sea there is no precipitation; it is a rainless region; not a river runs down to it, not a brook empties into it; therefore there is no process by which the salts and washings of the earth, which are taken up and held in solution by rain or river water, can be brought down into the Red Sea. Its salts come from the ocean, and the air takes up from it, in the process of evaporation, fresh water, leaving behind, for the currents to carry away, the solid matter which, as sea water, it held in solution. On the other hand, numerous rivers discharge themselves into the Mediterranean, some of which are filtered through soils and among minerals which yield one kind of salts or soluble matter, another river runs through a limestone or volcanic region of country, and brings down in solution solid matter — it may be common salt, sulphate or carbonate of lime, magnesia, soda, potash, or iron — either or all may be in its waters. Still, the constituents of sea water from the Mediterranean and of sea water from the Red Sea are quite the same. But the waters of the Dead Sea have no connection with those of the ocean; they are cut off from its channels of circulation, and are therefore quite different, as to their components, from any arm, frith, or gulf of the broad ocean. Its inhabitants are also different from those of

the high seas. "The water which evaporates from the sea is nearly pure, containing but very minute traces of salts. Falling as rain upon the land, it washes the soil, percolates through the rocky layers, and becomes charged with saline substances, which are borne seaward by the returning currents. The ocean, therefore, is the great depository of every thing that water can dissolve and carry down from the surface of the continents; and, as there is no channel for their escape, they of course consequently accumulate." [3] They would constantly accumulate, as this very shrewd author remarks, were it not for the shells and insects of the sea and other agents mentioned.

467. *A general system of circulation required for the ocean.* How, therefore, shall we account for this sameness of compound, this structure of coral (§ 465), this stability as to animal life in the sea, but upon the supposition of a general system of circulation in the ocean, by which, in process of time, water from one part is conveyed to another part the most remote, and by which a general interchange and commingling of the waters take place? In like manner, the constituents of the atmosphere, whether it be analyzed at the equator or the poles, are the same. By cutting off and shutting up from the general channels of circulation any portion of sea water, as in the Dead Sea, or of atmospheric air, as in mines or wells, we can easily fill either with gases or other matter that shall very much affect its character, or alter the proportion of its ingredients, and affect the health of its inhabitants; but in the open sea or open air we can do no such thing.

468. *Dynamical agents.* The principal agents that are supposed to be concerned in giving circulation to the atmosphere, and in preserving the ratio among its components, are light, heat, electricity, and perhaps magnetism. But with regard to the sea, it is not known what office, if any, is performed by electricity, in giving dynamical force to its system of circulation. The chief motive power from which marine currents derive their velocity has been ascribed to heat; but a close study of the agents concerned has suggested that an important — nay, a powerful and active agency in the system of oceanic circulation is derived from the salts of the sea water, through the instrumentality of the winds, of marine plants, and animals. These give the ocean great dynamical force. Let us, for the sake of illustrating and explaining the nature of this force, suppose the sea in all its parts — in its depths and at the surface, at the equator and about the poles — to be of one uniform temperature, and to be all of fresh water; and, moreover, that there be neither wind to disturb its surface, nor tides nor rains to raise the level in this part, or to depress it in that. In this case, there would be nothing of heat to disturb its equilibrium, and there would be no motive power (§ 461) to beget currents,

[3] Youmans' Chemistry.

or to set the water in motion by reason of the difference of level or of specific gravity due to water at different densities and temperatures. Now let us suppose the winds, for the first time since the creation, to commence to blow upon this quiescent sea, and to ruffle its surface; they, by their force, would create partial surface currents, and thus agitating the waters, as they do, but only for a little way below the surface, would give rise to a feeble circulation in the supposed sea of fresh water. The surface currents thus created would set with the wind, giving rise to counter currents in the shape of under-tows and eddies. This, then, is one of the sources whence power is given to the system of oceanic circulation; but, though a feeble one, it is one which exists in reality, and, therefore, need not be regarded as hypothetical. Some (§ 79) think it the "*sole cause!*" Let us next call in evaporation and precipitation, with heat and cold — more powerful agents still. Suppose the evaporation to commence from this imaginary fresh-water ocean, and to go on as it does from the seas as they are. In those regions, as in the trade-wind regions, where evaporation is in excess of precipitation (§ 545), the general level of this supposed sea would be altered, and immediately as much water as is carried off by evaporation would commence to flow in from north and south toward the trade-wind or evaporating region to restore the level. On the other hand, the winds would have taken this vapor, borne it off to the extra-tropical regions, and precipitated it, we will suppose, where precipitation is in excess of evaporation. Here is another alteration of sea level by elevation instead of by depression; and hence we have the motive power for a surface current from each pole toward the equator, the object of which is only to supply the demand for evaporation in the trade-wind regions — demand for evaporation being taken here to mean the difference between evaporation and precipitation for any part of the sea. Now imagine this sea of uniform temperature to be suddenly stricken with the invisible wand of heat and cold, bringing its waters to the various temperatures at which they at this instant are standing. This change of temperature would make a change of specific gravity in the waters, which would destroy the equilibrium of the whole ocean; upon this a set of currents would immediately commence to flow, namely, a current of cold and heavy water to the place of the warm, and a current of warm and lighter to the place of the cold. The motive power of the currents thus created would be difference of specific gravity arising from difference of temperature in fresh water. We have now traced the effect of two agents, which, in a sea of fresh water, would tend to create currents, and to beget a system of aqueous circulation; but a set of currents, and a system of circulation which, it is readily perceived, would be quite feeble in comparison with those which we find in the salt sea. One of these agents would be employed in restoring, by means of one or more polar currents, the water that is taken from one part of the ocean by evaporation, and deposited in another by precipitation. The other agent would be employed in restoring, by the forces due difference of specific gravity, the equilibrium, which has been disturbed by heating, and of course expanding, the waters of the torrid zone on one hand, and by cooling, and consequently contracting, those of the frigid zone on the other. This agency would, if it were not modified by others, find expression in a system of currents and counter currents, or rather in a set of surface currents of warm and lighter water, from the equator toward the poles, and in another set of under currents of cooler, dense, and heavy water from the poles toward the equator.

469. *Currents without wind.* Such, keeping out of view the influence of the winds, which we may suppose would be the same whether the sea were salt or fresh, would be the system of oceanic circulation were the sea all of fresh water. But fresh water, in cooling, begins to expand near the temperature [4] of 40°, and expands more and more till it reaches the freezing point, and ceases to be fluid. This law of expansion by cooling would impart a peculiar feature to the system of oceanic circulation were the waters all fresh, which it is not necessary here to notice farther than to say it can not exist in seas of salt water, for salt water (§ 405) contracts as its temperature is lowered, and until it passes its freezing point. Hence, in consequence of its salts, changes of temperature derive increased power to disturb the equilibrium of the ocean. If this train of reasoning be good, we may infer that, in a system of oceanic circulation, the dynamical force to be derived from difference of temperature, where the waters are all fresh, would be quite feeble; and that, were the sea not salt, we should (§ 462) probably have no such current in it as the Gulf Stream. So far we have been reasoning hypothetically, to show what would be the chief agents, exclusive of the winds, in disturbing the equilibrium of the ocean were its waters fresh and not salt. And whatever disturbs equilibrium there may be regarded as the *primum mobile* in any system of marine currents.

470. *Influence of salts and evaporation.* Let us now proceed another step in the process of explaining and illustrating the effect of the salts of the sea in the system of oceanic circulation. To this end, let us suppose the imaginary ocean of fresh water suddenly to become that which we have, namely, an ocean of salt water, which contracts as its temperature is lowered (§ 441) till it reaches 25°.6. Let evaporation now commence in the trade-wind region, as it was supposed to do (§ 468) in the case of the fresh-water seas, and as it actually goes on in nature — and what takes place? Why, a lowering of the sea level, as before. But as the vapor of salt water is fresh, or nearly so, fresh water only is taken up from the ocean;

[4] 39°.5.

that which remains behind is therefore more salt. Thus, while the level is lowered in the *salt* sea, the equilibrium is destroyed because of the saltness of the water; for the water that remains after the evaporation takes place is, on account of the solid matter held in solution, specifically heavier than it was before any portion of it was converted into vapor. The vapor is taken from the surface water; the surface water thereby becomes more salt (§ 463), and, under certain conditions, heavier; when it becomes heavier, it sinks; and hence we have, due to the salts of the sea, a vertical circulation, namely, a descent of heavier — because salter and cooler — water from the surface, and an ascent of water that is lighter — because it is not so salt, or, being as salt, is not so cool (§ 404) — from the depths below. This vapor, then, which is taken up from the evaporating regions (§ 293), is carried by the winds through their channels of circulation, and poured back into the ocean where the regions of precipitation are; and by the regions of precipitation I mean those parts of the ocean, as in the polar basins, where the ocean receives more fresh water in the shape of rain, snow, etc., than it returns to the atmosphere in the shape of vapor. In the precipitating regions, therefore, the level is destroyed, as before explained, by elevation; and in the evaporating regions, by depression; which, as already stated (§ 468), gives rise to a system of surface currents, moving on an inclined plane, from the poles toward the equator. But we are now considering the effects of evaporation and precipitation in giving impulse to the circulation of the ocean where its waters are *salt*. The fresh water that has been taken from the evaporating regions is deposited upon those of precipitation, which, for illustration merely, we will locate in the north polar basin. Among the sources of supply of fresh water for this basin, we must include not only the precipitation which takes place over the basin itself, but also the amount of fresh water discharged into it by the rivers of the great hydrographical basins of Arctic Europe, Asia, and America. This fresh water, being emptied into the Polar Sea and agitated by the winds, becomes mixed with the salt; but as the agitation of the sea by the winds is supposed to extend to no great depth (§ 468), it is only the upper layer of salt water, and that to a moderate depth, which becomes mixed with the fresh. The specific gravity of this upper layer, therefore, is diminished just as much as the specific gravity of the sea water in the evaporating regions was increased. And thus we have a surface current of saltish water from the poles toward the equator, and an under current of water salter and heavier from the equator to the poles. This under current supplies, in a great measure, the salt which the upper current, freighted with fresh water from the clouds and rivers, carries back.

471. *The under currents owing entirely to the salts of sea water.* Thus it is to the salts of the sea that we owe that feature in the system of

oceanic circulation which causes an under current to flow from the Mediterranean into the Atlantic (§ 385), and another (§ 377) from the Red Sea into the Indian Ocean. And it is evident, since neither of these seas is salting up, that just as much, or nearly just as much salt as the under current brings out, just so much the upper currents carry in. We now begin to perceive what a powerful impulse is derived from the salts of the sea in giving effective and active circulation to its waters. Hence we infer (§ 461) that the currents of the sea, by reason of its saltness, attain their maximum of volume and velocity. Hence, too, we infer that the transportation of warm water from the equator toward the frozen regions of the poles, and of cold water from the frigid toward the torrid zone, is facilitated; and consequently here, in the dynamical power which the sea derives from its salts, have we not an agent by which climates are mitigated — by which they are softened and rendered much more salubrious than it would be possible for them to be were the waters of the ocean deprived of their property of saltness?

472. *A property peculiar to seas of salt water.* This property of saltness imparts to the waters of the ocean another peculiarity, by which the sea is still better adapted for the regulation of climates, and it is this: by evaporating fresh water from the salt in the tropics, the surface water becomes heavier than the average of sea water (§ 427). This heavy water is also warm water; it sinks, and being a good retainer, but a bad conductor of heat, this warm water is employed in transporting through under currents heat for the mitigation of climates in far-distant regions. Now this also is a property which a sea of fresh water could not have (§ 430). Let the winds take up their vapor from a sheet of fresh water, and that at the bottom is not disturbed, for there is no change in the specific gravity of that at the surface by which that at the bottom may be brought to the top; but let evaporation go on, though never so gently, from salt water, and the specific gravity of that at the top will soon be so changed as (§ 404) to bring that from the very lowest depths of the sea to the top.

473. *Quantity of salt in the sea.* If all the salts of the sea were precipitated and spread out equally over the northern half of this continent, it would, it has been computed, cover the ground one mile deep. What force could move such a mass of matter on the dry land? Yet the machinery of the ocean, of which it forms a part, is so wisely, marvelously, and wonderfully compensated, that the most gentle breeze that plays on its bosom, the tiniest insect that secretes solid matter for its sea-shell, is capable of putting it instantly in motion. Still, when solidified and placed in a heap, all the mechanical contrivances of man, aided by the tremendous forces of all the steam and water power of the world, could not, in centuries of time, move even so much as an inch this matter which

the sunbeam, the zephyr, and the infusorial insect keep in perpetual motion and activity.

474. *Deductions.* If these inferences as to the influences of the salts upon the currents of the sea be correct, the same cause which produces an under current from the Mediterranean (§ 471), and an under current from the Red Sea into the ocean, should produce an under current from the ocean into the north polar basin; for it may be laid down as a law, that whenever two oceans, or two arms of the sea, or two sheets of water, differing as to saltness, are connected with each other, there are currents between them, viz., a surface current from, and an under current into the sea of lightest water. In every case, the hypothesis with regard to the part performed by the salt, in giving vigor to the system of oceanic circulation, requires that, counter to the surface current of water with less salt, there should be an under current of water with more salt in it. That such is the case with regard both to the Mediterranean and the Red Sea has been amply shown in other parts of this work (§ 471), and abundantly proved by other observers. That, in obedience to this law, there is a constant current setting out of the Arctic Ocean through Davis' and other straits thereabout, which connect it with the Atlantic Ocean, is generally admitted. Lieutenant De Haven, United States Navy, when in command of the American expedition in search of Sir John Franklin, was frozen up with his vessels — the Advance and the Rescue — in mid-channel near Wellington Straits; and during the nine months that he was so frozen, his vessels, like H. B. M. ship Resolute and the Fox (§ 431), each holding its place in the ice, were drifted with it bodily for more than a thousand miles toward the south.

475. *Drift of the Resolute.* The drift of these vessels is sufficient, were there no other evidence, to establish the existence of an open sea in the Arctic Ocean; for this drift can not be accounted for upon any other hypothesis, as a slight examination of the arctic regions on a terrestrial globe, and a careful study of the facts (§§ 457–460), and other phenomena of the case, will show.

476. *De Haven's drift.* About the middle of September, 1850, being in latitude 74°40', and in the fair way of Wellington Channel, De Haven found himself, with the Advance, frozen in her tracks, as M'Clintock did the Fox,[5] in August, 1857, who tried to reach the shore, but he was fast bound, and drifting to the west. De Haven, after having been carried as far as 75° 25', and M'Clintock as far as 75° 30', say within nine hundred miles of the pole, their northerly course was arrested, and then commenced with each that celebrated drift of a thousand miles to the south, and which from December lasted, the one till June, the other till April 25th. These vessels were not drifted *through* the ice, but *with* the ice; for

[5] A screw yacht of 177 tons.

in lat. 65° 30', when De Haven was liberated on the 9th of June, he had the same "hummocks," the same snow-drifts, and the same icy landscape which set out with him on December 2d, when he commenced his drift from the parallel [6] of 75° 25'.

477. *An anti-polynian view.* Now, upon the theory of no open water, and upon the supposition of an ice-covered sea that seals up in winter all the unexplored regions of the north, let us, in imagination, take a survey of that sea just as the anti-polynians, according to their theory, would have it. Let the time of the survey be at the beginning of winter, when De Haven commenced his southwardly drift. From the Advance to the pole — a distance of 900 miles — no water is to be seen; the frost has bridged it all over. From the pole to the distance of 900 miles beyond, and all around, it is one field of thick-ribbed ice. The flat, and tame, and dreary landscape may be relieved here and there, perhaps, by islands, capes, and promontories dotting the surface, but nevertheless it is now at least as cold — being winter — from the pole all around to the parallel of 75°, as it was in early fall when De Haven, being near that parallel in Wellington Channel, found his vessel fast bound with fetters by the frost-king. Wherefore we may suppose that these theorists would admit the whole to be frozen by December. So that, according to the anti-polynian view, we have, measuring from the pole as a centre, a disc of ice more than five thousand miles in circumference, and extending quite down to the shores of Arctic America and Asia. Such is the aspect presented by the polar sea without an open water in winter; and on the 2d of December — the moment before this remarkable drift commenced; and when it did commence — was the entire sheet of ice with which we have supposed the Arctic Ocean to be covered put in motion, or was that only put in motion which drifted out? By the hypothesis there is no open water in all the circumference of this sea into which the ice might drift. We therefore may well ask the anti-polynians again, How did this drift commence? for commence it did; its movement was out of that sea, and from the pole toward the equator, and so it continued for six months at the average rate of 5½ miles a day. But whence — on what parallel — did it commence? Was the whole disc in motion from the shores of Siberia over across by way of the north pole toward Wellington Channel? If one part of this disc be put in motion, either the whole must be, or there must be a split or a rent with open water between. If, during the winter and spring — the coldest period — the edge of this ice-disc nearest Wellington Channel be carried by the currents a thousand miles toward the south, the edge

---

[6] De Haven was frozen in lat. 74° 40', long. 92° 55'; was carried up to 75° 25' N., and thence down to 66° 15' N., 58° 35' W., when he was liberated. The Fox was frozen in 75° 30' N., 64° W.; was carried west to 69° in the same latitude, and thence down to 63° 50' N., and 57° W., when she was liberated. The Resolute was abandoned in lat. 74° 40', long. 101° 20', and was picked up afloat off Cape Mercy in 65° N.

along the Russian shores on the opposite side must have been drifted toward the north a thousand miles also, and so leave an open water behind. Now we simply know there was no such drifting up from the Siberian shores, and the case is put simply to show that in any case the northerly edge of the drifting ice must have come from open water; for if we deny the existence of an open water in that direction, then we must go back and admit that at the beginning of the drift there was ice all the way from Wellington Channel to the north pole, and thence all the way from the north pole to the nearest land beyond, which is supposed to be the Siberian shores of the Old World. But, on the other hand, we must also admit the fact — for the Advance, the Rescue, the Fox, and the Resolute are witnesses of it — that a tongue of this ice 1000 miles long was in each of these winters thrust out of the polar basin down through Baffin's Bay into Davis' Straits. These ships came down upon it. It would be difficult for those who oppose the existence of an open water here in the Arctic Ocean to discover a force there which, during the extreme cold months of the northern night, when the ice is making all the time, could tear from its fastenings and move 5½ miles a day all through the winter and spring a disc of ice seven feet thick [7] and 1800 geographical miles in diameter. Yet such seem to be the conditions which the absence of open water would require; for, when the Advance was thawed out, there was a thousand miles of ice to the northward of her, and between her and Wellington Channel. This 1000 miles of ice had drifted out of the polar basin during her journey to the south; for when she was liberated there was doubtless a continuous sheet of ice between her in lat. 65°, and Wellington Channel in lat. 75°. This tongue of ice is what the whalemen call the middle ice of Baffin's Bay. When the Advance was at Wellington Channel, this thousand miles of ice must, according to the anti-polynians, have been to the north of her; or, according to the other school, it must, as it drifted toward the south, have been forming toward the north at the edge of an open sea (§ 459). And toward the north De Haven saw a water-sky, and toward the north Penny found an open sea and sailed upon it.

478. *The drift explained.* Upon the supposition that the ice which drifts out of the Arctic Ocean in the dead of winter is formed on the edge of an open water not far from the channel through which it drifts, we can account for all the known facts which attended the celebrated drifts of De Haven, M'Clintock,[8] and the Resolute. Upon no other theory can these well-known and well-authenticated facts be reconciled. If there be no open water during this winter drift, which there is reason to believe takes place annually, both the Advance, Fox, and the Resolute indicate

---

[7] De Haven found the ice upon which his vessel was brought out 7 feet 2 inches thick.

[8] In the Fox, 1857–1858.

that the whole icy covering — the frost-shell of the polar sea in winter — must have drifted bodily far enough, on these three several occasions at least, to set each vessel a thousand miles on her way toward the south. And thus, without bringing in again the long chain of evidence from Chapter IX, the physical necessity of an open sea in the Arctic Ocean is proved.[9]

[9] "The Fox accomplished another of those remarkable drifts which can be explained upon no other hypothesis but that of an open water in the Arctic Ocean, and that, too, not far from the entrance into it of some of the channels which connect it with Baffin's Bay on the polar side of 75°. The Fox was attempting to pass from Melville Bay over to Lancaster Sound, in August, 1857, when, on the 18th day of that month she fell in with ice, in which she was finally frozen up, and remained so for 242 days, during which time she was drifted to the southward 1194 miles, which gives an average rate of five miles a day.

"This drift — the drift of the Resolute, of the Advance, and Rescue, each upward of a thousand miles — appears to indicate that a similar drift takes place every year. They show the existence of a polynia, and indicate that the open sea is to be sought for at no great distance from Kennedy's Channel on the one hand, and Maury's on the other. This conclusion is reached by a process of reasoning of this sort:

"When each one of these vessels was released from her cold fetters, there was doubtless behind her, and between her place of release and her place of original imprisonment, an uninterrupted reach of a thousand miles covered with ice; which ice, during the fall, the winter, and early spring, drifted out of the Arctic Ocean. Now we have the choice of two suppositions, and of only two, in explanation of this phenomenon, and they are: Either that the great body of all the winter-formed ice of the Arctic Ocean must have drifted in an unbroken mass over toward Baffin's Bay; for these vessels were brought out upon a tongue of ice thrust through that bay down into Davis' Straits; or that this tongue must have been separated from the main mass, leaving behind that from which it had been severed.

"By the latter supposition all the known facts of the case may be reconciled; by the former, not one.

"If we suppose this drifting field of ice to be formed upon the very verge of an open sea, and to drift to the south as fast as it is formed, then the whole phenomenon becomes one of easy solution. At any rate, we are now possessed of a physical fact which probably would have returned Captain Crozier and his companions to us all safe and sound had they been aware of its existence; and that fact is in this oft-occurring, if not regular and annual, southward drift of ice from the Arctic Ocean down through Baffin's Bay into Davis' Strait. Captain Franklin, being ignorant of it, placed his vessels out of its reach on the south, where he was frozen in and died, and where Captain Crozier, his successor, remained imprisoned for eighteen months, and then abandoned his ships; their drift in the mean time, and for obvious reasons, being almost, if not quite insensible, except as influenced by the summer thaw and 'winter wedgings.' Now if those vessels, with their scurvy-riddled, frostworn, and disabled crews, could have been placed farther to the north, as in Barrow's Strait, or in the fairway of any of those channels connecting with it from the northward and westward, or with Baffin's Bay, the probabilities are that this regularly-recurring winter drift would have brought them down safely into milder climates, and into the glad waters of the Atlantic Ocean, as it did those four other vessels.

"The frequent, if not the regular annual occurrence of this drift down through Baffin's Bay is a fact which will be considered by all future arctic explorers as one of great importance, for it affords the means of escaping from the Arctic Ocean in the severest winter." — ["The Progress of Marine Geography, Compiled by the General Secretary from Data Furnished by the Hydrographical Office, Washington," *Jour.*] *Amer. Geogr.* [*and Stat.*] *Soc.*, [*1859*], (1860) [pp. 1–12, esp. pp. 4–5].

479. *Thickness of a winter's ice.* On the first of April De Haven measured the ice, and found it seven feet two inches thick. It was formed probably mostly of rain and river water, which, like our own littoral waters (§ 426), protect the salter and heavier waters below from the cold, for De Haven invariably found the temperature of the water under the ice 28°, which is the temperature that average sea water invariably assumes during the process of congelation (§ 442). Moreover, the specific gravity of the surface water which Rodgers measured in the Arctic Ocean was (§ 427) less than that of average sea water — a fact in confirmation of this conjecture as to the office of rain and river water in the polar seas. The freezing-point of the strongest brine is 4°; consequently, the freezing-point of water in the sea varies according to the proportion of salts from 4° all the way up to just below 32°. Thus the salts of the sea impart to its waters an elasticity, as it were, giving a law, a sort of sliding-scale both for the thermal dilatation and of congelation, which varies between that of fresh water and the saltest sea waters according to the degree of their saltness.

480. *Layers of water of different temperature in the Arctic Ocean.* Rodgers tried with his hydrometer and thermometer the waters of the Arctic Ocean at the surface, below, and at the bottom, and as often as he tried he found this arrangement: warm and light water on the top, cool in the middle, "hot and heavy" at the bottom. His experiments were made near Behring's Straits in August, 1855, between the parallels of 71°–2° [and yielded the results given in Table 9]. Assuming the surface

Table 9

[Layers of water of different temperature in the Arctic Ocean]

Date	Depth	Temp.	Sp. Grav.	Place			
Aug. 13	surface	43.8	.0264	Lat. 72° 2′	Long. 174° 37′ W.		
"	20 fath.	33.5	.0266	"	"	"	"
"	40 fath.[a]	40.5	.0266	"	"	"	"
Aug. 15	surface	42.5	.0258	" 71° 21′	"	175° 22′	
"	12 fath.	39.8	.0264	"	"	"	"
"	24 fath.[a]	40.2	.0264				

[a] Near bottom

water which Rodgers used for these experiments to be a fair average of arctic surface waters generally, this table affords data that show the proportion of rain and river water that the Arctic Ocean receives annually. The quantity may be inferred from the fact that average sea water has ten per cent. more salt than Rodgers's arctic.

481. *The ice-bearing drift from the Arctic like the ordinary drift from the Baltic.* Returning now to the drift of the ice, and the drift of the Advance and her followers, we see that, so far as currents are concerned,

we have in the Arctic Ocean a repetition merely of the more familiar phenomenon that is seen in the Baltic, where (§ 383, *note*) an under current of salt water runs in, and an upper current of brackish water runs out. Then, since there is salt always flowing out of the north polar basin, we infer that there must be salt always flowing into it, else it would either become fresh, or the whole Atlantic Ocean would become more and more briny, and be finally silted up with salt. It might be supposed, were there no evidence to the contrary, that this salt was supplied to the polar seas from the Atlantic around North Cape, and from the Pacific through Behring's Straits, and through no other channels. But, fortunately, arctic voyagers, who have cruised in the direction of Davis' Straits, have confirmed by their observations a law of nature (§ 474), and afforded us proof positive as to the fact of this other source for supplying the polar seas with salt. They tell us of an under current setting from the Atlantic toward the polar basin. They describe huge icebergs, with tops high up in the air, and of course the bases of which extend far down into the depths of the ocean, ripping and tearing their way with terrific force and awful violence through the surface ice or against a surface current, on their way into the polar basin.

482. *Icebergs drifting north.* Passed Midshipman S. P. Griffin, who commanded the brig Rescue in the American searching expedition after Sir John Franklin, informs me that, on one occasion, the two vessels were endeavoring, when in Baffin's Bay, to warp up to the northward against a strong surface current, which of course was setting to the south; and that, while so engaged, an iceberg, with its top many feet above the water, came "drifting up" from the south, and passed by them "like a shot." Although they were stemming a surface current against both the berg and themselves, such was the force and velocity of the under current that it carried the berg to the northward faster than the crew could warp the vessel against a surface but counter current. They hooked on to it, and were towed to the north by it. Captain Duncan,[10] master of the English whale-ship Dundee, says, at page 76 of his interesting little narrative: "*December 18th* (1826). It was awful to behold the immense icebergs working their way to the northeast from us, and not one drop of water to be seen; they were working themselves right through the middle of the ice." And again, at page 92, etc.: "*February 23d.* Latitude 68° 37′ north, longitude about 63° west. The dreadful apprehensions that assailed us yesterday by the near approach of the iceberg were this day most awfully verified. About three P.M. the iceberg came in contact with our floe, and in less than one minute it broke the ice; we were frozen in quite close to the shore; the floe was shivered to pieces for several miles, causing an explosion like an earthquake, or one hundred pieces of heavy ordnance

[10] David Duncan, *Arctic Regions; Voyage to Davis' Strait* [London, 1827].

fired at the same moment. The iceberg, with awful but majestic grandeur (in height and dimensions resembling a vast mountain), came almost up to our stern, and every one expected it would have run over the ship . . . The iceberg, as before observed, came up very near to the stern of the ship; the intermediate space between the berg and the vessel was filled with heavy masses of ice, which, though they had been previously broken by the immense weight of the berg, were again formed into a compact body by its pressure. The berg was drifting at the rate of about four knots, and by its force on the mass of ice was pushing the ship before it, as it appeared, to inevitable destruction. *Feb. 24th.* The iceberg still in sight, but drifting away fast to the northeast. *Feb. 25th.* The iceberg that so lately threatened our destruction had driven completely out of sight to the northeast from us."

483. *Temperature of the under current.* Now, then, whence, unless from the difference of specific gravity due sea water of different degrees of saltness and temperature, can we derive a motive power in the depths of the sea, with force sufficient to give such tremendous masses of ice such a velocity? What is the temperature of this under current? Rodgers's observations (§ 480) would seem to indicate that at the depth of 150 feet it is not below 40°. Assuming the water of the surface current which runs out with the ice to be all at 28°, as De Haven found it (§ 479), we observe that it is not unreasonable to suppose that the water of the under current, inasmuch as it comes from the south, and therefore from warmer latitudes, is not so cold; and if it be not so cold, its temperature, before it comes out again, must be reduced to 28°, or whatever be the average temperature of the outer but surface current. Dr. Kane found the temperature of the open sea in the Arctic Ocean (§ 429) as high as 36°. Can water in the depths below flow from the mild climate of the temperate zones to the severer climates of the frigid zone without falling below 36°? To what, in the depths of the sea, can a warm current of large volume impart its heat? The temperature of sea water from the tropics in which ice is forming is invariably (§ 442) 28°. Does not the circumstance of De Haven's *invariably* finding this to be the temperature below the ice on which he drifted tend to confirm the conjecture (§ 479) about the ice and the river water?

484. *It comes to the surface.* This under polar current water, then, as it rises to the top, and is brought to the surface by the agitation of the sea in the arctic regions, gives out its surplus heat to warm the atmosphere there till the temperature of this warm under current water is lowered to the requisite degree for going out on the surface. Hence the water-sky of those regions. And the heat that it loses in falling from its normal temperature, be that what it may, till it reaches the temperature of 28°, is so much caloric set free in the polar regions, to temper the air and mitigate the climate there. Now is not this one of those modifications

of climate which may be fairly traced back to the effect of the saltness of the sea in giving energy to its circulation? Moreover, if there be a deep sea in the polar basin, which serves as a receptacle for the waters brought into it by this under current, which, because it comes from toward the equatorial regions, comes from a milder climate, and is therefore warmer, we can easily imagine why there might be an open sea in the polar regions — why Lieutenant De Haven, in his instructions (§ 428), was directed to look for it; and why both he and Captain Penny, of one of the English searching vessels, and afterward Dr. Kane, found it there. And in accounting for this polynia, we see that its existence is not only consistent with the hypothesis with which we set out, touching a perfect system of oceanic circulation, but that it may be ascribed, in a great degree at least, if not wholly, to the effect produced by the salts of the sea upon the mobility and circulation of its waters. Here, then, is an office which the sea performs in the economy of the universe by virtue of its saltness, and which it could not perform were its waters altogether fresh. And thus philosophers have a clew placed in their hands which will probably guide them to one of the many hidden treasures that are embraced in the true answer to the question, "Why is the sea salt?"

485. *Sea Shells — their influence upon currents.* We find in sea water other matter besides common salt. Lime is dissolved by the rains and the rivers, and emptied in vast quantities into the ocean. Out of it, coral islands and coral reefs of great extent — marl-beds, shell-banks, and infusorial deposits of enormous magnitude, have been constructed by the inhabitants of the deep. These creatures are endowed with the power of secreting, apparently for their own purposes only, solid matter, which the waters of the sea hold in solution. But this power was given to them that they also might fulfill the part assigned them in the economy of the universe. For to them, probably, has been allotted the important office of assisting to give circulation to the ocean, of helping to regulate the climates of the earth, and of preserving the purity of the sea. The better to comprehend how such creatures may influence currents and climates, let us again suppose the ocean to be perfectly at rest — that throughout, it is in a state of complete equilibrium — that, with the exception of those tenants of the deep which have the power of extracting from it the solid matter held in solution, there is no agent in nature capable of disturbing that equilibrium — and that all these fish, etc., have suspended their secretions, in order that this state of a perfect aqueous equilibrium and repose throughout the sea might be attained. In this state of things — the waters of the sea being in perfect equilibrium — a single mollusk or coralline, we will suppose commences his secretions, and abstracts from the sea water (§ 465) solid matter for his cell. In that act this animal has destroyed the equilibrium of the whole ocean, for the specific gravity of

that portion of water from which this solid matter has been abstracted is altered. Having lost a portion of its solid contents, it has become specifically lighter than it was before; it must, therefore, give place to the pressure which the heavier water exerts to push it aside and to occupy its place, and it must consequently travel about and mingle with the waters of the other parts of the ocean until its proportion of solid matter is returned to it, and until it attains the exact degree of specific gravity due sea water generally.

486. *Solid matter secreted by them.* How much solid matter does the whole host of marine plants and animals abstract from sea water daily? Is it a thousand pounds, or a thousand millions of tons? No one can say. But, whatever be its weight, it is so much of the power of gravity applied to the dynamical forces of the ocean. And this power is derived from the salts of the sea, through the agency of sea-shells and other marine animals, that of themselves scarcely possess the power of locomotion. Yet they have power to put the whole sea in motion, from the equator to the poles, and from top to bottom. But we have yet to inquire how far may currents be due to the derangement of equilibrium arising from the change of specific gravity caused by the secretions of the myriads of marine animals that are continually at work in various parts of the ocean. These little creatures abstract from sea water solid matter enough to build continents of. And, also, we have to remember as to the extent to which equilibrium in the sea is disturbed by the salts which evaporation leaves behind. Thus, when we consider the salts of the sea in one point of view, we see the winds and the marine animals operating upon the waters, and, in certain parts of the ocean, developing by their action upon the solid contents of the same those very principles of antagonistic forces which hold the earth in its orbit, and preserve the harmonies of the universe.

487. *Dynamical force derived from.* In another point of view, we see the sea-breeze and the sea-shell, in performing their appointed offices, acting so as to give rise to a reciprocating motion in the waters; and thus they impart to the ocean dynamical forces also for its circulation. The sea-breeze plays upon the surface; it converts only fresh water into vapor, and leaves the solid matter behind. The surface water thus becomes specifically heavier, and sinks. On the other hand, the little marine architect below, as he works upon his coral edifice at the bottom, abstracts from the water there a portion of its solid contents; it therefore becomes specifically lighter, and up its goes, ascending to the top with increased velocity, to take the place of the descending column, which, by the action of the winds, has been sent down loaded with fresh food and materials for the busy little mason in the depths below. Seeing, then, that the inhabitants of the sea, with their powers of secretion, are competent to exercise at least *some* degree of influence in disturbing equilibrium, are not these

creatures entitled to be regarded as agents which have their offices to perform in the system of oceanic circulation, and do they not belong to its physical geography? Their influences upon the economy of the sea are like those outstanding quantities which the astronomer finds in the periods of heavenly bodies. He calls them perturbations; for short, or even during considerable intervals, their effects may be inappreciable, but, unless there was a balance provided somewhere — a pendulum that requires ages for a vibration — they would, during the progress of time, accumulate so as to produce disorder, and finally cause the destruction of worlds. So, too, with the salts of the sea, and those little microscopic inhabitants of its waters. They take care of its outstanding quantities of solid matter, and by their influence preserve harmony in the ocean. It is immaterial how great or how small that influence may be supposed to be; for, be it great or small, it is *cumulative;* and we therefore may rest assured it is not a chance influence, but it is an influence exercised by design, and according to the commandment of Him whose "voice the winds and the sea obey." Thus God speaks through sea-shells to the ocean.

488. *Their physical relations.* It may therefore be supposed that the arrangements in the economy of nature are such as to require that the various kinds of marine animals, whose secretions are calculated to alter the specific gravity of sea water, to destroy its equilibrium, to beget currents in the ocean, and to control its circulation, should be distributed according to order. Upon this supposition — the like of which nature warrants throughout her whole domain — we may conceive how the marine animals of which we have been speaking may impress other features upon the physical relations of the sea by assisting also to regulate climates, and to adjust the temperature of certain latitudes. For instance, let us suppose the waters in a certain part of the torrid zone to be 90°, but, by reason of the fresh water which has been taken from them in a state of vapor, and consequently by reason of the proportionate increase of salts, these waters are heavier than waters that may be cooler, but not so salt (§ 105). This being the case, the tendency would be for this warm, but salt and heavy water, to flow off as an under current toward the polar or some other regions of lighter water. Moreover, if the sea were not salt, there would be no coral islands to beautify its landscape and give variety to its features; sea-shells and marine insects could not operate upon the specific gravity of its waters, nor give diversity to its climates; neither could evaporation give dynamical force to its circulation; its waters, ceasing to contract as their temperature falls below 39°, would give but little impulse to its currents, and impart no motion (§ 404) to its waters in the depths below: thus its circulation would be torpid, and its bosom lack animation. This under current may be freighted with heat to temper

some hyperborean region or to soften some extra-tropical climate, for we know that such is among the effects of marine currents. At starting, it might have been, if you please, so loaded with solid matter that, though its temperature were 90°, yet, by reason of the quantity of such matter held in solution, its specific gravity might have been greater even than that of extra-tropical sea water generally at 28°. Notwithstanding this, it may be brought into contact, by the way, with those kinds and quantities of marine organisms that shall abstract solid matter enough to reduce its specific gravity, and, instead of leaving it greater than common sea water at 28°, make it less than common sea water at 40; consequently, in such a case, this warm sea water, when it comes to the cold latitudes, would be brought to the surface through the instrumentality of shell-fish, and various other tribes that dwell far down in the depths of the ocean. Thus we perceive that these creatures, though they are regarded as beings so low in the scale of creation, may nevertheless be regarded as agents of much importance in the terrestrial economy; for we now comprehend how they are capable of spreading over certain parts of the ocean those benign mantles of warmth which temper the winds, and modify, more or less, all the marine climates of the earth.

489. *The regulators of the sea.* The makers of nice astronomical instruments, when they have put the different parts of their machinery together, and set it to work, find, as in the chronometer, for instance, that it is subject in its performance to many irregularities and imperfections; that in one state of things there is expansion, and in another state contraction among cogs, springs, and wheels, with an increase or diminution of rate. This defect the makers have sought to overcome; and, with a beautiful display of ingenuity, they have attached to the works of the instrument a contrivance which has had the effect of correcting these irregularities by counteracting the tendency of the instrument to change its performance with the changing influences of temperature. This contrivance is called a *compensation;* and a chronometer or clock that is well regulated and properly compensated will perform its office with certainty, and preserve its rate under all the vicissitudes of heat and cold to which it may be exposed. In the clock-work of the ocean and the machinery of the universe, order and regularity are maintained by a system of compensations. A celestial body, as it revolves around its sun, flies off under the influence of centrifugal force; but immediately the forces of compensation begin to act; the planet is brought back to its elliptical path, and held in the orbit for which its mass, its motions, and its distances were adjusted. Its compensation is perfect. So, too, with the salts and the shells of the sea in the machinery of the ocean; from them are derived principles of compensation the most perfect; through their agency the undue effects of heat and cold, of storm and rain, in disturbing the

equilibrium, and producing thereby currents in the sea, are compensated, regulated, and controlled. The dews, the rains, and the rivers are continually dissolving certain minerals of the earth, and carrying them off to the sea. This is an accumulative process; and if it were not *compensated*, the sea would finally become, as the Dead Sea is, saturated with salt, and therefore unsuitable for the habitation of many fish of the sea. The sea-shells and marine insects afford the required *compensation*. They are the conservators of the ocean. As the salts are emptied into the sea, these creatures secrete them again and pile them up in solid masses, to serve as the bases of islands and continents, to be in the process of ages upheaved into dry land, and then again dissolved by the dews and rains, and washed by the rivers away into the sea again.

490. *Whence does the sea derive its salts?* The question as to whence the salts of the sea were originally derived, of course has not escaped the attention of philosophers. I once thought with Darwin and those other philosophers who hold that the sea derived its salts originally from the washings of the rains and rivers. I now question that opinion; for, in the course of the researches connected with the "Wind and Current Charts," I have found evidence, from the sea and in the Bible, which seems to cast doubt upon it. The account given in the first chapter of Genesis, and that contained in the hieroglyphics which are traced by the hand of Nature on the geological column as to the order of creation, are marvelously accordant. The Christian man of science regards them both as true; and he never overlooks the fact that, while they differ in the mode and manner as well as in the things they teach, yet they never conflict; and they contain no evidence going to show that the sea was ever fresh; on the contrary, they both afford circumstantial evidence sufficient for the belief that the sea was salt as far back as the morning of creation, or at least as the evening and the morning of the day when the dry land appeared. That the rains and the rivers do dissolve salts of various kinds from the rocks and soil, and empty them into the sea, there is no doubt. These salts can not be evaporated, we know; and we also know that many of the lakes, as the Dead Sea, which receive rivers and have no outlet, are salt. Hence the inference by some philosophers that these inland water-basins received their salts wholly from the washings of the soil; and consequently the conjecture arose that the great sea derived its salts from the same source and by the same process. But, and per contra, though these solid ingredients can not be taken out of the sea by evaporation, they can be extracted by other processes. We know that the insects of the sea do take out a portion of them, and that the salt ponds and arms which, from time to time in the geological calendar, have been separated from the sea, afford an escape by which the quantity of chloride of sodium in its waters — the most abundant of its solid ingredients — is regulated. The insects of the

sea can not build their structures of this salt, for it would dissolve again, and as fast as they could separate it. But here the ever-ready atmosphere comes into play, and assists the insects in regulating the salts. It can not take them up from the sea, it is true, but it can take the sea away from them; for it pumps up the water from these pools that have been barred off, transfers it to the clouds, and they deliver it back to the sea as fresh water, leaving the salts it contained in a solid state behind. These are operations that have been going on for ages; proof that they are still going on is continually before our eyes; for the "hard water" of our fountains, the marl-banks of the valleys, the salt-beds of the plains, Albion's chalky cliffs, and the coral islands of the sea, are monuments in attestation. These masses of solid matter have been secreted from the sea waters; they express the ability of these creatures to prevent the accumulation of salts in the sea.

491. *Their antiquity.* There is no proof, nor is there any reason for the belief, that the sea is growing salter or fresher. Hence we infer that the operations of addition and extraction are reciprocal and equal; that the effect of rains and rivers in washing down is compensated by the processes of evaporation and secretion in taking out. If the sea derived its salts originally from the rivers, the geological records of the past would show that river beds were scored out in the crust of our planet before the sea had deposited any of its fossil shells and infusorial remains upon it. If, therefore, we admit the Darwin theory, we must also admit that there was a period when the sea was without salt, and consequently without shells or animals either of the silicious or calcareous kind. If ever there were such a time, it must have been when the rivers were collecting and pouring in the salts which now make the brine of the ocean. But while the palæontological records of the earth, on one hand, afford no evidence of any such fresh-water period, the Mosaic account is far from being negative with its testimony on the other. According to it, we infer that the sea was salt as early, at least, as the fifth day, for it was on that day of creation that the waters were commanded to "bring forth abundantly the moving creature that hath life." It is in obedience to that command that the sea now teems with organisms; and it is marvelous how abundantly the obedient waters do bring forth, and how wonderful for variety as well as multitude their progeny is. All who pause to look are astonished to see how the prolific ocean teems and swarms with life. The moving creatures in the sea constitute in their myriads of multitudes one of the "wonders of the deep."

492. *Insects of the sea — their abundance.* It is the custom of Captain Foster, of the American ship Garrick, who is one of my most patient of observers, to amuse himself by making drawings in his abstract log of the curious animalculæ which, with the microscope, he finds in the surface

water alongside; and though he has been following the sea for many years, he never fails to express his wonder and amazement at the immense numbers of living creatures that the microscope reveals to him in sea water. Hitherto his examinations related only to the surface waters, but in the log now before me he went into the depths, and he was more amazed than ever to see how abundantly the waters even there bring forth. "*January 28th*, 1855. In examining animalculæ in sea water, I have," says he, "heretofore used surface water. This afternoon, after pumping for some time from the stern pump seven feet below the surface, I examined the water, and was surprised to find that the fluid was literally alive with animated matter, embracing beautiful varieties." Of some he says, "Numerous heads, purple, red, and variegated." There is wonderful meaning in that word ABUNDANTLY, as it stands recorded in that Book, and as it is even at this day repeated by the great waters, a striking instance of which has been furnished by Piazzi Smyth, the Astronomer Royal of Edinburgh, during his voyage in 1856 on an astronomical expedition to Teneriffe. On that occasion he fell in with the annual harvest of medusæ that are sent by the Gulf Stream to feed the whales. [His description [11] of them is quoted in § 161.] Imagine how deep and thickly the bottom of the sea must, during the process of ages, have become covered with the flinty shells of these little creatures. And then recollect the command which was given to the waters of the sea on the fifth day, and we may form some idea of how literally they have obeyed this order, bringing forth most abundantly even now "the moving creature that hath life," and doing it in obedience to that command.

493. *Ditto, calcareous in the Pacific, silicious in the Atlantic.* In the waters of the Pacific Ocean, the calcareous matter seems to be in excess, for the microscopic shells there, as well as the conch and the coral, are built mostly of lime. In contemplating this round of compensations, the question may be asked, Where is the agent that regulates the supply of solid materials for the insects of the sea to build their edifices of? Answer: The rivers. They bring down, and pour into the sea continually, the pabulum which those organisms require. This amount again depends upon the quantity and power of the rains to wash out from the solid rock; and the rains depend upon the amount of vapor that the sea delivers to the winds, which, as Chapman's observations show, depends directly upon the salts of the sea.

494. *The records of the sea and of revelation agree.* So far the two records agree, and the evidence is clear that the sea was salt when it received this command. Do they afford any testimony as to its condition previously? Let us examine. On the second day of creation the waters

[11] [A passage approximately 340 words in length, repeated from §§ 161 and 162, is omitted here.]

were gathered together unto one place, and the dry land appeared. Before that period, therefore, there were no rivers, and consequently no
washings of brine by mists, nor dew, nor rains from the valleys among the
hills. The water covered the earth. This is the account of revelation; and
the account which Nature has written, in her own peculiar characters, on
the mountain and in the plain, on the rock and in the sea, as to the early
condition of our planet, indicates the same. The inscriptions on the geological column tell that there was a period when the solid parts of the
earth's crust which now stand high in the air were covered by water.
The geological evidence that it was so, with perhaps the exception of a
solitary mountain peak here and there, is conclusive; and when we
come to examine the fossil remains that are buried in the mountains and
scattered over the plains, we have as much reason to say that the sea was
salt when it covered or nearly covered the earth, as the naturalist, when
he sees a skull or bone whitening on the wayside, has to say that it was
once covered with flesh. Therefore we have reason for the conjecture that
the sea was salt "in the beginning," when "the waters under heaven were
gathered together into one place," and the dry land first appeared; for, go
back as far as we may in the dim records which young Nature has left
inscribed upon the geological column of her early processes, and there we
find the fossil shell and the remains of marine organisms to inform us
that when the foundations of our mountains were laid with granite, and
immediately succeeding that remote period when the primary formations
were completed, the sea was, as it is now, salt; for had it not been salt,
whence could those creeping things which fashioned the sea-shells that
cover the tops of the Andes, or those madrepores that strew the earth with
solid matter that has been secreted from briny waters, or those infusorial deposits which astound the geologist with their magnitude and extent,
or those fossil remains of the sea which have astonished, puzzled, and be-
wildered man in all ages — whence, had not the sea been salt when its
metes and bounds were set, could these creatures have obtained solid
matter for their edifices and structures? Much of that part of the earth's
crust which man stirs up in cultivation, and which yields him bread, has
been made fruitful by these "salts," which all manner of marine insects,
aqueous organisms, and sea-shells have secreted from the ocean. Much
of this portion of our planet has been filtered through the sea, and
its insects and creeping things are doing now precisely what they were set
about when the dry land appeared, namely, preserving the purity of the
ocean, and regulating it in the due performance of its great offices. As
fast as the rains dissolve the salts of the earth, and send them down
through the rivers to the sea, these faithful and everlasting agents of the
Creator elaborate them into pearls, shells, corals, and precious things;
and so, while they are preserving the sea, they are also embellishing the

land by imparting new adaptations to its soil, fresh beauty and variety to its landscapes. Whence came the salts of the sea originally is a question which perhaps never will be settled satisfactorily to every philosophic mind, but it is sufficient for the Christian philosopher to recollect that the salts of the sea, like its waters and the granite of the hills, are composed of substances which, when reduced to their simple state, are found for the most part to be mere gaseous or volatile matter of some kind or other. Thus we say that granite is generally composed of feldspar, mica, and quartz, yet these three minerals are made of substances more or less volatile in combination with oxygen gas. Iron, of which there is merely a trace, is the only ingredient which, in its uncombined and simple state, is not gaseous or volatile. Now was the feldspar of the granite originally formed in one heap, the mica in another, and the quartz in a third, then the three brought together by some mighty power, and welded into the granite rock for the everlasting hills to stand upon? or were they, as they were formed of the chaotic matter, made into rock? Sea water is composed of oxygen and hydrogen, and its salts, like the granite, also consist of gases and volatile metals. But whether the constituents of sea water, like those of the primitive rocks, were brought together in the original process of formation, and united in combination as we now find them in the ocean, or whether the sea was fresh "in the beginning" and became salt by some subsequent process, is not material to our present purpose. Some geologists suppose that in the chalk period, when the ammonites, with their huge chambered shells, lived in the sea, the carbonaceous material required by these creatures for their habitations must have been more abundant in its waters than it now is; but, though the constituents of sea water may have varied as to proportions, they probably were never, at least since "its waters commenced to bring forth," widely different from what they now are. It is true, the strange cuttle-fish, with its shell twelve feet in circumference, is no longer found alive in the sea: it died out with the chalk period; but then its companion, the tiny nautilus, remains to tell us that even in that remote period the proportion of salt in sea water was not unsuited to its health, for it and the coral insect have lived through all the changes that our planet has undergone since the sea was inhabited, and they tell us that its waters were salt as far back, at least, as their records extend, for they now build their edifices and make their habitations of the same materials, collected in the same way that they did then, and, had the sea been fresh in the interim, they too would have perished, and their family would have become extinct, like that of the great ammonite, which perhaps ceased to find the climates of the sea, not the proportion of its salts, suited to its well-being.

495. *Cubic miles of sea salt.* Did any one who maintains that the salts of the sea were originally washed down into it by the rivers and the rains

ever take the trouble to compute the quantity of solid matter that the sea holds in solution as salts? Taking the average depth of the ocean at three miles, and its average saltness at 3½ per cent., it appears that there is salt enough in the sea to cover to the thickness of one mile an area of ten millions of square miles. These ten millions of cubic miles of crystal salt have not made the sea any fuller. All this solid matter has been received into the interstices of sea water without swelling the mass; for chemists tell us that water is not increased in volume by the salt it dissolves. Here we have therefore displayed before us an economy of space calculated to surprise even the learned author himself of the "Plurality of Worlds." [12]

496. *The saltness of water retards evaporation.* There has been another question raised which bears upon what has already been said concerning the offices which, in the sublime system of terrestrial arrangements, have been assigned to the salts of the sea. On the 20th of January, 1855, Professor Chapman, of the University College, Toronto, communicated to the Canadian Institute a paper on the "Object of the Salt Condition of the Sea," which, he maintains, is *"mainly intended to regulate evaporation."* [13] To establish this hypothesis, he shows by a simple but carefully conducted set of experiments that, the salter the water, the slower the evaporation from it; and that the evaporation which takes place in 24 hours from water about as salt as the average of sea water is 0.54 per cent. less in quantity than from fresh water. "This suggestion and these experiments give additional interest to our investigations into the manifold and marvelous offices which, in the economy of our planet, have been assigned by the Creator to the salts of the sea. It is difficult to say what, in the Divine arrangement, was the *main* object of making the sea salt and not fresh. Whether it was to assist in the regulation of climates, or in the circulation of the ocean, or in re-adapting the earth for new conditions by transferring solid portions of its crust from one part to another, and giving employment to the corallines and insects of the sea in collecting this solid matter into new forms, and presenting it under different climates and conditions; or whether the *main* object was, as the distinguished professor suggests, to regulate evaporation, it is not necessary now or here to discuss. I think we may regard all the objects of the salts of the sea as *main* objects. But we see in the professor's experiments the dawn of more new beauties, and the appearance of other exquisite compensations, which, in studying the 'wonders of the deep,' we have so often paused to contemplate and admire. As the trade-wind region feeds the air with the vapor of fresh water, the process of evaporation from the sea is checked, for the water which remains, being salter,

---

[12] [William Whewell. The book was published anonymously.]
[13] [E. J. Chapman, "Note on the Object of the Salt Condition of the Sea," *The Canadian Journal,* III(1855), 186–187, esp., p. 187.]

parts with its vapor less readily; and thus, by the salts of the sea, floods may be prevented. But again, if the evaporating surface were to grow salter and salter, whence would the winds derive vapor duly to replenish the earth with showers; for the salter the surface, the more scanty the evaporation. Here is compensation, again, the most exquisite; and we perceive how, by reason of the salts of the sea, drought and flood, if not prevented, may be, and probably are, regulated and controlled; for that compensation which assists to regulate the amount of evaporation is surely concerned in adjusting also the quantity of rain. Were the salts of the sea lighter instead of heavier than the water, they would, as they feed the winds with moisture for the cloud and the rain, remain at its surface, and become more niggardly in their supplies, and finally the winds would howl over the salt-covered sea in very emptiness, and instead of cool and refreshing sea-breezes to fan the invalid and nourish the plants, we should have our gentle trade-winds coming from the sea in frightful blasts of parched, and thirsty, and blighting air. But sea salts, with their manifold and marvelous adaptations, come in here as a counterpoise, and, as the waters attain a certain degree of saltness, they become too heavy to remain longer in contact with the thirsty trade-winds, and are carried down, because of their weight, into the depths of the ocean; and thus the winds are *dieted* with vapor in due and wholesome quantities." — Maury's *Sailing Directions*, 7th ed., p. 862.

497. *The harmonies of the ocean.* Since the offices which, in the operations of the physical machinery of the earth, have been assigned to the salts of the sea, are obviously so important and manifold, it is fair FOR US to presume that, as for the firmament above, so with that below, the principles of conservation were in the beginning provided for each alike, for the world in the sky and the drop in the sea; that when the Creator gathered the waters together into one place, and pronounced his handiwork "GOOD," some check or regulator had already been provided for the one as well as the other — checks which should keep the sea up to its office, preventing it from growing, in the process of ages, either larger or smaller, fresher or salter. As we go down into the depths of the sea, we find that we are just beginning to penetrate the chambers of its hidden things, and to comprehend its wonders. The heart of man was never rightly attuned to the music of the spheres until he was permitted to stand with his eye at the telescope, and then, for the first time, the song of the morning stars burst upon him in all their glory. And so it is with the harmonies of old Ocean when contemplated through the microscope; then every drop of water in the sea is discovered to be in tune with the hosts of heaven, for each stands forth a peopled world.

498. *The microscope and the telescope.* Catching, as we contemplate the hosts of heaven through the telescope and the moving creatures of

the sea through the microscope, the spirit of Chalmers, and borrowing his fine imagery, let us draw a contrast between the glories of the heavens and the wonders of the insect world of earth and sea, as they are presented through these instruments to the mind of a devout philosopher: 'one leads him to see a world in every atom, the other a system for every star. One shows him that this vast globe, with its mighty nations and multitudinous inhabitants, is but a grain of sand in the immensity of space; the other, that every particle of clay that lies buried in the depths of the sea has been a living habitation, containing within it the workshops of a busy population. One tells him of the insignificance of the world we inhabit; the other redeems it from that insignificance by showing in the leaves of the forest, in the flowers of the field, and in every drop of water in the sea, worlds as numberless as the sands on its shores, all teeming with life, and as radiant as the firmament with glories. One suggests that, beyond and above all that is visible to man, there are fields of creation which sweep immeasurably along, and carry to the remotest regions of space the impress of the Almighty hand; the other reminds us that, within and beneath all that minuteness which the eye of man has been able to explore, there may be a region of invisibles, and that, could we draw aside the veil that hides it from our senses, we should behold a theatre of as many worlds as astronomy has unfolded — a universe within the compass of a point so small as to elude the highest power of the microscope, but where the wonder-working finger of the Almighty finds room for the exercise of his attributes — where He can raise another mechanism of worlds, filling and animating them all with the evidences of His glory.' When we lay down the microscope, and study the organisms of the sea by the light of reason, we find grounds for the belief that the sea was made salt in the beginning, for the marine fossils that are found nearest the foundation of the geological column remind us that in their day the sea was salt; and then, when we take up the microscope again to study the foraminiferæ, the diatomes, and corallines, and examine the structure of the most ancient inhabitants of the deep, comparing their physiology with that of their kindred in the fossil state, we are left to conjecture no longer, but are furnished with evidence and proof the most convincing and complete that the sea is salt from a physical necessity.

499. *Sea-shells and animalculæ in a new light.* Thus we behold sea-shells and animalculæ in a new light. May we not now cease to regard them as beings which have little or nothing to do in maintaining the harmonies of creation? On the contrary, do we not see in them the principles of the most admirable compensation in the system of oceanic circulation? We may even regard them as regulators, to some extent, of climates in parts of the earth far removed from their presence. There is something suggestive, both of the grand and the beautiful, in the idea

that, while the insects of the sea are building up their coral islands in the perpetual summer of the tropics, they are also engaged in dispensing warmth to distant parts of the earth, and in mitigating the severe cold of the polar winter. Surely an hypothesis which, being followed out, suggests so much design, such perfect order and arrangement, and so many beauties for contemplation and admiration as does this, which, for want of a better, I have ventured to offer with regard to the solid matter of the sea water, its salts and its shells — surely such an hypothesis, though it be not based entirely on the results of actual observation, can not be regarded as wholly vain or as altogether profitless.

CHAPTER XI

§ 501–526. THE CLOUD REGION, THE EQUATORIAL CLOUD-RING, AND SEA FOGS

501. *Cloud region — highest in the calm belts.* To simplify the discussion of these phenomena, let us consider fogs at sea to be in character like clouds in the sky. So treating them, and confining our attention to them as they appear to the mariner, we discover that the cloud region in the main is highest in the trade-wind and calm belts, lowest in extratropical regions.

502. *Fogless regions.* At sea, beyond "the offings," fogs are rarely seen between the parallels of 30° N. and S. Sea fogs, therefore, may be considered a rare phenomenon over one half of the surface of the globe. These fogless regions, though certain parts of them are not unfrequently visited by tempests, tornadoes, and hurricanes, are nevertheless much less frequented by gales of wind, as all furious winds are called, than are the regions on the polar side of these two parallels.

503. *The most stormy latitudes.* Taking the Atlantic Ocean, north and south, as an index of what takes place on other waters, the abstract logs of the Observatory show, according to the records of 265,304 observing days contained therein, that for every gale of wind that seamen encounter on the equatorial side of these two parallels of 30° N. and S., they encounter 10.4 on the polar side, and that for every fog on the equatorial they encounter 83 on the polar side. As a rule, fogs and gales increase both in numbers and frequency as you recede from the equator. The frequency of these phenomena between the parallels of 5° N. and 5° S., compared with their frequency between the parallels of 45° and 50° N. and S., is as 1 to 103 for gales, and as 1 to 102 for fogs. The observations do not extend beyond the parallels of 60°. It appears from these, however, that both the most stormy and foggy latitudes in the North Atlantic are between the parallels of 45° and 50°; that in the South Atlantic the most stormy latitudes are between the parallels of 55° and 60°, the most foggy between 50° and 55°.

504. *Influences of the Gulf Stream and the ice-bearing currents of the south.* How suggestively do these two groups of phenomena remind us, on the one hand, of the Gulf Stream and the ice-bearing currents of the north, and, on the other, of Cape Horn and the antarctic icebergs which cluster off the Falkland Islands! [1]

[1] Captain Chadwick reports, by letter of 30th April, 1860, an iceberg, first seen by

505. *Sea fogs rare within 20° of the equator — red fogs.* Sea fogs within 20° on either side of the equator are things of rare occurrence. Within this distance, however, on the north side, red fogs of "sea-dust" (§ 322) are not unfrequently encountered by navigators. These can scarcely be considered as coming within the category of sea fogs. The falling of this dust in the form of fog is no doubt owing to those influences (§ 331), the effects of which are so often observable morning and evening in the settling smoke from neighboring chimneys. The fogs which at early dawn are discovered hovering over our cities or skirting the base of the hills, are of the same sort. These particles of dust, like the atoms of smoke, are brought into conditions favorable for radiation on occasions when the air in which they are floating happens to have a high dew point. Thus each one of these innumerable little atoms of smoke and microscopic particles of sea-dust become loaded with dew, and being made visible, have the appearance of fog. Red fogs, therefore, do not properly come under our classification of sea fogs.

506. *Cloudless regions and height of clouds at sea.* On the polar side of 40° at sea the weather is for the most part cloudy. On the equatorial side, and especially within the trade-wind region, it is for the most part clear until we approach the cloud-ring, where clouds again indicate the normal state of the sky at sea. What is the height of the cloud region at sea? for vapor *plane* it can scarce be called. As yet our sailor observers have not turned their attention either to the height or the velocity of clouds. It is to be hoped that they will. Observations here are to be made rather under the direction of the commander of a fleet or squadron than of a single ship, and it is hoped that some of the distinguished admirals and brave old commodores who cruise about the world, with willing hearts and ready hands for the cause we advocate, may signalize their flag by contributing, for the advancement of human knowledge touching the physics of the sea and the machinery of the air, a series of well-conducted observations upon the force of the trade-winds,[2] upon the height and velocity of the clouds, the height and velocity of the waves, etc., in different parts of the ocean.

507. *Height and velocity of waves — plan for determining.* Commodore Wüllerstorf, of the Austrian frigate Novara, made an interesting series of observations upon the height and velocity of the waves during

---

him 14th September, 1859, in S. lat. 52° 25′, long. 57° 8′ W.; next, October 10th, in 47° 15′ S., 59° 30′ W., by the Wild Pigeon. Five days later he fell in with it in lat. 45° 40′, long. 58° 40′. It was last seen 7th November, in lat. 43° 44′ S., long. 57° 14′ W., by the British ship "City of Candy." Whether this were the same "berg" or not, it shows that icebergs are not unknown to the north of the Falkland Islands, as, indeed, the aqueous isotherm of 60°, Fig. 9, indicates by its sharp curve about those islands.

[2] *Sailing Directions*, 8th ed., II [80, 91].

his cruise in that vessel upon his last scientific mission. These no doubt will be published with the other important results of that admirably-conducted expedition. The most simple plan for determining the velocity of waves, and it may be hourly practiced on board of every vessel, is the plan which is followed by Captain Ginn, of the American ship John Knox, one of our co-operators. When he heaves the log with the seas following, instead of hauling in the line immediately, he leaves the chip to tow, watching till he observes it on the crest of a wave; he then turns the glass, or notes his watch, and marks the time it takes the wave to reach the ship. The usual velocity of the waves in the Atlantic is 22–3 miles an hour, off Cape Horn 26–8.[3]

508. *Determining the height of clouds at sea.* It would afford a pleasant and agreeable diversion for a squadron of men-of-war, as they pursue their voyage at sea, to amuse themselves and instruct their friends at home with observations upon all such phenomena. Those who are willing to undertake the clouds will have no difficulty in devising a plan both for the upper and the lower strata.

509. *Cloud region at sea in the shape of a double inclined plane.* Over the land the cloud region is thought to vary from three to five miles in height; there the height of clouds is known to be very variable. At sea it is no doubt less so. Here the cloud region is somewhat in the form of a double inclined plane, stretching north and south from the equatorial cloud-ring as a sort of ridge-pole. In the balloon ascents which have taken place from the Kew Observatory in England, it has been ascertained that there the cloud region is from 2000 to 6500 feet high, with a thickness varying from 2000 to 3000 feet, and that its temperature at the top is not lower than it is at the bottom of the cloud, notwithstanding its thickness. We are also indebted to Piazzi Smyth for interesting observations on the cloud region in the belt of northeast trades and of the upper counter current there. They were made from the Peak of Teneriffe, at

---

[3] *From Captain Ginn's Abstract Log.* "Saturday, September 11th, 1858, doubling Cape Horn. The long regular swell during this part of the day afforded me another opportunity of trying the velocity of the waves. This I did by paying out the log-line enough to be equal to 13 knots with the 14-second glass; then by watching the chip — to which I had fixed a piece of white rag to render it more distinguishable — as it appeared on the crest of a well-defined wave, and turning the glass at the same time, and then noting where the crest of the wave is at the moment the glass is 'out.'

"I have several times before tried the experiment in this way with the same length of line out astern, and have always found about the same rate for the velocity, namely, 22 to 23 miles an hour; but to-day I found it to be considerably more, namely, 26 to 28 miles an hour. Thus the crest of a wave would pass, while the 14-second glass ran out, from the place where the log-chip was towing astern (13 knots) to just ahead of the ship. The length of the ship is equal to about 6½ knots; the ship's speed at the time was 8 knots; thus, $13 + 6\frac{1}{2} + 8 = 27\frac{1}{2}$. A few days ago I tried the same experiment, and found the velocity to be 22 to 23. What has accelerated the velocity of these waves? Has the soundings any thing to do with it?"

the height of 12,200 feet, during the months of August and September, 1856.[4] The cloud region of the trades was between 3000 and 5000 feet high; of the upper or southwest current, it was above the mountain. Islands only a few hundred feet high are generally cloud-capped in the trade-wind regions at sea; another indication that, with a given amount of moisture in the wind, the cloud region is higher at sea than it is over the land. For most of the time during his sojourn on the Peak, the sea was concealed from view by the cloud stratum below, though the sky was clear overhead. Farther to the north, in the Atlantic, however, as in the fog region about the meeting of the cool and warm currents near the Grand Banks, the look-out at the masthead often finds himself above the fog or cloud in which the lower parts of the ship are enveloped. Going still farther toward the north and reaching the ice, the cloud region would again, for obvious reasons, mount up until you reached the open sea there, when again it would touch the earth with its smoke.

510. *Fogs in the harbor of Callao.* In the harbor of Callao, in Peru, which is filled with the cool waters of Humboldt's current, I have seen the bay covered with a fog only a few inches high. I have seen fogs there so dense, and with outlines so sharp, as to conceal from view the row-boats approaching the ship's side. These fogs, especially early in the morning, are sometimes so thick as to conceal from view not only the boat, but the persons of the crew up to the neck, so as to leave nothing visible but two rows of trunkless heads nodding catenaries at the oars, skimming through the air and dancing on the fog in a manner at once both magical and fantastic. At other times the cloud stratum is thicker and higher. Then may be seen three masts coming into port with top-gallant-sails and royals set, but no ship. These sails, nicely trimmed and swelling to the breeze in the sky, skim along over the clouds, and seem like things in a fairy scene. However, there are influences exerted in the formation of clouds and fogs over and near the land which appear not to be felt at sea.

511. *The cloudy latitudes.* In the extra-tropical north, the cloud region is high over the land, low over the water; and, as a rule, the farther inland, the dryer the air and the higher the cloud region. In the circum-antarctic regions, where all is sea, the rising vapors form themselves into clouds low down, and keep the face of the sky almost uninterruptedly obscured. The southern *eaves* of the cloud plane (§ 509), like the calm belts, vary their latitude as the sun does its declination, though their place is generally found between the parallels of 50° and 70° S. — farther or nearer according to the season; but under this edge, wherever it be, the mariner's heart is seldom made glad by the cheering influences of a clear sky. If not wrapped in mist, or covered with snow,

[4] *Teneriffe* [pp. 102, 169, and *passim*].

or pelted with hail, or drenched with rain, as he sails through these latitudes, he is dispirited under the influences of the gloomy and murky
weather which pervades those regions. His hope in the "brave west
winds" and trust in the prowess of a noble ship are then his consolation
and his comfort.

512. *Why there should be less atmosphere in the southern than in the
northern hemisphere.* Such are the quantities of vapor rising up from the
engirdling ocean about those austral regions, that it keeps permanently
expelled thence a large portion of the atmosphere. The specific gravity of
dry air being 1, that of aqueous vapor is 0.6 (§ 252). According to the
table (§ 362), the mean height of the barometer at sea, between the
equator and 78° 37' north, is 30.01; while its mean height in lat. 70° S.
is 29.0. To explain the great and grand phenomena of nature by illustrations drawn from the puny contrivances of human device is often a feeble
resort, but nevertheless we may, in order to explain this expulsion of air
from the watery south, where all is sea, be pardoned for the homely
reference. We all know, as the steam or vapor begins to form in the tea-
kettle, it expels air thence, and itself occupies the space which the air
occupied. If still more heat be applied, as to the boiler of a steam-engine,
the air will be entirely expelled, and we have nothing but steam above
the water in the boiler. Now at the south, over this great waste of circum-
fluent waters, we do not have as much heat for evaporation as in the
boiler or the tea-kettle; but, as far as it goes, it forms vapor which has
proportionally precisely the same tendency that the vapor in the tea-
kettle has to drive off the air above and occupy the space it held. Nor is
this all. This austral vapor, rising up, is cooled and condensed. Thus a
vast amount of heat is liberated in the upper regions, which goes to heat
the air there, expand it, and thus, by altering the level, causing it to flow
off. This unequal distribution of atmosphere between the two halves of
the globe is represented in *barometric* profile by Plate I (§ 215) — the
shading around the periphery of the circle being intended to represent
the relative height, and the scales standing up in it the barometric
column.

513. *Influence of antarctic icebergs in expelling the air from austral
regions.* This part of the southern ocean where the barometer shows
diminished pressure is frequented by icebergs, many of them very large
and high, and some of them sending up towers, minarets, and steeples,
which give them the appearance in the distance of beautiful cities afloat.
Each one of them is a centre of condensation. Could an eye from aloft
look down upon the scene, the upper side of the cloud stratum would
present somewhat the appearance of an immense caldron, boiling, and
bubbling, and intumescing in the upper air. These huge bergs condense
the vapor, and the liberated heat causes the air above them to swell out,

and to stand like so many curiously-shaped fungi above the general cloud level. And thus, where the icebergs are thick, the clouds are formed low down. Icebergs, like islands, facilitate the formation of clouds and promote precipitation.

514. *The horse latitudes — the doldrums.* Turn we now to the equatorial cloud-ring. Seafaring people have, as if by common consent, divided the ocean off into regions, and characterized them according to the winds; *e. g.*, there are the "trade-wind regions," the "variables," the "horse latitudes," the "doldrums," etc. The "horse latitudes" are the belts of calms and light airs (§ 210) which border the polar edge of the northeast trades. They were so called from the circumstance that vessels formerly bound from New England to the West Indies, with a deck-load of horses, were often so delayed in this calm belt of Cancer, that, for the want of water for their animals, they were compelled to throw a portion of them overboard. The "equatorial doldrums" is another of these calm places (§ 212). Besides being a region of calms and baffling winds, it is a region noted for its rains and clouds, which make it one of the most oppressive and disagreeable places at sea. The emigrant ships from Europe for Australia have to cross it. They are often baffled in it for two or three weeks; then the children and the passengers who are of delicate health suffer most. It is a frightful grave-yard on the way-side to that golden land. A vessel bound into the southern hemisphere from Europe or America, after clearing the region of variable winds and crossing the "horse latitudes," enters the northeast trades. Here the mariner finds the sky sometimes mottled with clouds, but for the most part clear. Here, too, he finds his barometer rising and falling under the ebb and flow of a regular atmospherical tide, which gives a high and low barometer every day with such regularity that the hour within a few minutes may be told by it. The rise and fall of this tide, measured by the barometer, amounts to about one tenth (0.1) of an inch, and it occurs daily and every where between the tropics: the maximum about 10h. 30m. A.M., the minimum between 4h. and 5h. P.M., with a second maximum and minimum about 10 P.M. and 5 A.M.[5] The diurnal variation of the needle (§ 344) changes also with the turning of these invisible tides. Continuing his course toward the equinoctial line, and entering the region of equatorial calms and rains, the navigator feels the weather to become singularly close and oppressive; he discovers here that the elasticity of feeling which he breathed from the trade-wind air has forsaken him; he has entered the doldrums, and is under the "cloud-ring."

515. *A frigate under the cloud-ring.* I find in the journal of the late Commodore Arthur Sinclair, kept on board the United States frigate

---

[5] [W. H.] Sykes ["Discussion of Meteorological Observations taken in India . . ."], *Philos. Trans.* [CXL] (1850) [297–378, esp., pls. 17–20].

Congress during a cruise to South America in 1817–18, a picture of the weather under this *cloud-ring* that is singularly graphic and striking. He encountered it in the month of January, 1818, between the parallel of 4° north and the equator, and between the meridians of 19° and 23° west. He says of it, "This is certainly one of the most unpleasant regions in our globe. A dense, close atmosphere, except for a few hours after a thunderstorm, during which time torrents of rain fall, when the air becomes a little refreshed; but a hot, glowing sun soon heats it again, and but for your awnings, and the little air put in circulation by the continual flapping of the ship's sails, it would be almost insufferable. No person who has not crossed this region can form an adequate idea of its unpleasant effects. You feel a degree of lassitude unconquerable, which not even the sea-bathing, which every where else proves so salutary and renovating, can dispel. Except when in actual danger of shipwreck, I never spent twelve more disagreeable days in the professional part of my life than in these calm latitudes. I crossed the line on the 17th of January, at eight A.M., in longitude 21° 20′, and soon found I had surmounted all the difficulties consequent to that event; that the breeze continued to freshen and draw round to the south-southeast, bringing with it a clear sky and most heavenly temperature, renovating and refreshing beyond description. Nothing was now to be seen but cheerful countenances, exchanged as by enchantment from that sleepy sluggishness which had borne us all down for the last two weeks."

516. *Subjects which at sea present themselves for contemplation.* One need not go to sea to perceive the grand work which the clouds perform in collecting moisture from the crystal vaults of the sky, in sprinkling it upon the fields, and making the hills glad with showers of rain. Winter and summer, "the clouds drop fatness upon the earth." This part of their office is obvious to all, and I do not propose to consider it now. But the sailor at sea observes phenomena and witnesses operations in the terrestrial economy which tell him that, in the beautiful and exquisite adjustments of the grand machinery of the atmosphere, the clouds have other important offices to perform besides those merely of dispensing showers, of producing the rains, and of weaving mantles of snow for the protection of our fields in winter. As important as are these offices, the philosophical mariner, as he changes his sky, is reminded that the clouds have commandments to fulfill, which, though less obvious, are not therefore the less benign in their influences, or the less worthy of his notice. He beholds them at work in moderating the extremes of heat and cold, and in mitigating climates. At one time they spread themselves out; they cover the earth as with a mantle; they prevent radiation from its crust, and keep it warm. At another time they interpose between it and the sun; they screen it from his scorching rays, and protect the tender plants from his

heat, the land from the drought; or, like a garment, they overshadow the sea, defending its waters from the intense forces of evaporation. Having performed these offices for one place, they are evaporated and given up to the sunbeam and the winds again, to be borne on their wings away to other places which stand in need of like offices. Familiar with clouds and sunshine, the storm and the calm, and all the phenomena which find expression in the physical geography of the sea, the right-minded mariner, as he contemplates "the cloud without rain," ceases to regard it as an empty thing; he perceives that it performs many important offices; he regards it as a great moderator of heat and cold — as a "compensation" in the atmospherical mechanism which makes the performance perfect. Marvelous are the offices and wonderful is the constitution of the atmosphere. Indeed, I know of no subject more fit for profitable thought on the part of the truth-loving, knowledge-seeking student, be he seaman or landsman, than that afforded by the atmosphere and its offices. Of all parts of the physical machinery, of all the contrivances in the mechanism of the universe, the atmosphere, with its offices and its adaptations, appears to me to be the most wonderful, sublime, and beautiful. In its construction, the grandeur of knowledge is displayed. The perfect man of Uz, in a moment of inspiration, thus bursts forth in laudation of this part of God's handiwork, demanding of his comforters, "But where shall wisdom be found, and where is the place of understanding? The depth saith, it is not in me; and the sea saith, it is not with me. It can not be gotten for gold, neither shall silver we weighed for the price thereof. No mention shall be made of coral or of pearls, for the price of wisdom is above rubies. Whence, then, cometh wisdom, and where is the place of understanding? Destruction and Death say, we have heard the fame thereof with our ears. God understandeth the way thereof, and he knoweth the place thereof; for he looketh to the ends of the earth, and seeth under the whole heaven; *to make the weight for the winds*; and he weigheth the waters by measure. When he made a decree for the rain, and a way for the lightning of the thunder, then did he see it and declare it; he prepared it, yea, and searched it out." [6] When the pump-maker came to ask Galileo to explain how it was that his pump would not lift water higher than thirty-two feet, the philosopher thought, but was afraid to say, it was owing to "the weight of the winds;" and though the fact that the air has weight is here so distinctly announced, philosophers never recognized the fact until within comparatively a recent period, and then it was proclaimed by them as a great discovery. Nevertheless, the fact was set forth as distinctly in the book of nature as it is in the book of revelation; for the infant, in availing itself of atmospherical pressure to draw milk from its mother's breast, unconsciously proclaimed it.

[6] Job, chap. xxviii.

517. *The barometer under the cloud-ring.* The barometer [7] stands lower under this cloud-ring than on either side of it (§ 362). After having crossed it, the attentive navigator may perceive how this belt of clouds, by screening the parallels over which he may have found it to hang from the sun's rays, not only promotes the precipitation which takes place within these parallels at certain periods, but how, also, the rains are made to change the places upon which they are to fall; and how, by traveling with the calm belt of the equator up and down the earth, this cloud-ring shifts the surface from which the heating rays of the sun are to be excluded; and how, by this operation, tone is given to the atmospherical circulation of the world, and vigor to its vegetation.

518. *Its motions.* Having traveled with the calm belt to the north or south, the cloud-ring leaves a clear sky about the equator; the rays of the torrid sun then pour down upon the solid crust of the earth there, and raise its temperature to a scorching heat. The atmosphere dances (§ 356), and the air is seen trembling in ascending and descending columns, with busy eagerness to conduct the heat off and deliver it to the regions aloft, where it is required to give dynamical force to the air in its general channels of circulation. The dry season continues; the sun is vertical; and finally the earth becomes parched and dry; the heat accumulates faster than the air can carry it away; the plants begin to wither, and the animals to perish. Then comes the mitigating cloud-ring. The burning rays of the sun are intercepted by it: the place for the absorption and reflection, and the delivery to the atmosphere of the solar heat, is changed; it is transferred from the upper surface of the earth to the upper surface of the clouds.

519. *Meteorological processes.* Radiation from land and sea below the cloud-belt is thus interrupted, and the excess of heat in the earth is delivered to the air, and by absorption carried up to the clouds, and there transferred to their vapors to prevent excess of precipitation. In the mean time, the trade-winds north and south are pouring into this cloud-covered receiver, as the calm and rain belt of the equator may be called, fresh supplies in the shape of ceaseless volumes of heated air, which, loaded to saturation with vapor, has to rise above and get clear of the clouds before it can commence the process of cooling by radiation. In the mean time, also, the vapors which the trade-winds bring from the north and the south, expanding and growing cooler as they ascend, are being condensed on the lower side of the cloud stratum, and their latent heat is set free, to check precipitation and prevent a flood. While this process and these operations are going on upon the nether side of the

[7] Observations *now* show that the thermometer stands highest *under* the cloud-ring. Indeed, the indications are that it coincides with the thermal equator.

cloud-ring, one not less important is, we may imagine, going on upon the upper side. There, from sunrise to sunset, the rays of the sun are pouring down without intermission. Every day, and all day long, they play with ceaseless activity upon the upper surface of the cloud stratum. When they become too powerful, and convey more heat to the cloud vapors than the cloud vapors can reflect and give off to the air above them, then, with a beautiful elasticity of character, the clouds absorb the surplus heat. They melt away, become invisible, and retain, in a latent and harmless state, until it is wanted at some other place and on some other occasion, the heat thus imparted. We thus have an insight into the operations which are going on in the equatorial belt of precipitation, and this insight is sufficient to enable us to perceive that exquisite indeed are the arrangements which Nature has provided for supplying this calm belt with heat, and of pushing the snow-line there high up above the clouds, in order that the atmosphere may have room to expand, to rise up, overflow, and course back into its channels of healthful circulation. As the vapor is condensed and formed into drops of rain, a two-fold object is accomplished; coming from the cooler regions of the clouds, the rain-drops are cooler than the air and earth below; they descend, and by absorption take up the heat which has been accumulating in the earth's crust during the dry season, and which can not now escape by radiation.

520. *Snow-line mounts up as it crosses the equatorial calm belt.* In the process of condensation, these rain-drops, on the other hand, have set free a vast quantity of latent heat, which has been gathered up with the vapor from the sea by the trade-winds and brought hither. The caloric thus liberated is taken by the air and carried up aloft still farther, to keep, at the proper distance from the earth, the line of perpetual congelation. Were it possible to trace a thermal curve in the upper regions of the air to represent this line, we should no doubt find it mounting sometimes at the equator, sometimes on this side, and sometimes on that of it, but always so mounting as to overleap this cloud-ring. This thermal line would not ascend always over the same parallels: it would ascend over those between which this ring happens to be; and the distance of this ring from the equator, north or south, is regulated according to the seasons. If we imagine the atmospherical equator to be always where the calm belt is which separates the northeast from the southeast trade-winds, then the loop in the thermal curve, which should represent the line of perpetual congelation in the air, would be always found to stride this equator; and it may be supposed that a thermometer, kept sliding on the surface of the earth so as always to be in the middle of this rain-belt, would show very nearly the same temperature all the year round; and so, too, would a barometer the same pressure, though the

height of the atmosphere over this calm belt would, in consequence of so much heat and expansion, be very much greater than it is over the trade-winds or tropical calms.

521. *Offices of the cloud-ring.* Returning and taking up the train of contemplation as to the office which this belt of clouds, as it encircles the earth, performs in the system of oceanic adaptations, we may see how the cloud-ring and calm zone which it overshadows perform the office both of ventricle and auricle in the immense atmospherical heart, where the heat and the forces which give vitality and power to the system are brought into play — where dynamical strength is gathered, and an impulse given to the air sufficient to send it thence through its long and tortuous channels of circulation.

522. *It acts as a regulator.* Thus this ring, or band, or belt of clouds is stretched around our planet to regulate the quantity of precipitation in the rain-belt beneath it; to preserve the due quantum of heat on the face of the earth; to adjust the winds; and send out for distribution to the four corners vapors in proper quantities to make up to each river-basin, climate, and season its quota of sunshine, cloud, and moisture. Like the balance-wheel of a well-constructed chronometer, this cloud-ring affords the grand atmospherical machine the most exquisitely-arranged *self-compensation.* If the sun fail in his supply of heat to this region, more of its vapors are condensed, and heat is discharged from its latent store-houses in quantities just sufficient to keep the machine in the most perfect compensation. If, on the other hand, too much heat be found to accompany the rays of the sun as they impinge upon the upper circumference of this belt, then again on that side the means of self-compensation are ready at hand; so much of the cloud-surface as may be requisite is then resolved into invisible vapor — for of invisible vapor are made the vessels wherein the surplus heat of the sun is stored away and held in the latent state until it is called for, when it is instantly set free, and becomes a palpable and an active agent in the grand design.

523. *The latent heat liberated in the processes of condensation from and under the cloud-ring, true cause of the trade-winds.* Ceaseless precipitation goes on under this cloud-ring. Evaporation under it is suspended almost entirely. We know that the trade-winds encircle the earth; that they blow perpetually; that they come from the north and the south, and meet each other near the equator; therefore we infer that this line of meeting extends around the world. By the rainy seasons of the torrid zone, except where it may be broken by the continents, we can trace the declination of this cloud-ring, stretched like a girdle around our planet, up and down the earth; it travels after the sun up and down the ocean, as from north to south and back. It is broader than the belt of calms out of which it rises. As the air, with its vapors, rises up in this calm belt and

ascends, these vapors are condensed into clouds, and this condensation is followed by a turgid intumescence, which causes the clouds to overflow the calm belt, as it were, both to the north and the south. The air flowing off in the same direction assumes the character of winds that form the upper currents that are counter (Plate I) to the trade-winds. These currents carry the clouds still farther to the north and south, and thus make the cloud-ring broader. At least we infer such to be the case, for the rains are found to extend out into the trade-winds, and often to a considerable distance both to the north and the south of the calm belt.

524. *Imagined appearance of the cloud-ring to a distant observer.* Were this cloud-ring luminous, and could it be seen by an observer from one of the planets, it would present to him an appearance not unlike the rings of Saturn do to us. Such an observer would remark that this cloud-ring of the earth has a motion contrary to that of the axis of our planet itself — that while the earth was revolving rapidly from west to east, he would observe the cloud-ring to go slowly, but only relatively, from east to west. As the winds which bring this cloud-vapor to this region of calms rise up with it, the earth is slipping from under them; and thus the cloud-ring, though really moving from west to east with the earth, goes relatively slower than the earth, and would therefore appear to require a longer time to complete a revolution. But, unlike the rings of Saturn through the telescope, the outer surface, or the upper side to us, of this cloud-ring would appear exceedingly jagged, rough, and uneven.

525. *Thunder.* The rays of the sun, playing upon this peak and then upon that of the upper cloud-surface, melt away one set of elevations and create another set of depressions. The whole stratum is, it may be imagined, in the most turgid state; it is in continued throes when viewed from above; the heat which is liberated from below in the process of condensation, the currents of warm air ascending from the earth, and of cool descending from the sky, all, we may well conceive, tend to keep the upper cloud-surface in a perpetual state of agitation, upheaval, and depression. Imagine in such a cloud-stratum an electrical discharge to take place; the report, being caught up by the cloud-ridges above, is passed from peak to peak, and repeated from valley to valley, until the last echo dies away in the mutterings of the distant thunder. How often do we hear the voice of the loud thunder rumbling and rolling away above the cloud-surface, like the echo of artillery discharged among the hills! Hence we perceive or infer that the clouds intercept the progress of sound, as well as of light and heat, through the atmosphere, and that this upper surface is often like Alpine regions, which echo back and roll along with rumbling noise the mutterings of the distant thunder.

526. *Exceeding interest attached to physical research at sea.* It is by trains of reasoning like this that we are continually reminded of the in-

terest which attaches to the observations which the mariner is called on to make. There is no expression uttered by nature which is unworthy of our most attentive consideration — for no physical fact is too bald for study — and mariners, by registering in their logs the kind of lightning, whether sheet, forked, or streaked, and the kind of thunder, whether rolling, muttering, or sharp, may be furnishing facts which will throw much light on the features and character of the clouds in different latitudes and seasons. Physical facts are the language of Nature, and every expression uttered by her is worthy of our most attentive consideration, for it is the voice of WISDOM.

CHAPTER XII

§ 531–555. THE GEOLOGICAL AGENCY OF THE WINDS

531. *The sea and air regarded as parts of the same machine.* Properly to appreciate the various offices which the winds and the waves perform, we must regard nature as a whole, for all the departments thereof are intimately connected. If we attempt to study in one of them, we often find ourselves tracing clews which insensibly lead us off into others, and, before we are aware, we discover ourselves exploring the chambers of some other department. The study of drift takes the geologist out to sea, and reminds him that a knowledge of waves, winds, and currents, of navigation and hydrography, are closely and intimately connected with his specialty. The astronomer directs his telescope to the most remote star or to the nearest planet in the sky, and makes an observation upon it. He can not reduce this observation, nor make any use of it, until he has availed himself of certain principles of optics — until he has consulted the thermometer, gauged the atmosphere, and considered the effect of heat in changing its powers of refraction. In order to adjust the pendulum of his clock to the right length, he has to measure the water of the sea and weigh the earth. He, too, must therefore go into the study of the tides; he must examine the earth's crust, and consider the matter of which it is composed, from pole to pole, circumference to centre; and in doing this, he finds himself, in his researches, alongside of the navigator, the geologist, and the meteorologist, with a host of other good fellows, each one holding on by the same thread, and following it up into the same labyrinth — all, it may be, with different objects in view, but, nevertheless, each one feeling sure that he is to be led into chambers where there are stores of knowledge and instruction especially for him. And thus, in undertaking to explore the physical geography of the sea, I have found myself standing side by side with the geologist on the land, and with him, far away from the sea-shore, engaged in considering marine fossils, changes of climates, the effects of deserts upon the winds, or the influence of mountains upon rains, or some of the many phenomena which the inland basins of the earth — those immense indentations on its surface that have no sea-drainage — present for contemplation and study.

532. *The level of the Dead Sea.* Among the most interesting of these last is that of the Dead Sea. Lieutenant Lynch, of the United States Navy, has run a level from that sea to the Mediterranean, and finds the former to be about one thousand three hundred feet below the general

sea-level of the earth. In seeking to account for this great difference of water level, the geologist examines the neighboring region, and calls to his aid the forces of elevation and depression which are supposed to have resided in the neighborhood; he then points to them as the agents which did the work. Truly they are mighty agents, and they have diversified the surface of the earth with the most towering monuments of their power. But is it necessary to suppose that they resided in the vicinity of this region? May they not have come from the sea, and been, if not in this case, at least in the case of other inland basins, as far removed as the other hemisphere? This is a question which I do not pretend to answer definitely. But the inquiry as to the geological agency of the winds in such cases is a question which my investigations have suggested. It has its seat in the sea, and therefore I propound it as one which, in accounting for the formation of this or that inland basin, is worthy, at least, of consideration.

533. *An ancient river from it.* Is there any evidence that the annual amount of precipitation upon the water-shed of the Dead Sea, at some former period, was greater than the annual amount of evaporation from it now is? If yea, from what part of the sea did the vapor that supplied the excess of that precipitation come, and what has cut off that supply? The mere elevation of the rim and depression of the lake basin would not cut it off. If we establish the fact that the Dead Sea at a former period did send a river to the ocean, we carry along with this fact the admission that when that sea overflowed into that river, then the water that fell from the clouds over the Dead Sea basin was more than the winds could convert into vapor and carry away again; the river carried off the excess to the ocean whence it came (§ 260).

534. *Precipitation and evaporation in the Dead Sea valley.* In the basin of the Dead Sea, in the basin of the Caspian, of the Sea of Aral, and in the other inland basins of Asia, we are entitled to infer that the precipitation and evaporation are at this time exactly equal. Were it not so, the level of these seas would be rising or sinking. If the precipitation were in excess, these seas would be gradually becoming fuller; and if the evaporation were in excess, they would be gradually drying up; but observation does not show, nor history tell us, that either is the case. As far as we know, the level of these seas is as permanent as that of the ocean, and it is difficult to realize the existence of subterranean channels between them and the great ocean. Were there such a channel, the Dead Sea being the lower, it would be the recipient of ocean waters; and we can not conceive how it should be such a recipient without ultimately rising to the level of its feeder.

535. *Whence come its rains?* It may be that the question suggested by my researches has no bearing upon the Dead Sea; that local elevations

and subsidences alone were concerned in placing the level of its waters where it is. But is it probable that, throughout all the geological periods, during all the changes that have taken place in the distribution of land and water surface over the earth, the winds, which in the general channels of circulation pass over the Dead Sea, have alone been unchanged? Throughout all ages, periods, and formations, is it probable that the winds have brought us just as much moisture to that sea as they now bring, and have just taken up as much water from it as they now carry off? Obviously and clearly not. The salt-beds, the water-marks, the geological formations, and other facts traced by Nature's own hand upon the tablets of the rock, all indicate plainly enough that not only the Dead Sea, but the Caspian also, had upon them, in former periods, more abundant rains than they now have. Where did the vapor for those rains come from? and what has stopped the supply? Surely not the elevation or depression of the Dead Sea basin. My researches with regard to the winds have suggested the probability (§ 290) that the vapor which is condensed into rains for the lake valley, and which the St. Lawrence carries off to the Atlantic Ocean, is taken up by the southeast trade-winds of the Pacific Ocean. Suppose this to be the case, and that the winds which bring this vapor arrive with it in the lake country at a mean dew-point of 50°. Let us also admit the southwest winds to be the rain winds for the lakes generally, as well as for the Mississippi Valley; they are also, speaking generally, the rain winds of Europe, and, I have no doubt, of extra-tropical Asia also.

536. *The influence of mountain ranges.* Now suppose a certain mountain range, hundreds of miles to the southwest of the lakes, but across the path of these winds, with their dew-point at 50°, were to be suddenly elevated, and its crest pushed into the regions of snow, having a mean temperature at its summit of 30° Fahrenheit. The winds, in passing that range, would be subjected to a mean dew-point of 30°; and, not meeting (§ 297) with any more evaporating surface between such range and the lakes, they would have no longer any moisture to deposit at the supposed lake temperature of 50°; for they could not yield their moisture to any thing above 30°. Consequently, the amount of precipitation in the lake country would fall off; the winds which feed the lakes would cease to bring as much water as the lakes now give to the St. Lawrence. In such a case, that river and the Niagara would drain them to the level of their own beds; evaporation would be increased by reason of the dryness of the atmosphere and the want of rain, and the lakes would sink to that level at which, as in the case of the Caspian Sea, the precipitation and evaporation would finally become equal.

537. *How the level of Caspians is reduced.* There is a self-regulating principle that would bring about this equality; for as the water in the lakes becomes lower, the area of its surface would be diminished, and

the amount of vapor taken from it would consequently become less and less as the surface was lowered, until the amount of water evaporated would become equal to the amount rained down again, precisely in the same way that the amount of water evaporated from the sea is exactly equal to the whole amount poured back into it by the rains, the fogs, and the dews.[1] Thus the great lakes of this continent would remain inland seas at a permanent level; the salt brought from the soil by the washings of the rivers and rains would cease to be taken off to the ocean as it now is; and finally, too, the great American lakes, in the process of ages, would become first brackish, and then briny. Now suppose the water basins which hold the lakes to be over a thousand fathoms (six thousand feet) deep. We know they are not more than four hundred and twenty feet deep; but suppose them to be six thousand feet deep. The process of evaporation, after the St. Lawrence has gone dry, might go on until one or two thousand feet or more were lost from the surface, and we should then have another instance of the level of an inland water-basin being far below the sea-level, as in the case of the Dead Sea; or it would become a rainless district, when the lakes themselves would go dry. Or let us take another case for illustration. Corallines are at work about the Gulf Stream; they have built up the Florida Reefs on one side, and the Bahama Banks on the other. Suppose they should build up a dam across the Florida Pass, and obstruct the Gulf Stream; and that, in like manner, they were to connect Cuba with Yucatan by damming up the Yucatan Pass, so that the waters of the Atlantic should cease to flow into the Gulf of Mexico. What should we have? The depth of the marine basin which holds the waters of that Gulf is, in the deepest part, about a mile. We should therefore have, by stopping up the channels between the Gulf and the Atlantic, not a sea-level in the Gulf, but we should have a mean level between evaporation and precipitation. If the former were in excess, the level of the Gulf waters would sink down until the surface exposed to the air would be just sufficient to return to the atmosphere, as vapor, the amount of water discharged by the rivers — the Mississippi and others, into the Gulf. As the waters were lowered, the extent of evaporating surface would grow less and less, until Nature should establish the proper ratio between the ability of the air to take up and the capacity of the clouds to let down. Thus we might have a sea whose level would be much farther below the water-level of the ocean than is the Dead Sea.

538. *The formation of inland basins — a third process.* There is still another process, besides the one already alluded to, by which the drainage of these inland basins may, through the agency of the winds, have

---

[1] The quantity of dew in England is about five inches during a year. — *Glaisher.* [Cited by George Buist, Report of Secretary, *Trans. Bombay Geogr. Soc.*, vol IX, p. cv.]

been cut off by the great salt seas, and that is by the elevation of continents from the bottom of the sea in distant regions of the earth, and
the substitution caused thereby of dry land instead of water for the
winds to blow upon. Now suppose that a continent should rise up in that
part of the ocean, wherever it may be, that supplies the clouds with the
vapor that makes the rain for the hydrographic basin of the great
American lakes. What would be the result? Why, surely, fewer clouds
and less rain, which would involve a change of climate in the lake
country; an increase of evaporation from it, because a decrease of precipitation upon it; and, consequently, a diminution of cloudy screens to protect the waters of the lakes from being sucked up by the rays of the sun;
and consequently, too, there would follow a low stage for water-courses,
and a lowering of the lake-level would ensue.

539. *Examples.* So far, I have instanced these cases only hypothetically; but, both in regard to the hydrographical basins of the Mexican
Gulf and American lakes, I have confined myself strictly to analogies.
Mountain ranges have been upheaved across the course of the winds,
and continents have been raised from the bottom of the sea; and, no
doubt, the influence of such upheavals has been felt in remote regions
by means of the winds, and the effects which a greater or less amount of
moisture brought by them would produce. In the case of the Salt Lake of
Utah, we have an example of drainage that has been cut off, and an illustration of the process by which Nature equalizes the evaporation and precipitation. To do this, in this instance, she is salting up the basin which
received the drainage of this inland water-shed. Here we have the appearance, I am told, of an old channel by which the water used to flow
from this basin to the sea. Supposing there was such a time and such a
water-course, the water returned through it to the ocean was the amount
by which the precipitation used to exceed the evaporation over the whole
extent of country drained through this now dry bed of a river. The winds
have had something to do with this; they are the agents which used to
bring more moisture from the sea to this water-shed than they carried
away; and they are the agents which now carry off from that valley more
moisture than is brought to it, and which, therefore, are making a salt-
bed of places that used to be covered by water. In like manner, there is
evidence that the great American lakes formerly had a drainage with the
Gulf of Mexico; for boats or canoes have been actually known, in former
years, and in times of freshets, to pass from the Mississippi River over into
the lakes. At low water, the bed of a dry river can be traced between
them. Now the Salt Lake of Utah is to the southward and westward of
our northern lake basin; that is the quarter (§ 357) whence the rain-
winds have been supposed to come. May not the same cause which lessened the preciptation or increase the evaporation in the Salt Lake water-

shed, have done the same for the water-shed of the great American system of lakes? If the mountains to the west — The Sierra Nevada, for instance — stand higher now than they formerly did, and if the winds which
feed the Salt Lake valley with precipitation formerly had, as I suppose
they now have, to pass the summits of these mountains, it is easy to perceive why the winds should not convey as much vapor across them now as
they did when the summit of the range was lower and not so cool. The
Andes, in the trade-wind region of South America, stand up so high, that
the wind, in order to cross them, has to part with all its moisture (§ 297),
and consequently there is, on the west side, a rainless region. Now suppose a range of such mountains as these to be elevated across the track of
the winds which supply the lake country with rains; it is easy to perceive how the whole country to the leeward of such range, and now
watered by the vapor which such winds bring, would be converted into a
rainless region. I have used these hypothetical cases to illustrate a position which any philosopher, who considers the geological agency of the
winds, may with propriety consult, when he is told of an inland basin the
water-level of which, it is evident, was once higher than it now is; and
that position is that, though the evidences of a higher water-level be un-
mistakable and conclusive, it does not follow therefore that there has been
a subsidence of the lake basin itself, or an upheaval of the water-shed
drained by it. The cause which has produced this change in the water-
level, instead of being local and near, may be remote; it may have its
seat in the obstructions to "the wind in his circuits," which have been in-
terposed in some other quarter of the world, which obstructions may prevent the winds from taking up or from bearing off their wonted supplies
of moisture for the region whose water-level has been lowered.

540. *The influence of the South American continent upon the climate
of the Dead Sea.* Having therefore, I hope, made clear the meaning of
the question proposed, by showing the manner in which winds may become important geological agents, and having explained how the upheav-
ing of a mountain range in one part of the world may, through the winds,
bear upon the physical geography of the sea, affect climates, and produce geological phenomena in another, I return to the Dead Sea and the
great inland basins of Asia, and ask, How far is it possible for the elevation of the South American continent, and the upheaval of its mountains,
to have had any effect upon the water-level of those seas? There are
indications (§ 535) that they all once had a higher water-level than they
now have, and that formerly the amount of precipitation was greater than
it now is; then what has become of the sources of vapor? What has diminished its supply? Its supply would be diminished (§ 538) either by the
substitution of dry land for water-surface in those parts of the ocean
which used to supply that vapor; or the quantity of vapor deposited in the

hydrographical basins of those seas would have been lessened if a snow-capped range of mountains (§ 536) had been elevated across the path of these winds, between the places where they were supplied with vapor and these basins. A chain of evidence which it would be difficult to set aside is contained in Chapters [IV, VI, and VII], going to show that the vapor which supplies the extra-tropical regions of the north with rains comes, in all probability, from the trade-wind regions of the southern hemisphere.

541. *The path of the S.E. trade-winds over into the northern hemisphere.* Now if it be true that the trade-winds from that part of the world take up there the water which is to be rained in the extra-tropical north, the path ascribed to the southeast trades of Africa and America, after they descend and become the prevailing southwest winds of the northern hemisphere, should pass over a region of less precipitation generally than they would do if, while performing the office of southeast trades, they had blown over water instead of land. The southeast trade-winds, with their load of vapor, whether great or small, take, after ascending in the equatorial calms, a northeasterly direction; they continue to flow in the upper regions of the air in that direction until they cross the tropic of Cancer. The places of least rain, then, between this tropic and the pole, should be precisely those places which depend for their rains upon the vapor which the winds that blow over southeast trade-wind Africa and America convey. Now, if we could trace the path of the winds through the extra-tropical regions of the northern hemisphere, we should be able to identify the track of these Andean winds by the droppings of the clouds; for the path of the winds which depend for their moisture upon such sources of supply as the dry land of Central South America and Africa can not overshadow a country that is watered well. It is a remarkable fact that the countries in the extra-tropical regions of the north that are situated to the northeast of the southeast trade-winds of South Africa and America — that these countries, over which theory makes these winds to blow, include all the great deserts of Asia, and the districts of least precipitation in Europe. A line from the Galapagos Islands through Florence in Italy, another from the mouth of the Amazon through Aleppo in Holy Land (Plate IV), would, after passing the tropic of Cancer, mark upon the surface of the earth the route of these winds; this is that "lee country" (§ 298) which, if such be the system of atmospherical circulation, ought to be scantily supplied with rains. Now the hyetographic map of Europe, in Johnston's beautiful *Physical Atlas*, places the region of least precipitation between these two lines (Plate IV).

542. *Relays for supplying them with vapor by the way.* It would seem that Nature, as if to reclaim this "lee" land from the desert, had stationed by the wayside of these winds a succession of inland seas to serve them

as relays for supplying them with moisture. There is the Mediterranean, with its arms, the Caspian Sea and the Sea of Aral, all of which are situated exactly in this direction, as though these sheets of water were designed, in the grand system of aqueous arrangements, to supply with fresh vapor winds that had already left rain enough behind them to make an Amazon and an Orinoco of. Now that there has been such an elevation of land out of the water we infer from the fact that the Andes were once covered by the sea, for their tops are now crowned with the remains of marine animals. When they and their continent were submerged — admitting that Europe in general outline was then as it now is — it can not be supposed, if the circulation of vapor were then such as it is supposed now to be, that the climates of that part of the Old World which is under the lee of those mountains were then as scantily supplied with moisture as they now are. When the sea covered South America, nearly all the vapor which is now precipitated upon the Amazonian water-shed was conveyed thence by the winds, and distributed, it may be supposed, among the countries situated along the route (Plate IV) ascribed to them.

543. *Adjustments in this hygrometry of Caspians.* If ever the Caspian Sea exposed a larger surface for evaporation than it now does — and no doubt it did; if the precipitation in that valley ever exceeded the evaporation from it, as it does in all valleys drained into the open sea, then there must have been a change of hygrometrical conditions there. And admitting the vapor-springs for that valley to be situated in the direction supposed, the rising up of a continent from the bottom of the sea, or the upheaval of a range of mountains in certain parts of America, Africa, or Spain, across the route of the winds which brought the rain for the Caspian water-shed, might have been sufficient to rob them of the moisture which they were wont to carry away and precipitate upon this great inland basin. See how the Andes have made Atacama a desert, and of Western Peru a rainless country: these regions have been made rainless simply by the rising up of a mountain range between them and the vapor-springs in the ocean which feed with moisture the winds that blow over those now rainless regions.

544. *Countries in the temperate zone of this hemisphere that are under the lee of land in the trade-wind regions of the other are dry countries.* That part of Asia, then, which is under the lee of southern trade-wind Africa, lies to the north of the tropic of Cancer, and between two lines, the one passing through Cape Palmas and Medina, the other through Aden and Delhi. Being extended to the equator, they will include that part of it which is crossed by the continental southeast trade-winds of Africa after they have traversed the greatest extent of land surface (Plate IV). The range which lies between the two lines which represent the course of the American winds with their vapors, and the two lines which

represent the course of the African winds with their vapors, is the range which is under the lee of winds that have, for the most part, traversed water surface or the ocean in their circuit as southeast trade-winds. But a bare inspection of Plate IV will show that the southeast trade-winds which cross the equator between longitude 15° and 50° west, and which are supposed to blow over into this hemisphere between these two ranges, have traversed land as well as water; and the Trade-wind Chart [2] shows that it is precisely those winds which, in the summer and fall, are converted into southwest monsoons for supplying the whole extent of Guinea with rains to make rivers of. Those winds, therefore, it would seem, leave much of their moisture behind them, and pass along to their channels in the grand system of circulation, for the most part, as dry winds. Moreover, it is not to be supposed that the channels through which the winds blow that cross the equator at the several places named are as sharply defined in nature as the lines suggested, or as Plate IV would represent them to be.

545. *Their situation, and the range of dry winds.* The whole region of the extra-tropical Old World that is included within the ranges marked is the region which has most land to windward of it in the southern hemisphere. Now it is a curious *coincidence*, at least, that all the great extra-tropical deserts of the earth, with those regions in Europe and Asia which have the least amount of precipitation upon them, should lie within this range. That they are situated under the lee of the southern continents, and have but little rain, may be a coincidence, I admit; but that these deserts of the Old World are placed where they are is no coincidence — no accident: they are placed where they are, and as they are, by design; and in being so placed, it was intended that they should subserve some grand purpose in the terrestrial economy. Let us see, therefore, if we can discover any other marks of that design — any of the purposes to be subserved by such an arrangement — and trace any connection between that arrangement and the supposition which I maintain as to the place where the winds that blow over these regions derive their vapors. It will be remarked at once that all the inland seas of Asia, and all those of Europe except the semi-fresh-water gulfs of the north, are within this range. The Persian Gulf and the Red Sea, the Mediterranean, the Black, and the Caspian, all fall within it. And why are they planted there? Why are they arranged to the northeast and southwest under this lee, and in the very direction in which theory makes this breadth of thirsty winds to prevail? Clearly and obviously, one of the purposes in the divine economy was, that they might replenish with vapor the winds that are almost vaporless when they arrive at these regions in the general system of circulation. And why should these winds be almost vaporless? They are almost

[2] Maury [*Trade-wind Chart of the Atlantic Ocean*].

vaporless because their route, in the general system of circulation, is such, that they are not brought into contact with a water-surface from which the needful supplies of vapor are to be had; or, being obtained, the supplies have since been taken away by the cool tops of mountain ranges over which these winds have had to pass.

546. *The Mediterranean within it.* In the Mediterranean, the evaporation is greater than the precipitation. Upon the Red Sea there never falls a drop of rain; it is all evaporation. Are we not, therefore, entitled to regard the Red Sea as a make-weight, thrown in to regulate the proportion of cloud and sunshine, and to dispense rain to certain parts of the earth in due season and in proper quantities? Have we not, in these two facts, evidence conclusive that the winds which blow over these two seas come, for the most part, from a dry country — from regions which contain few or no pools to furnish supplies of vapor?

547. *Heavy evaporation.* Indeed, so scantily supplied with vapor are the winds which pass in the general channels of circulation over the water-shed and sea-basin of the Mediterranean, that they take up there more water as vapor than they deposit. But, throwing out of the question what is taken up from the surface of the Mediterranean itself, these winds deposit more water upon the water-shed whose drainage leads into the sea than they take up from it again. The excess is to be found in the rivers which discharge themselves into the Mediterranean; but so thirsty are the winds which blow across the bosom of that sea, that they not only take up again all the water that those rivers pour into it, but they are supposed by philosophers to create a demand for an immense current from the Atlantic to supply the waste. It is estimated that three [3] times as much water as the Mediterranean receives from its rivers is evaporated from its surface. This may be an overestimate, but the fact that evaporation from it is in excess of the precipitation, is made obvious by the current which the Atlantic sends into it through the Straits of Gibraltar; and the difference, we may rest assured, whether it be much or little, is carried off to modify climate elsewhere — to refresh with showers and make fruitful some other parts of the earth.

548. *The winds that give rains to Siberian rivers have to cross the steppes of Asia.* The great inland basin of Asia, which contains the Sea of Aral and the Caspian, is situated on the route which this hypothesis requires these thirsty winds from southeast trade-wind Africa and America to take; and so scant of vapor are these winds when they arrive in this basin, that they have no moisture to leave behind; just as much as they pour down they take up again and carry off. We know (§ 260) that

---

[3] [Herschel] "Physical Geography" [*Encyclop. Britan.* XVII, 573. Herschel estimates that 508 cubic miles are evaporated, but only 173 cubic miles contributed by rivers.]

the volume of water returned by the rivers, the rains, and the dews, into the whole ocean, is exactly equal to the volume which the whole ocean gives back to the atmosphere; as far as our knowledge extends, the level of each of these two seas is as permanent as that of the great ocean itself. Therefore, the volume of water discharged by rivers, the rains, and the dews, into these two seas, is exactly equal to the volume which these two seas give back as vapor to the atmosphere. These winds, therefore, do not begin permanently to lay down their load of moisture, be it great or small, until they cross the Oural Mountains. On the steppes of Issam, after they have supplied the Amazon and the other great equatorial rivers of the south, we find them first beginning to lay down more moisture than they take up again. In the Obi, the Yenesi, and the Lena is to be found the volume which contains the expression for the load of water which these winds have brought from the southern hemisphere, from the Mediterranean, and the Red Sea; for in these almost hyperborean river-basins do we find the first instance in which, throughout the entire range assigned these winds, they have, after supplying the Amazon, etc., left more water behind them than they have taken up again and carried off. The low temperatures of Siberian Asia are quite sufficient to extract from these winds the remnants of vapor which the cool mountain-tops and mighty rivers of the southern hemisphere have left in them.

549. *How climates in one hemisphere depend upon the arrangement of land in the other, and upon the course of the winds.* Here I may be permitted to pause, that I may call attention to another remarkable coincidence, and admire the marks of design, the beautiful and exquisite adjustments that we here see provided, to insure the perfect workings of the great aqueous and atmospherical machine. This coincidence — may I not call it cause and effect? — is between the hygrometrical conditions of all the countries within, and the hygrometrical conditions of all the countries without, the range included within the lines which I have drawn (Plate IV) to represent the route in the northern hemisphere of the southeast trade-winds *after* they have blown their course over the land in South Africa and America. Both to the right and left of this range are countries included between the same parallels in which it is, yet these countries all receive more water from the atmosphere than they give back to it again; they all have rivers running into the sea. On the one hand, there is in Europe the Rhine, the Elbe, and all the great rivers that empty into the Atlantic; on the other hand, there are in Asia the Ganges, and all the great Chinese rivers; and in North America, in the latitude of the Caspian Sea, is our great system of fresh-water lakes; all of these receive from the atmosphere immense volumes of water, and pour it back into the sea in streams the most majestic. It is remarkable that none of these copiously-supplied water-sheds have, to the southwest of

them in the trade-wind regions of the southern hemisphere, any considerable body of land; they are, all of them, under the lee of evaporating surfaces, of ocean waters in the trade-wind regions of the south. Only those countries in the extra-tropical north which I have described as lying under the lee of trade-wind South America and Africa are scantily supplied with rains. Pray examine Plate IV in this connection. It tends to confirm the views taken in Chapter VII. The surface of the Caspian Sea is about equal to that of our lakes; in it, evaporation is just equal to the precipitation. Our lakes are between the same parallels, and about the same distance from the western coast of America that the Caspian Sea is from the western coast of Europe; and yet the waters discharged by the St. Lawrence give us an idea of how greatly the precipitation upon it is in excess of the evaporation. To windward of the lakes, and in the trade-wind regions of the southern hemisphere, is no land; but to windward of the Caspian Sea, and in the trade-wind region of the southern hemisphere, there is land. Therefore, supposing the course of the vapor-distributing winds to be such as I maintain it to be, ought they not to carry more water from the ocean to the American lakes than it is possible for them to carry from the land — from the interior of South Africa and America — to the valley of the Caspian Sea? In like manner (§ 365), extra-tropical New Holland and South Africa have each land — not water — to the windward of them in the trade-wind regions of the northern hemisphere, where, according to this hypothesis, the vapor for their rains ought to be taken up: they are both countries of little rain; but extra-tropical South America has, in the trade-wind region to windward of it in the northern hemisphere, a great extent of ocean, and the amount of precipitation (§ 299) in extra-tropical South America is wonderful. The coincidence, therefore, is remarkable, that the countries in the extra-tropical regions of this hemisphere, which lie to the northeast of large districts of land in the trade-wind regions of the other hemisphere, should be scantily supplied with rains; and likewise, that those so situated in the extra-tropical south, with regard to land in the trade-wind region of the north, should also be scantily supplied with rains.

550. *Terrestrial adaptations.* Having thus remarked upon the coincidence, let us turn to the evidences of design, and contemplate the beautiful harmony displayed in the arrangement of the land and water, as we find them along this conjectural "wind-road." (Plate IV) Those who admit design in terrestrial adaptations, or who have studied the economy of cosmical arrangements, will not be loth to grant that by design the atmosphere keeps in circulation a certain amount of moisture; that the water of which this moisture is made is supplied by the aqueous surface of the earth, and that it is to be returned to the seas again through rivers and the process of precipitation; for were it not so, there would be a

permanent increase or decrease of the quantity of water thus put and kept in circulation by the winds, which would be followed by a corresponding change of hygrometrical conditions, which, in turn, would draw after it permanent changes of climate; and permanent changes of climate would involve the ultimate well-being of myriads of organisms, both in the vegetable and animal kingdoms. The quantity of moisture that the atmosphere keeps in circulation is, no doubt, just that quantity which is best suited to the well-being, and most adapted to the proper development of the vegetable and animal kingdoms; and that quantity is dependent upon the arrangement and the proportions that we see in nature between the land and the water — between mountain and desert, river and sea. If the seas and evaporating surfaces were changed, and removed from the places they occupy to other places, the principal places of precipitation probably would also be changed: whole families of plants would wither and die for want of cloud and sunshine, dry and wet, in proper proportions and in due season; and, with the blight of plants, whole tribes of animals would also perish. Under such a chance arrangement, man would no longer be able to rely upon the early and the latter rain, or to count with certainty upon the rains being sent in due season for seed-time and harvest. And that the rain will be sent in due season we are assured from on high; and when we recollect who it is that "sendeth" it, we feel the conviction strong within us, that He that sendeth the rain has the winds for his messengers; and that they may do his bidding, the land and the sea were arranged, both as to position and relative proportions, where they are, and as they are.

551. *The Red Sea and its vapors.* It should be borne in mind that, by this hypothesis, the southeast trade-winds, after they rise up at the equator (Plate I), have to overleap the northeast trade-winds. Consequently, they do not touch the earth until near the tropic of Cancer (see the bearded arrows, Plate IV), more frequently to the north than to the south of it; but for a part of every year, the place where these vaulting southeast trades first strike the earth, after leaving the other hemisphere, is very near this tropic. On the equatorial side of it, be it remembered, the northeast trade-winds blow; on the polar side, what were the southeast trades, and what are now the prevailing southwesterly winds of our hemisphere, prevail. Now examine Plate IV, and it will be seen that the upper half of the Red Sea is north of the tropic of Cancer; the lower half is to the south of it; that the latter is within the northeast trade-wind region; the former, in the region where the southwest passage winds are the prevailing winds. The River Tigris is probably evaporated from the upper half of this sea by these winds; while the northeast trade-winds take up from the lower half those vapors which feed the Nile with rain, and which the clouds deliver to the cold demands of the Mountains of the

Moon. Thus there are two "wind-roads" crossing this sea: to the windward of it, each road runs through a rainless region; to the leeward there is, in each case, a river rained down. The Persian Gulf lies, for the most part, in the track of the southwest winds; to the windward of the Persian Gulf is a desert; to the leeward, the River Indus. This is the route by which theory would require the vapor from the Red Sea and Persian Gulf to be conveyed, and this is the direction in which we find indications that it is conveyed. For to leeward do we find, in each case, a river, telling to us, by signs not to be mistaken, that it receives more water from the clouds than it gives back to the winds.

552. *Certain seas and deserts considered as counterpoises in the terrestrial machinery.* Is it not a curious circumstance, that the winds which travel the road suggested from the southern hemisphere should, when they touch the earth on the polar side of the tropic of Cancer, be so thirsty, more thirsty, much more, than those which travel on either side of their path, and which are supposed to have come from southern seas, not from southern lands? The Mediterranean has to give those winds three times as much vapor as it receives from them (§ 547); the Red Sea gives them as much as they can take, and receives nothing back in return but a little dew (§ 376); the Persian Gulf also gives more than it receives. What becomes of the rest? Doubtless it is given to the winds, that they may bear it off to distant regions, and make lands fruitful, that but for these sources of supply would be almost rainless, if not entirely arid, waste, and barren. These seas and arms of the ocean now present themselves to the mind as counterpoises in the great hygrometrical machinery of our planet. — As sheets of water placed where they are to balance the land in the trade-wind region of South America and South Africa, they now present themselves. When the foundations of the earth were laid, we know who it was that "measured the waters in the hollow of his hand, and meted out the heavens with a span, and comprehended the dust of the earth in a measure, and weighed the mountains in scales, and the hills in a balance;" and hence we know also that they are arranged both according to proportion and to place. Here, then, we see harmony in the winds, design in the mountains, order in the sea, arrangement for the dust, and form for the desert. Here are signs of beauty and works of grandeur; and we may now fancy that, in this exquisite system of adaptations and compensations, we can almost behold, in the Red and Mediterranean Seas, the very waters that were held in the hollow of the Almighty hand when he weighed the Andes and balanced the hills of Africa in the comprehensive scales. In that great inland basin of Asia which holds the Caspian Sea, and embraces an area of one million and a half of geographical square miles, we see the water-surface so exquisitely adjusted, that it is just sufficient, and no more, to return to the atmosphere as vapor exactly as

much moisture as the atmosphere lends in rain to the rivers of that basin — a beautiful illustration of the fact that the span of the heavens was meted out according to the measure of the waters. Thus we are entitled to regard (§ 542) the Mediterranean, the Red Sea, and Persian Gulf as relays, distributed along the route of these thirsty winds from the continents of the other hemisphere, to supply them with vapors, or to restore to them that which they have left behind to feed the sources of the Amazon, the Niger, and the Congo.

553. *Hypothesis supported by facts.* The hypothesis that the winds from South Africa and America do take the course through Europe and Asia which I have marked out for them (Plate IV), is supported by so many coincidences, to say the least, that we are entitled to regard it as probably correct, until a train of coincidences at *least* as striking can be adduced to show that such is not the case. Returning once more to a consideration of the geological agency of the winds in accounting for the depression of the Dead Sea, we now see the fact palpably brought out before us, that if the Straits of Gibraltar were to be barred up, so that no water could pass through them, we should have a great depression of water-level in the Mediterranean. Three times as much water (§ 547) is evaporated from that sea as is returned to it through the rivers. A portion of water evaporated from it is probably rained down and returned to it through the rivers; but, supposing it to be barred up: as the demand upon it for vapor would exceed the supply by rains and rivers, it would commence to dry up; as it sinks down, the area exposed for evaporation would decrease, and the supplies to the rivers would diminish, until finally there would be established between the evaporation and precipitation an equilibrium, as in the Dead and Caspian Seas. But, for aught we know, the water-level of the Mediterranean might, before this equilibrium were attained, have to reach a stage far below that of the Dead Sea level. The Lake Tadjura is now in the act of attaining such an equilibrium. There are connected with it the remains of a channel by which the water ran into the sea; but the surface of the lake is now five hundred feet below the sea-level, and it is salting up. If not in the Dead Sea, do we not, in the valley of this lake, find outcropping some reason for the question, What have the winds had to do with the phenomena before us?

554. *How, by the winds, the age of certain geological phenomena in our hemisphere may be compared with the age of those in the other.* The winds, in this sense, are geological agents of great power. It is not impossible but that they may afford us the means of comparing, directly, geological events which have taken place in one hemisphere, with geological events in another: *e. g.*, the tops of the Andes were once at the bottom of the sea. — Which is the oldest formation, that of the Dead Sea or the Andes? If the former be the older, then the climate of the Dead Sea must

have been hygrometrically very different from what it now is. In regarding the winds as geological agents, we can no longer consider them as the type of instability. We should rather treat them in the light of ancient and faithful chroniclers, which, upon being rightly consulted, will reveal to us truths that nature has written upon their wings in characters as legible and enduring as any with which she has ever engraved the history of geological events upon the tablet of the rock.

555. *The Andes older than the Dead Sea as an inland water.* The waters of Lake Titicaca, which receives the drainage of the great inland basin of the Andes, are only brackish, not salt. Hence we may infer that this lake has not been standing long enough to become briny, like the waters of the Dead Sea; consequently, it belongs to a more recent period. On the other hand, it will also be interesting to hear that my friend, Captain Lynch, informs me that, in his exploration of the Dead Sea, he saw what he took to be the dry bed of a river that once flowed from it. And thus we have two more links, stout and strong, to add to the chain of circumstantial evidence going to sustain the testimony of this strange and fickle witness which I have called up from the sea to testify in this presence concerning the works of Nature, and to tell us which be the older — the Andes, watching the stars with their hoary heads, or the Dead Sea, sleeping upon its ancient beds of crystal salt.

CHAPTER XIII

§ 560–575. THE DEPTHS OF THE OCEAN

560. *Submarine scenery.* "We dive," says Schleiden,[1] "into the liquid crystal of the Indian Ocean, and it opens to us the most wondrous enchantments, reminding us of fairy tales in childhood's dreams. The strangely branching thickets bear living flowers. Dense masses of meandrinas and astræas contrast with the leafy, cup-shaped expansions of the explanarias, the variously-ramified Madrepores, which are now spread out like fingers, now rise in trunk-like branches, and now display the most elegant array of interlacing branches. The coloring surpasses every thing; vivid green alternates with brown or yellow; rich tints of purple, from pale red-brown to the deepest blue. Brilliant rosy, yellow, or peach-colored Nullipores overgrow the decaying masses, and are themselves interwoven with the pearl-colored plates of the Retipores, resembling the most delicate ivory carvings. Close by wave the yellow and lilac fans, perforated like trellis-work, of the Gorgonias. The clear sand of the bottom is covered with the thousand strange forms and tints of the sea-urchins and star-fishes. The leaf-like flustras and escharas adhere like mosses and lichens to the branches of the corals; the yellow, green, and purple-striped limpets cling like monstrous cochineal insects upon their trunks. Like gigantic cactus-blossoms, sparkling in the most ardent colors, the sea anemones expand their crowns of tentacles upon the broken rocks, or more modestly embellish the flat bottom, looking like beds of variegated ranunculuses. Around the blossoms of the coral shrubs play the humming-birds of the ocean, little fish sparkling with red or blue metallic glitter, or gleaming in golden green, or in the brightest silvery lustre. Softly, like spirits of the deep, the delicate milk-white or bluish bells of the jelly-fishes float through this charmed world. Here the gleaming violet and gold-green Isabelle, and the flaming yellow, black, and vermilion-striped coquette, chase their prey; there the band-fish shoots, snake-like, through the thicket, like a long silver ribbon, glittering with rosy and azure hues. Then come the fabulous cuttle-fish, decked in all colors of the rainbow, but marked by no definite outline, appearing and disappearing, intercrossing, joining company and parting again, in most fantastic ways; and all this in the most rapid change, and amid the most wonderful play of light and shade, altered by every breath of wind, and

---

[1] [Matthias Jacob] Schleiden, *The Plant; [a Biography. In a Series of Thirteen Popular Lectures.* 2nd ed., with additions (London, 1853)], pp. 403–406.

every slight curling of the surface of the ocean. When day declines, and the shades of night lay hold upon the deep, this fantastic garden is lighted up in new splendor. Millions of glowing sparks, little microscopic medusas and crustaceans, dance like glow-worms through the gloom. The sea-feather, which by daylight is vermilion-colored, waves in a greenish, phosphorescent light. Every corner of it is lustrous. Parts which by day were perhaps dull and brown, and retreated from the sight amid the universal brilliancy of color, are now radiant in the most wonderful play of green, yellow, and red light; and, to complete the wonders of the enchanted night, the silver disk, six feet across, of the moon-fish,[2] moves, slightly luminous, among the cloud of little sparkling stars. The most luxuriant vegetation of a tropical landscape can not unfold as great wealth of form, while in the variety and splendor of color it would stand far behind this garden landscape, which is strangely composed exclusively of animals, and not of plants; for, characteristic as the luxuriant develop-ment of vegetation of the temperate zones is of the sea bottom, the full-ness and multiplicity of the marine Fauna is just as prominent in the regions of the tropics. Whatever is beautiful, wondrous, or uncommon in the great classes of fish and Echinoderms, jelly-fishes and Polypes, and the Mollusks of all kinds, is crowded into the warm and crystal waters of the tropical ocean — rests in the white sands, clothes the rough cliffs, clings, where the room is already occupied, like a parasite, upon the first comers, or swims through the shallows and depths of the elements — while the mass of the vegetation is of a far inferior magnitude. It is pe-culiar in relation to this that the law valid on land, according to which the animal kingdom, being better adapted to accommodate itself to outward circumstances, has a greater diffusion than the vegetable kingdom — for the polar seas swarm with whales, seals, sea-birds, fishes, and countless numbers of the lower animals, even where every trace of vegetation has long vanished in the eternally frozen ice, and the cooled sea fosters no sea-weed — that this law, I say, holds good also for the sea, in the direc-tion of its depth; for when we descend, vegetable life vanishes much sooner than the animal, and, even from the depths to which no ray of light is capable of penetrating, the sounding-lead brings up news at least of living infusoria."

561. *Ignorance concerning the depth of "blue water."* Until the com-mencement of the plan of deep-sea soundings, as they have been con-ducted in the American navy, the bottom of what the sailors call "blue water" was as unknown to us as is the interior of any of the planets of our system. Ross and Dupetit Thouars, with other officers of the English, French, and Dutch navies, had attempted to fathom the deep sea, some with silk threads, some with spun-yarn (coarse hemp threads twisted

[2] *Orthagoriscus mola.*

together), and some with the common lead and line of navigation. All of these attempts were made upon the supposition that when the lead reached the bottom, either a shock would be felt, or the line, becoming slack, would cease to run out.

562. *Early attempts at deep-sea soundings — unworthy of reliance.* The series of systematic experiments recently made upon this subject shows that there is no reliance to be placed on such a supposition, for the shock caused by striking bottom can not be communicated through very great depths. Furthermore, the lights of experience show that, as a general rule, the under currents of the deep sea have force enough to take the line out long after the plummet has ceased to do so. Consequently, there is but little reliance to be placed upon deep-sea soundings of former methods, when the depths reported exceeded eight or ten thousand feet.

563. *Various methods tried or proposed.* Attempts to fathom the ocean, both by sound and pressure, had been made, but out in "blue water" every trial was only a failure repeated. The most ingenious and beautiful contrivances for deep-sea soundings were resorted to. By exploding petards, or ringing bells in the deep sea, when the winds were hushed and all was still, the echo or reverberation from the bottom might, it was held, be heard, and the depth determined from the rate at which sound travels through water. But, though the concussion took place many feet below the surface, echo was silent, and no answer was received from the bottom. Ericsson and others constructed deep-sea leads having a column of air in them, which, by compression, would show the aqueous pressure to which they might be subjected. This was found to answer well for ordinary purposes, but in depths of the sea, where the pressure would be equal to several hundred atmospheres, the trial was more than this instrument could stand. Mr. Baur, an ingenious mechanician of New York, constructed, according to a plan which I furnished him, a deep-sea sounding apparatus. To the lead was attached, upon the principle of the screw propeller, a small piece of clock-work for registering the number of revolutions made by the little screw during the descent; and, it having been ascertained by experiment in shoal water that the apparatus, in descending, would cause the propeller to make one revolution for every fathom of perpendicular descent, hands provided with the power of self-registration were attached to a dial, and the instrument was complete. It worked beautifully in moderate depths, but failed in blue water, from the difficulty of hauling it up if the line used were small, and from the difficulty of getting it down if the line used were large enough to give the requisite strength for hauling it up. An old sea-captain proposed a torpedo, such as is sometimes used in the whale fishery for blowing up the monsters of the deep, only this one was intended to explode on touching the bottom. It was proposed first to ascertain by actual experiment the rate

at which the torpedo would sink, and the rate at which the sound or the gas would ascend, and so, by *timing* the interval, to determine the depth. This plan would afford no specimens of the bottom, and its adoption was opposed by other obstacles. One gentleman proposed to use the magnetic telegraph. The wire, properly coated, was to be laid up in the sounding-line, and to the plummet was attached machinery, so contrived that at the increase of every 100 fathoms, and by means of the additional pressure, the circuit would be restored, somewhat after the method of Dr. Locke's electro-chronograph, and a message would come up to tell how many hundred fathoms up and down the plummet had sunk. As beautiful as this idea was, it was not simple enough in practical application to answer our purposes.

564. *Physical problems more difficult than that of measuring the depth of the sea have been accomplished.* Greater difficulties than any presented by the problem of deep-sea soundings had been overcome in other departments of physical research. These plans and attempts served to encourage, nor were they fruitless, though they proved barren of practical results. Astronomers had measured the volumes and weighed the masses of the most distant planets, and increased thereby the stock of human knowledge. Was it creditable to the age that the depths of the sea should remain in the category of an unsolved problem? Its "ooze and bottom" was a sealed volume, rich with ancient and eloquent legends, and suggestive of many an instructive lesson that might be useful and profitable to man. The seal which covered it was of rolling waves many thousand feet in thickness. Could it not be broken? Curiosity had always been great, yet neither the enterprise nor the ingenuity of man had as yet proved itself equal to the task. No one had succeeded in penetrating, and bringing up from beyond the depth of two or three hundred fathoms below the aqueous covering of the earth any specimens of solid matter for the study of philosophers.

565. *The deep-sea sounding apparatus of Peter the Great.* The honor of the first attempt to recover specimens of the bottom from great depths belongs to Peter the Great, of Russia. That remarkable man and illustrious monarch constructed a deep-sea sounding apparatus especially for the Caspian Sea. It was somewhat in the shape of a pair of ice-hooks, and such as are seen in the hands of the "ice-man" as, in his daily rounds, he lifts the blocks of ice from his cart in the street for delivery at the door. It was so contrived that when it touched the bottom the plummet would become detached, and the hook would bring up the specimen.

566. *A plan of deep-sea sounding devised for the American navy.* The sea, with its myths, has suggested attractive themes to all people in all ages. Like the heavens, it affords an almost endless variety of subjects for pleasing and profitable contemplation, and there has remained in the

human mind a longing to learn more of its wonders and to understand its mysteries. The Bible often alludes to them. Are they past finding out? How deep is it? and what is at the bottom of it? Could not the ingenuity and appliances of the age throw some light upon these questions? The government was liberal and enlightened; times seemed propitious; but when or how to begin, after all these failures, with this interesting problem, was one of the difficulties first to be overcome. It was a common opinion, derived chiefly from a supposed physical relation, that the depths of the sea are about equal to the heights of the mountains. But this conjecture was, at best, only a speculation. Though plausible, it did not satisfy. There were, in the depths of the sea, untold wonders and inexplicable mysteries. Therefore the contemplative mariner, as in mid-ocean he looked down upon its gentle bosom, continued to experience sentiments akin to those which fill the mind of the devout astronomer when, in the stillness of the night, he looks out upon the stars, and wonders. Nevertheless, the depths of the sea still remained as fathomless and as mysterious as the firmament above. Indeed, telescopes of huge proportions and of vast space-penetrating powers had been erected here and there by the munificence of individuals, and attempts made with them to gauge the heavens and sound out the regions of space. Could it be more difficult to sound out the sea than to gauge the blue ether and fathom the vaults of the sky? The result of the astronomical undertakings [3] lies in the discovery that what, through other instruments of less power, appeared as clusters of stars, were, by these of larger powers, separated into groups, and what had been reported as nebulæ could now be resolved into clusters; that, in certain directions, the abyss beyond these faint objects is decked with other nebulæ, which these great instruments may bring to light, but can not resolve; and that there are still regions and realms beyond, which the rays of the brightest sun in the sky have neither the intensity nor the force to reach, much less to penetrate. And what is more, these monster instruments have revealed to us, in those distant regions, forms or aggregations of matter which suggest to some the idea of the existence of physical forces there that we do not understand, and which raise the question in speculative minds, Is gravitation a universal thing, and do its forces penetrate every abyss of space? Could we not gauge the sea as well as the sky, and devise an instrument for penetrating the depths of the ocean as well as the depths of space? Mariners were curious concerning the bottom of the sea. Though nothing thence had been brought to light, exploration had invested the subject with additional interest, and increased the desire to know more. In this state of the case, the idea of a common twine thread for a sounding-line, and a cannon ball for a sinker, was suggested. It was a beautiful concep-

[3] See the work of Herschel and Rosse, and their telescopes.

tion; for, besides its simplicity, it had in its favor the greatest of recommendations — it could be readily put into practice.

567. *The great depths and failures of the first attempts.* Well-directed attempts to fathom the ocean began now to be made with such a line and plummet, and the public mind was astonished at the vast depths that were at first reported. Lieutenant Walsh, of the United States schooner "Taney," reported a cast with the deep-sea lead at thirty-four thousand feet without bottom. His sounding-line was an iron wire more than eleven miles in length. Lieutenant Berryman, of the United States brig "Dolphin," reported another unsuccessful attempt to fathom mid-ocean with a line thirty-nine thousand feet in length. Captain Denham, of her Britannic majesty's ship "Herald," reported bottom in the South Atlantic at the depth of forty-six thousand feet; and Lieutenant J. P. Parker, of the United States frigate "Congress," afterward, in attempting to sound near the same region, let go his plummet, and saw it run out a line fifty thousand feet long as though the bottom had not been reached. There are no such depths as these. The three last-named attempts were made with the sounding-twine of the American navy, which has been introduced in conformity with a very simple plan for sounding out the depths of the ocean. It involved for each cast only the expenditure of a cannon ball, and twine enough to reach the bottom. This plan was introduced as a part of the researches conducted at the National Observatory, and which have proved so fruitful and beneficial, concerning the winds and currents, and other phenomena of the ocean. These researches had already received the approbation of the Congress of the United States; for that body, in a spirit worthy of the representatives of a free and enlightened people, had authorized the Secretary of the Navy to employ three public vessels to assist in perfecting the discoveries, and in conducting the investigations connected therewith.

568. *The plan finally adopted.* The plan of deep-sea soundings finally adopted, and now in practice, is this: Every vessel of the navy, when she puts to sea, is, if she desires it, furnished with a sufficient quantity of sounding-twine, carefully marked at every length of one hundred fathoms — six hundred feet — and wound on reels of ten thousand fathoms each. It is made the duty of the commander to avail himself of every favorable opportunity to try the depth of the ocean, whenever he may find himself out upon "blue water." For this purpose he is to use a cannon ball of 32 or 68 pounds as a plummet. Having one end of the twine attached to it, the cannon ball is to be thrown overboard from a boat, and suffered to take the twine from the reel as fast as it will. The reel is made to turn easily. A silk thread, or the common wrapping-twine of the shops, would, it was thought, be strong enough for this purpose, for it was supposed there would be no strain upon the line except the very slight one re-

quired to drag it down, and the twine having nearly the specific gravity of sea water, this strain would, it was imagined, be very slight. Moreover, when the shot reached the bottom, the line, it was thought (§ 561), would cease to run out; then breaking it off, and seeing how much remained upon the reel, the depth of the sea could be ascertained at any place and time simply at the expense of one cannon ball and a few pounds of common twine.

569. *Discovery of currents in the depths of the sea.* But practical difficulties that were not expected at all were lurking in the way, and afterward showed themselves at every attempt to sound; and it was before these practical difficulties had been fairly overcome that the great soundings (§ 567) were reported. In the first place, it was discovered that the line, once started and dragged down into the depths of the ocean, never would cease to run out (§ 562), and, consequently, that there was no means of knowing when, if ever, the shot had reached the bottom. And, in the next place, it was ascertained that the ordinary twine (§ 566) would not do; that the sounding-line, in going down, was really subjected to quite a heavy strain, and that, consequently, the twine to be used must be strong; it was therefore subjected to a test which required it to bear a weight of at least sixty pounds freely suspended in the air. So we had to go to work anew, and make several hundred thousand fathoms of sounding-twine especially for the purpose. It was small, and stood the test required, a pound of it measuring about six hundred feet in length. The officers intrusted with the duty soon found that the soundings could not be made from the vessel with any certainty as to the depth. It was necessary that a boat should be lowered, and the trial be made from it; the men with their oars keeping the boat from drifting, and maintaining it in such a position that the line should be "up and down" the while. That the line would continue to run out after the cannon ball had reached bottom, was explained by the conjecture that there is in the ocean, as in the air, a system of currents and counter currents one above the other, and that it was one or more of these submarine currents, operating upon the bight of the line, which caused it to continue to run out after the shot had reached the bottom. In corroboration of this conjecture, it was urged, with a truth-like force of argument, that it was these under currents, operating with a swigging force upon the bights of the line — for there might be several currents running in different directions, and operating upon it at the same time — which caused it to part whenever the reel was stopped and the line held fast in the boat.

570. *Evidence in favor of a regular system of oceanic circulation.* A powerful train of circumstantial evidence was this (and it was derived from a source wholly unexpected), going to prove the existence of that system of oceanic circulation which the climates, and the offices, and the

adaptations of the sea require, and which its inhabitants (§ 465) in their mute way tell us of. This system of circulation commenced on the third day of creation, with the "gathering together of the waters," which were "called seas;" it will probably continue as long as sea water shall possess the properties of saltness and fluidity.

571. *Method of making a deep-sea sounding.* In making these deep-sea soundings, the practice is to time the hundred fathom marks (§ 568) as they successively go out; and by always using a line of the same size and "make," and a sinker of the same shape and weight, we at last established the law of descent. Thus the mean of our experiments gave us, for the sinker and twine used,

2 m. 21 s. as the average time of descent from 400 to 500 fathoms.
3 m. 26 s.   "   "   "   1000 to 1100   "
4 m. 29 s.   "   "   "   1800 to 1900   "

572. *The law of the plummet's descent.* Now, by aid of the law here indicated, we could tell very nearly when the ball ceased to carry the line out, and when, of course, it began to go out in obedience to the current and drift alone; for currents would sweep the line out at a uniform rate, while the cannon ball would drag it out at a decreasing rate. The development of this law was certainly an achievement, for it enabled us to show that the depth of the sea at the places named (§ 567) was not as great as reports made it. These researches were interesting; the problem in hand was important, and it deserved every effort that ingenuity could suggest for reducing it to a satisfactory solution.

573. *Brooke's sounding apparatus.* As yet, no specimens of the bottom had been brought up. The line was too small, the shot was too heavy, and it could not be weighed; and if we could reach the bottom, why should we not know its character? In this state of the case, Passed Midshipman J. M. Brooke, United States Navy, who, at the time, was associated with me on duty at the Observatory, proposed a contrivance by which the shot, on striking the bottom, would detach itself from the line, and send up a specimen of the bottom. This beautiful contrivance, called Brooke's Deep-sea Sounding Apparatus, is represented in Figure 4. A is a cannon ball, having a hole through it for the rod B. Figure 4, *a*, represents the rod, B, and the slings, D D, with the shot slung, ready for sounding. Figure 4, *b*, represents the apparatus in the act of striking the bottom; it shows how the shot is detached, and how specimens of the bottom are brought up, by adhering to a little soap or tallow,[4] called "arming," in the cup, C, at the lower end of the rod, B. With this contrivance specimens of the bottom have been brought up from the depth of nearly four miles.

[4] The barrel of a common quill attached to the rod has been found to answer better.

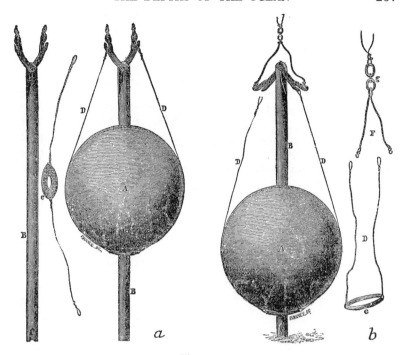

Figure 4

[Brooke's Deep-sea Sounding Apparatus for bringing up specimens of the bottom: *a*, ready for sounding; *b*, at moment of release on reaching bottom.]

574. *The deepest part of the Atlantic Ocean.* The greatest depths at which the bottom of the sea has been reached with the plummet are in the North Atlantic Ocean, and the places where it has been fathomed do not show it to be deeper than twenty-five thousand feet. The deepest place in this ocean (Plate VII) is probably between the parallels of 35° and 40° north latitude, and immediately to the southward of the Grand Banks of Newfoundland. The first specimens have been received from the coral sea of the Indian Archipelago and from the North Pacific. They were collected by the surveying expedition employed in those seas. A few soundings have been made in the South Atlantic, but not enough to justify deduction as to its depths or the precise shape of its floor.

575. *Deep-sea soundings by the English navy.* The friends of physical research at sea are under obligations to the officers of the English navy for much valuable information touching this interesting subject. Certain officers in that service have taken up the problem of deep-sea soundings with the most praiseworthy zeal, energy, and intelligence. Dayman in the Atlantic, Captains Spratt and Mansell in the Mediterranean, with

Captain Pullen in the Red Sea, have all made valuable contributions to the stock of human knowledge concerning the depths and bottom of the sea. To Mansell and Spratt we are indebted for all we know about deep-sea soundings in the Mediterranean, as we are to Pullen for those in the Red Sea. By their lines of soundings, their maps and profiles, they have enabled physical geographers to form, with some approach toward correctness, an idea as to the orography of the basins which hold the water for these two seas. We are also indebted to the French for deep-sea soundings in the Mediterranean. That sea appears to be about two miles deep in the deepest parts, which are in the isleless spaces to the west of Sardinia and to the east of Malta.

CHAPTER XIV

§ 580–618. THE BASIN AND BED OF THE ATLANTIC

580. *The wonders of the sea.* The wonders of the sea are as marvelous as the glories of the heavens; and they proclaim, in songs divine, that they too are the work of holy fingers. Among the revelations which scientific research has lately made concerning the crust of our planet, none are more interesting to the student of nature, or more suggestive to the Christian philosopher, than those which relate to the bed and bottom of the ocean.

581. *Its bottom and Chimborazo.* The basin of the Atlantic, according to the deep-sea soundings made by the American and English navies, is shown on Plate VII. This plate refers chiefly to that part of the Atlantic which is included within our hemisphere. In its entire length, the basin of this sea is a long trough, separating the Old World from the New, and extending probably from pole to pole. As to breadth, it contrasts strongly with the Pacific Ocean. From the top of Chimborazo to the bottom of the Atlantic, at the deepest place yet reached by the plummet in that ocean, the distance, in a vertical line, is nine miles.

582. *An orographic view.* Could the waters of the Atlantic be drawn off so as to expose to view this great sea-gash which separates continents, and extends from the Arctic to the Antarctic, it would present a scene the most rugged, grand, and imposing. The very ribs of the solid earth, with the foundations of the sea, would be brought to light, and we should have presented to us in one view, in the empty cradle of the ocean, "a thousand fearful wrecks," with that array of dead men's skulls, great anchors, heaps of pearl and inestimable stones, which, in the poet's eye, lie scattered on the bottom of the sea, making it hideous with sights of ugly death. To measure the elevation of the mountain-top above the sea, and to lay down upon our maps the mountain ranges of the earth, is regarded in geography as an important thing, and rightly so. Equally important is it, in bringing the physical geography of the sea regularly within the domains of science, to present its orology, by mapping out the bottom of the ocean so as to show the depressions of the solid parts of the earth's crust there, below the sea-level.

583. *Plate VII.* Plate VII presents the latest attempt at such a map. It relates exclusively to the bottom of that part of the Atlantic Ocean which lies north of 10° south. It is stippled with four shades: the darkest (that which is nearest the shore-line) shows where the water is less than

six thousand feet deep; the next, where it is less than twelve thousand feet deep; the third, where it is less than eighteen thousand; and the fourth, or lightest, where it is not over twenty-four thousand feet deep. The blank space south of Nova Scotia and the Grand Banks includes a district within which casts showing very deep water have been reported, but which subsequent investigation and discussion do not appear to confirm. The deepest part of the North Atlantic is probably somewhere between the Bermudas and the Grand Banks, but how deep it may be yet remains for the cannon ball and sounding-twine to determine. The waters of the Gulf of Mexico are held in a basin about a mile deep in the deepest part. THE BOTTOM OF THE ATLANTIC, or its depressions below the sea-level, are given, perhaps, on this plate with as much accuracy as the best geographers have been enabled to show, on a map, the elevations above the sea-level of the interior either of Africa or Australia.

584. *"What's the use" of deep-sea soundings?* "What is to be the use of these deep-sea soundings?" is a question that often occurs; and it is as difficult to be answered in categorical terms as Franklin's question, "What is the use of a new-born babe?" Every physical fact, every expression of nature, every feature of the earth, the work of any and all of those agents which make the face of the world what it is, and as we see it, is interesting and instructive. Until we get hold of a group of physical facts, we do not know what practical bearings they may have, though right-minded men know that they contain many precious jewels, which the experts of philosophy will not fail to bring out, polished and bright, and beautifully adapted, sooner or later, to man's purposes. Already we are obtaining practical answers to this question as to the use of deep-sea soundings; for, as soon as they were announced to the public, they forthwith assumed a practical bearing in the minds of men with regard to the question of a submarine telegraph across the Atlantic.

585. *The telegraphic plateau.* There is, at the bottom of this sea, between Cape Race, in Newfoundland, and Cape Clear, in Ireland, a remarkable steppe, which is already known as the telegraphic plateau, and has already been made famous by the attempts to run a telegraphic cable across the ocean upon it. In August, 1858, a cable was laid upon it from Valencia in Ireland to Trinity Bay in Newfoundland, and a few messages were passed through it, when it ceased to work. Whether messages can ever be *successfully* sent, in a commercial sense, through such a length of continuous submarine wire, is by no means certain; but that the wires of 1858 so soon ceased to pass any current at all was no doubt owing to the fact that the cable was constructed upon erroneous principles. Its projectors, in planning its construction, did not, unfortunately, avail themselves of the light which our deep-sea soundings had cast upon the bed of the ocean.

586. *The first specimens of deep-sea soundings.* It was upon this plateau that Brooke's sounding apparatus brought up its first trophies from the bottom of the sea. These specimens the officers of the Dolphin judged to be clay; but they took the precaution to label them, carefully to preserve them, and, on their return to the United States, to send them to the proper bureau. They were divided; a part was sent for examination to Professor Ehrenberg, of Berlin, and a part to the late Professor Bailey, of West Point — eminent microscopists both. The latter thus responded:

587. *Bailey's letter.* "I am greatly obliged to you for the deep soundings you sent me last week, and I have looked at them with great interest. They are exactly what I have wanted to get hold of. The bottom of the ocean at the depth of *more than two miles* I hardly hoped ever to have a chance of examining; yet, thanks to Brooke's contrivance, we have it clean and free from grease, so that it can at once be put under the microscope. I was greatly delighted to find that *all* these deep soundings are filled with microscopic shells; not a particle of sand or gravel exists in them. They are chiefly made up of perfect little calcareous shells (Foraminifera), and contain, also, a number of silicious shells (Diatomaceæ). It is not probable that these animals lived at the depths where these shells are found, but I rather think that they inhabit the waters near the surface; and when they die, their shells settle to the bottom. With reference to this point, I shall be very glad to examine bottles of water from various depths which were brought home by the Dolphin, and any similar materials, either 'bottom,' or water from other localities. I shall study them carefully. . . . The results already obtained are of very great interest, and have many important bearings on geology and zoology. . . . I hope you will induce as many as possible to collect soundings with Brooke's lead, in all parts of the world, so that we can map out the animalculæ as you have the whales. Get your whalers also to collect mud from pancake ice, etc., in the polar regions; this is always full of interesting microscopic forms."

588. *Specimens from the Coral Sea.* Lieutenant Brooke, of the North Pacific Exploring Expedition, procured specimens of the bottom from the depth of 2150 fathoms in the Coral Sea, lat. 13° S., long. 162° E. With regard to these, the admirable and lamented Bailey wrote in 1855, "You may be sure I was not backward in taking a look at the specimens you sent me, which, from their locality, promised to be so interesting. The sounding from 2150 fathoms, although very small in *quantity*, is not so bad in *quality*, yielding representatives of most of the great groups of microscopic organisms usually found in marine sediments. The predominant forms are silicious spicules of *sponges*. Various forms of these occur: some long and spindle-shaped, or acicular; others pin-headed; some three-spined, etc., etc. The Diatomes (silicious infusoria of Ehrenberg)

are very few in number, and mostly fragmentary. I found, however, some perfect valves of a coscinodiscus. The Foraminifera (Polythalamia of Ehrenberg) are very *rare*, only one perfect shell being seen, with a few fragments of others. The polycistineæ are present, and some species of haliomma were quite perfect. Fragments of other forms of this group indicate that various interesting species might be obtained if we had more of the material. You will see by the above that this deep sounding differs considerably from those obtained in the Atlantic. The Atlantic soundings were almost wholly composed of calcareous shells of the Foraminifera; these, on the contrary, contain very few Foraminifera, and are of a silicious rather than a calcareous nature. This only makes the condition of things in the Northern Atlantic the more interesting."

589. *They belong to the animal, not to the vegetable or mineral kingdom.* The first noticeable thing the microscope gives of these specimens is, that all of them are of the animal, not one of the mineral or vegetable kingdom. The ocean teems with life, we know. Of the four elements of the old philosophers — fire, earth, air, and water — perhaps the sea most of all abounds with living creatures. The space occupied on the surface of our planet by the different families of animals and their remains is inversely as the size of the individual. The smaller the animal, the greater the space occupied by his remains. Though not invariably the case, yet this rule, to a certain extent, is true, and will, therefore, answer our present purposes, which are simply those of illustration. Take the elephant and his remains, or a microscopic animal and his, and compare them. The contrast, as to space occupied, is as striking as that of the coral reef or island with the dimensions of the whale. The grave-yard that would hold the corallines is larger than the grave-yard that would hold the elephants.

590. *Quiet reigns in the depths of the sea.* We notice another practical bearing in this group of physical facts that Brooke's apparatus fished up from the bottom of the deep sea. Bailey, with his microscope (§ 587), could not detect a single particle of sand or gravel among these little mites of shells. They were from the great telegraphic plateau (§ 585), and the inference is that there, if any where, the waters of the sea are at rest. There was not motion enough to abrade these very delicate organisms, nor current enough to sweep them about and mix up with them a grain of the finest sand, nor the smallest particle of gravel torn from the loose beds of debris that here and there strew the bottom of the sea. This plateau is not too deep for the wire to sink down and rest upon, yet it is not so shallow that currents, or icebergs, or any abrading force can derange the wire after it is once lodged there.

591. *Is there life in them?* As Professor Bailey remarks (§ 587), the animalculæ, whose remains Brooke's lead has brought up from the bottom

of the deep sea, probably did not live or die there. They would have had no light there, and, had they lived there, their frail little texture would have been subjected, in its growth, to the pressure of a column of water twelve thousand feet high, equal to the weight of four hundred atmospheres. They probably lived and died near the surface, where they could feel the genial influences of both light and heat, and were buried in the lichen caves below after death.

592. *The ocean in a new light.* Brooke's lead and the microscope, therefore, it would seem, are about to teach us to regard the ocean in a new light. Its bosom, which so teems with animal life — its face, upon which time writes no wrinkles, makes no impression, are, it would now seem, as obedient to the great law of change as is any department whatever either of the animal or the vegetable kingdom. It is now suggested that henceforward we should view the surface of the sea as a nursery teeming with nascent organisms, its depths as the cemetery for families of living creatures that outnumber the sands on the sea-shore for multitude. Where there is a nursery, hard by there will be found also a grave-yard — such is the condition of the animal world. But it never occurred to us before to consider the surface of the sea as one wide nursery, its every ripple a cradle, and its bottom one vast burial-place.

593. *Leveling agencies.* On those parts of the solid portions of the earth's crust which are at the bottom of the atmosphere, various agents are at work, leveling both upward and downward. Heat and cold, rain and sunshine, the winds and the streams, all, assisted by the forces of gravitation, are unceasingly wasting away the high places on the land, and as perpetually filling up the low. But in contemplating the leveling agencies that are at work upon the solid portions of the crust of our planet which are at the bottom of the sea, one is led, at first thought, almost to the conclusion that these leveling agents are powerless there. In the deep sea there are no abrading processes at work; neither frosts nor rains are felt there, and the force of gravitation there is so paralyzed that it can not use half its power, as on dry land, in tearing the overhanging rock from the precipice and casting it down into the valley below.

594. *The offices of animalculæ.* When considering the bottom of the ocean, we have, in the imagination, been disposed to regard the waters of the sea as a great cushion, placed between the air and the bottom of the ocean, to protect and defend it from these abrading agencies of the atmosphere. The geological clock may, we thought, strike new periods; its hands may point to era after era; but, so long as the ocean remains in its basin, so long as its bottom is covered with blue water, so long must the deep furrows and strong contrasts in the solid crust below stand out bold, ragged, and grandly. Nothing can fill up the hollows there; no agent now at work, that we know of, can descend into its depths, and level off

the floors of the sea. But it now seems that we forgot the myriads of animalculæ that make the surface of the sea sparkle and glow with life: they are secreting from its surface solid matter for the very purpose of filling up those cavities below. These little marine insects build their habitations at the surface, and when they die, their remains, in vast multitudes, sink down and settle upon the bottom. They are the atoms of which mountains are formed — plains spread out. Our marl-beds, the clay in our river-bottoms, large portions of many of the great basins of the earth, are composed of the remains of just such little creatures as these, which the ingenuity of Brooke has enabled us to fish up from the depth of nearly four miles (twenty thousand feet) below the sea-level. These *Foraminifera*, therefore, when living, may have been preparing the ingredients for the fruitful soil of a land that some earthquake or up-heaval, in ages far away in the future, may be sent to cast up from the bottom of the sea for man's use.

595. *The study of them profitable.* The study of these "sunless treas-ures," recovered with so much ingenuity from the rich bottom of the sea, suggests new views concerning the physical economy of the ocean. It not only leads us into the workshops of the inhabitants of the sea — shows us through their nurseries and cemeteries, and enables us to study their economy — but it conducts us into the very chambers of the deep. Our investigations go to show that the roaring waves and the mightiest billows of the ocean repose, not upon hard or troubled beds, but upon cushions of still water; that every where at the bottom of the deep sea the solid ribs of the earth are protected, as with a garment, from the abrading action of its currents, and the cradle of its restless waves is lined by a stratum of water at rest, or so nearly at rest that it can neither wear nor move the lightest bit of drift that once lodges there.

596. *The abrasion of currents.* The tooth of running water is very sharp. See how the Hudson has eat through the Highlands, and the Niagara cut its way through layer after layer of the solid rock. But what are the Hudson and the Niagara, with all the fresh water-courses of the world, by the side of the Gulf Stream and other great "rivers in the ocean?" And what is the pressure of fresh water upon river-beds in comparison with the pressure of ocean water upon the bottom of the deep sea? It is not so great by contrast as the gutters in the streets are to the cataract. Then why have not the currents of the sea worn its bottom away? Simply because they are not permitted to get down to it. Suppose the currents which we see at and near the surface of the ocean were permitted to extend all the way to the bottom in deep as well as shallow water, let us see what the pressure and scouring force would be where the sea is only 3000 fathoms deep — for in many places the depth is even greater than that. It is equal there, in round numbers, to the

pressure of six hundred atmospheres. Six hundred atmospheres, piled up one above the other, would press upon every square foot of solid matter beneath the pile with the weight of 1,296,000 pounds, or 648 tons.

597. *Their pressure on the bottom.* The better to comprehend the amount of such a pressure, let us imagine a column of water just one foot square, where the sea is 3000 fathoms deep, to be frozen from the top to the bottom, and that we could then, with the aid of some mighty magician, haul this shaft of ice up, and stand it on one end for inspection and examination. It would be 18,000 feet high; the pressure on its pedestal would be more than a million and a quarter of pounds; and if placed on a ship of 648 tons burden, it would be heavy enough to sink her. There are currents in the sea where it is 3000 fathoms deep, and some of them — as the Gulf Stream — run with a velocity of four miles an hour, and even more. Every square foot of the earth's crust at the bottom of a four-knot current 3000 fathoms deep would have no less than 506,880 — in round numbers, half a million — of such columns of water daily dragging, and rubbing, and scouring, and chafing over it, under a continuous pressure of 648 tons. What would the bottom of the sea have to be made of to withstand such erosion? Water running with such a velocity, and with the friction upon the bottom which such a pressure would create, would in time wear away the thickest bed, though made of the hardest adamant. Why, then, has not the bottom of the sea been worn away? Why have not its currents cut through the solid crust in which its billows are rocked, and ripped out from the bowels of the earth the masses of incandescent, molten matter which geologists tell us lie pent up and boiling there?

598. *Why they can not chafe it.* If the currents of the sea, with this four-mile velocity at the surface, and this hundreds of ton pressure in its depths, were permitted to chafe against its bed, the Atlantic, instead of being two miles deep and 3000 miles broad, would, we may imagine, have been long ago cut down into a narrow channel that might have been as the same ocean turned up on edge, and measuring two miles broad and 3000 deep. But had it been so cut, the proportion of land and water surface would have been destroyed, and the winds, for lack of area to play upon, could not have sucked up from the sea vapors for the rains, and the face of the earth would have become as a desert without water. Now there is a reason why such changes should not take place, why the currents should not uproot nor score the deep bed of the ocean, why they should not throw out of adjustment any physical arrangement whatever in the ocean; it is because that in the presence of everlasting wisdom a *compass was set upon the face of the deep; because its waters were measured in the hollow of the Almighty hand; because bars and doors were set to stay its proud waves; and because, when He gave to the sea*

His decree that its waters should not pass His command, He laid the foundations of the world so fast that they should not be removed forever.

599. *What it consists of.* By bringing up specimens from the depths of the ocean, and studying them through the microscope, it has been ascertained that the bed of the ocean is lined with the microscopic remains of its own dead, with marine feculences which lie on the bottom as lightly as rests the gossamer in a calm at the bottom of the atmospherical ocean. How frail yet how strong, how light yet how firm are the foundations of the sea! Its waves can not fret them, its currents can not wear them, for the bed of the deep sea is protected from abrasion by a cushion of still and heavy water. There it lies — that beautiful arrangement — spread out over the bottom of the deep, and covering its foundations as with a garment, so that they may not be worn. If the currents chafe upon it now here, now there, as in shallow seas they sometimes do, this protecting cushion is self-adjusting; and the moment the unwonted pressure is removed the liquid cushion is restored, and there is again compensation.

600. *The causes that produce currents in the sea reside near its surface.* The discovery of this arrangement in the oceanic machinery suggests that the streams of running water in the sea play rather about its surface than in its depths; that the causes which produce currents reside at and near the surface; that they are changing heat and alternating cold with their powers of contraction and expansion — winds and sea-shells with evaporation and precipitation; and it is certain that none of these agents appear capable of reaching very far down into the depths of the great and wide sea with their influences. They go not much, if any, farther down than the light can reach. On the other hand, the most powerful agents in the atmosphere reside at and near its bottom; so that, where these two great oceans meet — the aqueous and the aerial — there we probably have the greatest conflict and the most powerful display of the forces that set and keep them in motion, making them to rage and roar.

601. *Their depth.* The greatest depth at which running water is to be found in the sea is probably in the narrowest part of the Gulf Stream, as, coming from its mighty fountain, it issues through the Florida Pass. The deep-sea thermometer shows that even here there is a layer of cold water in the depths beneath, so that this "river in the sea" may chafe not against the solid bottom. What revelations of the telescope, what wonders of the microscope, what fact relating to the physical economy of this terrestrial globe, is more beautiful or suggestive than some of the secrets which have been fished up from the caverns of the deep, and brought to light from the hidden paths of the sea?

602. *The cushion of still water — its thickness.* In my researches I have as yet found no marks of running water impressed upon the foundations of the sea beyond the depth of two or three thousand feet. Should

future deep-sea soundings establish this as a fact in other seas also, it will prove of the greatest value to submarine telegraphy. What may be the thickness of this cushion of still water that covers the bottom of the deep sea is a question of high interest, but we must leave it for future investigation.

603. *The conservators of the sea.* In Chapter X (*The Salts of the Sea*), I have endeavored to show how sea-shells and marine insects may, by reason of the offices which they perform, be regarded as compensations in that exquisite system of physical machinery by which the harmonies of nature are preserved. But the treasures of the lead and revelations of the microscope present the insects of the sea in a new and still more striking light. We behold them now, serving not only as compensations by which the motions of the water in its channels of circulation are regulated and climates softened, but acting also as checks and balances by which the equipoise between the solid and the fluid matter of the earth is preserved. Should it be established that these microscopic creatures live at the surface, and are only buried at the bottom of the sea, we may then view them as conservators of the ocean; for, in the offices which they perform, they assist to preserve its *status* by maintaining the purity of its waters.

604. *The anti-biotic view the most natural.* Does any portion of the shells which Brooke's sounding rod brings up from the bottom of the deep sea live there; or are they all the remains of those that lived near the surface in the light and heat of sun, and were buried at the bottom of the deep after death? Philosophers were divided in opinion upon this subject. The facts, as far as they went, seemed at first to favor the one conjecture nearly as well as the other. Under these circumstances, I inclined to the anti-biotic hypothesis, and chiefly because it would seem to conform better with the Mosaic account of creation. The sun and moon were set in the firmament before the waters were commanded to bring forth the living creature; and hence we infer that light and heat are necessary to the creation and preservation of marine life; and since the light and heat of the sun can not reach to the bottom of the deep sea, my own conclusion, in the absence of positive evidence upon the subject, has been, that the *habitat* of these mites of things hauled up from the bottom of the great deep is at and near the surface. On the contrary, others maintained, and perhaps with equal reason, the biotic side of the question. Professor Ehrenberg, of Berlin, is of this latter class.

605. *The question stated.* This is an interesting question. It is a new one; and it belongs to that class of questions which mere discussion helps to settle. It is therefore desirable to state both sides — present all the known facts; and then, provided with such lights as they afford, we may draw conclusions.

606. *The arguments of the biotics.* As soon as the deep-sea specimens were mounted on the slides of the microscope, the two great masters of that instrument — Bailey of West Point, and Ehrenberg of Berlin, discovered the greater part of the small calcareous *carapaces* to be filled with a soft pulp, which both admitted to be fleshy matter. From this fact the German argued that there is life at the bottom of the deep sea; the American (§ 587), that there is only death and repose there.

607. *Ehrenberg's statement of them.* "The other argument," says Ehrenberg, "for life in the deep which I have established is the surprising quantity of new forms which are wanting in other parts of the sea. If the bottom were nothing but the sediment of the troubled sea, like the fall of snow in the air, and if the biolithic curves of the bottom were nothing else than the product of the currents of the sea which heap up the flakes, similarly to the glaciers, there would *necessarily* be much less of unknown and peculiar forms in the depths. The surface and the borders of the sea are much more productive and much more extended than the depths; hence the forms peculiar to the depths should not be perceived. The great quantity of peculiar forms and of soft bodies existing in the innumerable carapaces, accompanied by the observation of the number of unknowns, *increasing with the depths* — these are the arguments which seem to me to hold firmly to the opinion of stationary life at the bottom of the deep sea." [1]

608. *The anti-biotic view.* The anti-biotics, on the other hand, quoted the observations of Professor Forbes, who has shown that, the deeper you go in the littoral waters of the Mediterranean, the fewer are the living forms.

609. *Their arguments based on the tides.* As for the number of unknowns increasing with the depth (§ 607), they contend that the tides, the currents, and the agitation of the waves all reach to the bottom in shallow water; that they sweep and scour from it the feculences of the sea, as these insect remains may be called, and bear them off into deep water. After reaching a certain depth, then this sediment passes into the stratum of quiet waters that underlie the roaring waves and tossing currents of the surface, and through this stratum these organic remains slowly find their way to the final place of repose as ooze at the bottom of the deep sea. Through such agencies the ooze of the deep sea ought, said the anti-biotics, to be richer than that of shallow water with infusorial remains; mud and all the light sedimentary matter of river waters are deposited in the deep pools, and not in the shoals and rapids of our freshwater streams; so we ought, reasoned they of this school, to have the most abundant deposits at the bottom of the deep sea.

[1] [Letter from C. G. Ehrenberg to Maury dated October, 1857, *Sailing Directions*, 8th ed., pp. 175–176.]

610. *On the antiseptic properties of sea water.* The anti-biotics referred to the antiseptic properties of sea water, and told how it is customary with mariners, especially with the masters of the sailing packets between Europe and America, to "corn" fresh meat by sinking it to great depths overboard. If they sink it too deep, or let it stay down too long, it becomes too salt. According to them, this process is so quick and thorough, because of the pressure and the affinity which not only forces the water among the fibres of the meat, but which also induces the salt to leave the water and take to the meat; and that the fleshy part of these microscopic organisms have been exposed to powerful antiseptic agents is proved by the fact that they are brought up in the middle of the ocean, and remain on board the vessel exposed to the air for months before they reach the hands of the microscopist; some of them have remained so exposed for more than a year, and then been found full of fleshy matter: a sure proof that it had been preserved from putrefaction and decay by processes which it had undergone in the sea, and before it was raised into the air.

611. *On pressure.* Thus the anti-biotics held that these little creatures were preserved for a while after death, and until they reached a certain depth, by salt, and afterward by pressure. They held that certain conditions are requisite in order that the decay of organic matter may take place; that the animal tissues of these shells during the process of decay are for the most part converted into gases; that these gases, in separating from the animal compound, are capable of exerting only a certain mechanical force, and no more; that this force is not very great; and, unless it were sufficient to overcome the pressure of deep-sea water, their separation could not go on, and that, consequently, there is a certain depth in the sea beyond which animal decomposition or vegetable decay can not take place. In support of this view, they referred to the well-known effects of pressure in arresting or modifying the energies displayed by certain chemical affinities; and in proof of the position that great compression in the sea prevents putrefaction, they referred to the fact well known to the fishermen of Nantucket and New Bedford, viz., that when a whale that they have killed sinks in shallow water, he, as the process of decay commences, is seen to swell and rise; but when he sinks in deep water, the pressure is such as to prevent the formation of the distending gases, and he never does rise. Some of these specimens have come from depths where the pressure is equal to that of 400 or 500 atmospheres. Specimens have been obtained by Lieutenant Brooke, in the Pacific, with "fleshy parts" among them, at the depth of 3300 fathoms, and where the pressure is nearly 700 atmospheres. We have brought up fleshy matter from the deep sea as deep down as we have gone; and we may infer that if we were to go to 4000 fathoms, we should still find pulpy matter among

the dead organisms there. At that depth, or a little over, common air, according to "Mariotte's *law*," would be heavier than water, and an air bubble down there, if one may imagine such a thing, would be heavy enough to sink. Under such conditions, and with the antiseptic agencies of the sea, the fleshy matter of these infusoria might be preserved at the bottom of the deep sea for a great length of time.

612. *Arguments from the Bible*. Moreover (§ 604), the anti-biotics pointed to the first chapter of Genesis to show that light and heat were ordained before the waters were commanded to bring forth. Hence they maintained that light and heat are necessary to marine life. In the depth of the sea there is neither light nor heat, wherefore they brought in circumstantial evidence from the Bible to sustain them in their view.

613. *A plan for solving the question*. This was an exceedingly interesting question, and we could suggest but one way of deciding it, which was this: Many of these little organisms of the sea are in the shape of plano-convex discs; all such, when alive, live with the convex side up, the flat side down; but when placed dead in the water and allowed freely to sink, the force of gravity always, and for obvious reasons, causes all such forms to sink with the convex side down. Brooke's lead will bring up these shells exactly as they lie on the bottom, and so he proposed to observe with regard to their manner of lying. Of course, if they lived at the bottom, they would die as they lived, and lie as they died, for (§ 590) there is nothing to turn them over after death at the bottom of the deep sea, consequently their skeletons would be brought up in the quills of the sounding machine flat side down, convex side up.

614. *An unexpected solution afforded*. But, before there was an opportunity of trying this plan, Ehrenberg himself afforded the solution in a most unexpected way; and in soundings from a great depth in the Mediterranean, among them he found many fresh-water shells with their fleshy parts still in them, though the specimens were taken from the middle of the Mediterranean. That savant, with his practiced eye, detected among them Swiss forms, which must have come down the Danube, and so out into the Mediterranean hundreds of miles, and on journeys which would require months, if not years, for these slowly-drifting creatures to accomplish. And so the anti-biotics maintain that their doctrine is established.[2]

---

[2] In a paper upon the organic life-forms from unexpected great depths of the Mediterranean, obtained by Captain Spratt from deep-sea soundings between Malta and Crete, in 1857, and read before the Berlin Academy, November 27, 1857, Ehrenberg said, "Especially striking among all the forms of the deep are the Phytolitharia, of which fifty-two in number are found. It would not be strange if these fifty-two forms were spongoliths, since we expect to find sponge in the sea. But a large number, not less than twenty kinds of Phytolitharia, are fresh-water and land forms. Hence the question arises, How came these forms into those depths in the middle of the sea?

615. *A discovery suggested by it.* Having thus discovered that the most frail and delicate organisms of the sea can remain in its depths for an indefinite length of time without showing a single trace of decay, we find ourselves possessed of a fact which suggests many beautiful fancies, some touching thoughts, and a few useful ideas; and among these last are found reasons for the conjecture that the gutta percha or other insulating material in which the conducting wires of the sub-Atlantic telegraph and other deep-sea lines are incased, becomes, when lodged beyond a certain depth, impervious to the powers of decay; that, with the weight of the sea upon them, the destructive agents which are so busy in the air can not find room for play. Curious that destruction and decay should be imprisoned and rendered inoperative at the bottom of the great deep!

616. *Specimens from the three oceans all tell the same story.* Specimens of the "ooze and bottom of the sea" have also been obtained by the ingenuity of Brooke from the depth of 2700 fathoms in the North Pacific, and examined by Professor Bailey.[3] We have now had specimens

---

"Naturally one looks at first to the Nile and the coasts; but the sea current carries the turbid Nile water eastward; for the current, according to Captain Smyth, especially in the middle of the sea, not only in the Levant, but also in the southern edge, is clearly a constant eastwardly one. Besides, there are among the forms some northern ones — e. g., *Eunotia triodon, Campylodiscus clypeus,* and many gallionella. This peculiarity may, perhaps, indicate a lower return current, hitherto observed only at Gibraltar, which probably brings into this basin the forms from the northern European rocks. Thus, for instance, the Danube may bring the Swiss forms in that circulation. But, on the other hand, a highly striking agreement with the forms of the 'trade-wind dust' is not to be overlooked.

"In reference to the question of permanent life in these most recent deep-sea materials, it may be observed, that the forms which we find are astonishingly well preserved, and in very large proportion, sometimes forming the principal mass of the earthy bottom.

"The striking fact, moreover, that every one who has the opportunity to compare accurately the microscopic forms of the whole land and sea under great variety of circumstance does, out of even the smallest specimens of the bottom, deduce so much that is new and peculiar to him, is no light testimony to show that the depth is not merely a collection of rubbish of the dead surface-life, however much there must be of fragments which naturally and undoubtedly deposit themselves there. I have considered this final remark necessary, because the distinguished sea-knower in Washington, often so kindly supplying and instructing me with material, has recently, in a report on Sub-oceanic Geography, New York, January 8, 1857, page 5 and yet more in detail in a late private letter, expressed a view opposite to that here laid down by me, in which, however, I can not coincide, for the reasons given above." — [Ehrenberg, "Die organischen Lebensformen in unerwartet grossen Tiefen des Mittelmeeres," *Monatsberichte der königlichen Preussischen Akademie der Wissenschaften, Berlin, a. d. J. 1857* (1858), pp. 538–570, esp., pp. 544–545, 570.]

[3]   "West Point, N.Y., January 29, 1856.

"MY DEAR SIR, — I have examined with much pleasure the highly interesting specimens collected by Lieutenant Brooke, of the United States Navy, which you kindly sent me for microscopic analysis, and I will now briefly report to you the results of general interest which I have obtained, leaving the enumeration of the organic contents and the description of new species for a more complete account, which I hope soon to publish. The specimens examined by me were as follows, viz.:

from the bottom of "blue water" in the narrow Coral Sea, the broad Pacific, and the long Atlantic, and they all tell the same story, namely,

"No. 1. Sea bottom, 2700 fathoms; lat. 56° 46′ N., long, 168° 18′ E.; brought up July 19, 1855, by Lieutenant Brooke, with Brooke's lead.

"No. 2. Sea bottom, 1700 fathoms; lat. 60° 15′ N., long. 170° 53′ E.; brought up as above, July 26, 1855.

"No. 3. Sea bottom, 900 fathoms; temperature (deep sea) 32°, Saxton; lat. 60° 30′ N., long. 175° E.

"A careful study of the above specimens gave the following results:

"1st. All the specimens contain some mineral matter, which diminishes in proportion to the depth, and which consists of minute angular particles of quartz, hornblende, feldspar, and mica.

"2d. In the deepest soundings (No. 1 and No. 2) there is the least mineral matter, the organic contents, which are the same in all, predominating, while the reverse is true of No. 3.

"3d. All these specimens are *very rich* in the silicious shells of the Diatomaceæ, which are in an admirable state of preservation, frequently with the valves united, and even retaining the remains of the soft parts.

"4th. Among the Diatomes the most conspicuous forms are the large and beautiful discs of several species of coscinodiscus. There is also, besides many others, a large number of a new species of rhizosolenia, a new syndendrium, a curious species of chætoceros, with furcate horns, and a beautiful species of asteromphalus, which I propose to call Asteromphalus Brookei, in honor of Lieutenant Brooke, to whose ingenious device for obtaining deep soundings, and to whose industry and zeal in using it, we are indebted for these and many other treasures of the deep.

"5th. The specimens contain a considerable number of silicious spicules of sponges, and of the beautiful silicious shells of the polycistineæ. Among the latter I have noticed Cornutella clathrata of Ehrenberg, a form occurring frequently in the Atlantic soundings. I have also noticed in all these soundings, and shall hereafter describe and figure, several species of eucyrtidium, halicalyptra, a perichlamidium, a stylodictya, and many others.

"6th. I have not been able to detect even a fragment of any of the calcareous shells of the polythalamia. This is remarkable, from the striking contrast it presents to the deep soundings of the Atlantic, which are chiefly made up of these calcareous forms. This difference can not be due to temperature, as it is well known that polythalamia are abundant in the Arctic Seas.

"7th. These deposits of microscopic organisms, in their richness, extent, and the high latitudes at which they occur, resemble those of the antarctic regions, whose existence has been proved by Ehrenberg, and the occurrence in these northern soundings of species of asteromphalus and chætoceros is another striking point of resemblance. These genera, however, are not exclusively polar forms, but, as I have recently determined, occur also in the Gulf of Mexico and along the Gulf Stream.

"8th. The perfect condition of the organisms in these soundings, and the fact that some of them retain their soft portions, indicate that they were very recently in a living condition, but it does not follow that they were living when collected at such immense depths. As among them are forms which are known to live along the shores as parasites upon the algæ, etc., it is certain that a portion, at least, have been carried by oceanic currents, by drift ice, by animals which have fed upon them, or by other agents, to their present position. It is hence probable that all were removed from shallower waters, in which they once lived. These forms are so minute, and would float so far when buoyed up by the gases evolved during decomposition, that there would be nothing surprising in finding them in any part of the ocean, even if they were not transported, as it is certain they often are, by the agents above referred to.

"9th. In conclusion, it is to be hoped that the example set by Lieutenant Brooke will be followed by others, and that, in all attempts to make deep soundings, the effort to bring up a portion of the bottom will be made. The soundings from any part

that the bed of the ocean is a vast cemetery. The ocean's bed has been found every where, wherever Brooke's sounding-rod has touched, to be soft, consisting almost entirely of the remains of infusoria. The Gulf Stream has literally strewed the bottom of the Atlantic with these microscopic shells; for the Coast Survey has caught up the same infusoria in the Gulf of Mexico and at the bottom of the Gulf Stream off the shores of the Carolinas, that Brooke's apparatus brought up from the bottom of the Atlantic off the Irish coast.

617. *Their suggestions.* The unabraded appearance of these shells, and the almost total absence among them of any detritus from the sea or foreign matter, suggest most forcibly the idea of perfect repose at the bottom of the deep sea. Some of the specimens are as pure and as free from the sand of the sea as the snow-flake that falls, when it is calm, upon the lea, is from the dust of the earth. Indeed, these soundings suggest the idea that the sea, like the snow-cloud with its flakes in a calm, is always letting fall upon its bed showers of these microscopic shells; and we may readily imagine that the "sunless wrecks," which strew its bottom, are, in the process of ages, hid under this fleecy covering, presenting the rounded appearance which is seen over the body of the traveler who has perished in the snow-storm. The ocean, especially within and near the tropics, swarms with life. The remains of its myriads of moving things are conveyed by currents, and scattered and lodged in the course of time all over its bottom. This process, continued for ages, has covered the depths of the ocean as with a mantle, consisting of organisms as delicate as the macled frost, and as light in the water as is down in air.

618. *The work of readaptation, how carried on.* The waters of the Mississippi and the Amazon, together with all the streams and rivers of the world, both great and small, hold in solution large quantities of lime, soda, iron, and other matter. They discharge annually into the sea an amount of this soluble matter, which, if precipitated and collected into one solid mass, would no doubt surprise and astonish the boldest speculator with its magnitude. This soluble matter can not be evaporated. Once in the ocean, there it must remain; and as the rivers are continually pouring in fresh supplies of it, the sea, it has been argued, must continue to become more and more salt. Now the rivers convey to the sea this solid matter mixed with fresh water, which, being lighter than that of the ocean, remains for a considerable time at or near the surface. Here the

---

of the ocean are sure to yield something of interest to microscopic analysis, and it is as yet impossible to tell what important results may yet flow from their study.

"The above is only a preliminary notice of the soundings referred to. I shall proceed without delay to describe and figure the highly interesting and novel forms which I have detected, and I hope soon to have them ready for publication.

"Yours very respectfully,   J. W. BAILEY.
"Lieutenant M. F. MAURY, National Observatory, Washington City, D. C."

microscopic organisms of the deep-sea lead are continually at work, secreting this same lime and soda, etc., and extracting from the sea water all this solid matter as fast as the rivers bring it down and empty it into the sea. Thus we haul up from the deep sea specimens of dead animals, and recognize in them the remains of creatures which, though invisible to the naked eye, have nevertheless assigned to them a most important office in the physical economy of the universe, viz., that of regulating the saltness of the sea (§ 489). This view suggests many contemplations. Among them, one, in which the ocean is presented as a vast chemical bath, in which the solid parts of the earth are washed, filtered, and precipitated again as solid matter, but in a new form, and with fresh properties. Doubtless it is only a readaptation — though it may be in an improved form — of old, and perhaps effete matter, to the uses and well-being of man. These are speculations merely; they may be fancies without foundation, but idle they are not, I am sure; for when we come to consider the agents by which the physical economy of this our earth is regulated, by which this or that result is brought about and accomplished in this beautiful system of terrestrial arrangements, we are utterly amazed at the offices which have been performed, the work which has been done, by the animalculæ of the water. But whence come the little silicious and calcareous shells which Brooke's lead has brought up, in proof of its sounding, from the depth of over two miles? Did they live in the surface waters immediately above? or is their *habitat* in some remote part of the sea, whence, at their death, the currents were sent forth as pall-bearers, with the command to deposit the dead corpses where the plummet found them?

CHAPTER XV

§ 621–680. SEA ROUTES, CALM BELTS, AND VARIABLE WINDS

621. *Practical results of physical researches at sea.* Plate V, so far as the winds are concerned, is supplemental to Plate I. The former shows the monsoon regions, and indicates the prevailing direction of the winds in every part of the ocean; the latter indicates it generally for any latitude, without regard to any particular sea. Plate V also exhibits the principal routes across the ocean. This plate indicates the great practical results of all the labor connected with this vast system of research; its aim is the improvement of navigation; its end, the shortening of voyages. Other interests and other objects, nay, the great cause of human knowledge, have been promoted by it; but the advancement that has been given to these do not, in this utilitarian age, and in the mind of people so eminently practical as mariners are, stand out in a relief half so grand and imposing as do those achievements by which the distant isles and marts of the sea have, for the convenience of commerce, been lifted up, as it were, and brought closer together by many days' sail.

622. *Time-tables.* So to shape the course on voyages as to make the most of the winds and currents at sea is the perfection of the navigator's art. How the winds blow and the currents flow along this route or that, is no longer matter of opinion or subject of speculation, but it is a matter of certainty determined by actual observation. Their direction has been determined for months and for seasons, along many of the principal routes, with all the accuracy of which results depending on the doctrine of chances are capable; and farther, these results are so certain that there is no longer any room for the mariner to be in doubt as to the best route. When a navigator undertakes a voyage now, he does it with the lights of experience to guide him. The winds and the weather daily encountered by hundreds who have sailed on the same voyage before him, and "the distance made good" by each one from day to day, have been tabulated in a work called Sailing Directions, and they are so arranged that he may daily see how much he is ahead of time, or how far he is behind time; nay, his path has been literally blazed through the winds for him on the sea; mile-posts have been set up on the waves, and finger-boards planted, and time-tables furnished for the trackless waste, by which the ship-master, on his first voyage to any port, may know as well as the most experienced trader whether he be in the right road or no.

623. *Close running.* From New York to the usual crossing of the equator on the route to Rio the distance, by an air line, is about 3400 miles; but the winds and currents are such as to force the Rio bound vessel out of its direct line. Nevertheless, they have been mapped down, studied, and discussed so thoroughly that we may, using as arguments the data obtained, compute with remarkable precision the detour that vessels sailing from New York or any other port, and attempting this route to the equator at the various seasons of the year, would have to make. This computation showed that, instead of 3400 miles, the actual distance to be accomplished through the water by vessels under canvas on this part of the voyage would be 4093 miles. More than a hundred sailing vessels have tried it by measuring and recording the distance actually sailed from day to day; their mean distance is 4099 miles, consequently their actual average differs only six miles from the computed average.[1]

624. *A desideratum on shipboard.* The best-navigated steam-ships do not sail closer than this, and a better proof of the accuracy of our knowledge concerning the prevailing direction of the winds at sea could not be afforded. Unfortunately, anemometers are not used on shipboard. Had they been in common use there, and had we been furnished with data for determining the force of the wind as well as its direction, we could compute the time as well as the distance required for the accomplishment of any given voyage under canvas. Thus the average time required to sail from New York to the equator might be computed within an hour, for it has been computed within an hour's sail — six miles (§ 623).

625. *How passages have been shortened.* By the knowledge thus elaborated from old and new log-books and placed before the nautical world, the average passage from Europe or the United States to all ports in the southern hemisphere has been shortened ten days, and to California a month and a half.[2] Between England and her golden colony in the South Seas the time required for the round voyage has been lessened fifty days or more, and from Europe to India and China the outward passage has been reduced ten days. Such are some of the achievements that

---

[1] *Sailing Directions*, II, 146.

[2] "During the last year [1859] the 8th edition of Maury's Sailing Directions, in two quarto volumes, has been published at the Observatory in Washington. It affords abundant evidence of the activity, to which allusion has already been made, in this field of research, and with regard to which all geographers feel the most lively interest.

"Official tables have been received from San Francisco, showing the vessels that have arrived at that port during the year, with the length of passage. Of those arriving direct, *via* Cape Horn, 124 were from the Atlantic ports of the United States, and 34 from Europe. Of these 124, 70 are known to have had the Wind and Current Charts on board; their average passage was 135 days, which is 11 days less than the average of those from the United States, and 24 days less than the average of those from Europe *without the Charts.* When these researches commenced, the general average was 180 days from the United States, and 183 from Europe to California." — ["The Progress of Marine Geography," *Jour. Amer. Geogr. Soc., 1859*, pp. 9–10.]

commend this beautiful system of research to the utilitarian spirit of the age.

626. *Fast sailing.* The route that affords the bravest winds, the fairest sweep, and the fastest running to be found among ships, is the route to and from Australia. But the route which most tries a ship's prowess is the outward-bound voyage to California. The voyage to Australia and back carries the clipper ship along a route which for more than 300° of longitude runs with the "brave west winds" of the southern hemisphere. With these winds alone, and with their bounding seas which follow fast, the modern clipper, without auxiliary power, has accomplished a greater distance in a day than any sea steamer has ever been known to reach. With these fine winds and heaving seas those ships have performed their voyages of circumnavigation in 60 days.

627. *The longest voyage.* The sea voyage to California, Columbia, and Oregon is the longest voyage in the world — longest both as to time and distance. Before these researches were extended to the winds and currents along that route, the average passage both from Europe and America to our northwest coast was not less than 180 days. It has been reduced so as to average only 135 days. This route is now so well established, and the winds of the various climates so well understood, that California bound vessels sailing about the same time from the various ports of Europe and America are, if they be at all of like prowess, almost sure to fall in with and speak each other by the way.

628. *Obstructions to the navigator.* The calm belts at sea, like mountains on the land, stand mightily in the way of the voyager, but, like mountains, they have their passes and their gaps. In the regions of light airs, of baffling winds, and deceitful currents, the seaman finds also his marshes and his "mud-holes" on the water. But these, these researches have taught him how best to pass or entirely to avoid. Thus the forks to his road, its turnings, and the crossings by the way, have been so clearly marked by the winds for him that there is scarcely a chance for him who studies the lights before him, and pays attention to the directions given, to miss his way.

629. *Plate V.* The arrows of Plate V are supposed to fly with the wind; the half-bearded and half-feathered arrows denote monsoons or periodic winds; the dotted bands, the regions of calm and baffling winds. Monsoons, properly speaking, are winds which blow one half of the year from one direction, and the other half from an opposite, or nearly an opposite direction. The time of the changing of these winds, and their boundaries at the various seasons of the year, have been discussed in such numbers and mapped down in such characters that the navigator who wishes to take advantage of them or to avoid them altogether is no longer in any doubt as to when and where they may be found.

630. *Deserts.* Let us commence the study of the calm belts as they are represented on Plates I and V. Both the monsoons and trade-winds are also represented on the latter. They often occupy the same region. But, turning to the trade-winds for a moment, we see that the belt or zone of the southeast trade-winds is broader than the belt or zone of northeast trades. This phenomenon is explained by the fact that there is more land in the northern hemisphere, and that most of the deserts of the earth — as the great deserts of Asia and Africa — are situated in the rear, or behind the northeast trades; so that, as these deserts become more or less heated, there is a call — a pulling back, if you please — upon these trades to turn about and restore the equilibrium which the deserts destroy. There being few or no such regions in the rear of the southeast trades, the southeast trade-wind force prevails, and carries them over into the northern hemisphere.

631. *Diurnal rotation.* We see by the plate that the two opposing currents of wind, called "the trades," are so unequally balanced that the one recedes before the other, and that the current from the southern hemisphere is larger in volume; *i. e.*, it moves a greater zone or belt of air. The southeast trade-winds discharge themselves over the equator — *i. e.*, across a great circle — into the region of equatorial calms, while the northeast trade-winds discharge themselves into the same region over a parallel of latitude, and consequently over a small circle. If, therefore, we take what obtains in the Atlantic as the type of what obtains entirely around the earth, as it regards the trade-winds, we shall see that the southeast trade-winds keep in motion more air than the northeast do, by a quantity at least proportioned to the difference between the circumference of the earth at the equator and at the parallel of latitude of 9° north. For if we suppose that those two perpetual currents of air extend the same distance from the surface of the earth, and move with the same velocity, a greater volume from the south should, as has already been shown (§ 343), flow across the equator in a given time than would flow from the north over the parallel of 9° in the same time; the ratio between the two quantities would be as radius to the secant of 9°. Besides this, the quantity of land lying within and to the north of the region of the northeast trade-winds is much greater than the quantity within and to the south of the region of the southeast trade-winds. In consequence of this, the mean level of the earth's surface within the region of the northeast trade-winds is, it may reasonably be supposed, somewhat above the mean level of that part which is within the region of the southeast trade-winds. And as the northeast trade-winds blow under the influence of a greater extent of land surface than the southeast trades do, the former are more obstructed in their course than the latter by the forests, the mountain ranges, unequally heated surfaces, and other such like inequalities.

632. *The land in the northern hemisphere.* That the land of the northern hemisphere does assist to turn these winds is rendered still more probable from this circumstance: All the great deserts are in the northern hemisphere, and the land surface is also much greater on our side of the equator. The action of the sun upon these unequally absorbing and radiating surfaces in and behind, or to the northward of the northeast trades, tends to check these winds, and to draw in large volumes of the atmosphere, that otherwise would be moved by them, to supply the partial vacuum made by the heat of the sun, as it pours down its rays upon the vast plains of burning sands and unequally heated land surfaces in our overheated hemisphere. The northwest winds of the southern are also, it may be inferred, stronger than the southwest winds of the northern hemisphere.

633. *Why the southeast trades are the stronger.* That the southeast trade-winds should, as observations (§ 343) have shown, be stronger than the northeast trade-winds, is due in part also to the well-established fact that the southern (§ 446) is cooler than the northern hemisphere. The isothermal lines of Dové show that the air of the southeast is also cooler than the air of the northeast trade-winds. Being cooler, the air from the cool side would, for palpable causes, rush with greater velocity into the equatorial calm belt than should the lighter air from the warmer or northern side. The fact that the air in the lower latitudes of the southern hemisphere is the cooler will assist to explain many other contrasts presented by the meteorological conditions on opposite sides of the equator. Plate IX shows that we have more calms and more fogs, more rains and more gales, with more thunder, on one side than the other, and that the atmosphere preserves its condition of unstable equilibrium with much more uniformity, being subject to changes less frequent and violent on the south side of the equator than on the north side.

634. *Their uniformity of temperature.* The highest summer temperature in the world is to be found in the extra-tropical countries of the north. The greatest extremes of temperature are also to be found among the valleys of the extra-tropical north. In the extra-tropical south there is but little land, few valleys, and much water; consequently the temperature is more uniform, changes are less sudden, and the consequent commotions in the air less violent.

635. *The mean place of the equatorial calm belt.* Following up these facts with their suggestions, we discover the key to many phenomena which before were locked up in "the chambers of the south." The belt of equatorial calms which separates the two systems of trade-winds is, as we know (§ 295), variable as to its position. It is also variable in breadth. Sometimes it covers a space of several degrees of latitude, sometimes not more than one. Its southern edge, in spring, sometimes goes

down to 5° S; its northern edge, in autumn, often mounts up to the parallel of 15° N. The key to these phenomena has been found; with it in hand, let us proceed to unlock, first remarking that the mean position of the equatorial calm belt in the Atlantic is between the equator and 9° N., and that as it is there, so I assume it to be in other oceans.

636. *Never at rest.* This calm belt is produced by the meeting of the two trade-winds, and it occupies strictly a medial position between them. It is in the barometric valley, between the two barometric ridges (§ 858, Figure 13) from which the trade-winds flow. If one "trade" be stronger than the other, the stronger will prevail so far as to force their place of meeting over and crowd it back upon the weaker wind. It is evident that this place of meeting will recede before the stronger wind, until the momentum of the stronger wind is so diminished by resistance, and its strength so reduced as exactly to be counterbalanced by the weaker wind. Then this calm place will become stationary, and so remain, until, for some cause, one or the other of the meeting winds gains strength or loses force; then the stronger will press upon the weaker, and the calm belt will change place and adjust itself to the new forces. The changes that are continually going on in the strength of the winds keep the calm belt in a trembling state, moving now to the north, now to the south, and always shifting its breadth or its place under the restless conditions of our atmosphere.

637. *The calm belts occupy medial positions.* The southern half of the torrid zone is cooler than the northern, and, parallel for parallel, the southeast trade-winds are consequently cooler than the northeast. They both flow into this calm belt, where the air, expanding, ascends, flows off above, produces a low barometer, and so makes room for the inflowing current below. Now if the trade-wind air which flows in on one side of this calm belt be heavier, whether from temperature or pressure, than the trade-wind air which flows in on the other, the wind from the heavy side will be the stronger. This is obvious, for it is evident that if the difference of temperature of the ascending column and the inflowing air were scarcely perceptible, the difference of specific gravity between the inflowing wind and the uprising air would be scarcely perceptible, and the movement of the inflowing wind would be very gentle; but if the difference of temperature were very great, the difference of specific gravity would be very great, and the violence of the inrushing wind proportionably great. Because the southern half of the torrid zone is the cooler, the difference in temperature between the air of the calm belt and the air of the trade-winds is greater, parallel for parallel, in the southeast than in the northeast trade-winds; consequently, the southeast trade-winds should be — as observations show them to be — stronger than the northeast, and consequently, also, their meeting should take place, not upon the equator,

but upon that side of it where the weaker winds prevail, and this is also in accordance (§ 343) with facts.

638. *Strength of the trade-winds varies with the seasons.* It follows from these premises that the winter trade-winds should be stronger than the summer. In our summer, the air which the northeast trade-winds put in motion has its temperature raised and brought more nearly up to that of the air in the calm belt. At the same time, the temperature of the air which the southeast trade-winds put in motion is proportionably lowered. Thus they increase in strength, while the northeast diminish; the consequence is, they push their place of meeting with the northeast trades far over on this side of the equator, and for two or three months of the year maintain the polar edge of the calm belt as high up as the parallel of 15° N. But with the change of seasons these influences are all transposed and brought into play on opposite sides — only in the southern summer the strength of the southeast and the temperature of the northeast trade-winds are diminished so as to admit of the edge of the calm belt being pressed down only as far as 5° instead of 15° S. The causes which produce this alternation of trade-wind strength are cumulative; consequently, the northeast trade-winds should be weakest in August or September, strongest in February or March, *after* the period of maximum heat in one case and of minimum in the other.

639. *Sailing through them in fall and winter.* In the other hemisphere, the period of strongest trades is coincident with that of the minimum in this. These deductions are also confirmed by observations; for such is the difference as to strength and regularity of the northeast trade-winds in September and February, that the average passage through them from New York to the line is 26.4 days in the winter against 38.8 in the fall month.

640. *A thermal adjustment.* Thus it appears that the equatorial calm belt is made to shift its place with the seasons, not by reason of the greater intensity of the solar ray in the latitude where the calm belt may be at that season, but by reason of the annual variations in the energy of each system of trades; which variations (§ 638) depend upon the changes in the temperature and barometric weight of the air which each system puts in motion. This calm belt, therefore, may be considered as a *thermal adjustment — the dynamical null-belt — between the trade-winds of the two hemispheres.*

641. *The barometer in the trade-winds and equatorial calms.* The observations on the barometer at sea (§ 855) shed light on this subject. According to the Dutch, that instrument stands higher by 0.055 inch in the southeast than it does in the northeast trade-winds. According to the observations of American navigators, it stands 0.050 inch higher.[3] The for-

[3] "Barometric Anomalies off Cape Horn," *Sailing Directions* [8th ed., II, 446–450].

mer determination is derived from 30,873, the latter from 1899 observations; therefore 0.055 inch is entitled to most weight. The trade-winds are best developed between the parallels of 5° and 20°. The mean barometric pressure between these parallels is 29.968 inches for the northeast, and 30.023 inches for the southeast trade-winds; while for the calm belt it is 29.915 inches. The pressure, therefore, upon the air in each of the trade-winds is greater than it is in the calm belt; and it is this *difference of pressure*, from whatever cause arising, that gives the wind in each system of trades its velocity. The difference between the calm belt and trade-wind pressure is 0.108 for the southeast and 0.053 for the northeast. According to the barometer, then, the southeast should be stronger than the northeast trade-winds, and according to actual observations they are.[4]

642. *Experiments in the French Navy.* Now if we liken the equatorial calm belt with its diminished pressure to a furnace, the northeast and the southeast trade-winds may be not inaptly likened to a pair of double bellows that are blowing into it. In excess of barometric pressure, the former is a bellows with a weight of 3.8 lbs., the latter with a weight of 7.8 lbs. to the square foot. It is this pressure which, like the weight upon the real bellows in the smithy, keeps up the steady blast; and as the effective weight upon the one system of trades is about double that upon the other, the one under the greatest pressure should blow with nearly double the strength of the other, and this appears, both from actual observations and calculations, as well as from direct experiments ordered in the French brig of war "Zebra," by Admiral Chabannes, to be the case.[5]

[4] "The Winds at Sea."

[5] Letter to Admiral Chabannes, with extract of his reply thereto:

"Observatory, Washington, 8th April, 1859.

"MY DEAR ADMIRAL, — My last was dated 15th January, ultimo. I hope the charts and vol. i., 8th ed. Sailing Directions, and part of vol. ii. in the sheets, came safely to hand. Vol. ii. is just out, and I hasten, in homage of my respect, and as a token of good-will, to lay a copy before you.

"Permit me, if you please, to call your attention to the chapter on the 'Average Force of the Trade-winds,' p. 857, and especially to the table of comparative speed (of sailing vessels) through the northeast and southeast trade-winds of the Atlantic, p. 865. The average speed, you observe, is nearly the same, notwithstanding that through the southeast trades the wind is aft, through the northeast just abaft the beam.

"In order to treat this question thoroughly, it is very desirable to know the difference in the speed of vessels when sailing with the same wind aft, with it quartering, with it a point or two abaft the beam, and with it close hauled. With a good series of experiments upon this subject, we should be able to arrive at definite conclusions with regard to the average difference in force not only of the two systems of trade-winds, but of the winds generally in various parts of the ocean.

"If we assume that a wind which, being dead aft, drives a vessel at the rate of six knots, will, when brought nearly abeam, drive her eight knots — as in this chapter I have supposed — and then, if we apply the dynamical law of the resistance increasing as the squares of the velocity of the ship, we should be led to the remarkable conclusion that the average velocity of the northeast to southeast trades of the Atlantic is as 36 to 64. Therefore, in conducting these experiments, it would be very desirable to know the area of canvas that fairly feels the wind when it is aft, and the area upon

643. *Difference in tons of the barometric pressure upon the northeast and southeast trade-winds.* — With these barometric observations, and the

which the wind blows when the ship is hauled up. Suffice it to say, that the facts which we already have indicate that the southeast trades, both of the Atlantic and Indian Oceans, are fresher than the northeast trades of the Atlantic. May we infer from this that the southeast trades of the Pacific are also fresher than the northeast trades of that ocean? If we may so infer and be right, then there is another step which we may take with great boldness, and pronounce the atmospherical circulation of the southern hemisphere to be much more active than that of the northern. And having reached this round in the ladder up which I am soliciting you to accompany me, we are prepared to pause and take a view of some of the new physical aspects which these facts and this reasoning spread out before us.

"That the atmospherical circulation is more active in the southern than in the northern hemisphere appears to be indicated also by the "brave west winds" of the extra-tropical south. (See also Plate IX and §§ 632 and 633.) If the air performs its circuit more rapidly through one system of trade-winds than the other, then it follows that it must perform its circuit more rapidly also along those regions through which it has to pass in order to reach such rapid trades. Consequently, there should be a great difference between the gales of the northern and those of the southern hemisphere. If we suppose the general circulation of the northern hemisphere to be sluggish, the air in its circuits there would have time to tarry by the way, as it were, and to blow gales of wind from all points of the compass. On the contrary, if the general circulation of the southern hemisphere be brisk and active, the air in its general circuits, like a fast train on the railway, would not have so much time to tarry by the way, because, like the cars, *it must be up to time.* Hence, admitting this view of the matter to be correct (and you perceive that for the want of the experiments already alluded to we are groping in the darkness of conjecture), though we might expect gales of wind in the extra-tropical regions of the south, yet they would, for the most part, blow *with* the prevailing direction of the wind, and not *against* it. Thus the gales on the polar side of Capricorn should, particularly at sea, have westing in them always — almost.

"In corroboration of this view, I may mention, on the authority of a paper just received from Lieutenant Van Gogh, of the Dutch Navy, that the gales of wind which take place between the meridians of 14° and 32° E., and between the parallels of 33° and 37° south, have been discussed at the Meteorological Institute of Utrecht. For this purpose he tabulated the results for the whole year of 17,810 observations — an observation comprehending a period of eight hours. According to these observations, it is blowing a gale of wind off the Cape of Good Hope 7.16 per cent. of the whole year, and from the following quarters: namely, between N.N.W. and S.S.W. 6.43 per cent.; from all other quarters, 0.73 per cent.

"Perhaps you may find it convenient to institute, with some of the vessels of your fleet, a regular series of experiments in the southeast trades upon speed, when sailing at various angles with the course of the wind. Besides answering our immediate purpose, the results might enable us to convert ships into very good anemometers for all winds except gales.

"Pardon me for being so tedious upon this subject. If you have felt me so, pray ascribe it to my desire to get by actual experiment an expression in the average speed of ships for the actual force and velocity of the winds.

"Wishing you all success and good luck in the investigation which you have in hand, pray believe me, my dear admiral, yours very truly, M. F. MAURY.
"Admiral C. DE CHABANNES, Commander-in-chief of the French
Naval Division of Brazil and La Plata, Rio de Janeiro."

Extract of a letter in reply to the foregoing:
"Montevideo, January 25, 1860.
"MY DEAR SIR, — * * * * As you have indicated to me in your letter of April, I have caused to be made, by a brig of my division, experiments upon the comparative

assumed fact that the mean pressure of the atmosphere is 15 lbs. upon the square inch, we may readily determine in tons the total by which the superincumbent pressure upon the southeast trade-winds between the parallels of 5° and 20° exceeds that upon the northeast between corresponding parallels. For the whole girdle of the earth, the excess of pressure upon the southeast trade-winds is 1,235,250 *millions* of tons. This is the superincumbent weight or pressure which is urging the southeast trade-winds forward faster than the northeast. It is inconceivably great; and to bring it within comprehensible terms, the mariner will be still astonished to hear that the weight of atmosphere which is bearing down upon the deck of a first class clipper ship is 15 or 20 tons greater when he is sailing in her through the southeast than it is when he is sailing in her through the northeast trade-winds.

644. *Why the barometer should stand higher in the southeast than in the northeast trade-winds.* The question now suggests itself, Why should the barometer stand higher in the southeast than it does in the northeast trade-winds? The theory of a crossing at the calm belts affords the answer. The air which the northeast trade-winds deliver into the calm belt is not as heavily laden with moisture as that of the southeast trades. It is not as heavily laden for two reasons; one is, the southeast trade-wind belt is broader than the northeast; consequently, in the former there is more air in contact with the evaporating surface. In the next place, the northeast trade-wind belt includes more land within it than the southeast; consequently, when the two winds arrive at the calm belt, they are, for this reason, also unequally charged with moisture. Now, when they rise up and precipitate this moisture, more heat is liberated from the southeast than from the northeast trade-wind air; the latter, therefore, is the cooler and the more compact; and as, by the theory of the crossings, it flows off to the south as an upper current, it presses upon the barometer with more weight than the warmer and more moist air that feeds the current which is above and counter to the northeast trades. There is not in the whole range of marine meteorology a single well-established fact that is inconsistent with the theory of a crossing at the calm belts.

645. *Cataclysms.* The geological record affords evidence that the climates of the earth were once very different from what they are now; that at one time intertropical climates extended far up toward the north; at another time polar climates reached, with their icebergs and their

---

velocities wind abaft and wind abeam with a given force of wind, but I have not yet been able to deduce any positive rule, the experiments not having been sufficiently multiplied. I can, however, give as a result that the increase of headway given by wind abeam over the headway with wind aft has been a little less than two knots; when the velocity with wind aft was from 6 to 8 knots, the force of the wind aft might be expressed by 4, and of the wind abeam by 6. * * * *

"C. de Chabannes."

drift, far down toward the equator; that in remote ages most of what we now call dry land was covered with water, for we find on the mountains and far away in the interior of continents deposits many feet thick, consisting of sea-shells, marine animals, and organic productions of many sorts. These fossils, marks, and traces indicate that since their day ages inconceivably great have elapsed. Not only so: the lines of drift, and boulders, and gashes with which the earth is scored and strewed, afford reason for the conjecture that there have been cataclysms, in which the waters have swept from north to south, and again from south to north, bearing with them icebergs, huge blocks of stone, rubble, drift, and sediment of various sorts. Lieutenant Julien, M. Le Hon, and M. Adhémar have, with marked ability, treated of this subject. They maintain that our earth has a "secular" as well as an annual summer and winter; that these "secular" seasons depend upon the precession of the equinoxes, and that the length of each is consequently 10,500 of our years; and that it is the melting of the polar ices in the "secular" season of one hemisphere, and their recongelation in the "secular" winter of the other, that causes a rush of the sea from one hemisphere into the other; and so cataclysms are produced at regular intervals of 10,500 years. In consequence of the inclination of the axis of the earth to the plane of its orbit, we have our change of seasons; and in consequence of the ellipticity of that orbit, the spring and summer of our hemisphere are at present longer than those of the southern. During the excess of time that the sun tarries on our side of the equator, the southern nights are prolonged, so that the night of the south pole — the antarctic winter — is annually a week longer (§ 366) than the arctic. Thus, during the period of 10,500 years, the antarctic regions will experience 142 years of night, or winter, in the aggregate, more than the arctic. Therefore it is manifest, say the *cataclysmatists*, that, though the two hemispheres do receive annually the same amount of solar heat, yet the amount dispensed by radiation is very much greater on one side of the equator than the other. The total effect of the alternate cooling down on each side of the equator causes an accumulation of ice at the pole — when the nights are longest — sufficient, say they, to disturb the centre of gravity of the earth, causing it to take up its position on the icy side of the equator. As the ice accumulates, so is the water drawn over from the opposite hemisphere. Such, briefly stated, is the theory which has found very ingenious and able advocates in the persons of MM. Julien [6] and Adhémar.[7]

646. *Are the climates of the earth changing?* This theory is alluded to here, not for the purpose of discussion, but for the purpose of directing

[6] Félix Julien, *Courants et Révolutions de l'Atmosphère et de la Mer, comprenant une théorie nouvelle sur les Déluges Périodiques* (Paris, 1860).

[7] J. Adhémar, *Révolutions de la Mer. Déluges Périodiques* (Paris, 1860).

attention to certain parts of this work in connection with it, as Chapters VII and XXI, for example, and of remarking upon the stability of terrestrial climates. Though the temperate regions be cooler in the southern than in the northern hemisphere, it does not appear certain that the climates of the earth are now changing. Observations upon the subject, however, are lacking. The question is one of widespread and exceeding interest; and it may be asked if we have not in the strength of the trade-winds a gauge, or in their barometric weight an index, or in the equatorial calm belt a thermometer — each one of the most delicate construction and sensitive character — which would, within the compass of human life, afford unerring indications of a change of climates, if any such change were going on? If the temperature of the S.E. trade-winds, or the barometric pressure upon the N.E. (§ 641), were to be diminished, the S.E. trades would force this calm belt still farther to the north, and we might have a regular rainy season in what is now the great desert of Sahara; for where this calm belt is (§ 517) there is the cloud-ring, with its constant precipitation. Therefore, if there be any indications that the southern edge of the great desert is gradually approaching the equator, it favors the supposition that the southern hemisphere is growing warmer; but if the indications be that the southern edge of the desert is receding from the equator, then the fact would favor the supposition that the southern hemisphere is growing still cooler. Nor are these the only latchets which a study of this calm belt and of the winds enables us to lift.

647. *Temperature of the trade-winds and calm belts.* Theory suggests, and observation, as far as it goes, seems to confirm the suggestion, that the N.E. and S.E. trade-winds enter the equatorial calm belt at the same temperature. I have followed 100 vessels with their thermometer across the equatorial calm belt of the Atlantic, and another 100 across it in the Pacific. Assuming its mean position to be as these observations indicate it to be — viz., between the parallels of 3° and 9° N. — the mean temperature is 81° at its northern, 81.4° at its southern edge, and 82° in the middle of it. These 200 logs were taken at random, and for all months. The temperature of the air was noted also in each trade at the distance of 5° from its edge of the calm belt. Thus the temperature of the N.E. trades, 5° from the north edge of the calm belt, or in 14° N., is 78.2°; at a like distance in the S.E. trades from the equatorial edge, or in 2° S., the mean temperature is 80.2°. From this it would seem that, in traversing this belt of 5°, the temperature of the N.E. is raised twice as much as the temperature of the S.E. trades; which is another indication that the velocity of the S.E. is nearly or quite double the velocity of the N.E. trades (§ 642). For if it be supposed that it takes the N.E. trades twice as long to traverse 5° of latitude as it does the S.E., it is evident that the former would be exposed twice as long to the solar ray, and receive twice

the amount of heat that is imparted to the S.E. trade-winds in traversing given differences of latitude. Thus the position of the calm belt, the barometer, the thermometer, and the rate of sailing, all indicate the S.E. trade-winds to be the stronger. It appears, moreover, that the temperature of the S.E. trade-wind is in 2° S. below the temperature of the N.E. in 9° N., the latter being 81°, the former 80.2°.

648. *The thermal equator.* The foregoing observations show that after these winds enter the calm belt, the air they bring into it continues to rise, and this also is what might well be anticipated, for the sun continues to pour down upon it. But while the temperature of the surface is kept down by the rain-drops from above, the temperature of the air in the whole belt is raised both by the direct heat of the sun and the latent heat which is set free by the constant (§ 514) and oftentimes heavy precipitation there. This latent heat is much more effective than is the direct heat of the sun in rarefying the air; consequently we here unmask the influences which place the thermal equator in the northern hemisphere.

649. *A natural actinometer in the trade-winds.* Nor is this the only chamber into which this calm belt key conducts us. Parallel for parallel (§ 446), the southern hemisphere is cooler than the northern; that is, the mean temperature for the parallel of 40° south, for example, is below the mean temperature for the parallel of 40° north, and so of corresponding parallels between 40° and the equator. It appears, moreover, that the mean temperature of the northeast trade-winds as they cross the parallel of 9° north, and the mean temperature of the southeast trade-winds as they cross the equator, is about the same (§ 647). The difference of temperature, then, between the southeast trades as they cross the parallel of 9° south, and as they cross the equator, expresses the *difference* in the thermal forces which give difference of energy to the dynamical power of the trade-winds. Not only so: it expresses the difference of temperature between the two corresponding parallels of 9° north and 9° south, and discovers to us a natural actinometer on a grand scale, and of the most delicate and beautiful kind.

650. *Heat daily received by the southeast trade-winds.* This actinometer measures for us the heat which the southeast trade-winds receive between the moment of crossing the parallel of 9° south and their arrival at the equator, for the heat thus received is just sufficient (§ 644) to bring so much of the southeast up to the temperature which the northeast trades have as they cross the parallel of 9° north. To complete this measurement of heat, we should know how long it takes the southeast trade-winds to get from the parallel of 9° south to the equator. According to the estimate, it takes about a day; but, knowing the exact time, we should have in the band of winds an actinometer which would disclose to us the average quantity of heat daily impressed by the sun upon the at-

mosphere at sea between the equator and 9° south. I say it takes about a day, and so infer from these data, viz.: The mean annual direction of the southeast trade-winds between 10° south and the line is south 40° east.[8] We suppose their average velocity to be (§ 343) about 25 miles an hour. At this rate it would take them 29h. 22m. 30s. to reach the equator. During this time they receive more heat than they radiate, and the excess is just sufficient to raise them from the normal temperature of the northeast trades as they enter the calm belt in 9° north. A series of observations on the temperature of the air in latitude 9° south at sea would, for the farther study of this subject, possess great value.[9]

651. *Equatorial calm belt never stationary.* If these views be correct, we should expect to find the equatorial calm belt changing its position with night and day, and yielding to all those influences, whether secular, annual, diurnal, or accidental, which are capable of producing changes in the thermal condition of the trade-winds. The great sun-swing of this calm belt from north to south is annual in its occurrence; it marks the seasons and divides the year (§ 296) into wet and dry for all those places that are within the arc of its majestic sweep. But there are other subordinate and minor influences which are continually taking place in the atmosphere, and which are also calculated to alter the place of this calm belt, and to produce changes in the thermal status of the air which the trade-winds move. These are, unusually severe winters or hot summers, remarkable spells of weather, such as long continuous rains or droughts over areas of considerable extent, either within or near the trade-wind belts. It is tremblingly alive to all such influences, and they keep it in continual agitation; accordingly we find that such is its state that within certain boundaries it is continually changing place and limits. This fact is abundantly proved by the speed of ships, for the log-books at the Observatory show that it is by no means a rare occurrence for one vessel, after she may have been dallying in the Doldrums for days in the vain effort to cross that calm belt, to see another coming up to her with fair winds, and crossing the belt after a delay in it of only a few hours instead of days.

652. *It varies with the strength of the trade-winds.* Hence we infer that the position of the equatorial calm belt is determined by the difference of strength between the northeast and southeast trade-winds, which difference, in turn, depends upon difference of barometric pressure (§ 642), and difference in temperature between them in corresponding latitudes north and south. In it the air which they bring ascends. Now if

---

[8] "The Winds at Sea."

[9] The mean temperature of sea water in the Atlantic is, for 9° north, 80°.26, 565 obs.; equator, 79°.63, 269 obs.; and 9° south, 78°.96, 223 obs. — [Maury, *Wind and Current Charts: Thermal Sheet*, Series D, nos, 2, 3, 4; *Wind and Current Chart of the South Atlantic*, Series D, nos. 2, 3.]

we liken this belt of calms to an immense atmospherical trough, extending, as it does, entirely around the earth, and if we liken the northeast and southeast trade-winds to two streams discharging themselves into it, we shall see that we have two currents perpetually running in at the bottom, and that, therefore, we must have as much air as these two currents bring in at the bottom to flow out at the top. What flows out at the top is carried back north and south by these upper currents, which are thus proved to exist and to flow counter to the trade-winds.

653. *Precipitation in it.* Captain Wilkes, of the Exploring Expedition, when he crossed this belt in 1838, found it to extend from 4° north to 12° north. He was ten days in crossing it, and during those ten days rain fell to the depth of 6.15 inches, or at the rate of eighteen feet and upward during the year. In its motions from south to north and back, it carries with it the rainy seasons of the torrid zone, always arriving at certain parallels at stated periods of the year; consequently, by attentively considering Plate V, one can tell what places within the range of this zone have, during the year, two rainy seasons, what one, and what are the rainy months for each locality.

654. *The appearance of the calm belts from a distant planet.* Were the northeast and the southeast trades, with the belt of equatorial calms, of different colors, and visible to an astronomer in one of the planets, he might, by the motion of these belts or girdles alone, tell the seasons with us. He would see them at one season going north, then appearing stationary, and then commencing their return to the south. But, though he would observe (§ 295) that they follow the sun in his annual course, he would remark that they do not change their latitude as much as the sun does his declination; he would therefore discover that their extremes of declination are not so far asunder as the tropics of Cancer and Capricorn, though in certain seasons the changes from day to day are very great. He would observe that these zones of winds and calms have their tropics or stationary nodes, about which they linger near three months at a time; and that they pass from one of their tropics to the other in a little less than another three months. Thus he would observe the whole system of belts to go north from the latter part of May till some time in August. Then they would stop and remain nearly stationary till winter, in December; when again they would commence to move rapidly over the ocean, and down toward the south, until the last of February or the first of March; then again they would become stationary, and remain about this, their southern tropic, till May again. Having completed his physical examination of the equatorial calms and winds, if the supposed observer should now turn his telescope toward the poles of our earth, he would observe a zone of calms bordering the northeast trade-winds on the north (§ 210), and another bordering the southeast trade-winds on the south (§ 213). These

calm zones also would be observed to vibrate up and down with the trade-wind zones, partaking (§ 296) of their motions, and following the declination of the sun. On the polar side of each of these two calm zones there would be a broad band extending up into the polar regions, the prevailing winds within which are the opposites of the trade-winds, viz., southwest in the northern and northwest in the southern hemisphere. The equatorial edge of these calm belts is near the tropics, and their average breadth is 10° or 12°. On one side of these belts (§ 210) the wind blows perpetually toward the equator; on the other, its prevailing direction is toward the poles. They are called (§ 210) the "horse latitudes" by seamen.

655. *Rainy seasons of the tropical calm belts.* Along the polar borders of these two calm belts (§ 296) we have another region of precipitation, though generally the rains here are not so constant as they are in the equatorial calms. The precipitation near the tropical calms is nevertheless sufficient to mark the seasons; for whenever these calm zones, as they go from north to south with the sun, leave a given parallel, the rainy season of that parallel, if it be in winter, is said to commence. Hence we may explain the rainy season in Chili at the south, and in California at the north.

656. *Their position.* We can now understand why the calm belts of Cancer and Capricorn occupy a medial position between the trades and the counter-trades; why, on one side of it, the prevailing direction of the wind should be poleward, on the other toward the equator; and we also discover the influences which determine their geographical position.

657. *A meteorological law.* An accumulation of atmosphere over one part of the earth's surface implies a depression over some other part, precisely as the piling up of water into a wave above the sea level involves a corresponding depression below; and in meteorology it may be regarded as a general law, that the tendency of all winds on the surface is to blow from the place where the barometer is higher to the place where the barometer is lower. This meteorological law is only a restatement of the dynamical truism about water seeking its own level.

658. *The barometer in the calm belts.* The mean height of the barometer in the calm belts of the tropics is greater (Plate I) than it is in any other latitude. The mean height of the barometer in the equatorial calm belt is less than it is on any other parallel between the tropical and equatorial calm belts. The difference for the calm belt of Cancer is 0.25 inch. This difference is permanent. It is sufficient to put both systems of trade-winds in motion, and to create an indraught of air flowing perpetually toward the equatorial calm belt from the distance of two thousand miles on each side of it.

659. *Winds with northing and winds with southing in them.* In like

manner, as we go from either tropical calm belt toward the nearest pole, the barometric pressure becomes less and less. The meteorological law just announced requires the prevailing wind on the polar side of these calm belts to be from them and in the direction of the poles; and observations (Plate I) show that such is the case. Dividing the winds in each hemisphere into winds with northing and winds with southing in them, actual observation shows that they balance each other in the southern hemisphere between the parallels of 35° and 40°, and in the northern between the parallels of 25° and 50°; that between these parallels the average annual prevalence of winds with northing and of winds with southing in them is the same, the difference being so small as to be apparently accidental; that, proceeding from the medial band toward the pole, polar-bound winds become more and more prevalent, and proceeding from it toward the equator, equatorial-bound winds become more and more prevalent. Now, in each case, the prevailing winds blow (§ 657) from the high to the low barometer (Plate I).

660. *The barometric ridges.* The fact of two barometric ridges encircling the earth, as the high barometer of the tropical calm belts do, and as they may be called (Plate I), suggests a place of low barometer on the polar side as naturally as the ascent of a hill on one side suggests to the traveler a descent on the other; and, had not actual observations revealed the fact, theory should have taught us (§ 654) the existence of a low barometer toward the polar regions as well as toward the equatorial.

661. *They make a depression in the atmosphere.* Let us contemplate for a moment this *accumulation* of air in the tropical belt about the earth in each hemisphere. Because it is an accumulation of atmospheric air about the calms — because the barometer stands higher under the calm belt of Capricorn, for instance, than it does on any other parallel between that calm belt and the pole on one side, or the equator on the other, it is not to be inferred that therefore there is a piling — a ridging up of the atmosphere there. On the contrary, were the upper surface of our atmosphere visible, and could we take a view of it from above, we should discover rather a valley than a ridge over this belt of greatest pressure; and over the belt of least pressure, as the equatorial calm belt, we should discover (§ 520), not a valley, but a ridge, and for these reasons: In the belts of low barometer, that is, in both the equatorial and polar calms, the air is expanded, made light, and caused to ascend chiefly by the latent heat that is liberated by the heavy precipitation which takes place there. This causes the air which ascends there to rise up and swell out far above the mean level of the great aerial ocean. This intumescence at the equatorial calm belt has been estimated to be several miles above the general level of the atmosphere. This calm belt air, therefore, as it boils up and flows off through the upper regions, north and south, to the trop-

ical calm belts, does not so flow by reason of any difference of barometric pressure, like that which causes the surface winds to blow, but it
so flows by reason of difference as to level.

662. *The upper surface of the atmosphere.* The tropical calm belts
(§ 278) are places where the mean amount of precipitation is small. The
air there is comparatively dry air. So far from being expanded by heat,
or swelled out by vapor, this air is contracted by cold, for the chief source
of its supply is through the upper regions, from the equatorial side,
where the cross section between any two given meridians is the larger;
and this upper current, while on its way from the equator, is continually
parting with the heat which it received at and near the surface, and
which caused it to rise under the equatorial cloud-ring. In this process
it is gradually contracted, thus causing the upper surface of the air to be
a sort of double inclined plane, descending from the equator and from
the poles to the place of the tropical calm belts.

663. *Winds in the southern stronger than winds in the northern hemisphere.* Observations show that the mean weight of the barometer in
high southern is much less (Plate I) than it is in corresponding high
northern latitudes; consequently, we should expect that the polar-bound
winds would be much more marked on the polar side of 40° S., than they
are on the polar side of 40° N. Accordingly, observations (Figure 2, §
352) show such to be the case; and they moreover show that the polar-
bound winds of the southern are much fresher than those of the northern
hemisphere.

664. *The waves and gales off the Cape of Good Hope.* To appreciate
the force and volume of these polar-bound winds in the southern hemisphere, it is necessary that one should "run them down" in that waste of
waters beyond the parallel of 40° S., where "the winds howl and the
seas roar." The billows there lift themselves up in long ridges with deep
hollows between them. They run high and fast, tossing their white caps
aloft in the air, looking like the green hills of a rolling prairie capped
with snow, and chasing each other in sport. Still, their march is stately
and their roll majestic. The scenery among them is grand, and the
Australian-bound trader, after doubling the Cape of Good Hope, finds
herself followed for weeks at a time by these magnificent rolling swells,
driven and lashed by the "brave west winds" most furiously. A sailor's
bride, performing this voyage with her gallant husband, thus alludes in
her "abstract log" to these rolling seas: "We had some magnificent gales
off the Cape, when the coloring of the waves, the transition from gray to
clear brilliant green, with the milky-white foam, struck me as most exquisite. And then in rough weather the moral picture is so fine, the
calmness and activity required is such an exhibition of the power of mind

over the elements, that I admired the sailors fully as much as the sea, and, *of course*, the sailor in command most of all; indeed, a sea voyage more than fulfills my expectations."

665. *Winds blow from a high to a low barometer.* It appears, therefore, that the low barometer about the poles and the low barometer of the equator cause an inrush of wind, and in each case the rushing wind comes from the high and blows toward the low barometer; that in one hemisphere the calm belt of Capricorn, and in the other the calm belt of Cancer, occupies the medial line between these places of low barometer.

666. *Polar rarefaction.* It appears, moreover, that the polar rarefaction is greater than the equatorial, for the mean height of the austral barometer is very much below that of the equatorial and, consequently, its influence in creating an indraught is felt at a greater distance — even at the distance of 50° of latitude from the south pole, while the influence of the equatorial depression is felt only at the distance of 30° in the southern, and of 25° in the northern hemisphere. The difference as to degree of rarefaction is even greater than this statement implies, for the influx into the equatorial calm belt is assisted also by temperature in this, that the trade-winds blow from cooler to warmer latitudes. The reverse is the case with the counter-trades; therefore, while difference of thermal dilatation assists the equatorial, it opposes the polar influx.

667. *The tropical calm belts caused by the polar and equatorial calms.* Thus we perceive that the tropical calm belts are simply an adjustment between the polar and equatorial calms; that the tropical calm belts assume their position and change their latitude in obedience to the energy with which the influence of the heated and expanding columns of air, as they ascend in the polar and equatorial calms, is impressed upon them.

668. *The meteorological power of latent heat.* This explanation of the calm places and of the movements of the low austral barometer shows, comparatively speaking, how much the latent heat of vapor, and how little the direct heat of the sun has to do in causing the air to rise up and flow off from these calm places, and, consequently, how little the direct action of the solar ray has to do either with the trades or the counter-trades. It regulates and controls them; it can scarcely be said to create them.

669. *The low barometer off Cape Horn.* The fact of a low barometer off Cape Horn was pointed out [10] as long ago as 1834. It was considered an anomaly peculiar to the regions of Cape Horn. It is now ascertained by the comparison of 6455 observations on the polar side of 40° south, and about 90,000 in all other latitudes, that the depression is not peculiar

[10] [Maury, "On the Navigation of Cape Horn,"] *Amer. Jour. Sci.*, XXVI(1834), 54–[63].

to the Cape Horn regions, but that it is general and alike in all parts of the austral seas, as [Table 10], compiled from the log-books of the Observatory by Lieutenants Warley and Young, shows.

670.   Table 10

Mean Height of the Barometer as observed between

the Parallels of	the Meridians of						Mean of all	
	20° W. and 140° E.		140° E. and 80° W.		Off Cape Horn			
	No. of Obs.	Barom. Inches	No. of Obs.	Barom. Inches	No. of Obs.	Barom. Inches	No. of Obs.	Barom. Inches
40° S. and 43° S.	1115	29.90	210	29.84	378	29.86	1703	29.88
43 " 45	738	.80	155	.73	237	.75	1130	.78
45 " 48	611	.58	226	.71	337	.68	1174	.63
48 " 50	175	.53	247	.56	250	.61	672	.62
50 " 53	108	.35	198	.45	359	.56	665	.48
53 " 55	6	.14	92	.35	377	.37	475	.36
	7	.27	64	.42	1055	.28	1126	.29

671. *Barometer at the poles.* These are the *observed* heights; for the want of data, no corrections have been applied to them; and for the want of numbers sufficient to give correct means, they lack that uniformity which larger numbers would doubtless give. They show, however, most satisfactorily, that a low barometer is not peculiar to Cape Horn regions alone; they show that it is common to all high southern latitudes; and other observations (§ 362) show that it is peculiar to these and not to northern latitudes. Projecting on a diagram, Figure 5, with parallels of latitude and the barometric scale as ordinates and abscissæ, a curve S, which will best represent the observations (§ 670), and continuing it to the south pole — also projecting another curve, N, which will best represent the observations (§ 362) on the polar side of 40° N., and continuing it to the north pole — we discover that if the barometric pressure in polar latitudes continues to decrease for the unknown as it does for the known regions, the mean height of the barometer would be at the north pole about 29.6, at the south about 28 inches. These lines, N and S, represent what may be called the *barometric descent* of the counter-trades.

672. *The "brave west winds" — their barometric descent.* The rarefaction of the air in the polar calms is, as we have seen (§ 667), sufficient to create an indraught all around to the distance of fifty degrees of latitude from the south pole; also (§ 662) the rarefaction in the belt of equatorial calms is sufficient to extend with its influence no farther than thirty degrees of latitude. This fact also favors the idea suggested by the diagram (§ 671), that the mean height of the barometer in the polar calms is very much less than it is in the equatorial. Moreover, the

counter-trades of the southern hemisphere are very much stronger (§ 626) than the counter-trades of the other. They are also stronger than the trade-winds of either; these facts likewise favor the idea of a greater exhaustion of air in the antarctic than in the arctic calm place; and it is manifest that actual observations also, as far as they go, indicate such to be the case. In other words, the "brave west winds" of the southern hemisphere have the greatest "barometric descent," and should therefore be, as they are, the strongest of the four winds.

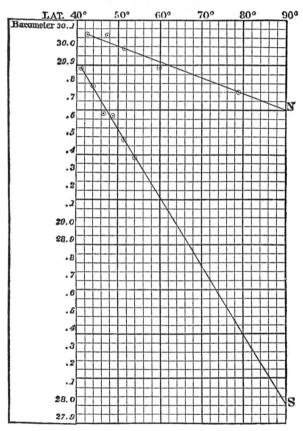

Figure 5

[Mean atmospheric pressure in high latitudes of the northern and southern hemispheres.]

673. *Study of the monsoons affords farther information touching the calm belts.* Farther information may be gained upon the subject of high and low barometers, of the "barometric declivity of winds," and of the

meteorological influence of diminished atmospheric pressure by studying the calm belts in connection with the monsoons.

674. *The southwest winds of the Atlantic.* Before, however, we proceed to these, let us take a hasty glance at the winds in certain other parts of the ocean. The winds which most prevail on the polar side of the calm belt of Cancer, and as far as 50° N. in the Atlantic, are the west winds. "Wind and weather in this part of the ocean," says Jansen, "are very unreliable and changeable; nevertheless, in the summer months, we find permanent north winds along the coast of Portugal. These north winds are worthy of attention, the more so from the fact that they occur simultaneously with the African monsoon, and because we then find northerly winds also in the Mediterranean, and in the Red Sea, and farther eastward to the north of the Indian monsoon. When, between the months of May and November, during which the African monsoon prevails, the Dutch ships, which have lingered in the calm belt of Cancer, run with the northeast trade and direct their course for the Cape Verd Islands, then it seems as if they were in another world. The sombre skies and changeable — alternately chilly and sultry — weather of our latitudes are replaced by a regular temperature and good settled weather. Each one rejoices in the glorious heavens, in which none save the little trade-clouds are to be seen — which clouds in the trade-wind region make the sunset so enchanting. The dark blue water, in which many and strange kinds of echinas sport in the sunlight, and, when seen at a distance, make the sea appear like one vast field adorned with flowers; the regular swellings of the waves with their silvery foam, through which the flying-fishes flutter; the beautifully-colored dolphins; the diving schools of tunnies — all these banish afar the monotony of the sea,[11] awake the love of life in the youthful seaman, and attune his heart to goodness. Every thing around him fixes his attention and increases his astonishment.

675. *Sailing through the trade-wind.* "If all the outbreathings of heartfelt emotion which the contemplation of nature forces from the sailor were recorded in the log-books, how much farther should we be advanced in the knowledge of the natural state of the sea! Once wandering over the ocean, he begins to be impressed by the grand natural tableau around him with feelings deep and abiding. The most splendid forecastle is lost in the viewless surface, and brings home to us the knowledge of our nothingness; the greatest ship is a plaything for the billows, and the slender keel seems to threaten our existence every moment. But when the eye of the mind is permitted to wander through space and into the depths of the ocean, and is able to form a conception

---

[11] When we, as our forefathers did, preserve in the journals all that we observe at sea, then we shall have abundant material with which to keep ourselves pleasantly occupied. — [*Bijdrage*, p. 285, n. 1.]

of Infinity and of Omnipotence, then it knows no danger; it is elevated — it comprehends itself. The distances of the heavenly bodies are correctly estimated; and, enlightened by astronomy, with the aid of the art of navigation, of which Maury's Wind and Current Charts form an important part, the shipmaster marks out his way over the ocean just as securely as any one can over an extended heath. He directs his course toward the Cape Verd Islands, and is carried there by the lively trade-wind. Yet beyond the islands, sooner or later, according to the month, the clear skies begin to be clouded, the trade-wind abates and becomes unsteady, the clouds heap up, the thunder is heard, heavy rains fall; finally, the stillness is deathlike, and we have entered the belt of calms. This belt moves toward the north from May to September. It is a remarkable phenomenon that the annual movements of the trades and calm belts from south to north, and back again, do not directly follow the sun in its declination, but appear to wait until the temperature of the sea-water puts it in motion.[12] If a ship which has come into the belt of calms between May and September could lie still in the place where it came into this belt — cast anchor, for example — then it would perceive a turning of the monsoon or of the trade-wind. It would see the belt of calms draw away to the north, and afterward get the southwest monsoon, or, standing more westerly, perhaps the southeast trade. On the contrary, later than September, this ship lying at anchor will see the northeast gradually awake. The belt of calms then moves toward the south, and removes from the ship, which remains there anchored on the north side."[13]

676. *The influence of the land upon the winds at sea.* The investigations that have taken place at the Observatory show that the influence of the land upon the normal directions of the wind at sea is an immense influence. It is frequently traced for a thousand miles or more out upon the ocean. For instance, the action of the sun's rays upon the great deserts and arid plains of Africa, in the summer and autumnal months, is such as to be felt nearly across the Atlantic Ocean between the equator and the parallel of 13° north. Between this parallel and the equator, the northeast trade-winds, during these seasons, are arrested in their course by the heated plains of Africa, as observation shows they are in India, and instead of "blowing home" to the equator, they stop and ascend over the desert sands of the continent. The southeast trade-winds, arriving at the equator during this period, and finding no northeast trades there to

[12] [*Bijdrage*, p. 286: *tot dat de isothermen van het zeewater zich in beweging stellen*; *i. e.*, until the isotherms of the sea water begin to move. Maury omits a passage following this sentence in which Jansen gives more details of the association between the motions of surface isotherms and of the wind belts. The remainder of the paragraph is much abridged from the passage in *Bijdrage*.]

[13] *Bijdrage* [pp. 285–287].

contest their crossing the line, continue their course, and blow *home* as a southwest monsoon to the deserts, where they ascend. These southwardly monsoons bring the rains which divide the seasons in these parts of the African coast. The region of the ocean embraced by these monsoons is cuneiform in its shape, having its base resting upon Africa, and its apex stretching over till within 10° or 15° of the mouth of the Amazon. Indeed, when we come to study the effects of South America and Africa (as developed by the Wind and Current Charts) upon the winds at sea, we should be led to the conclusion — had the foot of civilized man never trod the interior of these two continents — that the climate of one is humid; that its valleys are, for the most part, covered with vegetation, which protects its surface from the sun's rays; while the plains of the other are arid and naked, and, for the most part, act like furnaces in drawing the winds from the sea to supply air for the ascending columns which rise from its overheated plains. Pushing these facts and arguments still farther, these beautiful and interesting researches seem already sufficient almost to justify the assertion that, were it not for the great desert of Sahara and other arid plains of Africa, the western shores of that continent, within the trade-wind region, would be almost, if not altogether, as rainless and sterile as the desert itself.

677. *A "Gulf Stream" in the air.* Lieutenant Jansen has called my attention to a vein of wind which forms a current in the air as remarkable as that of the Gulf Stream is in the sea. This atmospherical Gulf Stream is in the southeast trade-winds of the Atlantic. It extends from near the Cape of Good Hope, in a direct line to the equator, on the meridian of Cape St. Roque (Plate V). The homeward route from the Cape of Good Hope lies in the middle of this vein; in it the winds are more steady than in any other part of the Atlantic. On the edges of this remarkable aerial current the wind is variable and often fitful; the homeward-bound Indiaman resorts to and uses this stream in the atmosphere as the European-bound American does the Gulf Stream. It is shaded on the plate.

678. *Counterpoises.* These investigations, with their beautiful developments, eagerly captivate the mind; giving wings to the imagination, they teach us to regard the sandy deserts, and arid plains, and the inland basins of the earth, as compensations in the great system of atmospherical circulation. Like counterpoises to the telescope, which the ignorant regard as incumbrances to the instrument, these wastes serve as makeweights, to give certainty and smoothness of motion — facility and accuracy to the workings of the machine.

679. *Normal state of the atmosphere.* When we travel out upon the ocean, and get beyond the influence of the land upon the winds, we find ourselves in a field particularly favorable for studying the general laws

of atmospherical circulation. Here, beyond the reach of the great equatorial and polar currents of the sea, there are no unduly heated surfaces, no mountain ranges, or other obstructions to the circulation of the atmosphere — nothing to disturb it in its natural courses. The sea, therefore, is the field for observing the operations of the general laws which govern the movements of the great aerial ocean. Observations on the land will enable us to discover the exceptions, but from the sea we shall get the rule. Each valley, every mountain range and local district, may be said to have its own peculiar system of calms, winds, rains and droughts. But not so the surface of the broad ocean; over it the agents which are at work are of a uniform character.

680. *Rain-winds.* Rain-winds are the winds which convey the vapor from the sea, where it is taken up, to other parts of the earth, where it is let down either as snow, hail, or rain. As a general rule, the trade-winds (§ 293) may be regarded as the evaporating winds; and when, in the course of their circuit, they are converted into monsoons, or the variables of either hemisphere, they then generally become also the rain-winds — especially the monsoons — for certain localities. Thus the southwest monsoons of the Indian Ocean are the rain-winds for the west coast of Hindostan (§ 298). In like manner, the African monsoons of the Atlantic are the winds which feed the springs of the Niger and the Senegal with rains. Upon every water-shed which is drained into the sea, the precipitation, for the whole extent of the shed so drained, may be considered as greater than the evaporation, by the amount of water which runs off through the river into the sea. In this view, all rivers may be regarded as immense rain-guages, and the volume of water annually discharged by any one as an expression of the quantity which is annually evaporated from the sea, carried back by the winds, and precipitated throughout the whole extent of the valley that is drained by it. Now, if we knew the rain-winds from the dry for each locality and season generally throughout such a basin, we should be enabled to determine, with some degree of probability at least, as to the part of the ocean from which such rains were evaporated. And thus, notwithstanding all the eddies caused by mountain chains and other uneven surfaces, we might detect the general course of the atmospherical circulation over the land as well as the sea, and make the general courses of circulation in each valley as obvious to the mind of the philosopher as is the current of the Mississippi, or of any other great river, to his senses.

CHAPTER XVI

§ 681–711. MONSOONS

681. The cause of. Monsoons are, for the most part, trade-winds deflected. When, at stated seasons of the year, a trade-wind is turned out of its regular course, as from one quadrant to another, it is regarded as a monsoon. The African monsoons of the Atlantic (Plate V), the monsoons of the Gulf of Mexico, and the Central American monsoons of the Pacific are, for the most part, formed of the trade-winds which are turned back or deflected to restore the equilibrium which the overheated plains of Africa, Utah, Texas, and New Mexico have disturbed. Thus with regard to the northeast and the southwest monsoons of the Indian Ocean, for example: a force is exerted upon the northeast trade-winds of that sea by the disturbance which the heat of summer creates in the atmosphere over the interior plains of Asia, which is more than sufficient to neutralize the forces which cause those winds to blow as trade-winds; it arrests them and turns them back; but, were it not for the peculiar conditions of the land about that ocean, what are now called the northeast monsoons would blow the year round; there would be no southwest monsoons there; and the northeast winds, being perpetual, would become all the year what in reality for several months they are, viz., northeast trade-winds.

682. The region of. Upon India and its seas the monsoon phenomena are developed on the grandest scale. They blow over all that expanse of northern water that lies between Africa and the Philippine Islands. Throughout this vast expanse, the winds that are known in other parts of the world as the N.E. trades, are here called N.E. monsoons, because, instead of blowing from that quarter for twelve months, as in other seas, they blow only for six. During the remaining six months they are turned back, as it were; for, instead of blowing toward the equator, they blow away from it, and instead of N.E. trades we have S.W. monsoons.

683. A low barometer in Northern India. If the N.E. trade-winds blow toward the equator by reason (§ 657) of the lower barometer of the calm belt there, we should — seeing them turned back and blowing in the opposite direction as the S.W. monsoon — expect to find toward the north, and at the place where they cease to blow, a lower barometer than that of the equatorial calm belt. The circumstances which indicate the existence of a low summer barometer — the period of the S.W. monsoon — in the regions about northern India are developed by the law which (§ 657) requires the wind to blow toward that place where there is least atmospheric pressure.

684. *The S.W. monsoons "backing down."* The S.W. monsoons commence at the north, and "back down," or work their way toward the south. Thus they set in earlier at Calcutta than they do at Ceylon, and earlier at Ceylon than they do at the equator. The average rate of travel, or "backing down to the south," as seamen express it, is from fifteen to twenty miles a day. It takes the S.W. monsoons six or eight weeks to "back down" from the tropic of Cancer to the equator. During this period there is a sort of barometric ridge in the air over this region, which we may call the monsoon wave. In this time it passes from the northern to the southern edge of the monsoon belt, and as it rolls along in its invisible but stately march, the air beneath its pressure flows out from under it both ways — on the polar side as the S.W. monsoon, on the equatorial as the N.E.

685. *How they begin.* As the vernal equinox approaches, the heat of the sun begins to play upon the steppes and deserts of Asia with power enough to rarefy the air, and cause an uprising sufficient to produce an indraught thitherward from the surrounding regions. The air that is now about to set off to the south as the N.E. monsoon is thus arrested, turned back, and drawn into this place of low barometer as the S.W. monsoon. These plains become daily more and more heated, the sun more and more powerful, and the ascending columns more and more active; the area of inrushing air, like a circle on the water, is widened, and thus the S.W. monsoons, "backing down" toward the equator, drive the N.E. monsoons from the land, replace them, and gradually extend themselves out to sea.

686. *The sun assisted by the latent heat of vapor.* Coming now from the water, they bring vapor, which, being condensed upon the hill-sides, liberates its latent caloric, and so, adding fuel to the flame, assists the sun to rarefy the air, to cause it to rise up and flow off more rapidly, and so to depress the barometer still more. It is not till the S.W. monsoons have been extended far out to sea that they commence to blow strongly, or that the rainy season begins in India. By this time the mean daily barometric pressure in this place of ascending air, which is also a calm place, has become less than it is in the equatorial calm belt; and the air which the S.E. trade-winds then bring to the equator, instead of rising up there in the calm belt, pass over without stopping, and flows onward to the calms of Central Asia as the S.W. monsoon. It is drawn over to supply the place of rarefaction over the interior of India.

687. *The rain-fall in India.* The S.W. monsoon commences to change at Calcutta, in 22° 34' N., in February, and extends thence out to sea at the rate of fifteen or twenty miles a day; yet these winds do not gather

vapor enough for the rainy season of Cherraponjie, in lat. 25° 16', to commence with until the middle or last of April, though this station, of all others in the Bengal Presidency, seems to be most favorably situated for wringing the clouds. Selecting from Colonel Sykes's report of the rain-fall of India those places nearest the same meridian, and about 2° of latitude apart, [Table 11 is compiled] with the view of showing, as far as such data can show, the time at which the rainy season commences in the interior.

Table 11

[*Time at which the rainy season commences in the interior of India*]

	Lat.	Long.	March	April	May	June	July	Aug.	Sept.	Oct.
	° '	° '	in.	in.	in.	in.	in.	in.	in.	in.
Poorie	19 48	85 49	—	1	1	5	14	7	4	—
Baitool	21 51	77 58	—	—	1	4	15	9	4	—
Sangor	23 50	78 47	—	—	2	2	15	12	13	1
Humeerpore	26 7	79 47	—	—	—	7	13	11	5	1
Bareilly	28 12	79 34	—	—	—	3	17	8	2	3
Ferozepore	30 57	74 41	—	—	—	1	19	—	—	—
Simla	31 6	77 11	1	—	1	4	18	12	—	—
Cherraponjie	25 16	91 43	1	28	115	147	99	104	72	40
			2	29	120	173	210	163	100	45

It is June before the S.W. monsoons have backed down as far as the equator and have regularly set in there.

688. *Its influences upon the monsoons.* These positions are selected without regard to elevation above the sea level. Of course, when the S.W. monsoon comes only from a short distance out to sea, as in April it does, it is but lightly loaded with moisture. The low country can not condense it, and it then remains for the mountain stations in the interior, such as Cherraponjie, to get the first rains of the season; and a most interesting physical problem may be here put on the road to solution by the question, Does not the rainy season of the S.W. monsoon commence at the high stations in the interior, as on the sides of the Himalaya, earlier than in the flat country along the sea-coast?

689. *The march of the monsoons.* With the view of investigating certain monsoon phenomena, recourse was had to our great magazine of undigested facts, the abstract logs; and after discussing not less than 11,697 observations on the winds at sea between the meridians of 80° and 85° E., and from Calcutta to the equator, results were obtained for Table 12, in which is stated in *days* the average monthly duration of the N.E. and S.W. winds at sea between the parallels [listed].

Table 12

[*Average monthly duration of N.E. and S.W. winds, Calcutta to the equator*]

	22° and 20° N.		20° and 15° N.		15° and 10° N.		10° and 5° N.		5° and 0° N.	
	Days	Days	Days	Days	Days	Days	Days	Days	Days	Days
	N.E.	S.W.	N.E.	S.W.	N.E.	S.W.	N.E.	S.W.	N.E.	S.W.
January	17	6	21	2	23	1	20	1	19	3[b]
February	11	11	13	6	19	3	22	1	16	2
March	4	18[a]	7	15	18	5	13	0	15	2
April	2	24	2	22[a]	6	12	6	11	4	14[a]
May	1	26	1	24	3	21[a]	1	23[a]	0	19
June	0	28	1	27	0	29	1	25	0	24
July	2	24	1	27	0	30	0	28	0	24
August	0	28	1	24	0	24	1	22	0	18
September	6	14[b]	1	18	0	23	0	26	1	18
October	9	6	12	6[b]	8	10	6	16	4	14
November	11	6	25	2	21	2[b]	10	6[b]	5	14
December	27	0	26	1	24	1	15	3	12	11

[a] Setting in of the S.W. monsoon.   [b] Ending of the S.W. monsoon.

It appears from this table that between Calcutta and the line the S.W. monsoons are the prevailing winds for seven months, the N.E. for five.

690. *Their conflict — it begins at the north.* Resorting to the graphic method of engraved squares for a farther discussion of these figures, it appears by Figure 6 that in February the northeast and southwest winds

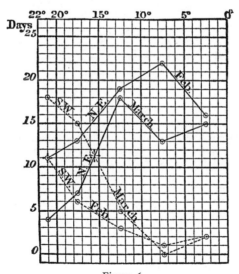

Figure 6

[Duration in days, in February and March, of the northeast and the southeast monsoons between Calcutta and the equator, 80° to 85° E. long.]

are in equal conflict between the parallels of 20° and 22°; that in March
the former have backed down as far as the parallel of 16°–15° — the
medial line between them *from* which each monsoon is blowing — and
where, again, the conflict of "back to back" is equally divided as to time
of mastery (12 days) on either side. By the month of June they (the
southwest) have fairly gained the ascendency, and so remain masters of
the field until October, when the bi-annual conflict is again commenced
at the north. The vanquished northeast trades now lead off in the attack,
and, as Figure 7 shows, the two combatants have force enough about the

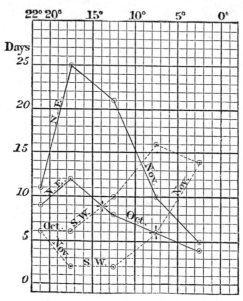

Figure 7

[Duration in days, in October and November, of the
northeast and the southeast monsoons between Calcutta and the equator, 80° to 85° E. long.]

parallel of 15° north to blow during this month 9 days each. The conflict,
instead of being "back to back," is now face to face; instead of blowing
away from the medial line, they blow toward it; instead of being a place
of high, the medial line is now (§ 657) a place of low barometer. By
November the northeast monsoon has pushed the place of equal contest
as far down as the parallel of 5° north.

691. *The barometric descent of the monsoons.* Each monsoon, like
the trade-winds, blows from a higher to a lower barometer. Taking up
the clew from this fact, and resorting again to the graphic method for

illustration, we may ascertain, with considerable accuracy, not only the relative strength of the northeast and southwest monsoons of the sea, but also the mean height of the barometer in the interior of India during the southwest monsoon, supposing that monsoon to go no farther than the mountain range, which may be taken at a mean to be about the parallel of 30° north. Now, taking the mean height of the barometer at the equatorial calm belt to be (§ 362) 29.92 inches; the mean height in the calm belt of Cancer to be 30.21 inches, the line N.E. of Figure 8 will

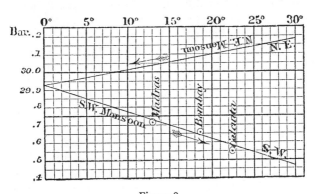

Figure 8

[Profile of atmospheric pressure along a N–S line across India during the northeast and the southwest monsoon.]

represent the average barometric declivity of the northeast monsoons generally. The mean height of the barometer during the three months of June, July, and August, when the southwest monsoons are at their height, is,

For Calcutta	29.55 inches
" Bombay	29.65 "
" Madras	29.73 "

The line S.W. represents the mean barometric declivity of the southwest monsoons at their height, and indicates that at their northern edge, supposed to be the parallel of 30° north, the barometer stands at about 29.45 inches. This barometric declivity indicates that the southwest are stronger than the northeast monsoons, and observation shows that they are.[1]

692. *The summer rains of Cherraponjie.* These are the winds – the southwest monsoons – which, coming from the sea, carry into the interior rains for the great water-shed of India. They bear with them an immense volume of vapor, as is shown by the rivers, and confirmed by the rain-fall

[1] Dr. Buist.

of Cherraponjie, and at 126 other stations. Cherraponjie is 4500 feet above the sea level. It reaches quite up to the cloud region, and receives a precipitation of 537⅓ inches during the southwest monsoon, from May to August inclusive. Col. Sykes reported to the British Association, at its meeting in 1852, the rain-fall at these 127 places, which are between the parallels of 20° and 34° in India.[2] According to this report, the southwest monsoons pour down during the three summer months upon this watershed 29¾ inches of rain. The latent heat that is liberated during the condensation of the vapor for all this rain expands the air, causing it to boil over, flow off, and leave a low barometer — a diminished atmospheric pressure throughout all the region south of the Himalaya.

693. *Dové and the monsoons.* As long ago as 1831, Dové maintained that the southwest monsoon was the southeast trade-wind rushing forward to fill the vacant places over the northern deserts. Dové admits the proofs of this to be indirect, and acknowledges the difficulty of finding out and demonstrating the problem.[3]

694. *The southeast trades passing into southwest monsoons.* But any navigator who, during the summer months, has occasion to traverse the Indian Ocean from north to south, may find that it is so. The outward-bound Indiaman, who, when on his way to Calcutta, crosses the equator in August, for example, will find the southeast trades, as he approaches the line, to haul more and more to the south. As he advances still farther north they get to the west of south. Finally, he discovers that he has got the regular southwest monsoons, and that he has passed from the southeast trades into them without any intervening calm. This in summer is the rule; it has its exceptions, but they are rare. Examining the logs of a number of vessels taken at random for the passage in August, we find, by 421 observations therein recorded, they had the wind thus:

Southeast from 10° to 5° south	0	calms
South " 5° south to equator	3	"
Southwest " equator to 5° north	3	"
Southwest " 5° to 10° north	0	"

695. *Lieutenant Jansen.* In like manner, and with like force, Jansen maintains that the northwest monsoon of Australia is the northeast trade-wind turned aside.

696. *Monsoons in the Pacific.* The influence exerted upon the winds by the deserts of Africa and the overheated plains of Asia is felt at sea for a thousand miles or more. Thus, though the desert of Cobi and the

---

[2] [W. H. Sykes, "Mean Temperature of the Day and Monthly Fall of Rain at 127 Stations under the Bengal Presidency, from Official Registers kept by Medical Officers for the Year 1851," *Rept. Brit. Assoc. Adv. Sci.,* 1852 (1853), pp. 252–261.]

[3] [The source of these statements is not Dove's article of 1831, but H. W. Dove, "On the Distribution of Rain in the Temperate Zone," *Amer. Jour. Sci.,* ser. 2, XX(1855), 397–402, esp. p. 400; and XXI(1856), 112–117.]

sun-burned plains of Asia are, for the most part, north of latitude 30°, their influence in causing monsoons (§ 692) is felt south of the equator (Plate V). So, too, with the great desert of Sahara and the African monsoons of the Atlantic; also with the Salt Lake country and the Mexican monsoons on one side, and those of Central America in the Pacific on the other. The influence (§ 298) of the deserts of Arabia upon the winds is felt in Austria and other parts of Europe, as the observations of Kriel, Lamont, and others show. So, also, do the islands, such as the Society and Sandwich, that stand far away from any extent of land, have a very singular but marked effect upon the wind. They interfere with the trades very often, and turn them back; for westerly and equatorial winds are common at both these groups in their winter time. Some hydrographers have even taken those westerly winds of the Society Islands to be an extension of the monsoons of the Indian Ocean.

697. *Influence of coral reefs upon winds.* It is a curious thing, is this influence of islands in the trade-wind region upon the winds in the Pacific. Every navigator who has cruised in those parts of that ocean has often turned with wonder and delight to admire the gorgeous piles of cumuli, heaped up and arranged in the most delicate and exquisitely beautiful masses that it is possible for fleecy matter to assume. Not only are these cloud-piles found capping the hills among the islands, but they are often seen to overhang the lowest islet of the tropics, and even to stand above coral patches and hidden reefs, "a cloud by day," to serve as a beacon to the lonely mariner out there at sea, and to warn him of shoals and dangers which no lead or seaman's eye has ever seen or sounded out. These clouds, under favorable circumstances, may be seen gathering above the low coral island, and performing their office in preparing it for vegetation and fruitfulness in a very striking manner. As they are condensed into showers, one fancies that they are a sponge of the most delicately elaborated material, and that he can see, as they "drop down their fatness," the invisible but bountiful hand aloft that is pressing it out.[4]

698. *Monsoons in miniature.* Land and sea breezes are monsoons in miniature, for they depend in a measure upon the same cause. In the monsoons, the latent heat of vapor which is set free over the land is a powerful agent. In the land and sea breezes, the heat of the sun by day and the radiation of caloric by night are alone concerned. In the monsoons the heat of summer and cold of winter are also concerned. But could the experiment be made with two barometers properly placed — one at sea and the other on land, but both within the reach of land and sea breezes — they would show regular alternations. In the sea breeze the land barometer would be low and the sea high, and *vice versa* in the

[4] *Sailing Directions*, 7th ed., p. 820.

land breeze; and when the barometer was highest and when it was lowest it would be calm.

699. *The changing of the monsoons.* It is these calm bands or "medial belts," as the crest and trough of the barometric wave may be called, which, with their canopy of clouds, follow the departing and herald the coming monsoon. They move to and fro, up and down the earth, like the sun in declination. As they have a breadth of 200 or 300 miles, they occupy several days in passing any given parallel, and while they overshadow it, then the monsoons are dethroned. During the interregnum, which lasts a week or two, the fiends of the storm hold their terrific sway in these bands. The changing of the monsoons is marked by storm and tempest. Becalmed in them, moanings are said by seamen to be heard in the air — a sign of the coming storm — a warning of impending danger to ship and crew. Then the props and stays are taken away from the air, and the wind seems ready to rush violently hither and thither, and whenever there is from any cause a momentary disturbance of the equilibrium. In such an atmosphere, the latent heat that is liberated by every heavy rain-shower has power to brew a storm. Throughout the monsoon region, the people know beforehand, almost to a day, the coming of this interregnum, which they call the changing of the monsoons, for the annual changing at the same place is very regular.

700. *How the calm belt of Cancer is pushed to the north.* Theory, therefore, points to a place in Northern India, which is near the northern limits of the southwest monsoon, where the mean height of the barometer during the rainy season (§ 691) is about 29.5 inches, the mean height at the equator being 29.92 inches. Into this monsoon place of low barometer over the land the wind rushes from the northeast as well as the southwest. The place of high pressure toward the north from which it rushes is under the calm belt of Cancer. Hence this belt is also pushed north, and made to occupy, in summer at least, the position over land somewhat like that assigned to it on Plate V. In the southwest monsoon the Malabar coast has its rainy season, so that the air over the peninsula is permanently kept more or less in a rarefied state, as actual observations abundantly show.

701. *The curved form of the equatorial calm belt in the Indian Ocean.* The equatorial calm belt in the Indian Ocean is a decided curve. The peculiar form may be ascribed to the meteorological influence of the Indian peninsula upon the calm belt, and in this way: The northeast monsoon brings the rainy season to the Coromandel coast and to the east coast of Ceylon. This rainy season embraces the land rather than the sea. The latent heat that is liberated during these rains, together with the effect of the solar ray upon this tongue of land, has the effect of expanding the air over it, and so deadening the northeast monsoon. In the mean

time, the meteorological influences from Africa on one side, and Australia on the other, tend to draw the wind in toward the land, and so retard the edges of the southeast trades, thus giving the calm belt the curved form shown in the plate.

702. *The winter monsoons.* In the winter time, and during the northeast monsoon, there is in the calm belt which intervenes between that monsoon and the southeast trades a belt of winter or westerly monsoons. It, too, is curved, as shown (Plate V) by the two lines drawn to represent its mean limits about the 1st of March. This is a most remarkable phenomenon, for which no satisfactory explanation has been suggested. It extends nearly, if not entirely, across the Pacific Ocean also, and the winds all the way prevail from the westward. The extreme breadth of this winter monsoon belt is about 9° or 10° of latitude. In the Indian Ocean, its middle is between the equator and 5° S.; in the Pacific, between the equator and 5° N.; in the Atlantic, between 5° and 10° N. In the Atlantic it is a summer monsoon easily to be accounted for. This belt of sub-monsoons, considering its great length and small breadth, is one of the most remarkable phenomena in marine meteorology.

703. *The monsoons of Australasia.* The northwest monsoons of Australia come from this belt; there it is widened, for these winds extend far down the west coast of that continent. The Malayan and Australasian archipelago have a complication of monsoons and sub-monsoons. The land and sea breezes impart to them peculiar features in many places, especially about the changing of the monsoons, as described by Jansen in his appendix to the Dutch edition of this work: "We have seen," says he, "that the calms which precede the sea-breeze generally continue longer, and are accompanied with an upward motion of the air; that, on the contrary, those which precede the land-breeze are, in the Java Sea, generally of shorter duration, accompanied by a heavy atmosphere, and that there is also an evident difference between the conversion of the land-breeze into the sea-breeze, and of the latter into the former. Even as the calms vary, so there appears to be a marked difference between the changing of the monsoons in the spring and in the autumn in the Java Sea. As soon as the sun has crossed the equator, and its vertical rays begin to play more and more perpendicularly upon the northern hemisphere, the inland plains of Asia, North Africa, and of North America are so heated as to give birth to the southwest monsoons in the China Sea, in the North Indian Ocean, in the North Atlantic, and upon the west coast of Central America: then the northwest monsoon disappears from the East Indian Archipelago, and gives place to the southeast trade-wind, which is known as the east monsoon, just as the northwest wind, which prevails during the southern summer, is called the west monsoon. This is the only northwest monsoon which is found in the southern hemisphere. While in the

northern hemisphere the northeast trade-wind blows in the China Sea and in the Indian Ocean, in the East Indian Archipelago the west monsoon prevails; and when here the southeast trade blows as the east monsoon, we find the southwest monsoon in the adjacent seas of the northern hemisphere. Generally the westerly monsoons blow during the summer months of the hemisphere wherein they are found.

704. *Thunder and lightning.* "In the Java Sea, during the month of February, the west monsoon blows strong almost continually; in March it blows intermittingly, and with hard squalls; but in April the squalls become less frequent and less severe. Now the changing commences; all at once gusts begin to spring up from the east: they are often followed by calms. The clouds which crowd themselves upon the clear sky give warning of the combat in the upper air which the currents there are about to wage with each other. The electricity, driven thereby out of its natural channels, in which, unobserved, it has been performing silently, but with the full consciousness of its power, the mysterious task appointed to it, now displays itself with dazzling majesty; its sheen and its voice fill with astonishment and deep reverence the mind of the sailor — so susceptible, in the presence of storm and darkness, to impressions that inspire feelings both of dread and anxiety, which by pretended occupations he strives in vain to conceal.[5] Day and night we now have thunder-storms. The clouds are in continual movement, and the darkened air, laden with vapor, flies in all directions through the skies. The combat which the clouds seem to court and to dread appears to make them more thirsty than ever. They resort to extraordinary means to refresh themselves; in funnel form, when time and opportunity fail to allow them to quench their thirst from the surrounding atmosphere in the usual manner, they descend near the surface of the sea, and appear to lap the water directly up with their black mouths. Water-spouts thus created are often seen in the changing season, especially among small groups of islands, which appear to facilitate their formation.[6] The water-spouts are not always accompanied by strong winds; frequently more than one is seen at a time, whereupon the clouds whence they proceed disperse in various directions, and the ends of the water-spouts bending over finally causes them to break in the middle, although the water which is now seen foaming around their base has suffered little or no movement laterally.

705. *Water-spouts.* "Yet often the wind prevents the formation of water-spouts. In their stead the wind-spout shoots up like an arrow, and the sea seems to try in vain to keep it back. The sea, lashed into fury,

---

[5] No phenomena in nature make a deeper impression upon the sailor than a dark [*Bijdrage*, p. 275, n. 1, *zwaar; i. e.,* heavy] thunderstorm in a calm sea.

[6] I never saw more water-spouts than in the archipelago of Riouw and Singen during the changing [*i. e.,* of the monsoon]. Almost daily we saw one or more. — [*Ibid.*, p. 275, n. 2.]

marks with foam the path along which the conflict rages, and roars with the noise of its water-spouts; and woe to the rash mariner who ventures therein! [7] The height of the spouts is usually somewhat less than 200 yards, and their diameter not more than 20 feet, yet they are often taller and thicker; when the opportunity of correctly measuring them has been favorable, however, as it generally was when they passed between the islands, so that the distance of their bases could be accurately determined, I have never found them higher than 700 yards, nor thicker than 50 yards. In October, in the Archipelago of Rio, they travel from southwest to northeast. They seldom last longer than five minutes; generally they are dissipated in less time. As they are going away, the bulbous tube, which is as palpable as that of a thermometer, becomes broader at the base, and little clouds, like steam from the pipe of a locomotive, are continually thrown off from the circumference of the spout, and gradually the water is released, and the cloud whence the spout came again closes its mouth.[8]

[7] The air-spouts near the equator always appear to me to be more dangerous than the water-spouts. I have once had one of the latter to pass a ship's length ahead of me, but I perceived little else than a waterfall in which I thought to come, yet no wind. Yet the water-spouts there also are not to be trusted. I have seen such spouts go up out of the water upon the shore, where they overthrew strong isolated frame houses. I have, however, never been in a situation to observe in what direction they revolved. — [*Ibid.*, n. 3.]

[8] Miniature water-spouts may be produced artificially by means of electricity, and those in nature are supposed to be caused by the display of electrical phenomena. "From the conductor of an electrical machine," says Dr. Bonzano, of New Orleans, "suspend by a wire or chain a small metallic ball (one of wood covered with tin-foil), and under the ball place a rather wide metallic basin containing some oil of turpentine, at the distance of about three quarters of an inch. If the handle of the machine be now turned slowly, the liquid in the basin will begin to move in different directions, and form whirlpools. As the electricity on the conductor accumulates, the troubled liquid will elevate itself in the centre, and at last become attached to the ball. Draw off the electricity from the conductor to let the liquid resume its position: a portion of the turpentine remains attached to the ball. Turn the handle again very slowly, and observe now the few drops adhering to the ball assume a conical shape, with the apex downward, while the liquid under it assumes also a conical shape, the apex upward, until both meet. As the liquid does not accumulate on the ball, there must necessarily be as great a current downward as upward, giving the column of liquid a rapid circular motion, which continues until the electricity from the conductor is nearly all discharged, silently, or until it is discharged by a spark descending into the liquid. The same phenomena take place with oil or water. Using the latter liquid, the ball must be brought much nearer, or a much greater quantity of electricity is necessary to raise it.

"If, in this experiment, we let the ball swing to and fro, the little water-spout will travel over its miniature sea, carrying its whirlpools along with it. When it breaks up, a portion of the liquid, and with it any thing it may contain, remains attached to the ball. The fish, seeds, leaves, etc., etc., that have fallen to the earth in rain-squalls, may have owed their elevation to the clouds to the same cause that attaches a few drops of the liquid, with its particles of impurities, to the ball."

By reference to Plate IX, we see that the phenomenon of thunder and lightning is of much more frequent occurrence in the North than in the South Atlantic; and I

706. *The east monsoon in the Java Sea.* "During the changing of the monsoons, it is mostly calm or cool, with gentle breezes, varied with rain-storms and light gales from all points of the compass. They are harassing to the crew, who, with burning faces under the clouded skies,[9] impatiently trim the sails to the changing winds. However, the atmosphere generally becomes clear, and, contrary to expectation, the northeast wind comes from a clear sky; about the coming of the monsoon it is northerly. Now the clouds are again packed together; the wind dies away, but it will soon be waked up to come again from another point. Finally, the regular land and sea breezes gradually replace rain, and tempests, calms, and gentle gales. The rain holds up during the day, and in the Java Sea we have the east monsoon. It is then May. Farther to the south than the Java Sea the east monsoon commences in April.[10] This monsoon prevails till September or October, when it turns to become the west monsoon. It has seemed to me that the east monsoon does not blow the same in every month, that its direction becomes more southerly, and its power greater after it has prevailed for some time.[11]

707. *Currents.* "It is sufficiently important to fix the attention, seeing that these circumstances have great influence upon the winds in the many straits of the Archipelago, in which strong currents run most of the time. Especially in the straits to the east of Java these currents are very strong. I have been unable to stem the current with eight-mile speed. However, they do not always flow equally strong, nor always in the same direction. They are probably the strongest when the tidal current and the equatorial current meet together. It is said that the currents in the straits during the east monsoon run eighteen hours to the north and six hours to the south, and the reverse during the west monsoon. The passing of the meridian by the moon appears to be the fixed point of time for the turning of the currents. It is probable that the heated water of the Archipelago is discharged to the north during the east monsoon, and to the south during the west monsoon.

---

infer that we have more electrical phenomena in the northern than in the southern hemisphere. Do water-spouts occur on one side of the equator more frequently than they do on the other? I have cruised a great deal on the southern hemisphere, and never saw a water-spout there. According to the log-books at the Observatory, they occur mostly on the north side of the equator. — M.

[9] At sea the face and hands burn (change the skin) [*Bijdrage*, p. 276, n. 1, *vervellen; i. e.,* peel] much quicker under a clouded than under a clear sky.

[10] In the northeast part of the Archipelago the east monsoon is the rainy monsoon. The phenomena in the northeast part are thus wholly different from those in the Java Sea. — [*Ibid.*, n. 2.]

[11] As is well known, the Strait of Soerabaya forms an elbow whose easterly outlet opens to the east, while the westerly outlet opens to the north. In the beginning of the east monsoon the sea-wind (east monsoon) blows through the westerly entrance as far as Grissee (in the elbow); in the latter part of this monsoon, the sea-wind blows, on the contrary, through the easterly entrance as far as Sambilangan (the narrow passage where the westerly outlet opens into the sea). — [*Ibid.*, n. 3.]

708. *Marking the seasons.* "As the sea makes the coming of the southern summer known to the inhabitants of the Java coast,[12] the turning of the east monsoon into the west monsoon commences. After the sun has finished its yearly task in the northern hemisphere, and brings its powerful influence to operate in the southern hemisphere, a change is at once perceived in the constant fine weather of the east monsoon of the Java Sea. As soon as it is at its height upon the Java Sea (6° south), then the true turning of the monsoon begins, and is accomplished much more rapidly than the spring turning. The calms then are not so continuous. The combat in the upper atmosphere appears to be less violent; the southeast trade, which has blown as the east monsoon, does not seem to have sufficient strength to resist the aggressors, who, with wild storms from the northwest and west, make their superiority known. Upon and in the neighborhood of the land, thunder-storms occur, but at sea they are less frequent.

709. *Conflicts in the air.* "The atmosphere, alternately clear and cloudy, moves more definitely over from the northwest, so that it appears as if no combat was there waged, and the southeast gives place without a contest. The land-breezes become less frequent, and the phenomena by day and night become, in a certain sense, more accordant with each other. Storms of wind and rain beneath a clouded sky alternate with severe gales and steady winds. In the last of November the west monsoon is permanent.

710. *Passing of the calm belts.* "Such are the shiftings. But what have they to do with the general system of the circulation of the atmosphere? Whenever we read attentively the beautiful meditations of the founder of the Meteorology of the Sea, and follow him in the development of his hypothesis, which lays open to view the wheels whereby the atmosphere performs its varied and comprehensive task with order and regularity, then it will not be necessary to furnish proof that these turnings are nothing else than the passing of a belt of calms which separates the monsoons from each other, and which, as we know, goes annually with the sun from the south to the north, and back over the torrid zone to and fro.

711. *Where they are, there the changing of the monsoons is going on.* "So also the calms, which precede the land and sea winds, are turned back.[13] If, at the coming of the land-wind in the hills, we go with it to the coast — to the sea, we shall perceive that it shoves away the calms which preceded it from the hills to the coast, and so far upon the sea as

[12] In the Archipelago we have generally high water but once a day, and, with the equinoxes, the tides also turn. The places which have high water by day in one monsoon get it at night in the other. — [*Bijdrage*, p. 277, n. 1.]

[13] [*Bijdrage*, p. 278, *Zoo ook zijn de stilten, die de land- en zeewinden vooraf gaan, kenteringen; i.e.,* The calms that precede the land and sea winds are comparable changes.]

the land-wind extends. Here, upon the limits of the permanent monsoon, the place for the calms remains for the night, to be turned back to the land and to the hills the following day by the sea-wind. In every place where these calms go, the land and sea winds turn back. If various observers, placed between the hills and the sea, and between the coast and the farthest limit of the land-wind, noted the moment when they perceived the calms, and that when they perceived the land-wind, then by this means they would learn how broad the belt of calms has been, and with what rapidity they are pushed over the sea and over the land. And even though the results one day should be found not to agree very well with those of another, they would at least obtain an average thereof which would be of value. So, on a larger scale, the belt of calms which separates the monsoons from each other presses in the spring from the south to the north, and in the fall from the north to the south, and changes the monsoons in every place where it presses." [14]

[14] [*Ibid.*, pp. 273–278. Jansen writes, in the last phrase quoted, . . . *in alle plaatsen, waarover hij heen trekt; i. e.*, . . . in all places over which it passes.]

CHAPTER XVII

§ 720–735. THE CLIMATES OF THE SEA

720. A *"milky way" in the ocean.* Thermal charts, showing the temperature of the surface of the Atlantic Ocean by actual observations made indiscriminately all over it, and at all times of the year, have been published by the National Observatory. The isothermal lines which these charts enable us to draw, and a few of which are traced on Figure 9, afford the navigator and the philosopher much valuable and interesting information touching the circulation of the oceanic waters, including the phenomena of their cold and warm currents; these lines disclose a thermal tide in the sea, which ebbs and flows but once a year; they also cast light upon the climatology of the sea, its hyetographic peculiarities, and the climatic conditions of various regions of the earth; they show that the profile of the coast-line of intertropical America assists to give expression to the mild climate of Southern Europe; they also increase our knowledge concerning the Gulf Stream, for they enable us to mark out, for the mariner's guidance, that "milky way" in the ocean, the waters of which teem, and sparkle, and glow with life and incipient organisms as they flow across the Atlantic. In them are found the clusters and nebulæ of the sea which stud and deck the great highway of ships on their voyage between the Old World and the New; and these lines assist to point out for the navigator their limits and his way. They show this *via lactea* to have a vibratory motion in the sea that calls to mind the graceful wavings of a pennon as it floats gently to the breeze. Indeed, if we imagine the head of the Gulf Stream to be hemmed in by the land in the Straits of Bemini, and to be stationary there, and then liken the tail of the Stream itself to an immense pennon floating gently in the current, such a motion as such a streamer may be imagined to have, very much such a motion, do my researches show the tail of the Gulf Stream to have. Running between banks of cold water (§ 71), it is pressed now from the north, now from the south, according as the great masses of sea water on either hand may change or fluctuate in temperature.

721. *The vibrations of the Gulf Stream.* In September, when the waters in the cold regions of the north have been tempered, and been made warm and light by the heat of summer, its limits on the left are as denoted by the line of arrows (Figure 9); but after this great sun-swing, the waters on the left side begin to lose their heat, grow cold, become heavy, and press the hot waters of this stream into the channel marked

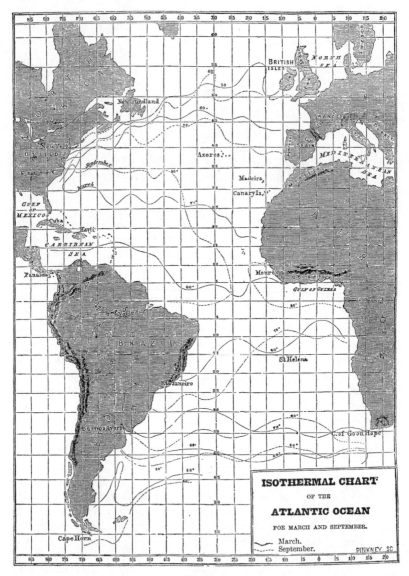

Figure 9

[The extreme movements of the isotherms 50°, 60°, 70°, etc., in the Atlantic Ocean during the year. The connection between the law of this motion and the climates of the sea is exceedingly interesting.]

out for them. Thus it acts like a pendulum, slowly propelled by heat on one side and repelled by cold on the other. In this view, it becomes a chronograph for the sea, keeping time for its inhabitants, and marking the seasons for the great whales; and there it has been for all time vibrating to and fro, once every year, swinging from north to south, and from south to north again, a great self-regulating, self-compensating liquid pendulum.

722. *Sea and land climates contrasted.* In seeking information concerning the climates of the ocean, it is well not to forget this remarkable contrast between its climatology and that of the land, namely: on the land February and August are considered the coldest and hottest months; but to the inhabitants of the sea, the annual extremes of cold and heat occur in the months of March and September. On the dry land, after the winter "is past and gone," the solid parts of the earth continue to receive from the sun more heat in the day than they radiate at night, consequently there is an accumulation of caloric, which continues to increase until August. The summer is now at its height; for, with the close of this month, the solid parts of the earth's crust and the atmosphere above begin to dispense with their heat faster than the rays of the sun can impart fresh supplies, and, consequently, the climates which they regulate grow cooler and cooler until the dead of winter again. But at sea a different rule seems to prevail. Its waters are the store-houses [1] in which the surplus heat of summer is stored away against the severity of winter, and its waters continue to grow warmer for a month after the weather on shore has begun to get cool. This brings the highest temperature to the sea in September, the lowest in March. Figure 9 is intended to show the extremes of heat and cold to which the *waters* — not the ice — of the sea are annually subjected, and therefore the isotherms of 40°, 50°, 60°, 70°, and 80° have been drawn for March and September, the months of extreme heat and extreme cold to the inhabitants of the "great deep." Corresponding isotherms for any other month will fall between these, taken by pairs. Thus the isotherm of 70° for July will fall nearly midway between the same isotherms (70°) for March and September.

723. *Figure 9.* A careful study of this figure, and the contemplation of the benign influence of the sea upon the climates which we enjoy, suggest many beautiful thoughts; for by such study we get a glimpse into the arrangements and the details of that exquisite machinery in the ocean which enables it to perform all its offices, and to answer with fidelity its marvelous adaptations. How, let us inquire, does the isothermal of 80°, for instance, get from its position in March to its position in September? Is it wafted along by currents, that is, by water which, after having been heated near the equator to 80°, then flows to the north with this tem-

[1] *Vide* Chap. XXII, "Actinometry of the Sea."

perature? Or is it carried there simply by the rays of the sun, as the snow-line is carried up the mountain in summer? We have reason to believe that it is carried from one parallel to another by each of these agents acting together, but mostly through the instrumentality of currents, for currents are the chief agents for distributing heat to the various parts of the ocean. The sun with his rays would, were it not for currents, raise the water in the torrid zone to blood heat; but before that can be done, they run off with it toward the poles, softening, and mitigating, and tempering climates by the way. The provision for this is as beautiful as it is benign; for, to answer a physical adaptation, it is provided by a law of nature that when the temperature of water is raised, it shall expand; as it expands, it must become lighter, and just in proportion as its specific gravity is altered, just in that proportion is equilibrium in the sea destroyed. Arrived at this condition, it is ordained that this hot water shall obey another law of nature, which requires it to run away, and hasten to restore that equilibrium. Were these isothermal lines moved only by the rays of the sun, they would slide up and down the ocean like so many parallels of latitude — at least there would be no break in them, like that which we see in the isotherm of 80° for September. It appears from this line that there is a part of the ocean near the equator, and about midway the Atlantic, which, with its waters, never does attain the temperature of 80° in September. Moreover, this isotherm of 80° will pass, in the North Atlantic, from its extreme southern to its extreme northern declination — nearly two thousand miles — in about three months. Thus it travels at the rate of about twenty-two miles a day. Surely, without the aid of currents, the rays of the sun could not drive it along that fast. Being now left to the gradual process of cooling by evaporation, atmospherical contact, and radiation, it occupies the other eight or nine months of the year in slowly returning south to the parallel whence it commenced to flow northward. As it does not cool as rapidly as it was heated, the disturbance of equilibrium by alteration of specific gravity is not so sudden, nor the current which is required to restore it so rapid. Hence the slow rate of movement at which this line travels on its march south. Between the meridians of 25° and 30° west, the isotherm of 60° in September ascends as high as the parallel of 56° N. In October it reaches the parallel of 50° north. In November it is found between the parallels of 45° and 47°, and by December it has nearly reached its extreme southern descent between these meridians, which it accomplishes in January, standing then near the parallel of 40°. It is all the rest of the year in returning northward to the parallel whence it commenced its flow to the south in September. Now it will be observed that this is the season — from September to December — immediately succeeding that in which the heat of the sun has been playing with greatest activity upon the polar ice. Its melted waters, which are thus put in motion in June, July, and August, would probably occupy the fall months in reaching the parallels indicated. These waters, though cold, and rising gradually in temperature as they flow south, are probably fresher, and if so, probably lighter than the sea water; and therefore it may well be that both the warmer and cooler systems of these isothermal lines are made to vibrate up and down the ocean principally by a gentle surface current in the season of quick motion, and in the season of the slow motion principally by a gradual process of calorific absorption on the one hand, and by a gradual process of cooling on the other. We have precisely such phenomena exhibited by the waters of the Chesapeake Bay as they spread themselves over the sea in winter. At this season of the year, the charts show that water of very low temperature is found projecting out and overlapping the usual limits of the Gulf Stream. The outer edge of this cold water, though jagged, is circular in its shape, having its centre near the mouth of the bay. The waters of the bay, being fresher than those of the sea, are therefore, though colder, yet lighter (§ 426) than the warmer waters of the ocean. And thus we have repeated here, though on a smaller scale, the phenomenon as to the flow of cold waters from the north, which force the surface isotherm of 60° from the latitude 56° to 40° during three or four months. Changes in the color or depth of the water, and the shape of the bottom, etc., would also cause changes in the temperature of certain parts of the ocean, by increasing or diminishing the capacities of such parts to absorb or radiate heat; and this, to some extent, would cause a bending, or produce irregular curves in the isothermal lines. After a careful study of this figure, and the Thermal Charts of the Atlantic Ocean, from which the materials for the former were derived, I am led to infer that the mean temperature of the atmosphere between the parallels of 56° and 40° north, for instance, and over that part of the ocean in which we have been considering the fluctuations of the isothermal line of 60°, is at least 60° of Fahrenheit, and upward, from January to August, and that the heat which the waters of the ocean derive from this source — atmospherical contact and radiation — is one of the causes which move the isotherm of 60° from its January to its September parallel. It is well to consider another of the causes which are at work upon the currents in this part of the ocean, and which tend to give the rapid southwardly motion to the isotherm of 60°. We know the mean dew-point must always be below the mean temperature of any given place, and that, consequently, as a general rule, at sea the mean dew-point due the isotherm of 60° is higher than the mean dew-point along the isotherm of 50°, and this, again, higher than that of 40°, this than 30°, and so on. Now suppose, merely for the sake of illustration, that the mean dew-point for each isotherm be 5° lower than the mean temperature, we should then have the atmosphere which crosses the isotherm of

60°, with a mean dew-point of 55°, gradually precipitating its vapors un-til it reaches the isotherm of 50°, with a mean dew-point of 45°; by which difference of dew-point the total amount of precipitation over the entire zone between the isotherms of 60° and 50° has exceeded the total amount of evaporation from the same surface. The prevailing direction of the winds to the north of the fortieth parallel of north latitude is from the southward and westward (Plate V); in other words, it is from the higher to the lower isotherms. Passing, therefore, from a higher to a lower tem-perature over the ocean, the total amount of vapor deposited by any given volume of atmosphere, as it is blown from the vicinity of the tropical toward that of the polar regions, is greater than that which is taken up again. The area comprehended on Plate V between the iso-therms of 40° and 50° Fahrenheit is less than the area comprehended between the isotherms 50° and 60°, and this, again, less than the area between this last and 70°, for the same reason that the area between the parallels of latitude 50° and 60° is less than the area between the par-allels of latitude 40° and 50°; therefore, more rain to the square inch ought to fall upon the ocean between the colder isotherms of 10° dif-ference, than between the warmer isotherms of the same difference. This is an interesting and an important view, therefore let me make myself clear. The aqueous isotherm of 50°, in its extreme northern reach, touches the parallel of 60° north. Now between this and the equator there are but three isotherms, 60°, 70°, and 80°, with the common difference of 10°. But between the isotherm of 40° and the pole there are at least five others, viz., 40°, 30°, 20°, 10°, 0°, with a common difference of 10°. Thus, to the north of the isotherm 50°, the vapor which would saturate the at-mosphere from zero, and perhaps far below, to near 40°, is deposited, while to the south of 50° the vapor which would saturate it from the tem-perature of 50° up to that of 80° can only be deposited. At least, such would be the case if there were no irregularities of heated plains, moun-tain ranges, land, etc., to disturb the laws of atmospherical circulation as they apply to the ocean.

724. *The effects of night and day upon the temperature of sea water.* Having therefore, theoretically, at sea more rain in high latitudes, we should have more clouds; and therefore it would require a longer time for the sun, with his feeble rays, to raise the temperature of the cold water which, from September to January, has brought the isotherm of 60° from latitude 56° to 40°, than it did for those cool surface currents to float it down. After this southward motion of the isotherm of 60° has been checked in December by the cold, and after the sources of the cur-rent which brought it down have been bound in fetters of ice, it pauses in the long nights of the northern winter, and scarcely commences its re-turn till the sun recrosses the equator, and increases its power as well in

intensity as in duration. Thus, in studying the physical geography of the sea, we must take cognizance of its actinometry also, for here we have the effects of night and day, of clouds and sunshine, upon its currents and its climates, beautifully developed. These effects are modified by the operations of certain powerful agents which reside upon the land; nevertheless, feeble though those of the former class may be, a close study of this plate will indicate that they surely exist.

725. *A belt of uniform temperature at sea.* Now, returning toward the south: we may, on the other hand, infer that the mean atmospherical temperature for the parallels between which the isotherm of 80° fluctuates is below 80°, at least for the nine months of its slow motion. This vibratory motion suggests the idea that there is probably, somewhere between the isotherm of 80° in August and the isotherm of 60° in January, a line or belt of invariable or nearly invariable temperature, which extends on the surface of the ocean from one side of the Atlantic to the other. This line or band may have its cycles also, but they are probably of a long and uncertain period.

726. *The western half of the Atlantic warmer than the eastern.* The fact has been pretty clearly established by the discoveries to which the wind and current charts have led, that the western half of the Atlantic Ocean is heated up, not by the Gulf Stream alone, as is generally supposed, but by the great equatorial caldron to the west of longitude 35°, and to the north of Cape St. Roque, in Brazil. The lowest reach of the 80° isotherm for September — if we except the remarkable equatorial flexure (Figure 9) which actually extends from 40° north to the line — to the west of the meridian of Cape St. Roque, is above its highest reach to the east of that meridian. And, now that we have the fact, how obvious, how beautiful, and striking is the cause! Cape St. Roque is in 5° 30′ south. Now study the configuration of the Southern American Continent from this cape to the Windward Islands of the West Indies, and take into account also certain physical conditions of these regions: the Amazon, always at a high temperature because it runs from west to east, is pouring an immense volume of warm water into this part of the ocean. As this water and the heat of the sun raise the temperature of the ocean along the equatorial sea-front of this coast, there is no escape for the liquid element, as it grows warmer and lighter, except to the north. The land on the south prevents the tepid waters from spreading out in that direction as they do to the east of 35° west, for here there is a space, about 18 degrees of longitude broad, in which the sea is clear both to the north and south. They must consequently flow north. A mere inspection of the figure is sufficient to make obvious the fact that the warm waters which are found east of the usual limits assigned the Gulf Stream, and between the parallels of 30° and 40° north, do not come from the Gulf

Stream, but from this great equatorial caldron, which Cape St. Roque blocks up on the south, and which dispenses its overheated waters up toward the fortieth degree of north latitude, not through the Caribbean Sea and Gulf Stream, but over the broad surface of the left bosom of the Atlantic Ocean.

727. *The warmest sides of oceans and the coldest shores of continents in juxtaposition.* Like the western half of the North Atlantic Ocean, the western half of every one of the three great oceans is the warmer. The great flow of warm water in the North Pacific is with the "Black Stream of Japan," on the Asiatic side; in the South Pacific it is with the Polynesian drift, on the Australian side: opposite to these, and on the eastern side, are the Humboldt current in one hemisphere, and the California current in the other — cold currents both. In the South Indian Ocean, the warm water is with the Mozambique current on the African side, and the cold drift on the Australian; and in the South Atlantic, Figure 9 shows that, parallel for parallel, the littoral waters of Brazil are many degrees warmer than those on the African side. Thus at sea the climatic conditions of the land are reversed, for the coldest side of the ocean is on the warmest side of the continent, and *vice versa.* The winds from extra-tropical seas temper the climates of the shores upon which they blow, not so much by the sensible heat they convey as by the latent heat which is liberated from the vapor they bring. This being condensed, as upon the British Islands and Western Europe, sets free heat enough not only to soften the climate, but to rarefy the air to such an extent as to be observed in the mean barometric pressure.

728. *The climates of Europe influenced by the shore-lines of Brazil.* Here we are again tempted to pause and admire the beautiful revelations which, in the benign system of terrestrial adaptations, these researches into the physics of the sea unfold and spread out before us for contemplation. In doing this, we shall have a free pardon from those at least who delight "to look through nature up to nature's God." What two things in nature can be apparently more remote in their physical relations to each other than the climate of Western Europe and the profile of a coast-line in South America? Yet this figure reveals to us not only the fact that these relations between the two are the most intimate, but makes us acquainted with the arrangements by which such relations are established. The barrier which the South American shore-line opposes to the escape, on the south, of the hot waters from this great equatorial caldron of St. Roque, causes them to flow north, and in September, as the winter approaches, to heat up the western half of the Atlantic Ocean, and to cover it, as far up as the parallel of 40° N., with a mantle of warmth above summer heat. Here heat to temper the winter climate of Western Europe is stored away as in an air-chamber to furnace-heated apart-

ments; and during the winter, when the fire of the solar rays sinks down, the westwardly winds and eastwardly currents are sent to perform their office in this benign arrangement. Though unstable and capricious to us they seem to be, they nevertheless "fulfill His commandments" with regularity and perform their offices with certainty. In tempering the climates of Europe with heat in winter that has been bottled away in the waters of the ocean during summer, these winds and currents are to be regarded as the flues and regulators for distributing it at the right time, and at the right places, in the right quantities. By March, when "the winter is past and gone," the furnace which had been started by the rays of the sun in the previous summer, and which, by autumn, had heated up the ocean in our hemisphere, has cooled down. The caldron of St. Roque, ceasing in activity, has failed in its supplies, and the chambers of warmth upon the northern sea, having been exhausted of their heated water (which has been expended in the manner already explained), have contracted their limits. The surface of heated water which, in September, was spread out over the western half of the Atlantic, from the equator to the parallel of 40° north, and which raised this immense area to the temperature of 80° and upward, is not to be found in early spring on this side of the parallel of 8° north. The isotherm of 80° in March, after quitting the Caribbean Sea, runs parallel with the South American coast toward Cape St. Roque, keeping some 8 or 10 degrees from it. Therefore the heat dispensed over Europe from this caldron falls off in March. But at this season the sun comes forth with fresh supplies; he then crosses the line and passes over into the northern hemisphere; observations show that the process of heating the water in this great caldron for the next winter is now about to commence. In the mean time, so benign is the system of cosmical arrangements, another process of raising the temperature of Europe commences. The land is more readily impressed than the sea by the heat of the solar rays; at this season, then, the summer climate due these transatlantic latitudes is modified by the action of the sun's rays directly upon the land. The land receives heat from them, but, instead of having the capacity of water for retaining it, it imparts it straightway to the air; and thus the proper climate, because it is the climate which the Creator has, for his own wise purposes, allotted to this portion of the earth, is maintained until the marine caldron of Cape St. Roque and the tropics is again heated and brought into the state for supplying the means of maintaining the needful temperature in Europe during the absence of the sun in the other hemisphere. Thus the equable climates of Western Europe are accounted for.

729. *The Gulf of Guinea and the climate of Patagonia.* In like manner, the Gulf of Guinea forms a caldron and a furnace, and spreads out over the South Atlantic an air-chamber for heating up in winter and

assisting to keep warm the extra-tropical regions of South America. Every traveler has remarked upon the mild climate of Patagonia and the Falkland Islands. "Temperature in high southern latitudes," says a very close observer, who is co-operating with me in collecting materials, "differs greatly from the temperature in northern. In southern latitudes there seem to be no extremes of heat and cold, as at the north. Newport, Rhode Island, for instance, latitude 41° north, longitude 71° west, and Rio Negro, latitude 41° south, and longitude 63° west, as a comparison: in the former, cattle have to be stabled and fed during the winter, not being able to get a living in the fields on account of snow and ice. In the latter, the cattle feed in the fields all winter, there being plenty of vegetation and no use of hay. On the Falkland Islands (latitude 51–2° south), thousands of bullocks, sheep, and horses are running wild over the country, gathering a living all through the winter." The water in the equatorial caldron of Guinea overflows to the south, as that of St. Roque does to the north; it carries to Patagonia and the Falkland Islands warmth, which, uniting with the heat set free by precipitation during the passage of the vapor-laden west winds across the Southern Andes, carries beyond latitude 50° in the other hemisphere the winter climate of South Carolina on one side of the North Atlantic, or of the "Emerald Island" on the other.

730. *Shore-lines.* All geographers have noticed, and philosophers have frequently remarked upon the conformity as to the shore-line profile of equatorial America and equatorial Africa. It is true, we can not now tell the reason, though explanations founded upon mere conjecture have been offered, why there should be this sort of jutting in and jutting out of the shore-line, as at Cape St. Roque and the Gulf of Guinea, on opposite sides of the Atlantic; but one of the purposes, at least, which this peculiar configuration was intended to subserve, is without doubt now revealed to us. We see that, by this configuration, two cisterns of hot water are formed in this ocean, one of which distributes heat and warmth to western Europe; the other, at the opposite season, tempers the climate of eastern Patagonia. Phlegmatic must be the mind that is not impressed with ideas of grandeur and simplicity as it contemplates that exquisite design, those benign and beautiful arrangements, by which the climate of one hemisphere is made to depend upon the curve of that line against which the sea is made to dash its waves in the other. Impressed with the perfection of terrestrial adaptations, he who studies the economy of the great cosmical arrangements is reminded that not only is there design in giving shore-lines their profile, the land and the water their proportions, and in placing the desert and the pool where they are, but the conviction is forced upon him also that every hill and valley, with the grass upon its sides, is a part of the wonderful mechanism, each having its offices to per-

form in the grand design. March is, in the southern hemisphere, the first month of autumn, as September is with us; consequently, we should expect to find in the South Atlantic as large an area of water at 80° and upward in March, as we should find in the North Atlantic for September. But do we? By no means. The area that is covered on this side of the equator with water at 80° and upward is nearly double that on the other. Thus we have the sea as a witness to the fact which the winds had proclaimed, viz., that summer in the northern hemisphere is hotter than summer in the southern.

731. *Sudden changes in the water thermometer.* Pursuing the study of the climates of the sea, let us now turn to Plate IV. Here we see at a glance how the cold waters, as they come down from the Arctic Ocean through Davis's Straits, press upon the warm waters of the Gulf Stream, and curve their channel into a horse-shoe. Navigators have often been struck with the great and sudden changes in the temperature of the water hereabouts. In the course of a single day's sail in this part of the ocean, changes of 15°, or 20°, and even of 30°, have been observed to take place in the temperature of the sea. The cause has puzzled navigators long, but how obvious is it now made to appear! This "bend" is the great receptacle of the icebergs which drift down from the north; covering frequently an area of hundreds of miles in extent, its waters differ as much as 20°, 25°, and in rare cases even as much as 30° of temperature from those about it. Its shape and place are variable. Sometimes it is like a peninsula, or tongue of cold water projected far down into the waters of the Gulf Stream. Sometimes the meridian upon which it is inserted into these is to the east of 40°, sometimes to the west of 50°.

732. *The fogs of Newfoundland.* By its discovery we have clearly unmasked the very seat of that agent which produces the Newfoundland fogs. It is spread out over an area frequently embracing several thousand square miles in extent, covered with cold water, and surrounded on three sides, at least, with an immense body of warm. May it not be that the proximity to each other of these two very unequally heated surfaces out upon the ocean would be attended by atmospherical phenomena not unlike those of the land and sea breezes? These warm currents of the sea are powerful meteorological agents. I have been enabled to trace in thunder and lightning the influence of the Gulf Stream in the eastern half of the Atlantic as far up as the parallel of 55° N., for there, in the dead of winter, a thunder-storm is not unusual.

733. *Aqueous isothermal lines.* These isothermal lines of 50°, 60°, 70°, 80°, etc., may illustrate for us the manner in which the climates in the ocean are regulated. Like the sun in the ecliptic, they travel up and down the sea in declination, and serve the monsters of the deep for signs and for seasons.

734. *The meeting of cool and warm waters.* It should be borne in mind that the lines of separation, as drawn on Plate VI, between the cool and warm waters, or, more properly speaking, between the channels representing the great polar and equatorial flux and reflux, are not so sharp in nature as this plate would represent them. In the first place, the plate represents the mean or average limits of these constant flows — polar and equatorial; whereas, with almost every wind that blows, and at every change of season, the line of meeting between their waters is shifted. In the next place, this line of meeting is drawn with a free hand on the plate, as if to represent an average; whereas there is reason to believe that this line in nature is variable and unstable as to position, and as to shape rough and jagged, and oftentimes deeply articulated. In the sea, the line of meeting between waters of different temperatures and density is not unlike the sutures of the skull-bone on a grand scale — very rough and jagged; but on the plate it is a line drawn with a free hand, for the purpose of showing the general direction and position of the channels in the sea, through which its great polar and equatorial circulation is carried on.

735. *The direction of aqueous isotherms on opposite sides of the sea.* Now, continuing for a moment our examination of Figure 9, we are struck with the fact that most of the thermal lines there drawn run from the western side of the Atlantic toward the eastern, in a northeastwardly direction, and that, as they approach the shores of this ocean on the east, they again turn down for lower latitudes and warmer climates. This feature in them indicates, more surely than any direct observations upon the currents can do, the presence, along the African shores in the North Atlantic, of a large volume of cooler waters. These are the waters which, having been first heated up in the caldron (§ 726) of St. Roque, in the Caribbean Sea, and Gulf of Mexico, have been made to run to the north, charged with heat and electricity to temper and regulate climates there. Having performed their offices, they have cooled down; but, obedient still to the "Mighty Voice" which the winds and the waves obey, they now return by this channel along the African shore to be again replenished with warmth, and to keep up the system of beneficent and wholesome circulation designed for the ocean.

CHAPTER XVIII

§ 740–772. TIDE–RIPS AND SEA DRIFT

740. *The glories of the sea.* We never tire of the sea; like the atmosphere, it is a laboratory in which the most exquisite processes are continually going on. Its flora and its fauna, its waves and its tides, its currents and its salts, all in themselves afford profitable subjects of study and charming themes for thought. But as interesting as they are individually, and as marvelous too, they are not half so marvelous, nor nearly so wonderful, as the offices which, with their aid, the sea performs in the physical economy of our planet. In this aspect the sea, with its insects, its salts, and its vapors, is a machine of the most beautiful construction. Its powers are vast, multitudinous, and varied. It is so stable and true in its work that nothing can throw it out of gearing, and yet its compensations are so delicate that the task of preserving them is assigned to the tiniest of its inhabitants, and to agents apparently the most subtle and fickle. They preserve its harmonies and make its adjustments, in beauty and sublimity of effect, to vie with the glories of the heavens.

741. *Drift described.* There is a movement of the waters of the ocean which, though it be a translation, yet it does not amount to what is known to the mariner as "current," for our nautical instruments and the art of navigation have not been brought to that state of perfection which will enable navigators generally to detect as currents the flow to which I allude as *drift.* If an object be set afloat in the ocean, as at the equator, it would, in the course of time, even though it should not be caught up by any of the known currents, find its way to the icy barriers about the poles, and again back among the tepid waters of the tropics. Such an object would illustrate the *drift of the sea,* and by its course would indicate the route which the surface-waters of the sea follow in their general channels of circulation to and fro between the equator and the poles.

742. *Plate VI.* The object of Plate VI, therefore, is to illustrate, as far as the present state of my researches enable me to do, the circulation of the ocean as influenced by *heat* and *cold,* and to indicate, on one hand, the routes by which the overheated waters of the torrid zone escape to cooler regions, and to point out, on the other, the great channel-ways through which the same waters, after having been deprived of this heat in the extra-tropical or polar regions, return again toward the equator; it being assumed that the drift or flow is from the poles when the temperature of the surface water is *below,* and from the equatorial regions when

it is *above* that due the latitude. Therefore, in a mere diagram, as this plate is, the numerous eddies and local currents which are found at sea are disregarded. Of all the currents in the sea, the Gulf Stream is the best defined; its limits, especially those of the left bank, are always well marked, and, as a rule, those of the right bank, as high as the parallel of the thirty-fifth degree of latitude, are quite distinct, being often visible to the eye. The Gulf Stream shifts its channel (§ 124), but nevertheless its banks are often very distinct. Ships, in crossing the edges of it, can sometimes know it by the color of the water; at other times they find, as they pass along, the temperature of the water to change 8° or 10° in the course of as many minutes; as an example of this, I quote from the abstract log of the "Herculean," in which Captain William M. Chamberlain, being in latitude 33° 39′ north, longitude 74° 56′ west (about one hundred and thirty miles east of Cape Fear), remarks: "Moderate breezes, smooth sea, and fine weather. At ten o'clock fifty minutes, entered into the southern (right) edge of the Stream, and in eight minutes the water rose six degrees; the edge of the stream was visible, as far as the eye could see, by the great rippling and large quantities of Gulf weed — more 'weed' than I ever saw before, and I have been many times along this route in the last twenty years." In this diagram, therefore, I have thought it useless to attempt a delineation of any of those currents, as the Rennell Current of the North Atlantic, the "connecting current" of the South, "Mentor's Counter Drift," "Rossel's Drift of the South Pacific," etc., which run now this way, now that, and which are frequently not felt by navigators at all. In overhauling the log-books for data for this chart, I have followed vessels with the water thermometer to and fro across the seas, and taken the registrations of it exclusively for my guide, without regard to the reported set of the currents. When, in any latitude, the temperature of the water has appeared too high or too low for the latitude, the inference has been that such water was warmed or cooled, as the case may be, in other latitudes, and that it has been conveyed to the place where found through the great channels of oceanic circulation. If too warm, it is supposed that it had its temperature raised in warmer latitudes, and therefore the channel in which it is found leads from the equatorial regions. On the other hand, if the water be too cool for the latitude, then the inference is that it has lost its heat in colder climates, and therefore is found in channels which lead from the polar regions. The arrow-heads point to the direction in which the waters are supposed to flow. Their rate, according to the best information that I have obtained, is, at a mean, only about four knots a day — rather less than more. Accordingly, therefore, as the immense volume of water in the antarctic regions is cooled down, it commences to flow north. As indicated by the arrow-heads, it strikes against Cape Horn, and is divided by the continent, one portion going along the west coast as Humboldt's Current (§ 398); the other, entering the South Atlantic, flows up into the Gulf of Guinea, on the coast of Africa. Now, as the waters of this polar flow approach the torrid zone, they grow warmer and warmer, and finally themselves become tropical in their temperature. They do not then, it may be supposed, stop their flow; on the contrary, they keep moving, for the very cause which brought them from the extra-tropical regions now operates to send them back. This cause is to be found in the difference of the specific gravity at the two places. If, for instance, these waters, when they commence their flow from the hyperborean regions, were at 30°, their specific gravity will correspond to that of sea water at 30°. But when they arrive in the Gulf of Guinea or the Bay of Panama, having risen by the way to 80°, or perhaps 85°, their specific gravity becomes such as is due sea water of this temperature; and, since fluids differing in specific gravity can no more balance each other on the same level than can unequal weights in the opposite scales of a true balance, this hot water must now return to restore that equilibrium which it has destroyed in the sea by rising from 30° to 80° or 85°. Hence it will be perceived that these masses of water which are marked as cold are not always cold. They gradually pass into warm; for in traveling from the poles to the equator they partake of the temperature of the latitudes through which they flow, and grow warm. Plate VI, therefore, is only introduced to give general ideas; nevertheless, it is very instructive. See how the influx of cold water into the South Atlantic appears to divide the warm water, and squeeze it out at the sides, along the coasts of South Africa and Brazil. So, too, in the North Indian Ocean, the cold water again compelling the warm to escape along the land at the sides, as well as occasionally in the middle. In the North Atlantic and North Pacific, on the contrary, the warm water appears to divide the cold, and to squeeze it out along the land at the sides. The impression made by the cold current from Baffin's Bay upon the Gulf Stream is strikingly beautiful.

743. *The great bend in the Gulf Stream.* Another feature of the sea expressed by this plate is a sort of reflection or recast of the shore-line in the temperature of the water. This feature is most striking in the North Pacific and Indian Ocean. The remarkable intrusion of the cool into the volume of warm waters to the southward of the Aleutian Islands is not unlike that which the cool waters from Davis's Straits make in the Atlantic upon the Gulf Stream. In sailing through this "horse-shoe," or bend in the Gulf Stream (§ 731), Captain N. B. Grant, of the American ship Lady Arbella, bound from Hamburg to New York, in May, 1854, passed, from daylight to noon, twenty-four large "bergs," besides several small ones, "the whole ocean, as far as the eye could reach, being literally covered with them. I should," he continues, "judge the average height of them

above the surface of the sea to be about sixty feet; some five or six of them were at least twice that height, and, with their frozen peaks jutting up in the most fantastic shapes, presented a truly sublime spectacle."

744. *The horse-shoe in the Japan current.* The "horse-shoe" of cold in the warm water of the North Pacific, though extending 5 degrees farther toward the south, can not be the harbor for such icebergs. The cradle of those of the Atlantic was perhaps in the Frozen Ocean, for they may have come thence through Baffin's Bay. But in the Pacific there is no nursery for them. The water in Behring's Strait is too shallow to let them pass from that ocean into the Pacific, and the climates of Russian America do not favor the formation of large bergs. But, though we do not find in the North Pacific the physical conditions which generate icebergs like those of the Atlantic, we find them as abundant with fogs. The line of separation between the warm and cold water assures us of these conditions.

745. *The animalculæ of the sea.* What beautiful, grand, and benign ideas do we not see expressed in that immense body of warm waters which are gathered together in the middle of the Pacific and Indian Oceans! It is the womb of the sea. In it coral islands innumerable have been fashioned, and pearls formed in "great heaps"; there multitudes of living things, countless in numbers and infinite in variety, are hourly conceived. With space enough to hold the four continents and to spare, the tepid waters of this part of the ocean teem with nascent organisms.[1] They sometimes swarm so thickly there that they change the color of the sea, making it crimson, brown, black, or white, according to their own hues. These patches of colored water sometimes extend, especially in the Indian Ocean, as far as the eye can reach. The question, "What produces them?" is one that has elicited much discussion in seafaring circles. The Brussels Conference deemed them an object worthy of attention, and recommended special observations with regard to them.

746. *Colored patches.* Capt. W. E. Kingman, of the American clipper ship the Shooting Star, reports in his abstract log a remarkable white patch, which he encountered in lat. 8° 46′ S., long. 105° 30′ E., and which, in a letter to me, he thus describes: "*Thursday*, July 27, 1854. At 7h. 45m. P.M., my attention was called to notice the color of the water, which was rapidly growing white. Knowing that we were in a much frequented part of the ocean, and having never heard of such

---

[1] "It is the realm of reef-building corals, and of the wondrously-beautiful assemblage of animals, vertebrate and invertebrate, that live among them or prey upon them. The brightest and most definite arrangements of color are here displayed. It is the seat of maximum development of the majority of marine genera. It has but few relations of identity with other provinces. The Red Sea and Persian Gulf are its off-sets." — [Edward] Forbes, "Distribution of Marine Life," *Physical Atlas*, [pp. 99–102, esp., p. 99].

an appearance being observed before in this vicinity, I could not account for it. I immediately hove the ship to and cast the lead; had no bottom at 60 fathoms. I then kept on our course, tried the water by thermometer, and found it to be 78½°, the same as at 8 A.M. We filled a tub, containing some 60 gallons, with the water, and found that it was filled with small luminous particles, which, when stirred, presented a most remarkable appearance. The whole tub seemed to be active with worms and insects, and looked like a grand display of rockets and serpents seen at a great distance in a dark night; some of the serpents appeared to be six inches in length, and very luminous. We caught, and could feel them in our hands, and they would emit light until brought within a few feet of a lamp, when, upon looking to see what we had, behold, nothing was visible; but, by the aid of a sextant's magnifier, we could plainly see a jelly-like substance without color. At last a specimen was obtained of about two inches in length, and plainly visible to the naked eye; It was about the size of a large hair, and tapered at the ends. By bringing one end within about one fourth of an inch of a lighted lamp, the flame was attracted toward it, and burned with a red light; the substance crisped in burning something like a hair, or appeared of a red heat before being consumed. In a glass of the water there were several small round substances (say $\frac{1}{16}$th of an inch in diameter), which had the power of expanding to more than twice their ordinary size, and then contracting again; when expanded, the outer rim appeared like a circular saw, only that the teeth pointed toward the centre. This patch of white water was about 23 miles in length, north and south, divided near its centre by an irregular strip of dark water half a mile wide; its east and west extent I can say nothing about. I have seen what is called white water in about all the known oceans and seas in the world, but nothing that would compare with this in extent or whiteness. Although we were going at the rate of nine knots, the ship made no noise either at the bow or stern. The whole appearance of the ocean was like a plain covered with snow. There was scarce a cloud in the heavens, yet the sky, for about ten degrees above the horizon, appeared as black as if a storm was raging; the star of the first magnitude shone with a feeble light, and the 'Milky Way' of the heavens was almost entirely eclipsed by that through which we were sailing. The scene was one of awful grandeur; the sea having turned to phosphorus, and the heavens being hung in blackness, and the stars going out, seemed to indicate that all nature was preparing for that last grand conflagration which we are taught to believe is to annihilate this material world. After passing through the patch, we noticed that the sky, for four or five degrees above the horizon, was considerably illuminated, something like a faint aurora borealis. We soon passed out of sight of the whole concern, and had a fine night, without any conflagration (except of midnight oil in trying to find out what was in the water). I send you this because I believe you request your corps of 'one thousand assistants' to furnish you with all such items, and I trust it will be acceptable. But as to its furnishing you with much, if any, information relative to the insects or animals that inhabit the mighty deep, time will only tell; I can not think it will."

747. *Whence the Red Sea derives its name.* These discolorations are no doubt caused by organisms of the sea, but whether wholly animal or wholly vegetable, or whether sometimes the one and sometimes the other, has not been satisfactorily ascertained. I have had specimens of the coloring matter sent to me from the pink-stained patches of the sea. They were animalculæ well defined. The tints which have given to the Red Sea its name may perhaps be in some measure due to agencies similar to those which, in the salt-makers' ponds, give a reddish cast (§ 71) to the brine just before it reaches that point of concentration when crystallization is to commence. Some microscopists maintain that this tinge is imparted by the shells and other remains of infusoria which have perished in the growing saltness of the water. The Red Sea may be regarded, in a certain light, as the scene of natural salt-works on a grand scale. The process is by solar evaporation. No rains interfere, for that sea (§ 376) is in a riverless district, and the evaporation goes on unceasingly, day and night, the year round. The shores are lined with incrustations of salt, and the same causes which tinge with red (§ 71) the brine in the vats of the salt-makers probably impart a like hue to the arms and ponds along the shore of this sea. Quantities, also, of slimy, red coloring matter are, at certain seasons of the year, washed up along the shores of the Red Sea, which Dr. Ehrenberg, after an examination under the microscope, pronounces to be a very delicate kind of sea-weed: from this matter that sea derives its name. So also the Yellow Sea. Along the coasts of China, yellowish-colored spots are said not to be uncommon. I know of no examination of this coloring matter, however. In the Pacific Ocean I have often observed these discolorations of the sea. Red patches of water are most frequently met with, but I have also observed white or milky appearances, which at night I have known greatly to alarm navigators by their being taken for shoals.

748. *The escape of warm waters from the Pacific.* These teeming waters bear off through their several channels the surplus heat of the tropics, and disperse it among the icebergs of the Antarctic. See the immense equatorial flow to the east of Australia, and which I have called the Polynesian Drift. It is bound for the icy barriers of that unknown sea, there to temper climates, grow cool, and return again, refreshing man and beast by the way, either as the Humboldt current, or the ice-bearing current which enters the Atlantic around Cape Horn, and changes into

warm again as it enters the Gulf of Guinea. It was owing to this great southern flow from the coral regions that Captain Ross was enabled to penetrate so much farther south than Captain Wilkes in his voyage to the Antarctic. The North Pacific, except in the narrow passage between Asia and America, is closed to the escape of these warm waters into the Arctic Ocean. The only outlet for them is to the south. They go down toward the antarctic regions to dispense their heat and get cool; and the cold of the Antarctic, therefore, it may be inferred, is not so bitter as is the extreme cold of the Frozen Ocean of the north.

749. *Ditto from the Indian Ocean.* The warm flow to the south from the middle of the Indian Ocean is remarkable. Masters who return their abstract logs to me mention sea-weed, which I suppose to be brought down by this current, as far as 45° south. There it is generally, but not always, about 5 degrees warmer than the ocean along the same parallel on either side.

750. *A wide current.* But the most unexpected discovery of all is that of the warm flow along the west coast of South Africa, its junction with the Lagulhas current, called, higher up, the Mozambique, and then their starting off as one stream to the southward. The prevalent opinion used to be that the Lagulhas current, which has its genesis in the Red Sea (§ 390), doubled the Cape of Good Hope, and then joined the great equatorial current of the Atlantic to feed the Gulf Stream. But my excellent friend, Lieutenant Marin Jansen, of the Dutch Navy, suggested that this was probably not the case. This induced a special investigation, and I found as he suggested, and as is represented on Plate VI. Captain N. B. Grant, in the admirably well-kept abstract log of his voyage from New York to Australia, found this current remarkably developed. He was astonished at the temperature of its waters, and did not know how to account for such a body of warm water in such a place. Being in longitude 14° east, and latitude 39° south, he thus writes in his abstract log: "That there is a current setting to the eastward across the South Atlantic and Indian Ocean is, I believe, admitted by all navigators. The prevailing westerly winds seem to offer a sufficient reason for the existence of such a current, and the almost constant southwest swell would naturally give it a northerly direction. But why the water should be *warmer* here (38° 40′ south) than between the parallels of 35° and 37° south, is a problem that, in my mind, admits not of so easy solution, especially if my suspicions are true in regard to the northerly set. I shall look with much interest for a description of the 'currents' in this part of the ocean." In latitude 38° south, longitude 6° east, he found the water at 56°. His course thence was a little to the south of east, to the meridian of 41° east, at its intersection with the parallel of 42° south. Here his water thermometer stood at 50°, but between these two places it ranged at 60° and

upward, being as high on the parallel of 39° as 73°. Here, therefore, was a stream — a mighty "river in the ocean" — one thousand six hundred miles across from east to west, having water in the middle of it 23° higher than at the sides. This is truly a Gulf Stream contrast. What an immense escape of heat from the Indian Ocean, and what an influx of warm water into the frozen regions of the south! This stream is not always as broad nor as warm as Captain Grant found it. At its mean stage it conforms more nearly to the limits assigned it in the diagram (Plate VI).

751. *Commotions in the sea.* Instances of commotions in the sea at uncertain intervals are not unfrequent. There are some remarkable disturbances of the sort which I have not been able wholly to account for. Near the equator, and especially on this side of it in the Atlantic, mention is made, in the "abstract log," by almost every observer that passes that way, of "tide-rips," which are a commotion in the water not unlike that produced by a conflict of tides or of other powerful currents. These "tide-rips" sometimes move along with a roaring noise, like rifts over rocks in rivers, and the inexperienced navigator always expects to find his vessel drifted by them a long way out of her course; but when he comes to cast up his reckoning the next day at noon, he remarks with surprise that no current has been felt.

752. *Humboldt's description of tide-rips.* Tide-rips present their most imposing aspect in the equatorial regions. Humboldt met some in 34° N., and thus describes them: "When the sea is perfectly calm, there appear on its surface narrow belts, like small rivulets, and in which the water runs with a noise very perceptible to the ear of an experienced pilot. On the 15th of June, in about 34° 36' N., we found ourselves in the midst of a great number of these belts of currents; we were able to determine their direction by the compass. Some were flowing to the N.E.; others E.N.E., although the general motion of the ocean, indicated by a comparison of the log and the longitude by chronometer, continued toward the S.E." [2] It is very common to see a mass of motionless water crossed by ridges of water which run in different directions. This phenomenon may be observed every day on the surface of our lakes; but it is more rare to find partial movements impressed by local causes on small portions of water in the midst of an oceanic river occupying an immense space, and moving in a constant direction, although with an inconsiderable velocity. In this conflict of currents, as in the oscillation of waves, our imagination is struck with these movements, which seem to penetrate each other, and by which the ocean is incessantly agitated.

[2] [Alexander von Humboldt, *Personal Narrative of Travels to the Equinoctial Regions of America* . . . translated . . . by Thomasina Ross, I(London, 1853), 25–26.]

753. *Horsburgh's.* Horsburgh, in his East India Directory, thus remarks on them, when speaking of the northeast monsoon about Java: "In the entrance of the Malacca Straits, near the Nicobar and Acheen Islands, and between them and Junkseylon, there are often very strong ripplings particularly in the southwest monsoon; these are alarming to persons unacquainted, for the broken water makes a great noise when the ship is passing through the ripplings in the night. In most places ripplings are thought to be produced by strong currents, but *here* they are frequently seen when there is no perceptible current. Although there is no perceptible current experienced so as to produce an error in the course and distance sailed, yet the surface of the water is impelled forward by some undiscovered cause. The ripplings are seen in calm weather approaching from a distance, and in the night their noise is heard a considerable time before they come near. They beat against the sides of a ship with great violence, and pass on, the spray sometimes coming on deck; and a small boat could not always resist the turbulence of these remarkable ripplings." [8]

754. *Tide-rips in the Atlantic.* Captain Higgins, of the "Maria," when bound from New York to Brazil, thus describes, in his abstract log, one of these "tide-rips," as seen by him, 10th October, 1855, in N. lat. 14°, W. long. 34°: "At 3 P.M. saw a tide-rip; in the centre, temp. air 80°, water 81°. From the time it was seen to windward, about three to five miles, until it had passed to leeward out of sight, it was not five minutes. I should judge it traveled at not less than sixty miles per hour, or as fast as the bores of India. Although we have passed through several during the night, we do not find they have the set the ship to the westward any; it may be that they are so soon passed that they have no influence on the ship, but they certainly beat very hard against the ship's sides, and jarred her all over. They are felt even when below, and will wake one out of sleep."

755. *Mock vigias.* Captain Wakeman, of the "Adelaide," in January, 1856, lat. 11° 21′ N., long. 33° 33′ W., encountered "tide-rips" which broke and foamed with such violence that he took them for breakers or a shoal. They sometimes are most alarming. Approaching through the stillness of the night with a roaring noise, and in the shape of tremendous rollers combing and foaming, they seem to threaten to overwhelm vessel and crew; but, breaking over the deck, they pass by, and in a few moments the sea is as smooth and as unruffled as before. Many of the "vigias" which disfigure our charts have no other foundation than the foam of a tide-rip. Captain Arquit's log of the "Comet" gives an account of many tide-rips which he encountered also in the northeast trade-wind region of the Atlantic. Thus, November 15, 1855, lat. 7° 34′ N., long.

[8] [*The India Directory*, II, 68.]

40° 30′ W.: "Many tide-rips, which we had a good opportunity of observing when becalmed. They came up in ridges as long as the eye could reach, from all parts of the compass, but mostly from the E. I examined the ridges very closely, but could not see any fine drift-matter of any kind, as you can on the ridges of currents in many parts of the ocean. We have had no currents unless they have been from different directions, and one counteracting the other. November 16th, lat. 6° 07′ N.: Light winds and pleasant. There has been no time since noon to midnight but there have been tide-rips either in sight or hearing, mostly tending N.E. and S.W. in long narrow ridges. From 8 P.M. to 9 P.M. the ocean appeared like a boiling caldron, which we sailed through for three miles. The bubbling made a loud noise, which we heard for a long time after we had sailed through it. The ship had a very singular motion, like striking her keel on a soft muddy bottom in a short rough seaway — the same as I have felt in the harbor of Montevideo. The motion was noticed by all on board. We have had a current of fifteen miles going west. I have often noticed tide-rips in this part of the ocean before, particularly when bound home (for I have never been where I am now, bound out, before), and have mentioned them in my abstract log, but they were different from what we had last night. The ship would come to and fall off three points without any regard to the rudder."

756. *Bores, eagres, and the earthquake wave of Lisbon.* But, besides tide-rips, bores, and eagres,[4] there are the sudden disruptions of the ice

---

[4] The bores of India, of the Bay of Fundy, and the Amazon are the most celebrated. They are a tremulous tidal wave, which at stated periods comes rolling in from the sea, threatening to overwhelm and ingulf every thing that moves on the beach. This wave is described, especially in the Bay of Fundy, as being many feet high; and it is said oftentimes to overtake deer, swine, and other wild beasts that feed or lick on the beach, and to swallow them up before the swiftest of foot among them have time to escape. The swine, as they feed on mussels at low water, are said to snuff the "bore," either by sound or smell, and sometimes to dash off to the cliffs before it rolls on.

The eagre is the bore of Tsien-Tang River. It is thus described by Dr. Macgowan, in a paper before the Royal Asiatic Society, 12th January, 1853, and as seen by him from the city of Hang-chow in 1848:

"At the upper part of the bay, and about the mouth of the river, the eagre is scarcely observable; but, owing to the very gradual descent of the shore, and the rapidity of the great flood and ebb, the tidal phenomena even here present a remarkable appearance. Vessels, which a few moments before were afloat, are suddenly left high and dry on a strand nearly two miles in width, which the returning wave as quickly floods. It is not until the tide rushes beyond the mouth of the river that it becomes elevated to a lofty wave constituting the eagre, which attains its greatest magnitude opposite the city of Hang-chow. Generally there is nothing in its aspect, except on the *third* day of the second month, and on the *eighteenth* of the eighth, or at the spring-tide about the period of the vernal and autumnal equinoxes, its great intensity being at the latter season. Sometimes, however, during the prevalence of easterly winds, on the *third* day after the sun and moon are in conjunction, or in opposition, the eagre courses up the river with hardly less majesty than when paying its ordinary periodical visit. On one of these unusual occasions, when I was traveling

which arctic voyagers tell of, the immense icebergs which occasionally appear in groups near certain latitudes, the variable character of all the currents of the sea — now fast, now slow (§ 401), now running this way,

in native costume, I had an opportunity of witnessing it, on December 14th, 1848, at about 2 P.M.

"Between the river and the city walls, which are a mile distant, dense suburbs extend several miles along the banks. As the hour of flood-tide approached, crowds gathered in the streets running at right angles with the Tsien-Tang, but at safe distances. My position was a terrace in front of the TRI-WAVE Temple, which afforded a good view of the entire scene. On a sudden all traffic in the thronged mart was suspended, porters cleared the front street of every description of merchandise, boatmen ceased lading and unlading their vessels, and put out in the middle of the stream, so that a few moments sufficed to give a deserted appearance to the busiest part of one of the busiest cities of Asia. The centre of the river teemed with craft, from small boats to huge barges, including the gay 'flower-boats.' Loud shouting from the fleet announced the appearance of the flood, which seemed like a glittering white cable, stretched athwart the river at its mouth, as far down as the eye could reach. Its noise, compared by Chinese poets to that of thunder, speedily drowned that of the boatmen; and as it advanced with prodigious velocity — at the rate, I should judge, of twenty-five miles an hour — it assumed the appearance of an alabaster wall, or, rather, of a cataract four or five miles across, and about thirty feet high, moving bodily onward. Soon it reached the advanced guard of the immense assemblage of vessels awaiting its approach. Knowing that the bore of the Hooghly, which scarcely deserves mention in connection with the one before me, invariably overturned boats which were not skillfully managed, I could not but feel apprehensive for the lives of the floating multitude. As the foaming wall of water dashed impetuously onward, they were silenced, all being intensely occupied in keeping their prows toward the wave which threatened to submerge every thing afloat; but they all vaulted, as it were, to the summit with perfect safety. The spectacle was of great interest when the eagre had passed about one half way among the craft. On one side they were quietly reposing on the surface of the unruffled stream, while those on the nether portion were pitching and heaving in tumultuous confusion on the flood; others were scaling with the agility of salmon the formidable cascade. This grand and exciting scene was but of a moment's duration; it passed up the river in an instant, but from this point with gradually diminishing force, size, and velocity, until it ceased to be perceptible, which Chinese accounts represent to be eighty miles distant from the city. From ebb to flood tide the change was almost instantaneous; a slight flood continued after the passage of the wave, but it soon began to ebb. Having lost my memoranda, I am obliged to write from recollection. My impression is that the fall was about twenty feet; the Chinese say that the rise and fall is sometimes forty feet at Hang-chow. The maximum rise and fall at spring-tides is probably at the mouth of the river, or upper part of the bay, where the eagre is hardly discoverable. In the Bay of Fundy, where the tides rush in with amazing velocity, there is at one place a rise of seventy feet; but there the magnificent phenomenon in question does not appear to be known at all. It is not therefore, where tides attain their greatest rapidity, or maximum rise and fall, that this wave is met with, but where a river and its estuary both present a peculiar configuration.

"Dryden's definition of an eagre, appended in a note to the verse above quoted from the *Threnodia Augustalis*, is, 'a tide swelling above another tide,' which he says he had himself observed in the River Trent. Such, according to Chinese oral accounts, is the character of the Tsien-Tang tides — a wave of considerable height rushes suddenly in from the bay, which is soon followed by one much larger. Other accounts represent three successive waves riding in; hence the name of the temple mentioned, that of the Three Waves. Both here and on the Hooghly I observed but one wave; my attention, however, was not particularly directed to this feature of the eagre. The term should, perhaps, be more comprehensive, and express 'the instantaneous rise

then that — all of which may be taken as so many signs of the tremendous throes which occur in the bosom of the ocean. Sometimes the sea recedes from the shore, as if to gather strength for a great rush against its barriers, as it did when it fled back to join with the earthquake and overwhelm Callao in 1746, and again Lisbon nine years afterward.

757. *Rains at sea and their effect upon its equilibrium.* Few persons have ever taken the trouble to compute (§ 402) how much the fall of a single inch of rain over an extensive region in the sea, or how much the change even of two or three degrees of temperature over a few thousand square miles of its surface, tends to disturb its equilibrium, and consequently to cause an aqueous palpitation that is felt from the equator to the poles. Let us illustrate by an example: The surface of the Atlantic Ocean covers an area of about twenty-five millions of square miles. Now, let us take one fifth of this area, and suppose a fall of rain one inch deep to take place over it. This rain would weigh three hundred and sixty thousand millions of tons; and the salt which, as water, it held in solution in the sea, and which, when that water was taken up as vapor, was left behind to disturb equilibrium, weighed sixteen millions more of tons, or nearly twice as much as all the ships in the world could carry at a cargo each. This rain might fall in an hour, or it might fall in a day; but, occupy what time it might in falling, this rain is calculated to exert so much force — which is inconceivably great — in disturbing the equilibrium of the ocean. If all the water discharged by the Mississippi River during the year were taken up in one mighty measure, and cast into the ocean at one effort, it would not make a greater disturbance in the equilibrium of the sea than would the supposed rain-fall. Now this is for but one fifth of the Atlantic, and the area of the Atlantic is about one fifth of the sea-area of the world; and the estimated fall of rain was but one inch, whereas the average for the year is (§ 757) sixty inches; but we will assume it for the sea to be no more than thirty inches. In the aggregate, and on an average, then, such a disturbance in the equilibrium of the whole ocean as is here supposed occurs seven hundred and fifty times a year, or at the rate of once in twelve hours. Moreover, when it is recollected that these rains take place now here, now there; that the vapor of which they were formed was taken up at still other places, we

---

and advance of a tidal wave;' the Indian barbarism 'bore' should be discarded altogether.

"A very short period elapsed between the passage of the eagre and the resumption of traffic. The vessels were soon attached to the shore again; women and children were occupied in gathering articles which the careless or unskillful had lost in the aquatic melee. The streets were drenched with spray, and a considerable volume of water splashed over the banks into the head of the grand canal, a few feet distant." — [D. J. Macgowan, "The Eagre of the Tsien-Tang River," *Trans. China Branch, Royal Asiastic Soc.*, IV(1855), 33–49, esp., pp. 37–39.]

shall be enabled to appreciate the better the force and the effect of these irregular movements in the sea.

758. *Ditto of cloud and sunshine.* Between the hottest hour of the day and the coldest hour of the night there is frequently a change of four degrees in the temperature of the sea.[5] Let us, therefore, the more thoroughly to appreciate those agitations of the sea which take place in consequence of the diurnal changes in its temperature, call in the sunshine, the cloud without rain, with day and night, and their heating and radiating processes. And to make the case as strong as to be true to nature we may, let us again select one fifth of the Atlantic Ocean for the scene of operation. The day over it is clear, and the sun pours down his rays with their greatest intensity, and raises the temperature two degrees. At night the clouds interpose, and prevent radiation from this fifth, whereas the remaining four fifths, which are supposed to have been screened by clouds, so as to cut off the heat from the sun during the day, are now looking up to the stars in a cloudless sky, and serve to lower the temperature of the surface-waters, by radiation, two degrees. Here, then, is a difference of four degrees, which we will suppose extends only ten feet below the surface. The total and absolute change made in such a mass of sea water by altering its temperature two degrees is equivalent to a change in its volume of three hundred and ninety thousand millions of cubic feet. And yet there are philosophers who maintain (§ 123) that evaporation and precipitation, changes of temperature and saltness, and the secretions of insects, are not to be reckoned among the current-producing agents of the sea — that the gentle trade-winds do it all!

759. *Day and night.* Do not the clouds, night and day, now present themselves to us in a new light? They are cogs, and rachets, and wheels in that grand and exquisite machinery which governs the sea, and which, amid all the jarring of the elements, preserves the harmonies of the ocean.

760. *Logs overhauled for kelp and ice.* The log-books of not less than 1843 vessels cruising on the polar side of 35° S. have, by the officers of the Observatory, been overhauled for kelp and ice. Of these, 367 — or one in five — mentioned kelp or sea-weed east of Cape Horn; 142 mention "rock-weed and drift matter" between the previous meridian and 10° W, and chiefly between 35° and 40° S. "Long kelp" is also found by Australian traders after passing the Cape of Good Hope; 146 logs make mention of it between the meridians of 40° and 120° E. It *most* abounds along this line, however, between the meridians of 45° and 65° E., and the parallels of 42° and 48° S. These sargassos are sketched with a free hand on Plate VI.

761. *A sargasso in the South Pacific.* Sea-weed is frequently mentioned also by the homeward-bound Australian traders on their way to

[5] Smyth, *The Mediterranean,* p. 125.

Cape Horn: this collection has (§ 139) already been alluded to. It now appears that instead of three, there are really five true sargassos, as shown on Plate VI.

762. *Sea-weed about the Falkland Islands.* The weedy space, marked as such, about the Falkland Islands, is probably not a true sargasso. The sea-weed reported there probably comes from the Straits of Magellan, where immense masses of it grow. These straits are so encumbered with sea-weed that steamers found great difficulty in making their way through it. It so encumbers their paddles as to make frequent stoppages necessary.

763. *The African sargasso.* The sargasso to the west of the Cape of Good Hope, though small, is perhaps the best defined of them all. Mention is generally made of it in the log as "rock-weed" and "drift matter." Now when it is recollected that weeds have been found as frequently *nearly* (§ 760) in this small space as they have been in the large space between the Cape and Australia, the reader will be able to form a more correct idea as to the relative abundance of weed in these seas of weeds.

764. *Icebergs.* By going far enough south, icebergs may be found on any meridian; but in searching for them, we can only look where commerce carries our colleagues of the sea. Out of the 1843 tracks traced on the polar side of 35° S., only 109 make mention of ice. Few of these went, except in doubling Cape Horn, beyond the parallel of 55° S., therefore we have not been able to trace the ice back into the "chambers of the frost." We can only say that north of 50° antarctic icebergs most abound between the meridians of 15° W. and 55° E.

765. *The largest drift farthest.* As a rule, the bergs which are the largest last longest, and approach nearest to the equator. Here, then, is the great line of antarctic drift; by studying it we may perhaps catch a glimmer of light from south polar shores. These icebergs, be it remembered, have drifted north through a belt of westerly winds. Their course, therefore, was probably not due north, but to the east of that rhomb.

766. *The line of antarctic drift.* Tracing this line of drift, then, backward in a southwesterly direction, it should guide us to that part of the southern continent where the icebergs have their principal nursery. This would take us to the sources of the Humboldt current, and seem to indicate that these glaciers are launched in its waters; but, as their motion is slow, the winds bear the bergs to the east, while the general drift sets them to the north.

767. *Necessity for, and advantages of an antarctic expedition.* Arrived at this point, fiords, deep bays, and capacious gulfs loom up before the imagination, reminding us to ask the question, Is there not embosomed in the antarctic continent a Mediterranean, the shores of which are favorable to the growth and the launching of icebergs of tremendous

size? and is not the entrance to this sea near the meridian of Cape Horn, perhaps to the west of it? Circumstances like these beget longings, and we sigh for fresh antarctic explorations. Surely, when we consider the advantages which the improvements of the age, the lights of the day, would afford an exploring expedition there now; when we reflect upon the drawbacks and difficulties with which former expeditions thither had to contend; when we call to mind the facilities with which one might be conducted *now*: surely, I say, when we thus reflect, no one can doubt as to the value and importance of the discoveries which a properly equipped expedition would *now* be sure to make.

768. *Commercial considerations.* In those regions there are doubtless elements of commercial wealth in the number of seals and abundance of whales, if in nothing else. It seems to be a physical law that cold-water fish are more edible than those of warm water. Bearing this fact in mind as we study Plate VI, we see at a glance the places which are most favored with good fish-markets. Both shores of North America, the east coast of China, with the west coasts of Europe and South America, are all washed by cold waters, and therefore we may infer that their markets abound with the most excellent fish. The fisheries of Newfoundland and New England, over which nations have wrangled for centuries, are in the cold water from Davis' Strait. The fisheries of Japan and Eastern China, which almost, if not quite, rival these, are situated also in the cold water. Neither India, nor the east coasts of Africa and South America, where the warm waters are, are celebrated for their fish.

769. *Value of the fisheries.* Three thousand American vessels, it is said, are engaged in the fisheries. If to these we add the Dutch, French, and English, we shall have a grand total, perhaps, of not less than six or eight thousand, of all sizes and flags, engaged in this one pursuit. Of all the industrial pursuits of the sea, however, the whale fishery is the most valuable. Wherefore, in treating of the physical geography of the sea, a map for the whales, it was thought, would be useful; it has so proved itself.

770. *Sperm whales.* The sperm whale is a warm-water fish. The *right* whale delights in cold water. An immense number of log-books of whalers have been discussed at the National Observatory with the view of detecting the parts of the ocean in which the whales are to be found at the different seasons of the year. Charts showing the result have been published; they form a part of the series of Maury's Wind and Current Charts.[6]

771. *A sea of fire to them.* In the course of these investigations, the discovery was made that the torrid zone is, to the right whale, as a sea of

[6] [Maury, "Whale Chart of the World," *Wind and Current Charts*, ser. F, nos. 1–4 (1852———).]

fire, through which he can not pass; that the right whale of the northern hemisphere and that of the southern are two different animals; and that the sperm whale has never been known to double the Cape of Good Hope — he doubles Cape Horn.

772. *Right whales.* With these remarks, and the explanation given on Plate VI, the parts of the ocean to which the right whale most resorts, and the parts in which the sperm are found, may be seen at a glance. The sargassos, or places of weed, are also represented on this plate.

CHAPTER XIX

§ 781–808. STORMS, HURRICANES, AND TYPHOONS

781. *Plate II.* Plate II is constructed from data furnished by the Pilot Charts, as far as they go, that are in process of construction at the National Observatory. For the Pilot Charts, the whole ocean is divided off into "fields" or districts of five degrees square, *i. e.*, five degrees of latitude by five degrees of longitude ["Explanation of the Plates, Plate II," *infra*]. Now, in getting out from the log-books materials for showing, in every district of the ocean, and for every month, how navigators have found the winds to blow, it has been assumed that, in whatever part of one of these districts a navigator may be when he records the direction of the wind in his log, from that direction the wind was blowing at that time all over that district; and this is the only assumption that is permitted in the whole course of the investigation. Now if the navigator will draw, or imagine to be drawn in any such district, twelve vertical columns for the twelve months, and then sixteen horizontal lines through the same for the sixteen points of the compass, *i. e.*, for N., N.N.E., N.E., E.N.E., and so on, omitting the *by*-points, he will have before him a picture of the "Investigating Chart" out of which the "Pilot Charts" are constructed. In this case the alternate points of the compass only are used, because, when sailing free, the direction of the wind is seldom given for such points as N. *by* E., W. *by* S., etc. Moreover, any attempt, for the present, at greater nicety would be over-refinement, for navigators do not always make allowance for the aberration of the wind; in other words, they do not allow for the apparent change in the direction of the wind caused by the rate at which the vessel may be moving through the water, and the angle which her course makes with the true direction of the wind. Bearing this explanation in mind, the intelligent navigator will have no difficulty in understanding the wind diagram (Plate II), and in forming a correct opinion as to the degree of credit due to the fidelity with which the prevailing winds of the year are represented on Plate V. As the compiler wades through log-book after log-book, and scores down in column after column, and upon line after line, mark upon mark, he at last finds that, under the month and from the course upon which he is about to make an entry, he has already made four marks or scores, thus (////). The one that he has now to enter will make the fifth, and he "scores and tallies," and so on until all the abstracts relating to that part of the ocean upon which he is at work have been

gone over, and his materials exhausted. These "fives and tallies" are exhibited on Plate II. Now, with this explanation, it will be seen that in the district marked A (Plate II) there have been examined the logs of vessels that, giving the direction of the wind for every eight hours, have altogether spent days enough to enable me to record the calms and the prevailing direction of the winds for eight hours, 2144 times: of these, 285 were for the month of September; and of these 285 observations for September, the wind is reported as prevailing for as much as eight hours at a time: from N., 3 times; from N.N.E., 1; N.E., 2; E.N.E., 1; E., 0; E.S.E., 1; S.E., 4; S.S.E., 2; S., 25; S.S.W., 45; S.W., 93; W.S.W., 24; W., 47; W.N.W., 17; N.W., 15; N.N.W., 1; Calms (the little 0's), 5; total, 285 for the month in this district. The number expressed in figures denotes the whole number of observations of calms and winds together that are recorded for each month and district. In C, the wind in May *sets* one third of the time from west. But in A, which is between the same parallels, the favorite quarter for the same month is from S. to S.W., the wind setting one third of the time from that quarter, and only 10 out of 221 times from the west; or, on the average, it blows from the west only 1⅓ day during the month of May. In B, notice the great "sun swing" of the winds in September, indicating that the change from summer to winter, in that region, is sudden and violent; from winter to summer, gentle and gradual. In some districts of the ocean, more than a thousand observations have been discussed for a single month, whereas, with regard to others, not a single record is to be found in any of the numerous logbooks at the National Observatory.

782. *Typhoons.* The China seas are celebrated for their furious gales of wind, known among seamen as typhoons and white squalls. These seas are included on the plate (V) as within the region of the monsoons of the Indian Ocean. But the monsoons of the China Sea are not five-month monsoons (§ 681); they do not prevail from the west of south for more than two or three months. Plate II exhibits the monsoons very clearly in a part of this sea. In the square between 15° and 20° north, 110° and 115° east, there appears to be a system of three monsoons; that is, one from northeast in October, November, December, and January; one from east in March and April, changing in May; and another from the southward in June, July, and August, changing in September. The great disturber of the atmospheric equilibrium appears to be situated among the plains and steppes of Asia; their influence extends to the China Seas, and about the changes of the monsoons these awful gales, called typhoons and white squalls, are experienced.

783. *The Mauritius hurricanes.* In like manner, the Mauritius hurricanes, or the cyclones of the Indian Ocean, occur during the unsettled state of the atmospheric equilibrium which takes place at that debatable

period during the contest between the trade-wind force and the monsoon force (§ 699), and which debatable period occurs at the changing of the monsoon, and before either force has completely gained or lost the ascendency. At this period of the year, the winds, breaking loose from their controlling forces, seem to rage with a fury that would break up the very fountains of the deep.

784. *The West India hurricanes.* So, too, with the West India hurricanes of the Atlantic. These winds are most apt to occur during the months of August and September. There is, therefore, this remarkable difference between these gales and those of the East Indies: the latter occur about the changing of the monsoons, the former during their height. In August and September, the southwest monsoons of Africa and the southeast monsoons of the West Indies are at their height; the agent of one drawing the northeast trade-winds from the Atlantic into the interior of New Mexico and Texas, the agent of the other drawing them into the interior of Africa. These two forces, pulling in opposite directions, assist now and then to disturb the atmospheric equilibrium to such an extent that the most powerful revulsions in the air are required to restore it. "The hurricane season in the North Atlantic Ocean," says Jansen, "occurs simultaneously with the African monsoon; and in the same season of the year in which the monsoons prevail in the North Indian Ocean, in the China Sea, and upon the Western coast of Central America, all the seas of the northern hemisphere have the hurricane season. On the contrary, the South Indian Ocean has its hurricane season in the opposite season of the year, and when the northwest monsoon prevails in the East Indian Archipelago." [1]

785. *The cyclone theory.* Under the head of hurricanes, typhoons, and tornadoes, I include all those gales of wind which are known as cyclones. These have been treated of by Redfield in America, Reid in England, Thom of Mauritius, and Piddington of Calcutta, with marked ability, and in special works. I refer the reader to them. The theory of this school is, that these are rotary storms; that they revolve against the hands of a watch in the northern, and with the hands of a watch in the southern hemisphere; that nearer the centre or vortex the more violent the storm, while the centre itself is a calm, which travels sometimes a mile or two an hour, and sometimes forty or fifty; that in the centre the barometer is low, rising as you approach the periphery of the whirl; that the diameter of these storms is sometimes a thousand miles, and sometimes not more than a few leagues; that they have their origin somewhere between the parallels of 10° and 20° north and south, traveling to the westward in either hemisphere, but increasing their distance from the equator, until they reach the parallel of 25° or 30°, when they turn to-

[1] [*Bijdrage*, p. 291.]

ward the east, or "recurvate," but continue to increase their distance from the equator — *i. e.*, they first travel westwardly, inclining toward the nearest pole; they then *recurve* and travel eastwardly, still inclining toward the pole; that such is their path in both hemispheres, etc.

786. *Puzzling questions.* The questions why these storms should recurve, and why they should travel as they do, and why they should turn *with* the hands of a watch in the southern, and *against* them in the northern hemisphere, are still considered by many as puzzles, though it is thought that their course to the westward in the trade-wind region, and to the eastward in the counter-trades, is caused by the general movement of the atmosphere, like the whirls in an angry flood, which, though they revolve, yet they are borne down stream with the currents as they do revolve. The motion polarward is caused, the conjecture is, by the fact that the equatorial edge of the storm has, in consequence of diurnal rotation, a greater velocity than its polar edge. There seems, however, to be less difficulty with regard to their turning than with regard to their course; the former is now regarded as the resultant of diurnal rotation and of these forces of translation which propel the winds along the surface of our planet. This composition of the forces of the revolving storm, and this resolution of them, are precisely such (§ 215) as to produce opposite rotation on opposite sides of the equator.

787. *Espy's theory.* Many of the phenomena connected with these storms still remain to be explained; even the facts with regard to them are disputed by some. The late Professor Espy, after having discussed for many years numerous observations that have been made chiefly on shore, maintained that the wind does not blow *around* the vortex or place of low barometer, but directly *toward* it. He held that the place of low barometer, instead of being a disc, is generally an oblong, in the shape of a long trough, between two atmospherical waves; that it is curved with its convex side toward the east; that it is sometimes nearly straight, and generally of great length from north to south, reaching in America from the Gulf of Mexico to the great lakes and beyond, and having but little breadth in proportion to its length; that it travels east, moving side foremost, requiring about two days to go from the Mississippi to St. John's, Newfoundland; that on either side of it, but many miles distant, there is a ridge of high barometer; that the wind on either side of the line of low barometer, in which there is little or no wind, blows toward it, etc., and, in support of these positions, he advanced this theory: "When the air in any locality acquires a higher temperature or a higher dew-point than that of surrounding regions, it is specifically lighter, and will ascend; in ascending, it comes under less pressure, and expands; in expanding from diminished pressure, it grows colder about a degree and a quarter for every hundred yards of ascent; in cooling as low as the dew-point (which

it will do when it rises as many hundred yards as the dew-point at the time is below the temperature of the air in degrees of Fahrenheit), it will begin to condense its vapor into cloud; in condensing its vapor into water or cloud, it will evolve its latent caloric; this evolution of latent caloric will prevent the air from cooling so fast in its farther ascent as it did in ascending below the base of the cloud now forming; the current of the air, however, will continue to ascend, and grow colder about half as much as it would do if it had no vapor in it to condense; and when it has risen high enough to have condensed, by the cold of expansion from diminished pressure, one hundredth of its weight of vapor, it will be about forty-eight degrees less cold than it would have been if it had no vapor to condense nor latent caloric to give out — that is, it will be about forty-eight degrees warmer than the surrounding air at the same height; it will, therefore (without making any allowance for the higher dew-point of the ascending current), be about one tenth lighter than the surrounding colder air, and, of course, it will continue to ascend to the top of the atmosphere, spreading out in all directions above as it ascends, overlapping the air in all the surrounding regions in the vicinity of the storm, and thus, by increasing the weight of the air around, cause the barometer to rise on the outside of the storm, and fall still more under the storm-cloud by the outspreading of air above, thus leaving less ponderable matter near the centre of the upmoving column to press on the barometer below. The barometer thus standing below the mean under the cloud in the central regions, and above the mean on the outside of the cloud, the air will blow on all sides from without, inward, under the cloud. The air, on coming under the cloud, being subjected to less pressure, will ascend and carry up the vapor it contains with it, and as it ascends will become colder by expansion from constantly diminishing pressure, and will begin to condense its vapor in cloud at the height indicated before, and thus the process of cloud-forming will go on. Now it is known that the upper current of air in the United States moves constantly, from a known cause, toward the eastward, probably a little to the south of east; and as the upmoving column containing the cloud is chiefly in this upper current of air, it follows that the storm-cloud must move in the same direction. And over whatever region the storm-cloud appears, to that region will the wind blow below; thus the wind must set in with a storm from some eastern direction, and, as the storm-cloud passes on toward the eastward, the wind must change to some western direction, and blow from that quarter till the end of the storm." [2]

788. *Dové's law.* According to Dové's "Law of Rotation," which is said to hold good in the northern hemisphere, and is supposed to obtain in the southern also, the wind being N.W. and veering, it ought to veer

[2] Espy, *Fourth Meteorological Report,* [pp. 11–12].

by W. to S.W., and so on, *against* the hands of a watch. This "law" is explained thus: Suppose a ship be in S. lat. off Cape Horn as at *a*, [Figure 10,] with a low barometer to the north of her, as at *C*, where the air ascends as fast as it comes pouring in from all sides. The ship, let it be supposed, is just on the verge of, but exterior to the vortex, or that place where the wind commences to revolve. The first rush of the air at *a* will be directly for the centre, *C*; consequently, a ship so placed would report the storm as commencing with the wind at south. For the sake of illustration, we will suppose this place of low barometer to be stationary, and the air, as it rushes in, to ascend at the disc, *C*. Thus the area of inrushing air will gradually enlarge itself by broad spreading, like a circle on the water, until it be compassed by a circle with a radius, *CS*, of indefinite length.

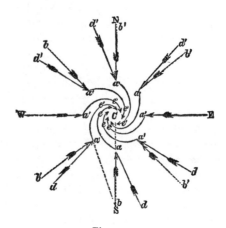

Figure 10
[Wind directions about a revolving storm in high southern latitudes.]

The air then, on the meridian, S C N, but to the south of *a*, will not blow along this meridian and pass over the ship; in consequence of the diurnal rotation of the earth, it will take a direction, S *a'*, to the westward; and the arrow *d a*, representing a S.S.E. wind, will now show the direction of the wind at *a*. Thus the ship will report that the wind commenced at south, and gradually hauled to S.S.E., *i. e.*, against the hands of a watch; and so the arrows *b' a'* will represent the direction of the wind at each station, *a' a' a'*, when the storm commenced, and the arrows *d' a'* the direction afterward, thus showing it to have veered *against* the hands of a watch. And this is the direction in which the forces of diurnal rotation, when not mastered by opposing forces, always require the wind, when not blowing round in spirals and a whirl, to haul in the southern hemisphere. Now, paradoxical as it may at first seem, it is also the forces of diurnal rotation that give that same wind, when it is blowing round in spirals, its first impulse to march round in the contrary direction, or *with the hands of a watch*; but this is as it should be — it *hauls* one way, and *marches* the other. After passing *a*, and each of the other stations, *a′ a′*, the rush of wind is sufficient, let us suppose, to create a whirl. The wind at *a′ a′ a′*, continuing on with a circular motion, is represented thenceforward in its course by the curved arrows *a e*, *a′ e′*. The wind coming from the east and the west has no direct impulse from diurnal rotation, but the wind on either side of it has, and hence the *prime vertical* wind is carried around with the rest. If, now, we imagine the disc *C* to be put in motion, and the storm to become a traveling one, we shall have to consider the composition and resolution of other forces also, such as those of traction, aberration, and the like, before we can resolve the whirlwind.

789. *Bernoulli's formula.* But the cyclonologists do not locate their storms in such high latitudes as the parallels of Cape Horn. Hence we might safely infer, one would suppose, that in high southern latitudes a north wind has a tendency to incline to the westward and a south wind to the eastward; and the cause of this tendency is in operation, whether the place of low barometer be a disc or an oblong, for it is in obedience to the trade-wind law, as expounded by Halley, that it so operates; and it will also be the case whether the wind be caused by an influx into the place of low, or the efflux from the place of high barometer; or, as is generally the case, by both together. If the distance between the place of high and low barometer were always the same, then a given difference of barometric pressure would always be followed by a wind of the same force or velocity. By expanding Bernoulli's formula for the velocity of gas jets under given pressures, Sir John F. W. Herschel has computed [3] the velocity and the force with which currents of air or winds would issue under certain differences of barometric pressure. Under the most favorable conditions, *i. e.*, when the places of high and of low barometer are in immediate juxtaposition, as on the inside and outside of an air-pump, an effective difference of 0.006 inch in the barometric pressure would create a breeze with a velocity of seven miles the hour. Such a wind is capable of exerting a horizontal pressure of 0.2 lb. the square foot [as appears from Table 13]. Changes, however, in the barometer, amounting to five or six times these differences, are observed to take place at sea without producing winds exceeding in velocity the rates above. This is because the places of high and low barometer at sea are far apart, and because, also of the obstructions to the winds afforded by the inequalities of the earth's surface.

[3] "Meteorology," *Encyclop. Britan.* [XIV, 647].

Table 13

[Velocity and force with which winds would issue under certain differences of barometric pressure]

Diff. barometric Pressure	Velocity of Wind	Horizontal Pressure	Strength of Wind
0.006 inch	7 miles per hour	0.2 lbs. per sq. foot	Gentle air
0.010 "	14 " "	0.9 " "	Light breeze
0.016 "	21 " "	1.9 " "	Good sailing breeze
0.06 "	41 " "	7.5 " "	A gale
0.14 "	61 " "	16.7 " "	Great storm
0.25 "	82 " "	30.7 " "	Tempest
0.41 "	92 " "	37.9 " "	Devastating hurricane

790. *Predicting storms.* But, in this view of the subject, the importance of a daily system of weather reports by telegraph on shore, and across the water between Europe and America when the sub-Atlantic cable is well laid, looms up and assumes all the proportions of one of the great practical questions of the age. We may conjecture, as the probable result of observation, that the greater the distance between the place of high and low barometer, the less the velocity of wind for a given barometric difference would be. Professor Buys Ballot has discovered, practically, the numerical relation between the force of the wind and given barometric differences for certain places in Holland. With the view of ascertaining like relations for this country, it has been proposed to establish a cordon of meteorological stations over the United States, each station being required to report daily to the Observatory in Washington, by telegraph, the height of the barometer, force of wind, etc. By such a plan, properly organized, we might expect soon to be able to give the ships, not only on the great lakes, but in our sea-port towns also, timely warning of many a gale, and to send by telegraph to Europe warning of many a one long before it could traverse the Atlantic. The contributions which the magnetic telegraph is capable of making for the advancement of meteorology will enable us to warn the ships in our Gulf ports, as well as those of Cuba, of the approach of every hurricane or tornado that visits those regions.

791. *The changing of the wind in a cyclone.* But, returning to the cyclone theory: though the wind be blowing around in spirals against the hands of the watch, yet, from the fact that the centre about which it is blowing is also traveling along, the changes of the wind, as observed by a vessel over which the storm is passing, will not, under all circumstances, be against the sun in the northern, or with the sun in the southern hemisphere. The reason is obvious. This point is worth studying, and any one who will resort to "*moving diagrams*" for illustration will be repaid with edification. Piddington's horn cards are the best; but let those who have

them not cut a disc of paper of any convenient diameter, say 2½ inches, and then cut out a circle of 2 inches from the middle; this will leave a ring half an inch broad upon which to draw arrows representing the course of the wind. Suppose them to be drawn for the northern hemisphere, as in [Figure 11]: lay the paper ring on the chart; suppose the ship to be in the N.E. quadrant of the storm, which is traveling north, the storm will pass to the west, but the wind will change from S.E. to S., and so on to the west, *with the hands of a watch,* though it be revolving about the centre *against the hands of a watch;* still the rule for finding the direction of the centre holds good: Face the wind, and the centre in the northern hemisphere will be to the right: in the southern, to the left.

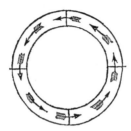

Figure 11
["Moving diagram" showing direction of wind in the four quadrants of a cyclone in the northern hemisphere.]

792. *The wind stronger on one side than the other.*   Suppose that in the case before us the storm is traveling to the north at the rate of 20 miles the hour, and that the wind is revolving around the centre also at the rate of 20 miles the hour: when the vortex bears west of the ship, the wind will be south. It is going 20 miles to the north with the body of the storm, and 20 miles around the centre; total force of the wind, 40 miles an hour on the *east* side. Now imagine yourself on the other side, that is, that you are in the northwest quadrant, and that the storm is traveling due north as before; the vortex will pass east of you, when the wind should have changed from N.E. to north, turning against the hands of a watch; but when the wind is north, it is, in the case supposed, traveling south at the rate of 20 miles an hour around the storm, while the progressive movement of the storm itself is north at the rate of 20 miles an hour. One motion exactly cancels the other, and there is, therefore, a line of calm and light, or moderate or not so heavy winds on one side of the centre, while on the other side there is a line of maximum violence; in other words, in every traveling cyclone the wind blows harder on one

side than the other. This is the case in both hemispheres; and by handling these moving diagrams for illustration, the navigator will soon become familiar with the various problems for determining the direction of the vortex, the course it is traveling, its distance, etc. Therefore, when it is optional with the navigator to pass the storm on either side, he should avoid the heavy side. These remarks apply to both hemispheres.

793. *The rainy quadrant of a cyclone.* Captain Toynbee asks if it rains more in one quarter of a cyclone than another? In cyclones that travel fast, I suppose there would be most rain in the *after* quarter; with those that have little or no progressive motion, I conjecture that the rainy quarter, if there be one, would depend upon the quarter whence that wind comes that brings most rain. The rain in a cyclone is supposed to come from the moisture of that air which has blown its round and gone up in the vortex; there it expands, grows cool, and condenses its vapor, which spreads out at top like a great mushroom in the air, the liberated heat adding fury to the storm.

794. *Erroneous theories.* Such, briefly stated, are the two theories. They appear to me, from such observation and study as I have been able to bestow, to be neither of them wholly right or altogether wrong. Both are instructive, and the suggestions of one will, in many instances, throw light upon the facts of the other. That rotary storms do frequently occur at sea we know, for vessels have sometimes, while scudding before the wind in them, sailed *round and round.* The United States brig "Perry" did this a few years ago in the West Indies; and so did the "Charles Heddle" in the East Indies: she went round and round a cyclone five times.

795. *The wind in a true cyclone blows in spirals.* From such observations as I have been able to obtain upon the subject, I am induced to believe, with Thom, that the wind in a cyclone does not blow round in a circle, but around in spirals. Nay, I go farther, and conjecture that it is only within a certain distance of the vortex that the wind gyrates, and that the gyrating column is never *hundreds of miles* in diameter, as the advocates of this theory make it: I shall allude to this again. The low barometer at the centre is owing, in part, to two causes; one is the condensation of vapor, with its liberated heat, as maintained by Espy; the other is the action of a real centrifugal force, which applies to all revolving bodies. In weighing the effect of this centrifugal force upon the low barometer, care should be taken not to give it an undue weight. It is not sufficient to cause the air to fly off in a tangent. The lateral atmospheric pressure would prevent that, if the centrifugal force were never so great; and the lower the barometer in the centre, the greater would be the pressure of the surrounding air. The proper weight, therefore, due the centrifugal force I hold to be not very great, though it is appreciable to this extent:

The storm having commenced revolving, the flow of air into the vortex is retarded, not prevented, by centrifugal tendency; and this retardation assists in causing the barometer to stand lower than it would if there were no revolution. Any one who has watched the little whirlwinds so often seen during summer and fall, or who can call to mind the whirls or "sucks" in a mill-pond, or at the lock in a canal when the water is drawn off at the bottom, may appreciate the extent to which the *centrifugal tendency* will help to make a low barometer at the centre of a cyclone.

796. *An illustration.* The low barometer, the revolving storm, and the ascending column require for a postulate the approach by spirals of the wind from circumference to centre. The wind blows toward the place of low barometer; that is admitted by all. It can only reach that place by a direct or by a curvilinear motion. If by the former, then there can be no revolution; but if there be revolution, then the air, while as wind it is revolving around the centre in the gyrations of the storm, is approaching the centre also. Hence we derive the elements of a spiral curve; and the physical necessity for spiral motion is demonstrated from the fact that there is circular motion and an uprising in the centre. This spiral movement and the uprising may be illustrated by familiar examples: The angles and corners of the Observatory, and its wings, are so arranged that at a certain place there is, with westerly winds, always a whirlwind. This whirlwind is six to eight feet in diameter; and when there is snow, there is a pile of it in the centre, with a naked path, in the shape of a ring, three or four feet broad about it. It is the spiral motion which brings the drift-snow to the centre or vortex, and the upward motion not being strong enough to carry the snow up, it is left behind, forming a sort of cone, which serves as a cast for the base of the vortex. If you throw chips or trashy matter into the lock of a canal and watch them, you will see that as they come within the influence of the "suck" they will approach the whirl by a spiral until they reach the centre, when, notwithstanding they may be lighter than the water, they will be "sucked" down. Here we see the effects of centrifugal force upon a fluid revolving within itself. The "suck" is funnel-shaped. As it goes down, the lateral pressure of the water increases; it counteracts more and more the effect of centrifugal force, and diminishes, by its increase, the size of the "suck."

797. *Dust whirlwinds and water-spouts.* So, too, with the little autumnal whirlwinds in the road and on the lawn: the dust, leaves, and trash will be swept in toward the centre at the bottom, whirl round and round, go up in the middle, and be scattered or spread out at the top. I recollect seeing one of these whirlwinds pass across the Potomac, raising from the river a regular water-spout, and, when it reached the land, it appeared as a common whirlwind, its course being marked, as usual, by a whirling column of leaves and dust. These little whirlwinds are, I take

it, the great storms of the sea in miniature; and a proper study of the miniature on land may give us an idea of the great original on the ocean.

798. A vera causa. The unequally heated plain is thought to be the cause of the one. But there are no unequally heated plains at sea; nevertheless, the *primum mobile* there is said, and rightly said, to be heat. Electricity, or some other imponderable, may be concerned in the birth of the whirlwind both ashore and afloat. But that is conjecture; the presence of heat is a fact. In the middle of the cyclone there is generally rain, or hail, or snow; and the amount of heat set free during the process of condensing the vapor for this rain, or hail, or snow, is sufficient to raise from the freezing to the boiling point more than five times the whole amount of water that falls. This vast amount of heat is set free, not at the surface of the sea, it is true, but in the cloud-region, and where the upward tendency of the indraught is still farther promoted. What sets the whirlwind a brewing is another question; but its elements being put in motion, there is a diminished barometric pressure, first, on account of *centrifugal tendency;* next, on account of the ascending column of air, which expands and ascends — ascends and expands on account of such diminished pressure — and next, though not least, on account of the heat which is set free by the condensation of the vapor which forms the clouds and makes the rain. This heat expands and pushes aside the upper air still more.

799. *Objections to the theory.* After much study, I find some difficulties about the cyclone theory that I can not overcome. They are of this sort: I can not conceive it possible to have a cyclone with a revolving and traveling disc 1000, 500, or even 100 miles in diameter, as the expounders of the theory have it. Is it possible for a disc of such an attenuated fluid as common air, having 1000 miles of diameter with its less than wafer-like thickness in comparison, to go traveling over the earth's surface and revolving about a centre with tornado violence? With the logbooks of several vessels before me that are *supposed* to have been in different parts of the same cyclone, I have a number of times attempted to project its path, but I always failed to bring out such a storm as the theory calls for. One or two vessels may do it; but is their testimony sufficient? I think not. Take as many as six or seven at sea, and their records never prove the existence of such a storm as the theory calls for. I make a distinction between the hauling of the wind in consequence of diurnal rotation of the earth, and the rotation of the wind in the cyclone in consequence of its centripetal force. For the sake of illustrating my difficulties a little farther, let us suppose a low barometer with a revolving storm to occur at A in the southern hemisphere [Figure 12]. Let the storm be traveling toward B. Let observers be at c'', d, and e, and let c'' and d be each several hundred miles from A. Now, then, will not the air at c'' and d

blow north and east as directly for the place of low barometer as it would
were that place an oblong, N A, instead of a disc, as per the arrows? And
why should it be a disc in preference to an ellipse, a square, an oblong, or
any figure, be it never so irregular? The trade-winds answer, and show
the equatorial calm belt — an oblong — as their place of meeting. But the
cyclonologists, instead of permitting the wind at the distance c'' to blow
to the east, and at d to blow to the north, merely because there is a low
barometer east of c'' and north of d, require it so to blow because, by
their *theory*, there is a low barometer east of d and south of c''! Thus, to
reach its theoretical place of destination, the wind must blow in a direction at right angles to that destination! It would require a rush of incon-
ceivable rapidity so to deflect currents of air while they are yet several

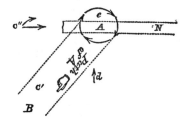

Figure 12
[Direction of wind about a storm in
southern latitudes.]

hundred miles from the centre of gyration. Moreover, the two cyclonologists, c'' and d, would differ with each other as to the centre of the storm.
The one in ship d would place it to the east of him, or in that direction, but
the other, c'', would place it to the south of him, as in the direction c'. The
gyrating disc of a cyclone can never, I apprehend, exceed a few miles in
diameter. On shore we seldom find it exceeding in breadth as many rods,
in most cases not as many fathoms, as its advocates give it miles at sea.
I think the dust-whirl in the street is a true type of the tornado (cyclone)
at sea. Imagine such a case to occur in nature as the one just supposed.
With the observations of d alone before him, the cyclonologist would say
the storm was traveling west, but passing north; with the log of c'', he
would say it travels west, but will pass south of us. By the rule, ship d
would be led toward the real track of the storm, *i. e.*, into danger, and ship
c away from it.[4]

800. *The three forces.* There are in the various parts of the storm at
least three forces at work in effecting a change of wind, as observed on
board ship at sea. (1) One is diurnal rotation: it alone can never work a

[4] See letter to Commodore Wüllerstorf, *Sailing Directions*, 8th ed., II, 457-[459].

change of direction exceeding 90°; (2) another is the varying position or traveling motion of the place of barometric depression: the change effected by it can not exceed 180°; (3) and the third is the whirling motion imparted by the rush to a common centre — the whirl of water at the floodgate of the mill, the whirlwind in the street.

801. *The effect of each.* Hence it appears that in a storm the wind may shift from any one of three causes, and we are not entitled to call it a cyclone unless the wind shift more than 180°. If the change of direction be less than 90°, the shifting may be due to diurnal rotation alone; if it be less than 180°, the shifting may, and is probably, due to (1) and (2). The sailor has therefore no proof to show that he has been in a cyclone unless the wind during the storm changed its direction more than 180°. Cyclones, there is reason to believe, are often whirlwinds in a storm. This may be illustrated by referring again to our miniature whirlwinds on the land; there we often see a number of them at one time and about the same place; and they often appear to skip, raging here, then disappearing for a moment, then touching the ground again, and pursuing the former direction.

802. *A storm within a storm.* Observations have proved that this is the case on land, and observations have not established that this is *not* the case at sea; observations are wanting upon this subject. Tornadoes on the land often divide themselves, sending out branches, as it were. It remains to be seen whether cyclones do not do the same at sea; and whether, in those widespread and devastating storms that now and then sweep over the ocean, there be only one vortex or several; and if only one, whether the whole storm partake of the cyclone character. In other words, may there not be at storm within a storm — that is, a cyclone traveling with the storm and revolving in it? I ask the question because the theory does not satisfy all the facts observed.

803. *The Black Sea storm of 1854.* The celebrated Black Sea storm of 1854, which did so much damage to the allied fleet, is still maintained by some to be a true cyclone; and by the observations of some of the vessels a cyclone may be made out. But if we take the observations of all of them, and discuss them upon the supposition that the whole storm was a cyclone, it will puzzle any one to make any thing of them. Admiral Fitzroy, in the Meteorological Papers of the Board of Trade, published diagrams of the winds as observed during that storm on board of various vessels in various parts of that sea. I have not been able to reconcile them with the cyclone theory. Espy maintains that they confirm his theory; and his (§ 787) is anti-cyclonic.

804. *Cyclones of the North Atlantic.* The cyclones of the North Atlantic take their rise generally (§ 785) somewhere between the parallels of 10° and 20° north. They take a westerly course until they fall in with

the Gulf Stream, when they turn about and run along upon it until their force is expended. The atmosphere over the Gulf Stream is generally well charged with moisture, and in this fact perhaps will be found the reason why the path of the storm is laid along the Gulf Stream.

805. *The hurricane season.* Table 14 is from Birt's [5] Handbook of Storms.

Table 14

Average Number of Cyclones or Hurricanes which have occurred in different Months of the Year, and in various Regions

Locality	Jan.	Feb.	Mar.	April	May	June	July	Aug.	Sep.	Oct.	Nov.	Dec.	Total
West Indies	—	1	2	—	—	4	15	36	25	27	1	2	113
S. Indian Ocean	9	13	10	8	4	—	—	—	1	1	4	3	53
Mauritius	9	15	15	8	—	—	—	—	—	—	—	6	53
Bay of Bengal	1	—	1	1	7	3	—	1	—	7	6	3	30
China Sea	—	—	—	—	—	2	5	5	18	10	6	—	46

806. *Cyclones in the Mississippi Valley.* The vortex of a cyclone is often and aptly compared to a meteor. I have often observed the paths of such through the forests of the Mississippi Valley, and the path has in no instance that has fallen under my observation been more than a few hundred yards broad. There the track of these tornadoes is called a "wind-road," because they make an avenue through the wood straight along, and as clear of trees as if the old denizens of the forest had been felled with an axe. I have seen trees three or four feet in diameter torn up by the roots, and the top, with its limbs, lying next the hole whence the roots came. Nevertheless, the passage of the meteor, whose narrow path was marked by devastation, would create a great commotion in the air, and there would be high winds raging for several miles on either side of the "wind-road." But (§ 799) let us consider for a moment the effect of the diurnal rotation of the earth upon one of these revolving discs 1000 miles in diameter: its height would scarcely be two miles, and its thickness would not be as great, in proportion to its diameter, as half the thickness of this leaf is to the length of an inch. Now the difference in rate of the diurnal rotation between the northern and southern limbs of the disc would be sufficient, irrespective of any other power, to break it up. Suppose its southern limb to be in 20° N., its northern limb would be 1000 miles, say 17°, farther north, that is, in 37°. Diurnal rotation would carry to the east the air in the southern limb at the rate of 845 miles an hour; but when this same air comes round on the northern limb, diurnal rotation would carry it eastward at the rate only of 720 miles. *Because* the wind hauls in a particular way, it does not, as by diagram (§ 799) it has been

[5] [Table 14 is not in William Radcliff Birt, *Handbook of the Law of Storms* . . . (Liverpool, 1853)].

shown, follow that it is blowing in a circle, or that the centre of the storm is at right angles to its line of direction.

807. *Extra-tropical gales.* In the extra-tropical regions of each hemisphere furious gales of wind also occur. One of these, remarkable for its violent effects, was encountered on the 24th of December, 1853, about three hundred miles from Sandy Hook, latitude 39° north, longitude 70° west, by the "San Francisco," steam-ship. That ship was made a complete wreck in a few moments, and she was abandoned by the survivors, after incredible hardships, exertions, and sufferings. Some months after this disaster I received by the California mail the abstract log of the fine clipper ship "Eagle Wing" (Ebenezer H. Linnell), from Boston to San Francisco. She encountered the ill-fated steamer's gale, and thus describes it: "*December 24th.* Latitude 39° 15' north, longitude 62° 32' west. First part threatening weather; shortened sail; at 4 P.M. close-reefed the top-sails and furled the courses. At 8 P.M. took in fore and mizzen top-sails; hove to under close-reefed main top-sail and spencer, the ship lying with her lee rail under water, nearly on her beam-ends. At 1 30 A.M. the fore and main top-gallant-masts went over the side, it blowing a perfect hurricane. At 8 A.M., moderated; a sea took away jib-boom and bowsprit cap. In my thirty-one years' experience at sea, I have never seen a typhoon or hurricane so severe. Lost two men overboard — saved one. Stove skylight, broke my barometer, etc., etc." Severe gales in this part of the Atlantic — *i. e.*, on the polar side of the calm belt of Cancer — rarely occur during the months of June, July, August, and September. This appears to be the time when the fiends of the storm are most busily at work in the West Indies. During the remainder of the year, these extra-tropical gales, for the most part, come from the northwest. But the winter is the most famous season for these gales. That is the time when the Gulf Stream has brought the heat of summer and placed it (§ 172) in closest proximity to the extremest cold of the north. And there would, therefore, it would seem, be a conflict between these extremes; consequently, great disturbances in the air, and a violent rush from the cold to the warm. In like manner, the gales that most prevail in the extra-tropics of the southern hemisphere come from the pole and the west, *i. e.*, southwest.

808. *Storm and Rain Charts.* Storm and Rain Charts for the Atlantic Ocean have already been published by the Observatory, and others for the other oceans are in process of construction. The object of such charts is to show the directions and relative frequency of calms, fogs, rain, thunder, and lightning. These charts are very instructive. They show that that half of the atmospherical coating of the earth which covers the northern hemisphere — if we may take as a type of the whole what occurs on either side of the equator in the Atlantic Ocean — is in a much less stable condition than that which covers the southern.

CHAPTER XX

§ 811–841. THE WINDS OF THE SOUTHERN HEMISPHERE

811. *Repetition often necessary.* A work of this sort, which is progressive, must necessarily bear with it more or less of repetition. It embodies the results of the most extensive system of philosophical observations, physical investigation, and friendly co-operation that has ever been set on foot. As facts are developed, theories are invented or expanded to reconcile them. As soon as this is done, or in a short time thereafter, some one or more of the fleets that are out reconnoitring the seas for us returns with additional facts for our store-house of knowledge. Whether these tend to confirm or disprove the theory, a restatement is often called for; hence the repetition, of which the case before us is an example.

812. *The S.E. and N.E. trade-winds put in a balance.* The facts stated in Chap. XV go to show that the southeast trade-winds are stronger than the northeast. The barometer tells us (§ 643) that between the parallels of 5° and 20° the southeast trade-winds bear a superincumbent pressure upon the square foot of nearly 4 pounds greater than that to which the northeast trades are subjected. Such an excess of superincumbent pressure upon a fluid so elastic and subtle as air ought to force the southeast trade-winds from under it more rapidly than the lighter pressure forces the northeast. Observations showing that such is or is not the case should not be ignored.

813. *Observations by 2235 vessels.* I have evidence from all the vessels in a fleet numbering no less than 2235 sail to show that the southeast are stronger than the northeast trades. Every one of these vessels passed through both systems of trade-winds. The knots run per hour by each one of them, as they passed through the southeast trades of the Indian Ocean and through both systems of the Atlantic, have been discussed from crossing to crossing. The average result in knots per hour is expressed in Table 15. The comparison is confined to the rate of sailing between the parallels of 10° and 25°, because this is the belt of steadiest trades.

814. *Ships used as anemometers.* It is well to observe, that on each of these three oceans, though the direction of the wind is the same, the course steered by each fleet is different; consequently, these *anemometers* are at different angles with the wind, through the southeast trades the wind is nearly aft in the Atlantic, and quartering in the Indian Ocean, giving an average sailing speed of 7 knots an hour in the latter, and of 6

Table 15

Average Speed through the Trade-winds of the North Atlantic and South Indian Oceans

Month	Knots per Hour from: 10° to 15° N.E. Trades	10° to 15° S.E. Trades	15° to 20° N.E. Trades	15° to 20° S.E. Trades	20° to 25° N.E. Trades	20° to 25° S.E. Trades	Average N.E. Trades	Average S.E. Trades	Course steered through: N.E. Trades (N. Atlantic)	S.E. Trades (S. I. Ocean)
January	8	6½	7½	7	6	6	7	6½	N. 49° W.	S. 69° W.
February	7½	6	6	6¾	5	6	6	6	N. 46 W.	"
March	8	7	7	7	5¼	6½	6¾	6¾	N. 47 W.	"
April	7¼	7	5¾	7¾	4	6	6	6¾	N. 48 W.	S. 70° W.
May	8¼	8	6¾	7½	6½	6	7	7	N. 46 W.	"
June	9	7½	8	7¾	5¼	7	7½	7½	N. 43 W.	"
July	5½	8	8	8¼	6¼	7	6½	7¾	N. 46 W.	"
August	4¾	7¾	6	8¼	4¾	7½	5	7¾	N. 40 W.	S. 69° W.
September	5½	8½	6	8	4	6¾	5	7¾	N. 50 W.	"
October	7½	8¼	6¾	8	6	5¾	6¾	7¼	N. 45 W.	"
November	6	8	6	7	4½	5¼	5½	6¾	N. 49 W.	"
December	6	6½	6¾	6¾	5½	5	6	6	N. 48 W.	"
Means	7	7¾	6¾	7½	5¼	6¼	6¼	7	N. 47 W.	S. 69¼° W.

Average course steered through the N.E. trades of the N. Atlantic Ocean, N.W. ¼ W.
" " " " S.E. trades of the S. Indian Ocean, W.S.W.

Average Speed through the Trade-winds of the North and South Atlantic Oceans

Month	Knots per Hour from: 25° to 20° S.E. Trades	25° to 20° N.E. Trades	20° to 15° S.E. Trades	20° to 15° N.E. Trades	15° to 10° S.E. Trades	15° to 10° N.E. Trades	Average S.E. Trades	Average N.E. Trades	Course steered through: S.E. Trades (S. Atlantic)	N.E. Trades (N. Atlantic)
January	6	5	5¾	5½	5¾	7½	6	6	N. 54° W.	S. 21° E.
February	4½	5	5½	6½	6	7	5¼	6	N. 55 W.	S. 25 E.
March	6½	4¾	6	6½	6	7½	6	6¼	N. 55 W.	S. 22½ E.
April	5½	5½	5½	6½	6¾	7¾	6	6½	N. 53 W.	S. 22½ E.
May	5¾	5	5¾	6½	6½	7	6	6	N. 55 W.	S. 24¾ E.
June	5¾	6	6	6	7	5	6¼	5¾	N. 55 W.	S. 28 E.
July	5¼	7	6	7½	6¼	4½	6	6	N. 55 W.	S. 27 E.
August	5	5½	5¾	6½	6	4	5¾	5¼	N. 55 W.	S. 28½ E.
September	5½	3¼	6¼	5½	7	5	6¼	4½	N. 55 W.	S. 22½ E.
October	5¼	4¼	6	4½	5¼	4¾	5½	4½	N. 56 W.	S. 20 E.
November	6½	4	6½	5¼	6¼	5¾	6½	5	N. 55 W.	S. 22½ E.
December	5½	4¾	5¾	6	6¼	7¾	5¾	5	N. 55 W.	S. 23 E.
Means	5½	5	5¾	6	6¼	6	6	5¾	N. 55 W.	S. 24° E.

Average course steered through the S.E. trades of the S. Atlantic, N.W. by W.
" " " " N.E. trades of the N. Atlantic, S.S.E.

in the former; while through the northeast trades the average speed is 6¼ knots an hour one way (N.W. ¼ W.), with the wind just abaft the beam, and 5¾ the other (S.S.E.), with the wind at a point not so favorable for speed. Indeed, most of the ships which average a S.S.E. course through this part of the northeast trade-wind belt are close hauled; therefore the average strength of the trades here can not be fairly compared with the average strength where the fleet have free winds. What is the difference in the strength of such winds, which, impinging upon the sails, each at the particular angle indicated above, imparts the aforesaid velocities? Moderate winds, such as these are, give a ship her highest speed generally when they are just abaft the beam, as they are for a northwest course through the northeast trades of the N. Atlantic. So, to treat these ships as *anemometers* which will really enable us to measure the comparative strength of the winds, we should reduce the average knots per hour to the average speed of a mean ship sailing through average "trades" in each ocean, with the wind impinging upon her sails at the same angle for all three, as, for example, just abaft the beam, as in the North Atlantic.

815. *Velocity of the trade-winds.* Let us apply to the average speed through the South Atlantic and Indian Oceans such a correction. Through the former the wind is aft; through the latter, quartering. If we allow two knots as a correction for the one, and one knot for the other, we shall not be greatly out. Applying this correction, we may state the speed of a mean ship sailing with average trades just abaft the beam to be as follows:

Through the N.E. of the Atlantic	6¼ knots per hour [1]
" " S.E. " "	8 " " "
" " S.E. " Indian Ocean	8 " " "

I do not take into this comparison the force of the N.E. trades on a S.S.E. course (§ 813), because the winds along this route are known not to be as steady as they are farther away from the African coast. Thus it is clearly established that the S.E. trades are stronger than the N.E., *and so they should be if there be a crossing of winds in the calm belt of Capricorn.*

816. *Ditto of the counter-trades.* The counter-trades of the southern hemisphere move, as before stated, toward their pole more steadily and briskly than do the counter-trades of the northern hemisphere. To give an idea of the difference of the strength of these two winds, I cite the fact that vessels sailing through the latter, as from New York to England, average 150 miles a day. Along the corresponding latitudes through the former, as on a voyage to Australia, the average speed is upward of 200 miles a day. Consequently, the counter-trades of the southern hemisphere transport in given times larger volumes of air toward the south than our counter-trades do toward the north. This air returns to the tropical calm

---

[1] The correctness of this estimate is sustained by experiments which Admiral Chabannes, in command of the French fleet on the coast of Brazil, has been so kind as to institute. See note [§ 642].

belts as an upper current. If, descending there, it feeds the trade-winds, then, the supply being more abundant for the S.E. trades than for the N.E., the S.E. trades must be the stronger; and so they are; observations prove them so to be. Thus the crossing of the air at the *tropical* calm belts, though it may not be proved, yet it is shown to be so very probable that the onus of proof is shifted. It now rests with those who dispute the crossing to prove their theory the true one.

817. *The waves they get up.* Arrived at this point, another view in the field of conjecture is presented, which it is proper we should pause to consider. The movements of the atmosphere on the polar side of 40° N. are, let it be repeated, by no means so constant from the west, nor is the strength of the westerly winds there nearly so great on the average as it is in the extra-tropical regions of the south. This fact is well known among mariners. Every one who has sailed in that southern girdle of waters which belt the earth on the polar side of 40°, has been struck with the force and trade-like regularity of the westerly winds which prevail there. The waves driven before these winds assume in their regularity of form, in the magnitude of their proportions, and in the stateliness of their march, an aspect of majestic grandeur that the billows of the sea never attain elsewhere. No such waves are found in the trade-winds; for, though the S.E. trades are quite as constant, yet they have not the force to pile the water in such heaps, nor to arrange the waves so orderly, nor to drive them so rapidly as those "brave" winds do. There the billows, chasing each other in stately march, look, with their rounded crests and deep hollows, more like mountains rolling over a plain than the waves which we are accustomed to see. Many days of constant blowing over a wide expanse of ocean are required to get up such waves. It is these winds and waves which, on the voyage to and from Australia, have enabled the modern clipper-ship to attain a speed, and, day after day, to accomplish runs which at first were considered, even by the nautical world, as fabulous, and are yet regarded by all with wonder and admiration.

818. *A meteorological corollary.* Seeing, therefore, that we can bring in such a variety of facts and circumstances, all tending to show that the southeast trade-winds are stronger than the northeast, and that the westerly winds which prevail on the polar side of 40° S. are stronger and more constant than their antœcian fellows of the north, we may consider it as a fact established, independently of Plate IX, that the general system of atmospherical circulation is more active in the southern than it is in the northern hemisphere. And, seeing that it blows with more strength and regularity from the west in the extra-tropical regions of the southern than it does in the extra-tropical regions of the northern hemisphere, we should deduce, by way of corollary, that the counter-trades of the south are not so easily arrested in their course, or turned back in their circuits, as

are those of the north. Consequently, moreover, we should not, either in the trades or the counter-trades of the southern hemisphere, look for as many calms as in those of the northern systems.

819. *Facts established.* Therefore, holding to this corollary, we may consider the following as established facts in meteorology: That the S.E. trade-winds are stronger than the N.E.; that the N.W. passage-winds — the counter-trades of the south — are stronger and less liable to interruption in their circuits than the S.W., the counter-trades of the north; that the atmospherical circulation is more regular and brisk in the southern than it is in the northern hemisphere; and, to repeat: since the wind moves in its circuits more briskly through the southern than it does through the northern hemisphere, it consequently has less time to tarry or dally by the way in the south than in the north; hence the corollary just stated. But observations also, as well as mathematically-drawn inferences, show that calms are much less prevalent in the southern hemisphere. For this inference observations are ample; they are grouped together by thousands and tens of thousands, both on the Pilot and the Storm and Rain Charts. These charts have not been completed for all parts of the ocean, but as far as they have been constructed the facts they utter are in perfect agreement with the terms of this corollary.

820. *Atmospherical circulation more active in the southern than in the northern hemisphere.* These premises being admitted, we may ascend another round on this ladder, and argue that, since the atmosphere moves more briskly and in more constant streams through its general channels of circulation in the southern than it does through them in the northern hemisphere; and that, since it is not arrested in its courses by calms as often in the former as it is in the latter, neither should it be turned back by the way, so as to blow in gales from the direction opposite to that in which the general circulation carries it. The atmosphere, in its movements along its regular channels of circulation, may be likened, that in the southern hemisphere to a fast railway train; that of the northern to a slow. The slow train may, when "steam is up," run as fast as the fast train, but it is not obliged to get through so quick; therefore it may dally by the way, stop, run back, and still be through in time. Not so the fast; it has not time to stop often or to run back far; neither have the counter-trades of the south time to blow backward; consequently, such being the conditions, we should also expect to find in the extra-tropical south a gale with easting in it much more seldom than in the extra-tropical north.

821. *Gales in the two hemispheres.* We shall appeal to observations for the correctness of this conjecture, and claim for it, also, as presently will appear, the dignity of an established truth [Table 16]. Thus the Storm and Rain Charts show that between the parallels of 40° and 55° there were in the northern hemisphere 33,515 observations, and that for every

Table 16

Average Number of Gales (to the 1000 Observations) with Easting and with Westing in them between the corresponding Parallels in the North and South Atlantic, as shown by the Storm and Rain Charts

		North	South
Between 40° and 45°	Number of observations	17,274	8756
	Gales in 1000 do. with easting	23	12
	"　　　"　do.　" westing	66	82
Between 45° and 50°	Number of observations	11,425	5548
	Gales in 1000 do. with easting	24	1
	"　　　"　do.　" westing	106	61
Between 50° and 55°	Number of observations	4,816	5169
	Gales in 1000 do. with easting	24	10
	"　　　"　do.　" westing	144	97

1000 observations there were 24 gales with easting and 105 with westing. In the southern, there were 19,473 observations, and for every 1000 of these there were 5 gales with easting and 80 with westing in them. Those for the southern hemisphere are only for that part of the ocean through which vessels pass on their way to and fro around Cape Horn. That part of this route which lies between 40° and 55° S. is under the lee of South America; and Patagonia, that lies east of the Andes, is almost a rainless region; consequently, we might expect to find more unsteady winds and fewer rains in that part of the ocean where the observations for the southern part of the tables were made than we should expect to meet with well out to sea, as at the distance of two or three thousand miles to the eastward of Patagonia. So that the contrast presented by the above statement would probably be much greater did our observations extend entirely across the South, as they do across the North Atlantic. But as it is, the contrast is very striking. In some aspects, the meteorological agents of the two hemispheres, especially those forces which control the winds and the weather, differ very much. The difference is so wide as to suggest greater regularity and rapidity of circulation on one side of the equator than on the other.

822. *Calms in the two hemispheres.* Each one of the observations [summarized in Table 17] embraces a period of eight hours; the grand total, if arranged consecutively, with the observations drawn out each to occupy its period separately, would be equal to 373 years. They exhibit several curious and suggestive facts concerning the difference of the atmospherical stability in the two hemispheres.

823. *The propelling power of the winds.* If we would discover the seat of those forces which produce this difference in the dynamical status of the two great aerial oceans that envelop our planet, we should search for them in the unequal distribution of land and water over the two

Table 17

Average Number of Calms to the 1000 *Observations between the Parallels of* 30° *and 55° in the North and South Atlantic, and between the Parallels of* 30° *and* 60° *in the North and South Pacific Oceans, as shown by the Pilot Charts*

	Atlantic		Pacific	
Between the Parallels of	North	South	North	South
30° and 35°, No. of observations	12,935	15,842	22,730	44,846
Calms to the 1000 do.	46	26	34	35
35° and 40°, No. of observations	22,136	23,439	13,939	66,275
Calms to the 1000 do.	37	24	31	23
40° and 45°, No. of observations	16,363	8,203	12,400	31,889
Calms to the 1000 do.	45	27	53	23
45° and 50°, No. of observations	8,907	4,183	15,897	4,940
Calms to the 1000 do.	38	25	35	21
50° and 55°, No. of observations	3,519	3,660	32,804	9,728
Calms to the 1000 do.	40	16	32	17
55° and 60°, No. of observations			15,470	9,111
Calms to the 100 do.			43	21
Total No. of observations	63,050	55,327	113,240	166,829
Average calms to the 1000 do.	41	24	39	25

hemispheres. In one the wind is interrupted in its circuits by the continental masses, with their wooded plains, their snowy mantles in winter, their sandy deserts in summer, and their mountain ranges always. In the other there is but little land and less snow. On the polar side of 40° S. especially, if we except the small remnant of this continent that protrudes beyond that parallel in the direction of Cape Horn, there is scarcely an island. All is sea. There the air is never dry; it is always in contact with a vapor-giving surface; consequently, the winds there are loaded with moisture, which, with every change of temperature, is either increased by farther evaporation or diminished by temporary condensation. *The propelling power of the winds in the southern hemisphere resides chiefly in the latent heat of the vapor which they suck up from the engirdling sea on the polar side of Capricorn.*

824. *Lt. Van Gogh's Storm and Rain Charts.* The Storm and Rain Charts show that within the trade-wind regions of both hemispheres the calm and rain curves are symmetrical; that in the extra-tropical regions the symmetry is between the calm and fog curves; and also, especially in the southern hemisphere, between the gale and rain curves. Lieutenant Van Gogh, of the Dutch Navy, in an interesting paper on the connection between storms near the Cape of Good Hope and the temperature of the sea,[2] presents a storm and rain chart for that region. It is founded on 17,810 observations, made by 500 ships, upon wind

[2] *De stormen nabij de Kaap de Goede Hoop in verband beschouwd met de temperatuur der zee.*

and weather, between 14° and 32° E., and 33° and 37° S. By that chart
the gale and rain curves are so symmetrical that the phenomena of rains
and gales in the extra-tropical seas present themselves suggestively as
cause and effect. The general storm and rain charts of the Atlantic Ocean,
prepared at the National Observatory, hold out the same idea. Let us
examine, expand, and explain this fact.

825. *The "brave west winds" caused by rarefaction in the antarctic
regions.* We ascribe the trade-winds to the equatorial calm-belt. But to
what shall we ascribe the counter-trades, particularly of the southern
hemisphere, which blow with as much regularity toward the pole as the
northeast trades of the Atlantic do toward the equator? Shall we say that
those winds are drawn toward the south pole by *heat*, which causes them
to expand and ascend in the antarctic regions? It sounds somewhat paradoxical to say that heat causes the winds to blow toward the poles as
well as toward the equator; but, after a little explanation, and the passing in review of a few facts and circumstances, perhaps the paradox may
disappear. It is held as an established fact by meteorologists that the
average amount of precipitation is greater in the northern than in the
southern hemisphere; but this, I imagine, applies rather to the land than
the sea. On the polar side of 40° it is mostly water in the southern, mostly
land in the northern hemisphere. It is only now and then, and on rare
occasions, that ships carry rain-gauges to sea. We can determine by
quantitive measurements the difference in amount of precipitation on the
land of the two hemispheres, and it is the result of this determination, I
imagine, that has given rise to the general remark that the rain-fall is
greater for the northern than it is for the southern hemisphere. But we
have few hyetographic measurements for quantity at sea; there the de-
terminations are mostly numerical. Our observers report the "times" of
precipitation, which, whether it be in the form of rain, hail, or snow, is
called by the charts, and in this discussion, *rain.* Among such a large
corps of observers, rain is sometimes, no doubt, omitted in the log; so
that, in all probability, the charts do not show as many "times" with rain
as there are "times" *actually* with rain at sea. This omission, however, is
as likely to occur in one hemisphere as in the other. Still, we may safely
assume that it rains oftener in all parts of the sea than our observations,
or the rain charts that are founded on them, indicate.

826. *Relative frequency of rains and gales at sea.* With the view of
comparing the rains at sea between the parallels of 55° and 60°, both in
the North and South Atlantic, we have taken from the charts the following figures:

        South — Observations, 8410; gales, 1228; rains, 1105
        North —   "   526;   "   135;   "   64
        Gales to the 1000 observations   S. 146;   N. 256
        Rains   "   "   S. 131;   N. 121

That is, for every 10 gales, there are in the southern hemisphere 9 rains, and in the northern 4.7. In which hemisphere does most water fall on the average during a rain at sea? Observations do not tell, but there seems to be a philosophical reason why it should rain not only oftener, but more copiously at sea, especially in the extra-tropical regions, in the southern hemisphere than in those of the northern. On the polar side of 40° N., for example, the land is stretched out in continental masses, upon the thirsty bosom of which, when the air drops down its load of moisture, only a portion of it can be taken up again; the rest is absorbed by the earth to feed the springs. On the polar side of 40° S. we have a water instead of a land surface, and as fast as precipitation takes place there, the ocean replenishes the air with moisture again. It may consequently be assumed that a high dew-point — at least one as high as the ocean can maintain in contact with winds blowing over it, and going from warmer to cooler latitudes all the time — is the normal condition of the air on the polar side of 40° S., whereas on the polar side of 40° N. a low dew-point prevails. The rivers to the north of 40° could not, I reckon, if they were all converted into steam, supply vapor enough to make up this average difference of dew-point between the two hemispheres. The symmetry of the rain and storm curves on the polar side of 40° S. suggests that it is the condensation of this vapor which, with the liberation of its latent heat, gives such activity and regularity to the circulation of the atmosphere in the other hemisphere.

827. *The rain-fall of Cape Horn and Cherraponjie.* On the polar side of 40° S., near Cape Horn, the gauge of Captains King and Fitzroy showed a rain-fall of 153.75 inches in 41 days. There is no other place except Cherraponjie where the precipitation approaches this in amount. Cherraponjie (§ 299) is a mountain station in India, 4500 feet high, which, in lat. 25° N., acts as a condenser for the monsoons fresh from the sea. But on the polar side of lat. 45°, in the northern hemisphere, it is, except along the American shores of the North Pacific, a physical impossibility that there should be a region of such precipitation as King and Fitzroy found on the western slopes of Patagonia — a physical impossibility, because the peculiar combination of conditions required to produce a Patagonian rain-fall are wanting on the polar side of 45° N. There is not in the North Atlantic water surface enough to afford vapor for such an amount of precipitation. In the North Pacific the water surface may be broad and ample enough to afford the vapor, but in neither of these two northern sheets of water are the winds continuous enough from the westward to bring in the requisite quantities of vapor from the sea. Moreover, if the westerly winds of the extra-tropical north were as steady and as strong as are those of the south, there is lacking in the north that continental relief — mountain ranges rising abruptly out of the

sea, or separated from it only by lowlands — that seems to be necessary to bring down the rain in such floods. Colonel Sykes [3] quotes the rain-fall of Cherraponjie at 605.25 inches for the 214 days from April to October, the season of the southwest monsoons. Computing the Cape Horn rains according to the ratio given by King and Fitzroy for their 41 days of observation, we should have a rain-fall in Patagonia of 825 inches in 214 days, or a yearly amount of 1368.7 inches. Neither the Cape Horn rains, nor the rains any where at sea on the polar side of 45° S., are periodical. They are continuous; more copious, perhaps, at some seasons than at others, but abundant at all.

828. *Influence of highlands upon precipitation.* Now, considering the extent of water surface on the polar side of the southeast trade-wind belt, we see no reason why, on these parallels, the engirdling air of that great watery zone of the south should not, entirely around the earth, be as heavily charged with vapor as was that which dropped this flood upon the Patagonian hills. If those mountains had not been there, the conden-sation and the consequent precipitation would probably not have been as great, because the *conditions at sea* are less apt to produce rain; but the quantity of vapor in the air would have been none the less, which vapor was being borne in the channels of circulation toward the antarctic re-gions for condensation and the liberation of its latent heat; and we make, as we shall proceed to show, no violent supposition if, in attempting to explain this activity of circulation south of the equator, we suppose a cloud region, a combination of conditions in the antarctic circle pecul-iarly favorable to heavy and almost incessant precipitation. But, before describing these conditions, let us turn aside to inquire how far precipi-tation in the supposed cloud region of the south may assist in giving force and regularity to the winds of the southern hemisphere.

829. *The latent heat of vapor.* If we take a measure, as a cubic foot, of ice at zero, and apply heat to it by means of a steady flame that will give off heat at a uniform rate, and in such quantities that just enough heat may be imparted to the ice to raise its temperature 1° a minute, we shall find that at the end of 32 minutes the ice will be at 32°. The ice will now begin to melt; but it and its water — the heat being continued — will remain at 32° for 140 minutes, when all the ice will have become water at 32°.[4] This 140° of heat, which is enough to raise the temperature of 140 cubic feet of ice one degree from any point below 32°, has been ren-dered latent in the process of liquefaction. Freeze this water again, and this latent heat will become sensible heat, for heat no more than ponder-

---

[3] [W. H. Sykes, "Mean Temperature of the Day and Monthly Fall of Rain at 127 Stations under the Bengal Presidency . . .,"] *Rept.* [22nd *Annual Meeting*], *Brit. Assoc. Adv. Sci.*, 1852 [pp. 252–261, esp. pp. 256–257].

[4] [James P.] Espy, [*The*] *Philosophy of Storms* [(Boston, 1841), p. viii].

able matter can be annihilated. But if, after the cubic foot of ice has been converted into water at 32°, we continue the uniform supply of heat as before and at the same rate, the water will, at the expiration of 180 minutes more, reach the temperature of 212° — the boiling point — and at this temperature it will remain for 1030 minutes, notwithstanding the continuous supply of heat during the interval. At the expiration of this 1030 minutes of boiling heat, the last drop of water will have been converted into steam; but the temperature of the steam will be that only of the boiling water; thus, in the evaporation of every measure of water, heat enough is rendered latent during the process to raise the temperature of 1030 such measures one degree. If this vapor be now condensed, this latent heat will be set free and become sensible heat again. Hence we perceive that every rain-drop that falls from the sky has, in its process of condensation, evolved heat enough to raise one degree the temperature of 1030 raindrops. But if, instead of the liquid state, as rain, it come down in the solid state, as hail or snow, then the heat of fluidity, amounting to enough to raise the temperature of 140 additional drops one degree, is also set free.

830. *The cause of the boisterous weather off Cape Horn.* We have in this fact a clew to the violent wind which usually accompanies hailstorms. In the hail-storm congelation takes place immediately after condensation, and so quickly that the heat evolved during the two processes may be considered as of one evolution. Consequently, the upper air has its temperature raised much higher than could be done by the condensing only. So also the storms which have made Cape Horn famous are no doubt owing, in a great measure, to this heavy Patagonian rain-fall. The latent heat which is liberated by the vapor as it is condensed into rain there, has the effect of producing a great intumescence in the air of the upper regions round about them, which in turn produces commotion in the air below. But this is digressive. Therefore let us take up the broken thread, and suppose, merely for illustration, such a rain-fall as King and Fitzroy encountered in Patagonia to have taken place under the supposed cloud region of the antarctic circle, and to have been hail or snow instead of rain, then the total amount of caloric set free among the clouds, in those 41 days of such a flood, would be enough to raise from freezing to boiling six and a half times as much water as fell. But if the supposed antarctic precipitation come down in the shape of rain, then the heat set free would be sufficient only to raise from freezing to boiling about 5¾ as much water as the flood brought down. We shall have, perhaps, a better idea of the amount of heat that would be set free, in the condensation and congelation in the antarctic regions of as much vapor as it took to make the Patagonian rain-fall, if we vary the illustration by supposing this rain-fall of 153.75 inches to extend over an area of 1000

square miles, and that it fell as snow or hail. The latent heat set free among the clouds during these 41 days would have been sufficient to raise from the freezing to the boiling point all the water in a lake 1000 square miles in area and 83¼ feet in depth. The unknown area of the antarctic is eight millions of square miles. We now see how the cold of the poles, by facilitating precipitation, is made to react and develop heat, to expand the air, and give force to the winds.

831. *Offices of icebergs in the meteorological machinery.* Thus we obtain another point of view from which we may contemplate, in a new aspect, the icebergs which the antarctic region send forth in such masses and numbers. They are a part of the meteorological machinery of our planet. The offices which they perform as such are most important, and oh, how exquisite! While they are in the process of congelation the heat of fluidity is set free, which, whether it be liberated by the freezing of water at the surface of the earth, or of the rain-drop in the sky, helps in either case to give activity and energy to the southern system of circulation by warming and expanding the air at its place of ascent. Thus the water, which, by parting with its heat of liquefaction, has expended its meteorological energy in giving dynamical force to the air, is like the exhausted steam of the engine; it has exerted its power and become inert. It is, therefore, to be got out of the way. In the grand meteorological engine which drives the wind through his circuits, and tempers it to beast, bird, and plant, this waste water is collected into antarctic icebergs, and borne away by the currents to more genial climes, where the latent heat of fluidity which they dispensed to the air in the frigid zone is restored, and where they are again resolved into water, which, approaching the torrid zone in cooling streams, again joins in the work and helps to cool the air of the trade-winds, to mitigate climate, and moderate the gale. For, if the water of southern seas were warmer, evaporation would be greater; then the S.E. trade-winds would deliver vapor more abundantly to the equatorial calm belts: this would make precipitation there more copious, and the additional quantity of heat set free would give additional velocity to the inrushing trade-winds. Thus it is, as has already been stated, that, parallel for parallel, trans-equatorial seas are cooler than cis-equatorial; thus it is that icebergs are employed to push forward the winds in the polar regions, to hold them back in the equatorial; and thus it is that, in contemplating the machinery of the air, we perceive how icebergs are "coupled on," and made to perform the work of a regulator, with adjustments the most beautiful, and compensations the most exquisite, in the grand machinery of the atmosphere.

832. *The antarctic calm place a region of constant precipitation.* With this illustration concerning the dynamical force which the winds derive

from the vapor taken up in one climate and transported to another, we may proceed to sketch those physical features which, being found in the antarctic circle, would be most favorable to heavy and constant precipitation, and, consequently, to the development of a system of aerial circulation peculiarly active, vigorous, and regular for the *aqueous* hemisphere, as the southern in contrast with the northern one may be called. These vapor-bearing winds which brought the rains to Patagonia are — I wish to keep this fact in the reader's mind — the counter-trades of the southern hemisphere. As such they have to perform their round in the grand system of aerial circulation, and as, in every system of aerial circulation there must be some point or place at which motion ceases to be direct and commences to be retrograde, so there must be a place somewhere on the surface of our planet where these winds cease to go forward, stop, and commence their return to the north; *and that place is, in all probability, within the antarctic regions.* Its precise locality has not been determined, but I suppose it to be a band or disc — an area — within the polar circle, which, could it be explored, would be found, like the equatorial calm belt, a place of light airs and calms, of ascending columns of air, a region of clouds, of variable winds, and constant precipitation.

833. *Also of a low barometer.* But, be that as it may, the air which these vapor-bearing winds — vapor-bearing because they blow over such an immense tract of ocean — pour into this stopping-place has to ascend and flow off as an upper current, to make room for that which is continually flowing in below. In ascending it expands and grows cool, and, as it grows cool, condensation of its vapor commences; with this, vast quantities of latent heat, which converted the water out at sea into vapor for these winds, are set free in the upper air. There it reacts by warming the ascending columns, causing them still farther to expand, and so to rise higher and higher, while the barometer sinks lower and lower. This reasoning is suggested not only by the facts and circumstances already stated as well known, but it derives additional plausibility for correctness by the low barometer of these regions. In the equatorial calm belts the mean barometric pressure is about 0.25 inch less than it is in the trade-winds, and this diminution of pressure is enough to create a perpetual influx of the air from either side, and to produce the trade-winds. Off Cape Horn the mean barometric pressure [5] is 0.75 inch less than in the trade-wind regions. This is for the parallel say of 57°–8° S. According to the mean of 2472 barometric observations made along that part only of the route to Australia which lies between the meridians of the Cape of Good Hope and Melbourne, the mean barometric pressure on the polar side of 42° S. has been shown by Lieutenant Van Gogh, of the Dutch

[5] *Sailing Directions*, 6th ed. (1854), p. 692; 8th ed., II, 450.

Navy, to be 0.33 inch less than it is in the trade-winds. The mean pressure in this part of the South Indian Ocean is, under winds with easting in them, 29.8 inches; ditto, under winds with westing, 29.6 inches. Plate I shows a supposed mean pressure in the polar calms of not more than 28.75 inches.

834. *Aqueous vapor the cause of both.* To what, if not to the effects of the condensation of vapor borne by those surcharged winds, and to the immense precipitation in the austral regions, shall we ascribe this diminution of the atmospherical pressure in high south latitudes? It is not so in high north latitudes, except about the Aleutian Islands of the Pacific, where the sea to windward is also wide, and where precipitation is frequent, but not so heavy. The steady flow of "brave" winds toward the south would seem to call for a combination of physical conditions about their stopping-place exceedingly favorable to rapid, and heavy, and constant precipitation. The rain-fall at Cherraponjie and on the slopes of the Patagonian Andes reminds us what those conditions are. There mountain masses seem to perform in the chambers of the upper air the office which the jet of cold water does for the exhausted steam in the condenser of the engine. The presence of land, not water, about this south polar stopping-place is therefore suggested; for the sea is not so favorable as the mountains are for aqueous condensation.

835. *The topographical features of the antarctic bands.* By the terms in which our proposition has been stated, and the manner in which the demonstration has been conducted, the presence in the antarctic regions of land in large masses is called for; and if we imagine this land to be relieved by high mountains and lofty peaks, we shall have in the antarctic continent a most active and powerful condenser. If, again, we tax imagination a little farther, we may, without transcending the limits of legitimate speculation, invest that unexplored land with numerous and active volcanoes. If we suppose this also to be the case, then we certainly shall be at no loss for sources of dynamical force sufficient to give that freshness and vigor to the atmospherical circulation which observations have abundantly shown to be peculiar to the southern hemisphere. Neither under such physical aspects need it be any longer considered paradoxical to ascribe the polar tendency of the "brave west winds" to rarefaction by heat in the antarctic circle. This heat is relative, and though it be imparted to air far below the freezing-point, raising its temperature only a few degrees, its expansive power for that change is as great when those few degrees are low down as it is when they are high up on the scale. If such condensation of vapor do take place, then liberation of heat and expansion of air must follow, and consequently the oblateness of the atmospherical covering of our planet will be altered; the flattening about the poles will be relieved by the intumescence of the

expanded and ascending air, which, protruding above the general level of the aerial ocean, will receive an impulse equatorially, as well from the mere derangement of equilibrium as from the centrifugal forces of the revolving globe. And so this air, having parted with its moisture, and having received the expansive force of all the latent heat evolved in the process of vaporous condensation, will commence its return toward the equator as an upper current of dry air.

836. *A perpetual cyclone.* Arrived at this point of the investigation, we may contemplate the whole system of these "brave west winds" in the light of an everlasting cyclone on a gigantic scale. The antarctic continent is in its vortex, about which the wind, in the great atmospherical ocean all around the world, from the pole to the edge of the calm belt of Capricorn, is revolving in spiral curves, continually going with the hands of a watch, and twisting from left to right.

837. *Discovery of design in the meteorological machinery.* In studying the workings of the various parts of the physical machinery that surrounds our planet, it is always refreshing and profitable to detect, even by glimmerings never so faint, the slightest tracings of the purpose which the Omnipotent Architect of the universe designed to accomplish by any particular arrangement among its various parts. Thus it is in this instance: whether the train of reasoning which we have been endeavoring to follow up, or whether the arguments which we have been adducing to sustain it be entirely correct or not, we may, from all the facts and circumstances that we have passed in review, find reasons sufficient for regarding in an instructive, if not in a new light, that vast waste of waters which surrounds the unexplored regions of the antarctic circle: it is a reservoir of dynamical force for the winds — a regulator in the grand meteorological machinery of the earth. The heat which is transported by the vapors with which that sea loads its superincumbent air is the chief source of the motive power which gives to the winds of the southern hemisphere, as they move through their channels of circulation, their high speed, great regularity, and consistency of volume. And this insight into the workings of the wonderful machinery of sea and air we obtain from comparing together the relative speed of vessels as they sail to and fro upon intertropical seas!

838. *Indications which the winds afford concerning the unexplored regions of the south.* Such is the picture which, after no little labor, much research, and some thought, the winds have enabled us to draw of certain unexplored portions of our planet. As we have drawn the picture, so, from the workings of the meteorological machinery of the southern hemisphere, we judge it to be. The evidence which has been introduced is meteorological in its nature, circumstantial in its character, we admit; but it shows the idea of land in the antarctic regions — of much land, and

high land — to be plausible at least. Not only so: it suggests that a group of active volcanoes there would by no means be inconsistent with the meteorological phenomena which we have been investigating. True, volcanoes in such a place may not be a meteorological necessity. We can not say that they are; yet the force and regularity of the winds remind us that their presence there would not be inconsistent with known laws. According to these laws, we may as well imagine the antarctic circle to encompass land as to encompass water. We know, ocularly, but little more of its topographical features than we do of those of one of the planets; but, if they be continental, we surely may, without any unwarrantable stretch of the imagination, relieve the face of nature there with snow-clad mountains, and diversify the landscape with flaming volcanoes. None of these features are inconsistent with the phenomena displayed by the winds. Let us apply to other departments of physics, and seek testimony from other sources of information. None of the evidence to be gathered there will appear contradictory — it is rather in corroboration. Southern explorers, as far as they have penetrated within the antarctic circle, tell us of high lands and mountains of ice; and Ross, who went farthest of all, saw volcanoes burning in the distance.

839. *Their extent; Plate X.* The unexplored area around the south pole is about twice as large as Europe. This untraveled region is circular in shape, the circumference of which does not measure less than 7000 miles. Its edges have been penetrated here and there, and land, wherever seen, has been high and rugged. Plate X shows the utmost reach of antarctic exploration. The unexplored area there is quite equal to that of our entire frigid zone. Navigators on the voyage from the Cape of Good Hope to Melbourne, and from Melbourne to Cape Horn, scarcely ever venture, except while passing Cape Horn, to go on the polar side of 55° S. The fear of icebergs deters them. These may be seen there drifting up toward the equator in large numbers and large masses all the year round. I have encountered them myself as high up as the parallel of 37° — 8° S. The belt of ocean that encircles this globe on the polar side of 55° S. is never free from icebergs. They are found in all parts of it the year round. Many of them are miles in extent and hundreds of feet thick. The area on the polar side of the 55th parallel of south latitude comprehends a space of 17,784,600 square miles. The nursery for the bergs, to fill such a field, must be an immense one; such a nursery can not be on the sea, for icebergs require to be fastened firmly to the shore until they attain full size. They therefore, in their mute way, are loud with evidence in favor of antarctic shore lines of great extent, of deep bays where they may be formed, and of lofty cliffs whence they may be launched.

840. *A physical law concerning the distribution of land and water.*

There is another physical circumstance which obtains generally with regard to the distribution of land and water over the surface of the earth, and which, as far as it goes, seems to favor the hypothesis of much land about the south pole; and that circumstance is this: It seems to be a physical necessity that land should not be antipodal to land. Except a small portion of South America and Asia, land is always opposite to water. Mr. Gardner has called attention to the fact that only one twenty-seventh part of the land is antipodal to land. The belief is, that on the polar side of 70° north we have mostly water, not land. The law of distribution, so far as it applies, is in favor of land in the opposite zone. Finally, geographers are agreed that, irrespective of the particularized facts and phenomena which we have been considering, the probabilities are in favor of an antarctic continent rather than of an antarctic ocean.

841. *Dr. Jilek.* "There is now no doubt," says Dr. Jilek, in his Lehrbuch der Oceanographie, "that around the south pole there is extended a great continent mainly within the polar circle, since, although we do not know it in its whole extent, yet the portions with which we have become acquainted, and the investigations made, furnish sufficient evidences to infer the existence of such with certainty. This southern or antarctic continent advances farthest northward in a peninsula S.S.E. of the southern end of America, reaching in Trinity Land almost to 62° south latitude. Outwardly these lands exhibit a naked, rocky, partly volcanic desert, with high rocks destitute of vegetation, always covered with ice and snow, and so surrounded with ice that it is difficult or impossible to examine the coasts very closely. * * *

842. *Antarctic expeditions.* "The principal discoverers of these coasts are (Wilkes), Dumont d'Urville, and Ross (the younger), of whom the latter in 1842 followed a coast over 100 miles between 72° and 79° south latitude, and 160° and 170° east longitude, to which he gave the name Victoria Land, and on which he discovered a volcano (Erebus) 10,200 feet high in 167° east longitude and 77° south latitude, as well as another extinct one (Terror) 10,200 feet high, and then discovered the magnetic south pole." [6]

[6] Text-book of Oceanography for the Use of the Imperial Naval Academy, by Dr. August Jilek, Vienna, 1857. [August von Jilek, *Lehrbuch der Ozeanographie zum Gebrauche der k.-k. Marine-Akademie.*]

CHAPTER XXI

§ 850–878. THE ANTARCTIC REGIONS AND THEIR CLIMA-
TOLOGY

850. *Indications of a mild climate about the south pole.* During our investigations of the winds and currents, facts and circumstances have been revealed which indicate the existence of a mild climate — mild by comparison — within the antarctic circle. They plead most eloquently the cause of exploration there. The facts and circumstances which suggest mildness of climate about the south pole are these: a low barometer, a high degree of aerial rarefaction, and strong winds from the north.

851. *The story of the winds.* The winds were the first to whisper of this strange state of things, and to intimate to us that the antarctic climates are very unlike the arctic for rigor and severity. In dividing the sea into wind-bands (§ 352) or longitudinal belts 5° of latitude broad each, I excluded from the subjoined table observations from those parts of the sea, such as the North Indian Ocean, the China Sea, and all those seas where monsoons prevail. The object was to investigate the *general* movements of the atmosphere, and therefore all regions which present exceptional cases to the general law were excluded as above. The grouping was not carried beyond 60° north and south, for the lack of observations on the polar side of those parallels. The number of observations thus remaining was 1,159,353. These were then divided simply into two classes for each belt, viz., polar winds [1] and equatorial winds. They were then reduced to terms of a year, and the average prevalence of each wind in days deduced therefrom, as per Figure 2 and Table 18.

852. *The* null *belts.* This table reveals a marked difference in the atmospherical movements north, as compared with the atmospherical movements south of the equator. The equatorial winds of the northern hemisphere are in excess only between the parallels of 10° and 30°; *i. e.,* they are dominant over a zone 20° of latitude in breadth, while the equatorial winds of the southern hemisphere hold the mastery from 35° S. to 10° N.; *i. e.,* they prevail over a belt 45° in breadth, while the others cover a space not half so broad. This table, moreover, shows that the debatable ground between the winds, or what may be called the *null belt,* in this general movement from poles toward the equator, and from equator

---

[1] Polar winds blow *toward* the pole, equatorial *toward* the equator. [These designations are directly opposite to common usage, in which winds are named according to the directions *from* which they blow.]

Table 18

Polar and Equatorial Winds

	Northern Hemisphere					Southern Hemisphere				
		Equa-torial	Polar	Excess in Days			Polar	Equa-torial	Excess in Days	
Belts	No. of Obsv'ns.	Days	Days	Equa-torial	Polar	No. of Obsv'ns.	Days	Days	Polar	Equa-torial
Between										
0° and 5°	67,829	79	268	—	189	72,945	83	269	—	186
5 " 10	36,841	158	183	—	25	54,648	72	283	—	211
10 " 15	27,339	278	73	205	—	43,817	82	275	—	193
15 " 20	33,103	273	91	182	—	46,604	91	266	—	175
20 " 25	44,527	246	106	140	—	66,395	128	227	—	99
25 " 30	68,777	185	163	22	—	66,635	147	208	—	61
30 " 35	62,514	155	195	—	40	76,254	150	204	—	54
35 " 40	41,233	173	179	—	6	107,231	178	178	0	0
40 " 45	33,252	163	186	—	23	63,669	202	155	47	—
45 " 50	29,461	164	189	—	25	29,132	209	148	61	—
50 " 55	41,570	148	203	—	55	14,286	208	151	57	—
55 " 60	17,874	142	213	—	71	13,617	224	132	92	—

toward the poles, is, in the northern hemisphere, between the parallels of 25° and 50°. In the southern the field of battle is narrowed down to a single belt (between 35° and 40°); here the two winds exactly counterbalance each other. As one proceeds from this medial belt, the winds increase belt for belt very nearly *pari passu;* on the polar side, the polar winds — on the equatorial, the equatorial winds, gaining more and more in days of annual duration, and more and more in average velocity each.

853. *Extent of the polar indraught.* The fact that the influence of the polar indraught upon the winds should extend from the antarctic to 40° S., while that from the arctic is so feeble as scarcely to be felt in 50° N., is indicative enough as to difference in degree of aerial rarefaction over the two regions. The significance of this fact is enhanced by the "brave west winds," which, being bound to the place of greatest rarefaction, rush more violently along to their destination than do the counter-trades of the northern hemisphere. Why should these polar-bound winds of the two hemispheres differ so much in strength and prevalence, unless there be a much more abundant supply of caloric, and, consequently, a higher degree of rarefaction, at one pole than the other?

854. *The rarefaction of the air over polar regions.* In the southern hemisphere — and our attention is now directed exclusively to that — the polar winds on the south side of 40° are very much stronger than are the equatorial winds on the north side of 35°: a fact indicative of a greater degree of rarefaction about the place of polar calms than we have in the equatorial calm belt.

855. *Barometrical observations.* That such is the case is also suggested

by the fact that the indraught into the antarctic calm place is felt (§ 854) at the distance of 50° from the pole all round, while the equatorial indraught is felt no farther than 35° from the equator; and that such is the case is proved by the barometer. Lieutenant Andrau, of the Meteorological Institute of Utrecht, has furnished us from the Dutch logs with 83,334 observations on the height of the barometer between the parallels of 50° N. and 36° S. at sea. Lieutenants Warley and Young have extracted from the log-books of this office, taken at random, 6945 observations on the barometer south of 40° at sea. Dr. Kane has furnished us with the mean height of the barometer in lat. 78° 37′ N., according to 12,000 hourly observations made during his imprisonment of 17 months in the ice there. The annals of Greenwich and St. Petersburg give us the mean height of the barometer in lat. 51° 29′ N. according to three years' observations, and in lat. 59° 51′ N. according to ten years of observation. Such are the sources of Table 5 § 362.

856. *The low austral barometer.* Captains Wilkes, U.S.N., and Clarke Ross, R.N., both, during the expeditions to the South Seas in 1839–41, had occasion to remark upon the apparent deficiency of atmosphere over the extra-tropical regions of the southern hemisphere; and the low barometer off Cape Horn had attracted the attention of navigators at an early day. I observed it in 1831 when doubling the Cape as master of the U.S.S. Falmouth, and wrote a paper on it, which was published in the American Journal of Science in 1833–4. The more abundant materials which the abstract logs have placed within my reach enable me to go more fully into this subject. To ascertain whether these "*barometric anomalies,*" as they are called, are peculiar to the regions about Cape Horn, or whether they are common to high southern latitudes in all longitudes, the observations about Cape Horn were arranged in one group; those between 20° W. and 140° E. in another; and those between 140° E. and 80° W. in another, with the results [shown in Table 10, § 670.]. (They are all on the polar side of 40° S.)

857. *Discussion of observations.* The instruments used for these observations were for the most part the old-fashioned marine barometer, to which no corrections have been applied. The discrepancies of this table evidently arise from the lack of numbers sufficient to mask these sources of error, or from the influence of the land, and not from any difference as to the mean height of the barometer along the same parallels *at sea* in any one of the three divisions. In this discussion, the observations of each group and every band were arranged according to the month. These monthly tables are not repeated here, but they do not indicate any decided change in the barometric pressure in high southern latitudes according to the season. The barometer there stands low the year round.

858. *Barometric curve at sea.* Resorting to the graphic method, and

using the table (§ 362) for the purpose, the barometric curve of Figure
13 has been projected from pole to pole.

859. *Ditto over the land.* Professor Schouw has given us the mean
height of the barometer for 32 places on the land between the parallels
of 33° S. and 75° 30′ N. They afford materials for Figure 14, and
show the exceptional character of the meteorological influences which
rule on shore when compared with those which rule at sea. There is
barely a resemblance between this profile of the atmosphere over the
land and the profile of it (Figure 13) over the sea, so different are these

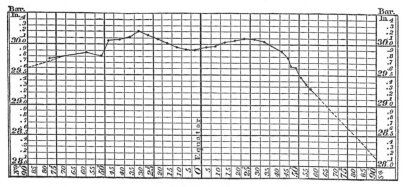

Figure 13

[Barometric curve from the North to the South Pole. The irregularities of the curve
in the higher northern latitudes probably owing to the observations at Greenwich,
St. Petersburg, *et al.*, which, being on the land, are more exceptional than the others,
all of which were made at sea.]

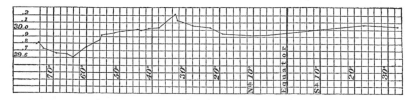

Figure 14

[Mean atmospheric pressure in different latitudes plotted from observations made at
stations on land. (From data compiled by J. F. Schouw).]

influences. The irregularities over the land are chiefly owing to the difference in the amount of precipitation at one station as compared with
the amount at another. Those islands, as the Sandwich and Society, which
are so situated as to bring down a heavy precipitation, are as *chimneys
to the atmosphere.* The latent heat which is liberated by the vapor they

condense has the effect of bringing down the barometer, and of causing, during the rainy season, an indraught thitherward from many miles at sea. Such is the rarefaction produced by the liberation of this heat, that its effect is, as the pilot charts show, felt and confessed by the winds at the distance out to sea of more than a thousand miles from the Sandwich Islands. Thus the land and the islands give us, in the circulation of the atmosphere, systems within systems. In the Mississippi and all great rivers, the general movement of the waters, notwithstanding the eddies and the whirlpools, is down stream with the current. So with the atmosphere: its general movements are indicated by observations at sea; its eddies and whirlpools are created by the mountains, and the islands, and other inequalities which obstruct its flow in the regular channels. The mean reading of the barometer when the rainy season in India is at its height is 0.4 inch less than it is in the midst of the dry.

860. *Agreement of observations at sea.* Figure 13 shows the observations in the southern hemisphere to be so accordant, and the curve itself so regular, that we feel no hesitation about projecting it into the vacant spaces of the south, and asserting, with all the boldness consistent with the true spirit of philosophical deduction, that, whether the actual barometric pressure at the south pole be as low as 28.14° or not, it is nevertheless very much lower in the antarctic than in the arctic regions.

861. *The question why the barometer should stand lower about the south than the north pole considered.* The question now arises, Whence this unequal distribution of atmosphere between the two hemispheres, and why should the mean height of the barometer in circumpolar regions be so much less for the austral than for the boreal? No one, it is submitted, will attempt to account for this difference by reason of any displacement of the geometrical centre of the earth with regard to its centre of attraction, in consequence of the great continental masses of the northern hemisphere; neither can it be ascribed to any difference in the forces of gravitation arising from the oblateness of our globe; neither can it be accounted for by the effects of diurnal rotation after the Halleyan theory: that would create as low a barometer at one pole as the other. The air, in its motions to the east and its motions to the west, is in equipoise between the parallels of 35° and 40° N., 25° and 30° S. There is near each pole and about the equator a place of permanently low barometer. The air from all sides is continually seeking to restore the equilibrium by rushing into these places of rarefaction and reduced pressure; consequently there ought to be between each pole and the equator a place of high barometer from which the air on one side flows toward the equator, on the other toward the pole. Observation (§ 362) shows this high place to be between the parallels of 25° and 40° in the north, and of 20° and 30° in the southern hemisphere: thus the barom-

eter as well as the winds (Figure 2) indicate a greater degree of rarefaction about the south than about the north pole. Were there no friction,
and were the atmosphere ordained to move without resistance, the air
from these *null belts* would carry with it to the polar calms the easterly
motion which it had acquired from the earth in its motion around its
axis at these null belts. Were this motion so impressed, the wind would
arrive, rushing with an hourly velocity about the polar calm places of 700
miles in the arctic, and 800 in the antarctic. Such a velocity would impart
a centrifugal force sufficient to keep the air away from the poles and produce almost a vacuum there. In this state of things, the *same* air would
continue to revolve about the poles were not some other agent, such as
heat, brought in to expand and drive it away. Being expanded and puffed
out above the general atmospherical level, but retaining its velocity — for
the supposition is that it moves without friction — and returning through
the upper regions, it would flow back as it went, viz., as a westerly wind,
and arrive at its null belt in the direction of the meridian. But the wind
has friction, and is resisted in every movement; the atmosphere partakes
of the spheroidal form, which has been impressed upon the earth itself
by its axial rotation. That form is to it the form of stability. The water at
the pole is 13 miles nearer to the centre of the earth than the water at the
equator; but there is not on that account any tendency in the sea to flow
back from the equator toward the poles; neither is there any tendency to
motion one way or the other in the atmospherical ocean by reason of the
oblateness of its surface. To produce the polar and equatorial movements
of the air, there must be an agent both at the equator and the poles to
prevent such stability by constantly disturbing equilibrium there, and
that agent is heat; therefore, whatever be the degree of polar depression
due the barometer in consequence of axial rotation, such depression
could, of itself, produce neither trade nor counter-trade wind; it could
no more produce currents in the air than in the sea, nor could axial rotation produce a high barometer at one pole, a low barometer at the
other; consequently, the difference in the pressure of the atmosphere
about the two poles, as shown by Figure 13, can not be ascribed to
the influence of axial rotation. It is doubtless due to the excess in antarctic regions of aqueous vapor and its latent heat.

862. *Psychrometry of polar winds.* The arctic circle lies chiefly on
the land, the antarctic on the water. As the winds enter one, they are
loaded with vapor; but on their way to the other they are desiccated
(§ 826). Northern mountains and the hills wring from them water for
the great rivers of Siberia and Arctic America. These winds, then, sweep
comparatively dry air across the arctic circle; and when they arrive at the
calm disc — the place of ascent there — the vapor which is condensed in
the act of ascending does not liberate heat enough to produce a rare-

faction sufficient to call forth a decided indraught from a greater distance in the surrounding regions than 40° (§ 852); and the rarefaction being not so great, the barometer is not so low there as in antarctic regions.[2]

863. *Aerial rarefaction about the north pole.* Nevertheless, there *is* rarefaction in the arctic regions. The winds show it, the barometer attests it, and the fact is consistent with the Russian theory of a polynia in polar waters. The presence within the arctic circle of a considerable body of comparatively warm water, which observation has detected going into it as an under current — which induction shows must rise up and flow as a surface current — would give forth vapor most freely. This vapor, being lighter than air, displaces a certain quantity of atmosphere. Rising up and being condensed, it liberates its latent heat in the cloud region, and so, by raising temperature, causes the moderate degree of rarefaction which the barometer at sea, at Greenwich, at St. Petersburg, and in the arctic ice indicates.

864. *Ditto about the south pole.* Within the antarctic circle, on the contrary, the winds bring air which has come over the water for the distance of hundreds of leagues all around; consequently, a large portion of atmospheric air is driven away from the austral regions by the force of vapor, which fills the atmosphere there. Now there must be a place — an immense disc, with irregular outlines, it may be, and probably is — where these polar winds (§ 855) cease to go forward, rise up, and commence to flow back as an upper current. If the physical aspects — the topographical features in and about this calm place — be such as to produce rapid condensation and heavy precipitation (Chap. XX), then we shall have, in the latent heat liberated from all this vapor, an agent sufficient not only to produce a low barometer and a powerful indraught, but quite adequate also to the mitigation of climate there.

865. *Influences favorable to heavy precipitation.* Mere altitude, with its consequent refrigeration, does not seem as favorable as mountain peaks and solid surfaces to the condensation and precipitation of vapor in the air. In the trade-wind regions out at sea it seldom rains; but let an island rise never so little above the water, and the precipitation upon it becomes copious. Colonel Sykes (§ 827) reports the rain-fall at Cherraponjie, a mountain station 4500 feet high in India, lat. 25° 16′ N., long. 91° 43′ E., to be 577.6 inches during the six months of S.W. monsoons — from May to October. Surely no one will maintain that this vapor, after rising from the sea, reached the height of 4500 feet for the *first* time when it was blown upon Cherraponjie. Islands in the South Sea are everlastingly cloud-capped. If it be mere refrigeration that condenses this vapor, why, one might ask, should not the clouds form at the same height above

---

[2] Captain M'Clintock, during his arctic explorations in the schooner Fox, records the barometer as high as 31 inches.

the sea, whether there be an island below or not, and why should not these clouds precipitate as copiously upon the water as they do upon the land? We only know that they do not.

866. *The climates of corresponding shores and latitudes north and south.* Captains King and Fitzroy exposed their rain-gauge on the western slopes of the Patagonian Andes, and it collected 153.75 inches in forty-one days; that is, at the rate, as already (§ 827) stated, of 1368.7 inches the year. The latent heat that is liberated during these rains gives to Eastern Patagonia its mild climate. It is this latent heat which causes the irregularity in the barometric curve (§ 858) between the parallels of 50°–55° S. Here the westerly winds prevail; they carry over to the eastern coasts the air that, in passing the mountains, is warmed by this liberated heat; and thus, as already (§ 729) explained, we have an exception to the rule under which meteorologists ascribe cold and severe climates to the windward or western, soft and mild to the leeward or eastern, shores of extra-tropical oceans. Labrador and the Falkland Islands [3] are in corresponding latitudes north and south. They are both on the windward side of the Atlantic; they occupy relatively the same position with regard to the wind. Labrador is almost uninhabitable on account of the severity of its climate, but in the Falkland Islands and their neighboring shores the cattle find pasturage always in winter. The *thermometrical* difference of climate at these two places, north and south, may be taken as a sort of index to the relative difference between the arctic and antarctic climates of our planet.

867. *Thermal difference between arctic and antarctic climates.* Along the eastern base of the Rocky Mountains the isotherms mount up [4] toward the north in consequence of the heavy winter precipitation upon the western slopes of these mountains. The heat which is required to convert the water of the Columbia and other rivers into vapor is set free on the mountain range, and the upper Missouri is by this heat kept open for navigation long after the lower and more southern portion of it is frozen up.

868. *How the temperature of air may be raised by crossing mountains.* In the preceding chapter the circumstances have been considered which favor the idea that most of the unknown surface of the antarctic circle is not only land, but that its coasts are probably highlands; than in its topographical features it presents all the conditions that are required for the rapid condensation of the vapor with which the impinging air from the sea is loaded, and that in the valleys beyond mild climates may be

[3] ["Captain W. H. Smyley to Lieut. M. F. Maury"], *Sailing Directions*, 6th ed., [548–553, esp.], p. 553.
[4] [Lorin] Blodget, *Climatology of the United States* [Philadelphia, 1857, map opp. p. 296].

expected. The aqueous vapor which the air carries along is one of the most powerful modifiers of climates. It is to the winds precisely what coals are to the steam-ship at sea — the source of motive power. The condensation of vapor is for one what the consumption of fuel is for the other; only with the winds the same heat may be used over and over again, and for many purposes. By simple sending moist air to the top of snow-capped mountains, condensing its moisture, and bringing it down to the surface again, it is made *hot*. Though by going up the air be cooled, it is expanded, and receives as sensible heat the latent heat of its vapor; being brought down to the surface again, and compressed by the whole weight of the barometric column, it is hotter than it was before by the amount of heat received from its vapor. That we may form some idea as to the modifying influences upon climate which *might* arise from this source, let us imagine the air as it impinges upon the antarctic continent to be charged with vapor at the temperature of 40°. In order to arrive at the place of polar calms, it has to cross a mountain range, we will suppose, the summits of which are pushed high up into the regions of perpetual snow. As this air, with its moisture, rises, it expands, cools, and liberates the latent heat of its vapor, which the air receives in the sensible form. Now suppose the expansion due the height of the mountain-top to be sufficient to lower the temperature of dry air to −50°; but, on account of the latent heat which is liberated from the vapor of the moist air, the temperature of the air that has ascended, instead of falling as it crosses the mountain to −50°, as dry air would do, falls, in consequence of the condensation of its vapor, no lower than −30°. Thus, in the case supposed, heat enough has been set free to raise the temperature of the newly-arrived air 20°. Consequently, when this air, which, at the temperature of 40°, came from the sea loaded with vapor, passes the mountain, it loses vapor, but receives heat; descending into the valleys beyond, it is again compressed by the weight of the barometric column, which, let us assume, is the same in the valley as at the sea level on the other side of the mountain. The temperature of this air now, instead of being 40°, will be 60°. A powerful modifier of climate is the latent heat of vapor in the air.

869. *Aurora australis.* There is not only reason to suppose that the topographical features and the climates of the antarctic regions differ greatly from the topographical features and climates of the arctic, but there is reason to suppose a difference in other physical aspects also. The aurora points to these. "On the morning of the second of September last," says Capt. B. P. Howes, in his abstract log of the "Southern Cross," lat. 58° S., long. 70° W., "at about half past one o'clock, the rare phenomenon of the aurora australis manifested itself in a most magnificent manner. Our ship was off Cape Horn in a violent gale, plunging furiously into a

heavy sea, flooding her decks, and sometimes burying her whole bows beneath the waves. The heavens were black as death: not a star was to be seen when the brilliant spectacle first appeared. I can not describe the awful grandeur of the scene; the heavens gradually changed from murky blackness till they became like livid fire, reflecting a lurid, glowing brilliancy over every thing. The ocean appeared like a sea of vermilion lashed into fury by the storm; the waves, dashing furiously over our sides, ever and anon rushed to leeward in crimson torrents. Our whole ship, sails, spars, and all, seemed to partake of the same ruddy hues. They were as if lighted up by some terrible conflagration. Taking all together, the howling, shrieking storm, the noble ship plunging fearlessly beneath the crimson-crested waves, the furious squalls of hail, snow, and sleet driving over the vessel and falling to leeward in ruddy showers, the mysterious balls of electric fire resting on our mast-heads, yard-arms, etc., and, above all, the awful sublimity of the heavens, through which coruscations of auroral light would often shoot in spiral streaks and with meteoric brilliancy, altogether presented a scene of terrible grandeur and awful sublimity surpassing the wildest dreams of fancy. Words fail to convey any just idea of the magnificence it presented. One must *see* it and *feel* it in order to realize it. I have written this because I believe it an unusual occurrence to see the 'southern lights' at all, and also because this was far superior, and, in fact, altogether different from our northern lights, as seen from the latitude of Boston."

870. *An erroneous opinion.* Some objections to these views respecting the comparative mildness of antarctic climates are suggested by common opinion. It is an opinion which is generally received among physicists that the southern is colder than the northern hemisphere, and that the austral are more severe than the boreal climates, and that the antarctic icebergs, both as to size and numbers, are witnesses of the fact. These objections have weight; they deserve consideration.

871. *Tropical regions of the southern hemisphere cooler, extra-tropical warmer, than those of the northern.* The answer is as follows: Between lat. 40° or lat. 45° and the equator, and parallel for parallel, the southern hemisphere is cooler than the northern. Reason teaches, and observations show that it is so. But beyond 45° S. observations are wanting, and we are left to reason and conjecture. That the southern hemisphere should, till within a certain distance of the pole, be warmer in winter and cooler in summer, and have a mean annual temperature less than that of our hemisphere, may be explained by the fact that the southern hemisphere has more water; that the vapor which is taken up from trans-equatorial seas and condensed into rains for cis-equatorial rivers conveys with it a vast amount of heat which the southern hemisphere receives from the sun. It is rendered latent by evaporation on one side of the

equator, and made sensible by precipitation on the other. Much of it is set free in the equatorial calm belt, and it is this heat which assists mightily to maintain the thermal equator in its northern position.

872. *Formation of southern icebergs.* So, in like manner, the vapor that is borne to the antarctic regions by the polar-bound winds transports immense volumes of heat from the more temperate latitudes of the south, and sets it free again in the polar regions there. And as for the southern icebergs, they are rather of fresh than of salt water; and they are the channels through which the water that the winds carry there as vapor finds its way back again. Being fresh water, and falling on the antarctic declivities of the land, it is by rills, and streams, and rains brought together, and by constant accretions formed into glaciers of a size and thickness that are almost impossible to be formed out of sea water unless it be dashed up as spray. Moreover, on the arctic ocean the rains are not so copious, and for that reason, though the climate be more severe, icebergs, or, rather, glaciers, are not formed on so grand a scale. Southern icebergs are true glaciers afloat. Arctic winds are dry enough to evaporate much of the ice and snow that fall and form in the north polar basin. As compared with arctic climates, antarctic are marine, arctic continental; and for the very reason that the English Channel is cooler in the summer and warmer in winter than the Canadian, so is winter at the south pole much less severe than that of the north. The difference between the two polar climates is, as the barometer indicates, even greater than is the difference between a Canadian and an English winter.

873. *Mild climate in 63° S.* As tending to confirm these views touching the mildness of unknown antarctic climates, the statement of Captain Smyley, an American sealer, may be mentioned. He planted a self-registering thermometer on the South Shetlands, lat. 63° S., and left it for several winters, during which time it went no lower than −5° Fahr.[5]

874. *Antarctic ice-drift.* The low barometer and the implied heavy precipitation in the antarctic regions are not the only witnesses that may be called up in favor of bluffs and bold shores to the antarctic continent. The icebergs, in their mute way, tell that the physical features of that unexplored land are such, in its northern slopes, as to favor the formation of glaciers on the shore, to become the huge icebergs that, on their journey to the milder climates of the north, are encountered far away at sea. After a somewhat attentive, but by no means a thorough examination and study of antarctic icebergs as they endanger the routes of navigation, the idea has been suggested that information may be gathered from them concerning antarctic regions which would be highly useful to any future expedition thitherward.

875. *Antarctic currents.* The conditions required for Gulf-Stream-like

[5] *Sailing Directions*, [7th ed. (1855), p. 488].

currents, or a rapid flow and reflow of equatorial and polar waters between the torrid and the frigid zones, as in the northern hemisphere, are not to be found about the antarctic regions. Of all the currents that come from those regions, Humboldt's current is by far the most majestic. It is believed also to be the least sluggish of them all. It certainly conveys the coldest water thence to the torrid zone; and yet it appears not to come from a nursery of icebergs, for in its line of march fewer icebergs are found than are encountered on the same parallels between other meridians, but where feebler currents flow. From the arctic regions the strongest currents bring down the most icebergs; not so from the antarctic. Hence the inference that, though icebergs have been encountered off the shores of the antarctic continent wherever they have been approached, yet it is only those which have been launched from particular points of that frost-bound coast which are stout enough to bear transportation to the parallel of 40° south. In Humboldt's current it is rare to see an iceberg as far from the pole as the parallel of the fifty-fifth degree of south latitude; but off the Cape of Good Hope on one side of the Atlantic, and Cape Corrientes on the other, antarctic icebergs are sometimes seen as far as the parallel of 35°, often as far as 40°. Lieutenants Warley and Young, after having examined the logs of 1843 ships cruising on the polar side of 35° S., report the great antarctic ice-drift to be toward the Falkland Islands on one hand, and the Cape of Good Hope on the other.

876. *Antarctic explorations demanded.* These facts and the stories of the icebergs are very suggestive. In mute eloquence and with great power they plead the cause of antarctic exploration. Within the periphery of that circle is included an area equal in extent to the one sixth part of the entire landed surface of our planet.[6] Most of this immense area is as unknown to the inhabitants of the earth as is the interior of one of Jupiter's satellites. With the appliances of steam to aid us, with the lights of science to guide us, it would be a reproach to the world to permit such a large portion of its surface any longer to remain unexplored. For the last 200 years the Arctic Ocean has been a theatre for exploration; but as for the antarctic, no expedition has attempted to make any *persistent* exploration, or even to winter there.

877. *Former expeditions.* England through Cook and Ross; Russia through Bellingshausen; France through D'Urville; and the United States through Wilkes, have sent expeditions to the South Sea. They sighted and sailed along the icy barrier, but none of them spent the winter or essayed to travel across and look beyond the first impediment. The expeditions which have been sent to explore unknown seas have contributed largely to the stock of human knowledge, and they have added renown to nations, lustre to diadems. Navies are not all for war. Peace

[6] The area of the antarctic circle is 8,155,600 square miles.

has its conquests, science its glories; and no navy can boast of brighter chaplets than those which have been gathered in the fields of geographical exploration and physical research.

878. *An appeal for others.* The great nations of the earth have all, with more or less spirit, undertaken to investigate certain phenomena touching the sea, and, to make the plans more effectual, they have agreed to observe according to a prescribed formula. The observations thus made have brought to light most of the facts and circumstances which indicate the existence within the antarctic circle of a mild climate — mild by comparison. The observations which have led to this conclusion were made by fellow-laborers under all flags. It is hoped that this circumstance may vindicate, in the eyes of all, the propriety of an appeal in this place for antarctic exploration, and plead for it favorable consideration among all nations.

CHAPTER XXII

§ 879–893. THE ACTINOMETRY OF THE SEA

879. *A new field.* One of the columns in the man-of-war log of the Brussels Conference calls for the temperature of the water below as well as at the surface of the sea. Only a few entries have been made in this column; but these, as far as they go, seem to indicate that the warmest water, especially in tropical seas, is not to be found at the top, but in a stratum a little way down. What is the depth of this stratum, and what may be the thermal difference between its waters and those of the surface, are questions for future observations to settle. Indeed, this subject opens a new field of inquiry; it is one from which much useful and instructive information is doubtless to be obtained by any one of our co-operators who patiently and with diligence will enter upon the investigation.[1]

880. *The warmest waters in the sea — where are they? at or below the surface?* The observations that we possess do not prove that the warmest water of intertropical seas is not at the surface: they go no farther than

[1] On the 26th of March, 1852, the late Passed Midshipman A. C. Jackson, U. S. N., being in the Gulf Stream, lat. 34° 55′ N., long. 74° 8′ W., found the temperature of the water 74.5° at the surface, 79° at the depth of six feet, and 86.5° at the depth of 16½ feet. Again, on the 30th, in lat. 24° 10′ N., long. 80° 11′ W. (near the edge of the Gulf Stream), he tried the temperature of the water by another carefully conducted set of observations, and found it 78° at the surface, and 79.5° at the depth of 16½ feet. The sea was rough, and he did not, for that reason, observe the temperature at six feet. The temperature of the air in the shade was 69.5° on the 26th, and 79° on the 30th. (*Sailing Directions*, 5th ed., 1853, p. 59).

Extract of a Letter from J. Bermingham, Chief Engineer of the American Steamer "Golden Age," dated Bay of Panama, June 29, 1860, and addressed to Lieut. John M. Brooke, U. S. N.

"On our late trip from San Francisco (June 5) to this port, we experienced the most remarkably fine weather and smooth sea that I have ever witnessed on the Pacific, or any where else.

"On the 14th, while crossing the Gulf of Tehuantepec, we found the temperature of the sea water on the surface (where it had not been disturbed by the progress of the vessel) 88°, and upon taking the temperature at the same time ten feet below the surface, the mean of three thermometers gave 90°. Temperature of atmosphere, 93°.

"I do not exactly understand why the temperature of the sea water should be so much greater at a distance of ten feet from the surface than it was immediately upon the surface.

"Mr. Agassiz (a son of Professor Agassiz) was on board, and he and myself made repeated tests of the temperature of the water during the four hours we were running through it — the warm belt.

"Ninety degrees is the highest temperature that I have ever known the water of the ocean to attain."

to show that it is *sometimes* not at the surface, and to suggest that, in all probability, it is generally below, especially in "blue water." Reason suggests it also. Supposing that, as a rule, the hottest water is below the surface, we may, in order to stimulate research, encourage investigation, and insure true progress, propound a theory in explanation of the phenomenon, looking to future observations to show how far it may hold good.

881. *The annual supply of solar heat uniform.* The flow of heat from the sun is held to be uniform, and the quantity of it that is annually impressed upon the earth is considered as a constant. The sun spots *may* make this "constant" a variable, but the amount annually received by the earth is so nearly uniform that our best instruments have not been able to show us any variation in its uniformity. Some maintain that climates are undergoing a gradual change as to temperature. However this may be as to certain localities, Baron Fourier, after a long and laborious calculation, claims to have shown that if the earth had been once heated, and, after having been brought to any given temperature, if it had then been plunged into a colder medium, it would not, in the space of 1,280,000 years, be reduced in temperature more than would, in one second of time, a 12-inch globe of like materials if placed under like conditions. It may be assumed that, for the whole earth, there has not been, since the invention of the thermometer, any appreciable change in the temperature of the crust of our planet.

882. *Quantity of heat daily impressed upon the earth.* The earth receives from the sun heat enough daily, it has been said (§ 271), to melt a quantity of ice sufficient to incase it in a film 1½ inch thick. What becomes of this heat after it is so impressed? how is it dispensed by the land? how by the sea? Let us inquire.

883. *How far below the surface does the heat of the sun penetrate?* The solar ray penetrates the solid parts of the earth's crust only to the depth of a few inches, but, striking its fluid parts with its light and heat, it penetrates the sea to depths more or less profound, according to the transparency of its waters. Let us, in imagination, divide these depths, whatever they may be, into any number of stratifications or layers of equal thickness. The direct heat of the sun is supposed to be extinguished in the lowest layer; the bottom layer, then, will receive and absorb the minimum amount of heat, the top the maximum; consequently, each layer, as we go from the top to the bottom, will receive less and less of the sun's heat.

884. *The stratum of warmest water.* Now, which will retain most heat and reach the highest temperature? Not the top layer, or that to which most heat is imparted, because by evaporation heat is carried off from the surface of the sea almost as fast as by the sun it is impressed

upon the surface of the sea; not the bottom layer, because that receives a minimum, which, though it can not escape by evaporation, may nevertheless fail to make any marked change in temperature — fail, not by reason of no evaporation, but by the ever-changing movements which, considering the length of time required to heat the lower stratum by such slow and gradual accumulation of heat, would alter its place and vary its condition, and, indeed, remove it beyond the reach of the observer.

885. *Its position.* The layer, therefore, which accumulates most heat and becomes warmest, should be neither at the bottom nor at the top, but intermediate, the exact temperature and depth of which it is for observation to determine. To encourage such determination and the investigations which it suggests is the main object of this chapter.

886. *The different subjects for observation.* In conducting such investigations, several questions are to be considered, such as the transparency and specific gravity of the water, its *phosphorescence;* the face of the sky, whether clear or cloudy; the state of the sea, whether smooth or rough; the condition of the weather, whether calm or windy. Then the temperature should be tried, at various depths and at various hours of the night and day, in order to ascertain not only the maximum temperature and average depth of the warmest stratum in the day, but the difference in its temperature and position by day and by night. These observations will afford the data, also, for computing the amount of solar heat that penetrates the bosom of the sea, as well as the amount that is radiated thence again. They will reveal to us knowledge concerning its actinometry in other aspects. We shall learn how absorption by, as well as radiation from, the under strata is affected by a rough sea, as when the waves are leaping and tossing their white caps, and how by its glassy surface, as when the winds are hushed and the sea smooth.

887. *Expected discoveries.* Here we are reminded, also, to anticipate the discovery of new beauties and fresh charms among the wonders of the sea. We have seen (§ 366) that while the heat of the sun is impressed alike upon sea and land, nevertheless the solid part of the earth's crust radiates much more freely than the fluid. On the land the direct heat of the sun operates only upon a mere shell a few inches in thickness; at sea it penetrates into the depths below, and operates upon a layer of water many feet thick. The solid land-crust has its temperature raised high by day and cooled low down by night; but the most powerful sun, after beating down all day with its fiercest intensity upon this liquid covering, has not power to raise its temperature more than three or four degrees. This covering serves as a reservoir for the solar heat. In the depths below it is concealed from the powers of intense radiation, and held by the obedient ocean in readiness to be brought to the surface from time to time, and as the winds and the clouds call for it. Here it is rendered

latent by the forces of evaporation, and in this form, having fulfilled its office in the economy of the ocean, it passes off into the air, there to enter, in mysterious ways, upon the performance of its manifold tasks.

888. *Actinic processes.* As evaporation goes on by day or night, the upper stratum is rendered heavier by reason of both the heat and the fresh water which are borne away by evaporation; the upper water, having been thus rendered both salter and cooler, has its specific gravity increased so much the more. On the other hand, the strata below, receiving more heat by day than they dispense again by radiation day and night, grow actually warmer and specifically lighter; and thus, by unseen hands and the "clapping of the waves," the waters below are brought to the surface, and those on the surface carried down to unknown depths; and thus, also, we discover new and strange processes which have been ordained for the waters of the ocean in their system of vertical circulation.

889. *The reservoirs of heat.* Thus we arrive at the conclusion that the ocean is the great reservoir of sensible, as the clouds are of latent heat; that in these two chambers it is innocuously stored, thence to be dispensed by processes as marvelous as they are benignant and wise, to perform its manifold offices in the economy of our planet. It is this heat which gives "his circuits" to the winds and circulation to the sea; it is it that fetches from the ocean the clouds that make "the earth soft with showers." Stored away in the depths of intertropical seas, it is conveyed along, by "secret paths," to northern climes, there to be brought to the surface in due season, given to the winds, and borne away to temper the climates of Western Europe, clothing the British Islands as they go, in green, and causing them to smile under the genial warmth even in the dead of winter.

890. *The radiating powers of earth, air, and water compared.* We may note, also, another peculiarity as to the difference in the direct heat-absorbing and radiating properties of sea, land, and air. It is one which presents the atmosphere in the light of a regulator between the two on one hand, and the heating powers of the sun on the other. It is suggestive, also, of other benign compensations and lovely offices in the physical machinery of our planet. Both land and water receive more heat from the sun than they radiate again; but the atmosphere receives less heat *direct* from the *sun* than it radiates off again into space. As the heat comes from the sun, part of it is absorbed by the atmosphere, but the largest portion of it is impressed upon the land and water. From these a portion passes off into the atmosphere by conduction, while another portion is radiated directly off into the realms of space. What becomes of the remainder? Let us inquire, for there is a remainder; and unless means for its escape were provided, the land and water, especially the latter, would continue to grow warmer and warmer, and so produce confusion in the terrestrial

economy. The remainder of this heat, being that which is neither radiated by sea and land directly off into space, nor imparted to the air by conduction from them, is absorbed in the process of evaporation. It is then delivered to the atmosphere, latent in the vesicles of vapor, to be set free in the cloud region, rendered sensible and imparted to the upper air, whence it is sent off by radiation into the "emptiness of space." Thus the air, with its actinometry, presents itself in the light of a thermal adjustment, by which the land and sea are prevented from becoming seething hot, and by which they are enabled to perform their wonderful offices with certainty and regularity.

891. *An office for waves in the sea.* Thus, perhaps, we discover a new office for the waves in the physical economy of the ocean. Is is not to them that has been assigned the task of bringing up, by their agitation of the surface, the layers of warm water that are spread out below? and are they not concerned also, as they draw up the genial waters, in regulating the supply of heat for the winds by night, as well as in cold or cloudy days, for the purposes of evaporation? Thus even the waves of the sea are made, by this beautiful study, to present themselves as parts, important parts, in the terrestrial machinery. We now view them, as it were, like balance-wheels in the complicated system of mechanism by which the climates of the earth are governed. If the waves did not stir up the heated waters from below (§ 881), the winds would evaporate slowly by night for the want of adequate supplies of caloric; the consequence would be less precipitation and a more scanty supply of latent heat for liberation in the cloud region. As a consequence of this, the winds would have less motive power, and the whole climatic arrangements of our planet would be different from what they are.

892. *A reflection concerning heat.* It is curious to think that this subtile thing, called heat, which we have been contemplating, now as latent in the clouds above, now as sensible in the waters below, comes from the same source whence originally came the heat which has been packed away in seams of coal and stored in the bowels of the earth for ages and ages, to be called forth by man at will for his own comfort, pleasure, and convenience; that this protean thing is the agent which controls sea and winds, and they it; that it is it which has lifted up the mountains, which clothes the world with beauty, and keeps the stupendous fabric of the universe in motion; and that, after all, this mighty agent is only that gentle thing that "warms in the sun!"

893. *Probable relation between the actinometry of the sea and its depth.* Pursuing this subject, the philosophical mariner, as he sails along and records observations for these purposes, may fancy — and perhaps rightly — that he has traced to the actinometry of the sea one of the physical conditions which, when the depths of the ocean were laid, had its weight with the ALMIGHTY ARCHITECT.

EXPLANATION OF THE PLATES

PLATE I — This plate combines in its construction the results of 1,159,353 separate observations on the force and direction of the wind, and a little upward of 100,000 observations on the height of the barometer at sea. The wind observations embrace a period of eight hours each, or three during the twenty-four hours. Each one of the barometric observations expresses the mean height of the barometer for the day; therefore each one of the 100,000 may itself be the mean of many, or it may be only one. Suffice it to say, that 83,334 of them were obtained by Lieutenant Andrau from the logs of Dutch ships during their voyages to and fro between the parallels of 50° N. and 36° S.; that nearly 6000 of them were made south of the parallel of 36° S., and obtained from the log-books at the Observatory in Washington; that for the others at sea I am indebted to the observations of Captain Wilkes, of the Exploring Expedition, of Sir James Clark Ross, on board the "*Erebus*" and "*Terror*," in high southern latitudes, and of Dr. Kane in the Arctic Ocean. Besides these, others made near the sea have been used, as those at Greenwich, St. Petersburg, Hobart Town, etc., making upward of 100,000 in all. This profile shows how unequally the atmosphere is divided by the equator.

The arrows *within* the circle fly with the wind. They represent its mean annual direction from each quarter, and by bands 5° of latitude in breadth, and according to actual observation at *sea*. They show by their length the annual duration of the wind in months. They are on a scale of one twentieth of an inch per month, except the half-bearded arrows, which are on a scale twice as great, or one tenth of an inch to the month. It will thus be perceived at a glance that the winds of the longest duration are the S.E. trades, between the parallels of 5° and 10° south, where the long feathered arrows represent an annual average of ten months.

The most prevalent winds in each band are represented by full-feathered arrows; the next by half-feathered, except between the parallels of 30° and 35° N., where the N.E. and S.W. winds, and between the parallels of 35° and 40°, where the N.W. and S.W. winds contend for the mastery as to average annual duration.

The rows of arrows on each side of the axis, and nearest to it, are projected with the utmost care as to direction, and length or duration.

The feathered arrows in the shading around the circle represent the crossing at the calm belts, and the great equatorial and polar movements by upper and lower currents of air in its general system of atmospherical circulation.

The small featherless and curved arrows, *n, q, r, s*, on the shading around the circle, are intended to suggest how the trade-winds, as they cross parallels of larger and larger circumference on their way to the equator, act as an undertow, and draw supplies of pure air down from the counter-current above; which supplies are required to satisfy the increasing demands of these winds; for, as they near the equator, they not only cross parallels of larger circumference, but, as actual observations show, they also greatly increase both their duration and velocity. In like manner, the counter-trades, as they approach the poles, are going from latitudes where the parallels are larger to latitudes where the parallels are smaller. In other words, they diminish, as they approach the poles, the area of their vertical section; consequently there is a crowding out — a sloughing off from the lower current, and a joining and a turning back with the upper current. This phenomenon is represented by the small featherless and curved arrows in the periphery on the polar side of the calm belts of Cancer and Capricorn.

This dotted or shaded periphery is intended to represent a profile view of the atmosphere as suggested by the readings of the barometer at sea. This method of delineating the atmosphere is resorted to in order to show the unequal distribution of the atmosphere, particularly on the polar side of lat. 40° S.; also the piling up over the calm belts, and the depression — barometrical — over the equatorial calms and cloud ring.

The engirdling seas of the extra-tropical south suggest at once the cause of this inequality in the arrangement over them of the airy covering of our planet. Excepting a small portion of South America, the belt between the parallels of 40° and 65° or 70° south may be considered to consist entirely of sea. This immense area of water surface keeps the atmosphere continually saturated with vapor. The specific gravity of common atmospheric air being taken as unity, that of aqueous vapor is about 0.6; consequently the atmosphere is expelled thence by the *steam*, if, for the sake of explanation, we may so call the vapor which is continually rising up from this immense boiler. This vapor displaces a certain portion of air, occupies its place, and, being one third lighter, also makes lighter the barometric column. Moreover, being lighter, it mounts up into the cloud region, where it is condensed either into clouds or rain, and the latent heat that is set free in the process assists still farther to lessen the barometric column; for the heat thus liberated warms and expands the upper air, causing it to swell out above its proper level, and so flow back toward the equator with the upper current of these regions.

Thus, though the barometer stands so low as to show that there is less atmosphere over high southern latitudes than there is in corresponding latitudes north, yet, if it were visible and we could see it, we should discover, owing to the effect of this vapor and the liberation of its latent heat, and the resulting intumescence of the lighter air over the austral regions, the actual height of this invisible covering to be higher there than it is in the boreal regions.

Taking the mean height of the barometer for the northern hemisphere to be 30 inches, and taking the 100,000 barometric observations used as data for the construction of this diagram to be correct, we have facts for the assertion that in the austral regions the quantity of air that this vapor permanently expels thence is from one twelfth to one fifteenth of the whole quantity that belongs to corresponding latitudes north — a curious, most interesting, and suggestive physical revelation.

PLATE II (§ 781) is a section taken from one of the manuscript charts at the Observatory. It illustrates the method adopted there for co-ordinating for the Pilot Charts the winds as reported in the abstract logs. For this purpose the ocean is divided into convenient sections, usually five degrees of latitude by five degrees of longitude. These parallelograms are then subdivided into a system of engraved squares, the months of the year being the ordinates, and the points of the compass being the abscissæ. As the wind is reported by a vessel that passes through any part of the parallelogram, so is it assumed to have been at that time all over the parallelogram. From such investigations as this the Pilot Charts are constructed.

PLATE III illustrates the position of the channel of the Gulf Stream (Chap. II) for summer and winter. The diagram A shows a thermometrical profile presented by cross-sections of the Gulf Stream, according to observations made by the hydrographical parties of the United States Coast Survey. The elements for this diagram were kindly furnished me by the superintendent of that work. They are from a paper on the Gulf Stream, read by him before the American Association for the Advancement of Science at its meeting in Washington, 1854. Imagine a vessel to sail from the Capes of Virginia straight out to sea, crossing the Gulf Stream at right angles, and taking the temperature of its waters at the surface and at various depths. The diagram shows the elevation and depression of the thermometer across this section as they were actually observed by such a vessel.

The black lines x, y, z, in the Gulf Stream, show the course which those threads of warm waters take (§ 130). The lines a, b show the computed drift route that the unfortunate steamer San Francisco would take after her terrible disaster in December, 1853.

PLATE IV is intended to show how the winds may become geological agents. It shows where the winds that, in the general system of atmospherical circulation, blow over the deserts and thirsty lands in Asia and Africa (where the annual amount of precipitation is small) are supposed to get their vapors from; where, as surface winds, they are supposed to condense portions of it; and whither they are supposed to transport the residue thereof through the upper regions, retaining it until they again become surface winds.

PLATE V shows the prevailing direction of the wind during the year in all parts of the ocean. It also shows the principal routes across the seas to various places. Where the cross-lines representing the yards are oblique to the keel of the vessel, they indicate that the winds are, for the most part, ahead; when perpendicular or square, that the winds are, for the most part, fair. The figures on or near the diagrams representing the vessels show the average length of the passage in days.

The arrows denote the prevailing direction of the wind; they are supposed to fly *with* it; so that the wind is going as the arrows point. The half-bearded and half-feathered arrows represent monsoons (§ 630), and the stippled or shaded belts the calm zones.

In the regions on the polar side of the calms of Capricorn and of Cancer, where the arrows are flying both from the northwest and the southwest, the idea intended to be conveyed is, that the prevailing direction of the wind is between the northwest and the southwest, and that their frequency is from these two quarters in proportion to the number of arrows.

PLATE VI is intended to show the present state of our knowledge with regard to the drift of the ocean, or, more properly, with regard to the great flow of polar and equatorial waters, and their channels of circulation as indicated by the thermometer (§ 742). Farther researches will enable us to improve this chart. The sargasso seas and the most favorite places of resort for the whale — *right* in cold, and *sperm* in warm water — are also exhibited on this chart.

PLATES VII and VIII speak for themselves. They are orographic for the North Atlantic Ocean, and exhibit completely the present state of our knowledge with regard to the elevations and depressions in the bed of that sea as derived from the deep-sea soundings taken by the American and English navies from the commencement of the system to Dayman's soundings in the Bay of Biscay, 1859; Plate VIII exhibiting a vertical section of the Atlantic, and showing the contrasts of its bottom with the sea-level in a line from Mexico across Yucatan, Cuba, San Domingo, and the Cape de Verds, to the coast of Africa, marked A on Plate VII.

PLATE IX — The data for this Plate are furnished by Maury's Storm and Rain Charts, including observations for 107,277 days in the North Atlantic, and 158,025 in the South; collated by Lieutenant J. J. Guthrie, at the Washington Observatory, in 1855.

The heavy vertical lines, 5°, 10°, 15°, etc., represent parallels of latitude; the other vertical lines, months; and the horizontal lines, per cents., or the number of days in a hundred.

The continuous curve line stands for phenomena in the North, and the broken curve line for phenomena in the South Atlantic. Thus the Gales' Curve shows that in every hundred days, and on the average, in the month of January of different years, there have been observed, in the northern hemisphere, 36 gales (36 per cent.) between the parallels of 50° and 55°; whereas during the same time and between the same parallels in the southern hemisphere, only 10 gales on the average (10 per cent.) have been reported.

The fact is here developed that the atmosphere is in a more unstable condition in the North than in the South Atlantic; that we have more calms, more rains, more fogs, more gales, and more thunder in the northern than in the southern hemisphere, particularly between the equator and the 55th parallel. Beyond that the influence of Cape Horn becomes manifest.

PLATE X (§ 839) shows the limits of the unexplored area about the south pole.

INDEX

PLATES

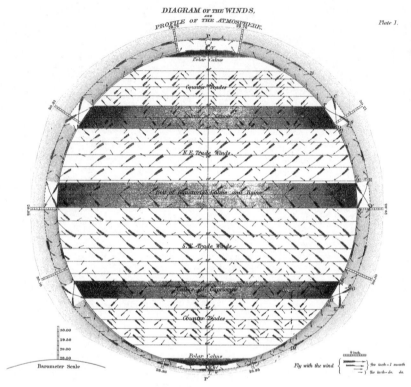

DIAGRAM OF THE WINDS,
AND
PROFILE OF THE ATMOSPHERE.

Plate 1.

Polar Calms

Counter Trades

Calms of Cancer

N.E. Trade Winds

Belt of Equatorial Calms and Rains

S.E. Trade Winds

Calms of Capricorn

Counter Trades

Polar Calms

30.00
29.50
29.00
28.50
Barometer Scale

Fly with the wind

Lieut. R. L. May, Delt.

Oliver J. Stuart, Sc.

110° East 115° 120° E.

20° N. 20° North

North

N.N.E.

N.E.

E.N.E. B

East 175

E.S.E. 164 D

S.E. 128

S.S.E. 78

South 102 180

S.S.W. 126 56

S.W. 72 18

W.S.W. 117 119 156

West 83 137 256

W.N.W. 195 76 160

N.W. 118 146 73

N.N.W. 168 76 15° North

Dec. Jan. Feb. March April May June July Aug. Sept. Oct. Nov. Dec. Jan. Feb. March April May June July Aug. Sept. Oct. Nov.

10° N. 10° North

North

N.N.E.

N.E.

E.N.E. 149

East 93

E.S.E. 28 70 C

S.E. 165 42 85

S.S.E. 131 75 64

South 121 18

S.S.W. 79 12

S.W. 193 19

W.S.W. 221 36

West 124 279

W.N.W. 132 239

N.W. 175 285

N.N.W. 5° North

105° East 110° 115° E.

D. McClelland Sc.

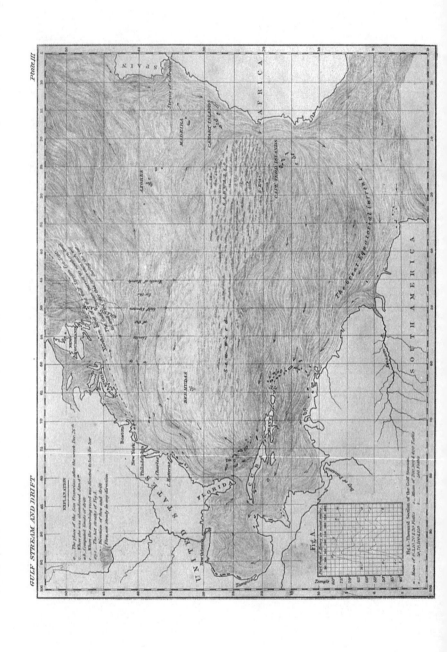

GULF STREAM AND DRIFT

Plate III

Fig. A.—Thermal Section of the Gulf Stream.

Plate IV

GEOLOGICAL AGENCY OF THE WINDS

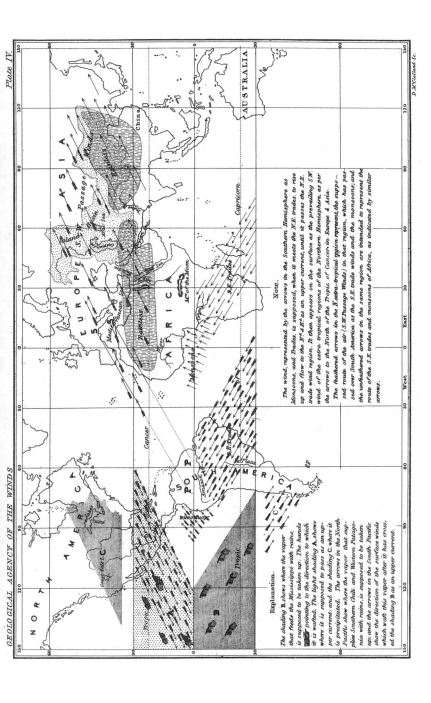

D. McClelland Sc.

Note.

The wind, represented by the arrows in the Southern Hemisphere as Monsoons, and Trades, is supposed, when it meets the N.E. trades, to rise up and flow to the N°. 48° as an upper current, until it passes the N.E. trade wind region. It then appears on the surface as the prevailing S.W. wind of the extra-tropical regions of the Northern Hemisphere, as per the arrows to the North of the Tropic of Cancer in Europe & Asia.

The feathered arrows in the Neutro-tropical region represent the supposed route of the air-(S.W.Passage Winds) in that region, which has passed over South America as the S.E. trade winds and the monsoons; and the unfeathered arrows in the same region are intended to represent the route of the S.E. trades and monsoons of Africa, as indicated by similar arrows.

Explanation.

The shading B. shows where the vapor that feeds the Mississippi with rains, is supposed to be taken up. The hands ☞ pointing to the direction to which it is wafted. The light shading A. shows where it is supposed to pass as an upper current: and the shading C. where it is precipitated. The arrows in the North Pacific show where the vapor that supplies Southern Chili and Western Patagonia with rains, is supposed to be taken up; and the arrows in the South Pacific show the direction of the surface winds which waft this vapor after it has crossed the shading B as an upper current.

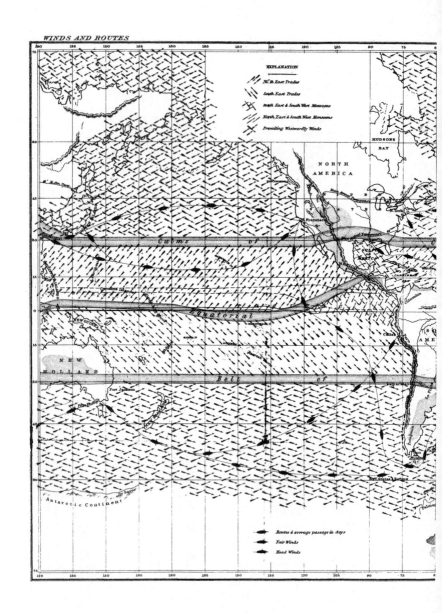

Plate V

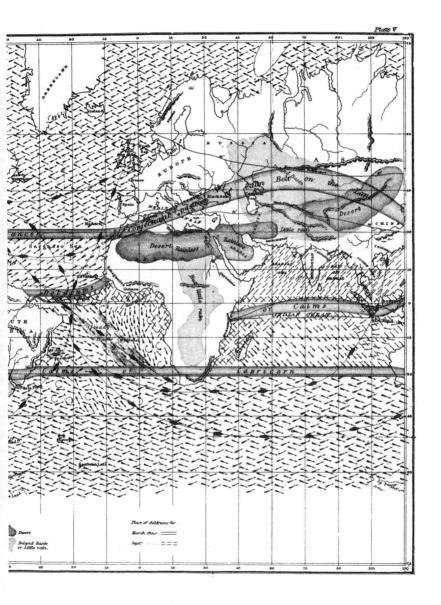

Desert

Inland Basin
or Little rain.

Place of doldrums for

March thus

Sept.

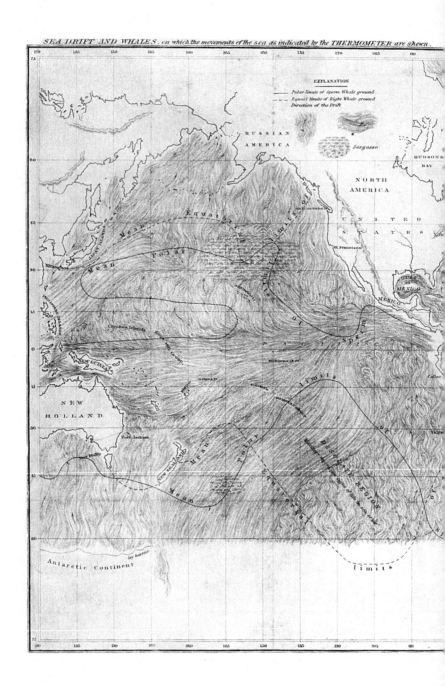

Plate VI

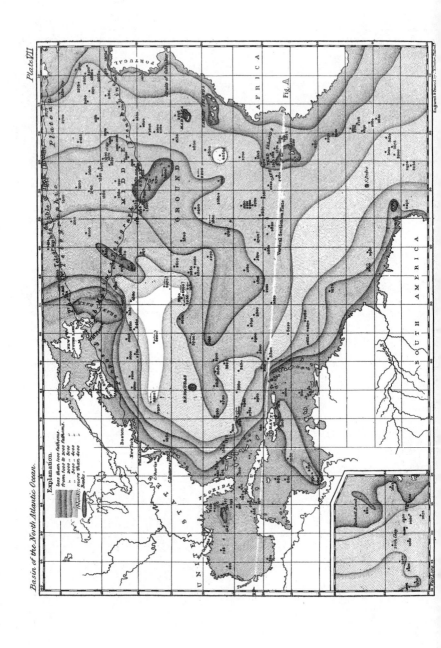

Plate VII.

Basin of the North Atlantic Ocean.

Explanation.

less than 1000 fathoms.
from 1000 to 2000 fathoms.
2000 .. 3000 ..
3000 .. 4000 ..
more than 4000.
Banks.

Fig. A.

UNITED STATES

FLORIDA

BERMUDAS

HAYTI

SOUTH AMERICA

AFRICA

PORTUGAL

MIDDLE GROUND

Plateau

Boston

Vertical Section Plate.

Engraved & Printed by J. Hullmandel. London.

PLATE VII

VERTICAL SECTION—NORTH ATLANTIC

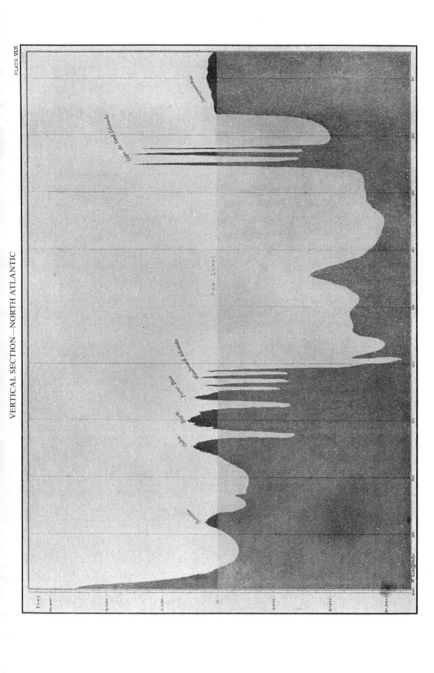

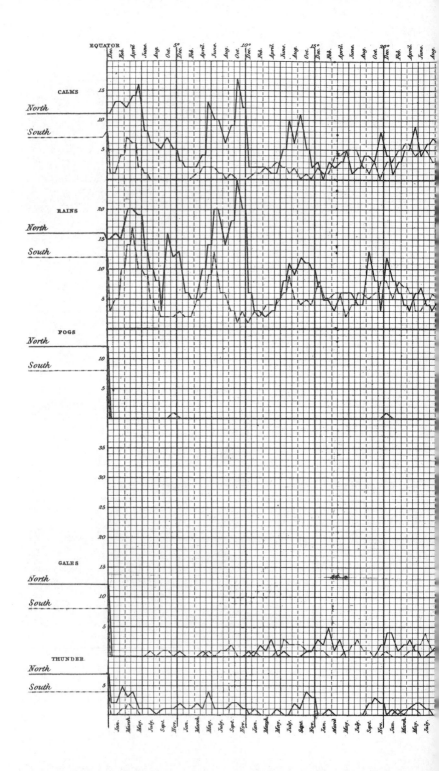

AND RAINS IN NORTH AND SOUTH ATLANTIC.
(West Indies not included.)

Plate IX.

PlateX.

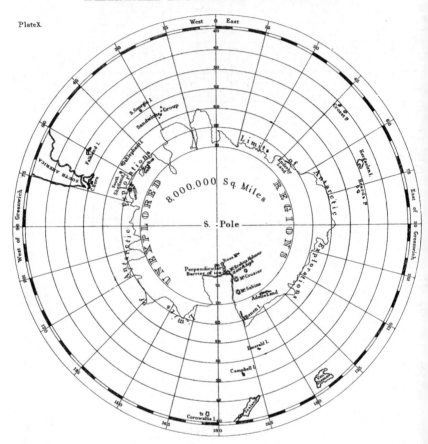

A CATALOG OF SELECTED
DOVER BOOKS
IN ALL FIELDS OF INTEREST

A CATALOG OF SELECTED DOVER
BOOKS IN ALL FIELDS OF INTEREST

CONCERNING THE SPIRITUAL IN ART, Wassily Kandinsky. Pioneering work by father of abstract art. Thoughts on color theory, nature of art. Analysis of earlier masters. 12 illustrations. 80pp. of text. 5⅜ x 8½.   23411-8

ANIMALS: 1,419 Copyright-Free Illustrations of Mammals, Birds, Fish, Insects, etc., Jim Harter (ed.). Clear wood engravings present, in extremely lifelike poses, over 1,000 species of animals. One of the most extensive pictorial sourcebooks of its kind. Captions. Index. 284pp. 9 x 12.   23766-4

CELTIC ART: The Methods of Construction, George Bain. Simple geometric techniques for making Celtic interlacements, spirals, Kells-type initials, animals, humans, etc. Over 500 illustrations. 160pp. 9 x 12. (Available in U.S. only.)   22923-8

AN ATLAS OF ANATOMY FOR ARTISTS, Fritz Schider. Most thorough reference work on art anatomy in the world. Hundreds of illustrations, including selections from works by Vesalius, Leonardo, Goya, Ingres, Michelangelo, others. 593 illustrations. 192pp. 7⅛ x 10¼.   20241-0

CELTIC HAND STROKE-BY-STROKE (Irish Half-Uncial from "The Book of Kells"): An Arthur Baker Calligraphy Manual, Arthur Baker. Complete guide to creating each letter of the alphabet in distinctive Celtic manner. Covers hand position, strokes, pens, inks, paper, more. Illustrated. 48pp. 8¼ x 11.   24336-2

EASY ORIGAMI, John Montroll. Charming collection of 32 projects (hat, cup, pelican, piano, swan, many more) specially designed for the novice origami hobbyist. Clearly illustrated easy-to-follow instructions insure that even beginning papercrafters will achieve successful results. 48pp. 8¼ x 11.   27298-2

THE COMPLETE BOOK OF BIRDHOUSE CONSTRUCTION FOR WOODWORKERS, Scott D. Campbell. Detailed instructions, illustrations, tables. Also data on bird habitat and instinct patterns. Bibliography. 3 tables. 63 illustrations in 15 figures. 48pp. 5¼ x 8½.   24407-5

BLOOMINGDALE'S ILLUSTRATED 1886 CATALOG: Fashions, Dry Goods and Housewares, Bloomingdale Brothers. Famed merchants' extremely rare catalog depicting about 1,700 products: clothing, housewares, firearms, dry goods, jewelry, more. Invaluable for dating, identifying vintage items. Also, copyright-free graphics for artists, designers. Co-published with Henry Ford Museum & Greenfield Village. 160pp. 8¼ x 11.   25780-0

HISTORIC COSTUME IN PICTURES, Braun & Schneider. Over 1,450 costumed figures in clearly detailed engravings–from dawn of civilization to end of 19th century. Captions. Many folk costumes. 256pp. 8⅜ x 11¾.   23150-X

STICKLEY CRAFTSMAN FURNITURE CATALOGS, Gustav Stickley and L. & J. G. Stickley. Beautiful, functional furniture in two authentic catalogs from 1910. 594 illustrations, including 277 photos, show settles, rockers, armchairs, reclining chairs, bookcases, desks, tables. 183pp. 6½ x 9¼.
23838-5

AMERICAN LOCOMOTIVES IN HISTORIC PHOTOGRAPHS: 1858 to 1949, Ron Ziel (ed.). A rare collection of 126 meticulously detailed official photographs, called "builder portraits," of American locomotives that majestically chronicle the rise of steam locomotive power in America. Introduction. Detailed captions. xi+ 129pp. 9 x 12.
27393-8

AMERICA'S LIGHTHOUSES: An Illustrated History, Francis Ross Holland, Jr. Delightfully written, profusely illustrated fact-filled survey of over 200 American lighthouses since 1716. History, anecdotes, technological advances, more. 240pp. 8 x 10¾.
25576-X

TOWARDS A NEW ARCHITECTURE, Le Corbusier. Pioneering manifesto by founder of "International School." Technical and aesthetic theories, views of industry, economics, relation of form to function, "mass-production split" and much more. Profusely illustrated. 320pp. 6⅛ x 9¼. (Available in U.S. only.)
25023-7

HOW THE OTHER HALF LIVES, Jacob Riis. Famous journalistic record, exposing poverty and degradation of New York slums around 1900, by major social reformer. 100 striking and influential photographs. 233pp. 10 x 7⅞.
22012-5

FRUIT KEY AND TWIG KEY TO TREES AND SHRUBS, William M. Harlow. One of the handiest and most widely used identification aids. Fruit key covers 120 deciduous and evergreen species; twig key 160 deciduous species. Easily used. Over 300 photographs. 126pp. 5⅜ x 8½.
20511-8

COMMON BIRD SONGS, Dr. Donald J. Borror. Songs of 60 most common U.S. birds: robins, sparrows, cardinals, bluejays, finches, more—arranged in order of increasing complexity. Up to 9 variations of songs of each species.
Cassette and manual 99911-4

ORCHIDS AS HOUSE PLANTS, Rebecca Tyson Northen. Grow cattleyas and many other kinds of orchids—in a window, in a case, or under artificial light. 63 illustrations. 148pp. 5⅜ x 8½.
23261-1

MONSTER MAZES, Dave Phillips. Masterful mazes at four levels of difficulty. Avoid deadly perils and evil creatures to find magical treasures. Solutions for all 32 exciting illustrated puzzles. 48pp. 8¼ x 11.
26005-4

MOZART'S DON GIOVANNI (DOVER OPERA LIBRETTO SERIES), Wolfgang Amadeus Mozart. Introduced and translated by Ellen H. Bleiler. Standard Italian libretto, with complete English translation. Convenient and thoroughly portable—an ideal companion for reading along with a recording or the performance itself. Introduction. List of characters. Plot summary. 121pp. 5¼ x 8½.
24944-1

TECHNICAL MANUAL AND DICTIONARY OF CLASSICAL BALLET, Gail Grant. Defines, explains, comments on steps, movements, poses and concepts. 15-page pictorial section. Basic book for student, viewer. 127pp. 5⅜ x 8½.
21843-0

THE CLARINET AND CLARINET PLAYING, David Pino. Lively, comprehensive work features suggestions about technique, musicianship, and musical interpretation, as well as guidelines for teaching, making your own reeds, and preparing for public performance. Includes an intriguing look at clarinet history. "A godsend," *The Clarinet*, Journal of the International Clarinet Society. Appendixes. 7 illus. 320pp. 5⅜ x 8½. 40270-3

HOLLYWOOD GLAMOR PORTRAITS, John Kobal (ed.). 145 photos from 1926-49. Harlow, Gable, Bogart, Bacall; 94 stars in all. Full background on photographers, technical aspects. 160pp. 8⅜ x 11¼. 23352-9

THE ANNOTATED CASEY AT THE BAT: A Collection of Ballads about the Mighty Casey/Third, Revised Edition, Martin Gardner (ed.). Amusing sequels and parodies of one of America's best-loved poems: Casey's Revenge, Why Casey Whiffed, Casey's Sister at the Bat, others. 256pp. 5⅜ x 8½. 28598-7

THE RAVEN AND OTHER FAVORITE POEMS, Edgar Allan Poe. Over 40 of the author's most memorable poems: "The Bells," "Ulalume," "Israfel," "To Helen," "The Conqueror Worm," "Eldorado," "Annabel Lee," many more. Alphabetic lists of titles and first lines. 64pp. 5$\frac{3}{16}$ x 8¼. 26685-0

PERSONAL MEMOIRS OF U. S. GRANT, Ulysses Simpson Grant. Intelligent, deeply moving firsthand account of Civil War campaigns, considered by many the finest military memoirs ever written. Includes letters, historic photographs, maps and more. 528pp. 6⅛ x 9¼. 28587-1

ANCIENT EGYPTIAN MATERIALS AND INDUSTRIES, A. Lucas and J. Harris. Fascinating, comprehensive, thoroughly documented text describes this ancient civilization's vast resources and the processes that incorporated them in daily life, including the use of animal products, building materials, cosmetics, perfumes and incense, fibers, glazed ware, glass and its manufacture, materials used in the mummification process, and much more. 544pp. 6⅛ x 9¼. (Available in U.S. only.) 40446-3

RUSSIAN STORIES/RUSSKIE RASSKAZY: A Dual-Language Book, edited by Gleb Struve. Twelve tales by such masters as Chekhov, Tolstoy, Dostoevsky, Pushkin, others. Excellent word-for-word English translations on facing pages, plus teaching and study aids, Russian/English vocabulary, biographical/critical introductions, more. 416pp. 5⅜ x 8½. 26244-8

PHILADELPHIA THEN AND NOW: 60 Sites Photographed in the Past and Present, Kenneth Finkel and Susan Oyama. Rare photographs of City Hall, Logan Square, Independence Hall, Betsy Ross House, other landmarks juxtaposed with contemporary views. Captures changing face of historic city. Introduction. Captions. 128pp. 8¼ x 11. 25790-8

AIA ARCHITECTURAL GUIDE TO NASSAU AND SUFFOLK COUNTIES, LONG ISLAND, The American Institute of Architects, Long Island Chapter, and the Society for the Preservation of Long Island Antiquities. Comprehensive, well-researched and generously illustrated volume brings to life over three centuries of Long Island's great architectural heritage. More than 240 photographs with authoritative, extensively detailed captions. 176pp. 8¼ x 11. 26946-9

NORTH AMERICAN INDIAN LIFE: Customs and Traditions of 23 Tribes, Elsie Clews Parsons (ed.). 27 fictionalized essays by noted anthropologists examine religion, customs, government, additional facets of life among the Winnebago, Crow, Zuni, Eskimo, other tribes. 480pp. 6⅛ x 9¼. 27377-6

CATALOG OF DOVER BOOKS

FRANK LLOYD WRIGHT'S DANA HOUSE, Donald Hoffmann. Pictorial essay of residential masterpiece with over 160 interior and exterior photos, plans, elevations, sketches and studies. 128pp. 9¼ x 10¾. 29120-0

THE MALE AND FEMALE FIGURE IN MOTION: 60 Classic Photographic Sequences, Eadweard Muybridge. 60 true-action photographs of men and women walking, running, climbing, bending, turning, etc., reproduced from rare 19th-century masterpiece. vi + 121pp. 9 x 12. 24745-7

1001 QUESTIONS ANSWERED ABOUT THE SEASHORE, N. J. Berrill and Jacquelyn Berrill. Queries answered about dolphins, sea snails, sponges, starfish, fishes, shore birds, many others. Covers appearance, breeding, growth, feeding, much more. 305pp. 5¼ x 8¼. 23366-9

ATTRACTING BIRDS TO YOUR YARD, William J. Weber. Easy-to-follow guide offers advice on how to attract the greatest diversity of birds: birdhouses, feeders, water and waterers, much more. 96pp. 5³⁄₁₆ x 8¼. 28927-3

MEDICINAL AND OTHER USES OF NORTH AMERICAN PLANTS: A Historical Survey with Special Reference to the Eastern Indian Tribes, Charlotte Erichsen-Brown. Chronological historical citations document 500 years of usage of plants, trees, shrubs native to eastern Canada, northeastern U.S. Also complete identifying information. 343 illustrations. 544pp. 6½ x 9¼. 25951-X

STORYBOOK MAZES, Dave Phillips. 23 stories and mazes on two-page spreads: Wizard of Oz, Treasure Island, Robin Hood, etc. Solutions. 64pp. 8¼ x 11. 23628-5

AMERICAN NEGRO SONGS: 230 Folk Songs and Spirituals, Religious and Secular, John W. Work. This authoritative study traces the African influences of songs sung and played by black Americans at work, in church, and as entertainment. The author discusses the lyric significance of such songs as "Swing Low, Sweet Chariot," "John Henry," and others and offers the words and music for 230 songs. Bibliography. Index of Song Titles. 272pp. 6½ x 9¼. 40271-1

MOVIE-STAR PORTRAITS OF THE FORTIES, John Kobal (ed.). 163 glamor, studio photos of 106 stars of the 1940s: Rita Hayworth, Ava Gardner, Marlon Brando, Clark Gable, many more. 176pp. 8⅜ x 11¼. 23546-7

BENCHLEY LOST AND FOUND, Robert Benchley. Finest humor from early 30s, about pet peeves, child psychologists, post office and others. Mostly unavailable elsewhere. 73 illustrations by Peter Arno and others. 183pp. 5⅜ x 8½. 22410-4

YEKL and THE IMPORTED BRIDEGROOM AND OTHER STORIES OF YIDDISH NEW YORK, Abraham Cahan. Film Hester Street based on *Yekl* (1896). Novel, other stories among first about Jewish immigrants on N.Y.'s East Side. 240pp. 5⅜ x 8½. 22427-9

SELECTED POEMS, Walt Whitman. Generous sampling from *Leaves of Grass*. Twenty-four poems include "I Hear America Singing," "Song of the Open Road," "I Sing the Body Electric," "When Lilacs Last in the Dooryard Bloom'd," "O Captain! My Captain!"—all reprinted from an authoritative edition. Lists of titles and first lines. 128pp. 5³⁄₁₆ x 8¼. 26878-0

THE BEST TALES OF HOFFMANN, E. T. A. Hoffmann. 10 of Hoffmann's most important stories: "Nutcracker and the King of Mice," "The Golden Flowerpot," etc. 458pp. 5⅜ x 8½. 21793-0

FROM FETISH TO GOD IN ANCIENT EGYPT, E. A. Wallis Budge. Rich detailed survey of Egyptian conception of "God" and gods, magic, cult of animals, Osiris, more. Also, superb English translations of hymns and legends. 240 illustrations. 545pp. 5⅜ x 8½. 25803-3

FRENCH STORIES/CONTES FRANÇAIS: A Dual-Language Book, Wallace Fowlie. Ten stories by French masters, Voltaire to Camus: "Micromegas" by Voltaire; "The Atheist's Mass" by Balzac; "Minuet" by de Maupassant; "The Guest" by Camus, six more. Excellent English translations on facing pages. Also French-English vocabulary list, exercises, more. 352pp. 5⅜ x 8½. 26443-2

CHICAGO AT THE TURN OF THE CENTURY IN PHOTOGRAPHS: 122 Historic Views from the Collections of the Chicago Historical Society, Larry A. Viskochil. Rare large-format prints offer detailed views of City Hall, State Street, the Loop, Hull House, Union Station, many other landmarks, circa 1904-1913. Introduction. Captions. Maps. 144pp. 9⅜ x 12¼. 24656-6

OLD BROOKLYN IN EARLY PHOTOGRAPHS, 1865-1929, William Lee Younger. Luna Park, Gravesend race track, construction of Grand Army Plaza, moving of Hotel Brighton, etc. 157 previously unpublished photographs. 165pp. 8⅞ x 11¾. 23587-4

THE MYTHS OF THE NORTH AMERICAN INDIANS, Lewis Spence. Rich anthology of the myths and legends of the Algonquins, Iroquois, Pawnees and Sioux, prefaced by an extensive historical and ethnological commentary. 36 illustrations. 480pp. 5⅜ x 8½. 25967-6

AN ENCYCLOPEDIA OF BATTLES: Accounts of Over 1,560 Battles from 1479 B.C. to the Present, David Eggenberger. Essential details of every major battle in recorded history from the first battle of Megiddo in 1479 B.C. to Grenada in 1984. List of Battle Maps. New Appendix covering the years 1967-1984. Index. 99 illustrations. 544pp. 6½ x 9¼. 24913-1

SAILING ALONE AROUND THE WORLD, Captain Joshua Slocum. First man to sail around the world, alone, in small boat. One of great feats of seamanship told in delightful manner. 67 illustrations. 294pp. 5⅜ x 8½. 20326-3

ANARCHISM AND OTHER ESSAYS, Emma Goldman. Powerful, penetrating, prophetic essays on direct action, role of minorities, prison reform, puritan hypocrisy, violence, etc. 271pp. 5⅜ x 8½. 22484-8

MYTHS OF THE HINDUS AND BUDDHISTS, Ananda K. Coomaraswamy and Sister Nivedita. Great stories of the epics; deeds of Krishna, Shiva, taken from puranas, Vedas, folk tales; etc. 32 illustrations. 400pp. 5⅜ x 8½. 21759-0

THE TRAUMA OF BIRTH, Otto Rank. Rank's controversial thesis that anxiety neurosis is caused by profound psychological trauma which occurs at birth. 256pp. 5⅜ x 8½. 27974-X

A THEOLOGICO-POLITICAL TREATISE, Benedict Spinoza. Also contains unfinished Political Treatise. Great classic on religious liberty, theory of government on common consent. R. Elwes translation. Total of 421pp. 5⅜ x 8½. 20249-6

MY BONDAGE AND MY FREEDOM, Frederick Douglass. Born a slave, Douglass became outspoken force in antislavery movement. The best of Douglass' autobiographies. Graphic description of slave life. 464pp. 5⅜ x 8½. 22457-0

FOLLOWING THE EQUATOR: A Journey Around the World, Mark Twain. Fascinating humorous account of 1897 voyage to Hawaii, Australia, India, New Zealand, etc. Ironic, bemused reports on peoples, customs, climate, flora and fauna, politics, much more. 197 illustrations. 720pp. 5⅜ x 8½. 26113-1

THE PEOPLE CALLED SHAKERS, Edward D. Andrews. Definitive study of Shakers: origins, beliefs, practices, dances, social organization, furniture and crafts, etc. 33 illustrations. 351pp. 5⅜ x 8½. 21081-2

THE MYTHS OF GREECE AND ROME, H. A. Guerber. A classic of mythology, generously illustrated, long prized for its simple, graphic, accurate retelling of the principal myths of Greece and Rome, and for its commentary on their origins and significance. With 64 illustrations by Michelangelo, Raphael, Titian, Rubens, Canova, Bernini and others. 480pp. 5⅜ x 8½. 27584-1

PSYCHOLOGY OF MUSIC, Carl E. Seashore. Classic work discusses music as a medium from psychological viewpoint. Clear treatment of physical acoustics, auditory apparatus, sound perception, development of musical skills, nature of musical feeling, host of other topics. 88 figures. 408pp. 5⅜ x 8½. 21851-1

THE PHILOSOPHY OF HISTORY, Georg W. Hegel. Great classic of Western thought develops concept that history is not chance but rational process, the evolution of freedom. 457pp. 5⅜ x 8½. 20112-0

THE BOOK OF TEA, Kakuzo Okakura. Minor classic of the Orient: entertaining, charming explanation, interpretation of traditional Japanese culture in terms of tea ceremony. 94pp. 5⅜ x 8½. 20070-1

LIFE IN ANCIENT EGYPT, Adolf Erman. Fullest, most thorough, detailed older account with much not in more recent books, domestic life, religion, magic, medicine, commerce, much more. Many illustrations reproduce tomb paintings, carvings, hieroglyphs, etc. 597pp. 5⅜ x 8½. 22632-8

SUNDIALS, Their Theory and Construction, Albert Waugh. Far and away the best, most thorough coverage of ideas, mathematics concerned, types, construction, adjusting anywhere. Simple, nontechnical treatment allows even children to build several of these dials. Over 100 illustrations. 230pp. 5⅜ x 8½. 22947-5

THEORETICAL HYDRODYNAMICS, L. M. Milne-Thomson. Classic exposition of the mathematical theory of fluid motion, applicable to both hydrodynamics and aerodynamics. Over 600 exercises. 768pp. 6⅛ x 9¼. 68970-0

SONGS OF EXPERIENCE: Facsimile Reproduction with 26 Plates in Full Color, William Blake. 26 full-color plates from a rare 1826 edition. Includes "The Tyger," "London," "Holy Thursday," and other poems. Printed text of poems. 48pp. 5¼ x 7. 24636-1

OLD-TIME VIGNETTES IN FULL COLOR, Carol Belanger Grafton (ed.). Over 390 charming, often sentimental illustrations, selected from archives of Victorian graphics—pretty women posing, children playing, food, flowers, kittens and puppies, smiling cherubs, birds and butterflies, much more. All copyright-free. 48pp. 9¼ x 12¼. 27269-9

PERSPECTIVE FOR ARTISTS, Rex Vicat Cole. Depth, perspective of sky and sea, shadows, much more, not usually covered. 391 diagrams, 81 reproductions of drawings and paintings. 279pp. 5⅜ x 8½. 22487-2

DRAWING THE LIVING FIGURE, Joseph Sheppard. Innovative approach to artistic anatomy focuses on specifics of surface anatomy, rather than muscles and bones. Over 170 drawings of live models in front, back and side views, and in widely varying poses. Accompanying diagrams. 177 illustrations. Introduction. Index. 144pp. 8⅜ x11¼. 26723-7

GOTHIC AND OLD ENGLISH ALPHABETS: 100 Complete Fonts, Dan X. Solo. Add power, elegance to posters, signs, other graphics with 100 stunning copyright-free alphabets: Blackstone, Dolbey, Germania, 97 more–including many lower-case, numerals, punctuation marks. 104pp. 8⅛ x 11. 24695-7

HOW TO DO BEADWORK, Mary White. Fundamental book on craft from simple projects to five-bead chains and woven works. 106 illustrations. 142pp. 5⅜ x 8. 20697-1

THE BOOK OF WOOD CARVING, Charles Marshall Sayers. Finest book for beginners discusses fundamentals and offers 34 designs. "Absolutely first rate . . . well thought out and well executed."–E. J. Tangerman. 118pp. 7¾ x 10⅝. 23654-4

ILLUSTRATED CATALOG OF CIVIL WAR MILITARY GOODS: Union Army Weapons, Insignia, Uniform Accessories, and Other Equipment, Schuyler, Hartley, and Graham. Rare, profusely illustrated 1846 catalog includes Union Army uniform and dress regulations, arms and ammunition, coats, insignia, flags, swords, rifles, etc. 226 illustrations. 160pp. 9 x 12. 24939-5

WOMEN'S FASHIONS OF THE EARLY 1900s: An Unabridged Republication of "New York Fashions, 1909," National Cloak & Suit Co. Rare catalog of mail-order fashions documents women's and children's clothing styles shortly after the turn of the century. Captions offer full descriptions, prices. Invaluable resource for fashion, costume historians. Approximately 725 illustrations. 128pp. 8⅜ x 11¼. 27276-1

THE 1912 AND 1915 GUSTAV STICKLEY FURNITURE CATALOGS, Gustav Stickley. With over 200 detailed illustrations and descriptions, these two catalogs are essential reading and reference materials and identification guides for Stickley furniture. Captions cite materials, dimensions and prices. 112pp. 6½ x 9¼. 26676-1

EARLY AMERICAN LOCOMOTIVES, John H. White, Jr. Finest locomotive engravings from early 19th century: historical (1804–74), main-line (after 1870), special, foreign, etc. 147 plates. 142pp. 11⅜ x 8¼. 22772-3

THE TALL SHIPS OF TODAY IN PHOTOGRAPHS, Frank O. Braynard. Lavishly illustrated tribute to nearly 100 majestic contemporary sailing vessels: Amerigo Vespucci, Clearwater, Constitution, Eagle, Mayflower, Sea Cloud, Victory, many more. Authoritative captions provide statistics, background on each ship. 190 black-and-white photographs and illustrations. Introduction. 128pp. 8⅞ x 11¾. 27163-3

LITTLE BOOK OF EARLY AMERICAN CRAFTS AND TRADES, Peter Stockham (ed.). 1807 children's book explains crafts and trades: baker, hatter, cooper, potter, and many others. 23 copperplate illustrations. 140pp. 4⅝ x 6.   23336-7

VICTORIAN FASHIONS AND COSTUMES FROM HARPER'S BAZAR, 1867–1898, Stella Blum (ed.). Day costumes, evening wear, sports clothes, shoes, hats, other accessories in over 1,000 detailed engravings. 320pp. 9⅜ x 12¼. 22990-4

GUSTAV STICKLEY, THE CRAFTSMAN, Mary Ann Smith. Superb study surveys broad scope of Stickley's achievement, especially in architecture. Design philosophy, rise and fall of the Craftsman empire, descriptions and floor plans for many Craftsman houses, more. 86 black-and-white halftones. 31 line illustrations. Introduction 208pp. 6½ x 9¼.   27210-9

THE LONG ISLAND RAIL ROAD IN EARLY PHOTOGRAPHS, Ron Ziel. Over 220 rare photos, informative text document origin (1844) and development of rail service on Long Island. Vintage views of early trains, locomotives, stations, passengers, crews, much more. Captions. 8⅞ x 11¾.   26301-0

VOYAGE OF THE LIBERDADE, Joshua Slocum. Great 19th-century mariner's thrilling, first-hand account of the wreck of his ship off South America, the 35-foot boat he built from the wreckage, and its remarkable voyage home. 128pp. 5⅜ x 8½.
40022-0

TEN BOOKS ON ARCHITECTURE, Vitruvius. The most important book ever written on architecture. Early Roman aesthetics, technology, classical orders, site selection, all other aspects. Morgan translation. 331pp. 5⅜ x 8½.   20645-9

THE HUMAN FIGURE IN MOTION, Eadweard Muybridge. More than 4,500 stopped-action photos, in action series, showing undraped men, women, children jumping, lying down, throwing, sitting, wrestling, carrying, etc. 390pp. 7⅞ x 10⅝.
20204-6 Clothbd.

TREES OF THE EASTERN AND CENTRAL UNITED STATES AND CANADA, William M. Harlow. Best one-volume guide to 140 trees. Full descriptions, woodlore, range, etc. Over 600 illustrations. Handy size. 288pp. 4½ x 6⅜.   20395-6

SONGS OF WESTERN BIRDS, Dr. Donald J. Borror. Complete song and call repertoire of 60 western species, including flycatchers, juncoes, cactus wrens, many more–includes fully illustrated booklet.   Cassette and manual 99913-0

GROWING AND USING HERBS AND SPICES, Milo Miloradovich. Versatile handbook provides all the information needed for cultivation and use of all the herbs and spices available in North America. 4 illustrations. Index. Glossary. 236pp. 5⅜ x 8½.
25058-X

BIG BOOK OF MAZES AND LABYRINTHS, Walter Shepherd. 50 mazes and labyrinths in all–classical, solid, ripple, and more–in one great volume. Perfect inexpensive puzzler for clever youngsters. Full solutions. 112pp. 8⅛ x 11.   22951-3

PIANO TUNING, J. Cree Fischer. Clearest, best book for beginner, amateur. Simple repairs, raising dropped notes, tuning by easy method of flattened fifths. No previous skills needed. 4 illustrations. 201pp. 5⅜ x 8½. 23267-0

HINTS TO SINGERS, Lillian Nordica. Selecting the right teacher, developing confidence, overcoming stage fright, and many other important skills receive thoughtful discussion in this indispensible guide, written by a world-famous diva of four decades' experience. 96pp. 5⅜ x 8½. 40094-8

THE COMPLETE NONSENSE OF EDWARD LEAR, Edward Lear. All nonsense limericks, zany alphabets, Owl and Pussycat, songs, nonsense botany, etc., illustrated by Lear. Total of 320pp. 5⅜ x 8½. (Available in U.S. only.) 20167-8

VICTORIAN PARLOUR POETRY: An Annotated Anthology, Michael R. Turner. 117 gems by Longfellow, Tennyson, Browning, many lesser-known poets. "The Village Blacksmith," "Curfew Must Not Ring Tonight," "Only a Baby Small," dozens more, often difficult to find elsewhere. Index of poets, titles, first lines. xxiii + 325pp. 5⅝ x 8¼. 27044-0

DUBLINERS, James Joyce. Fifteen stories offer vivid, tightly focused observations of the lives of Dublin's poorer classes. At least one, "The Dead," is considered a masterpiece. Reprinted complete and unabridged from standard edition. 160pp. 5³⁄₁₆ x 8¼. 26870-5

GREAT WEIRD TALES: 14 Stories by Lovecraft, Blackwood, Machen and Others, S. T. Joshi (ed.). 14 spellbinding tales, including "The Sin Eater," by Fiona McLeod, "The Eye Above the Mantel," by Frank Belknap Long, as well as renowned works by R. H. Barlow, Lord Dunsany, Arthur Machen, W. C. Morrow and eight other masters of the genre. 256pp. 5⅜ x 8½. (Available in U.S. only.) 40436-6

THE BOOK OF THE SACRED MAGIC OF ABRAMELIN THE MAGE, translated by S. MacGregor Mathers. Medieval manuscript of ceremonial magic. Basic document in Aleister Crowley, Golden Dawn groups. 268pp. 5⅜ x 8½. 23211-5

NEW RUSSIAN-ENGLISH AND ENGLISH-RUSSIAN DICTIONARY, M. A. O'Brien. This is a remarkably handy Russian dictionary, containing a surprising amount of information, including over 70,000 entries. 366pp. 4½ x 6⅛. 20208-9

HISTORIC HOMES OF THE AMERICAN PRESIDENTS, Second, Revised Edition, Irvin Haas. A traveler's guide to American Presidential homes, most open to the public, depicting and describing homes occupied by every American President from George Washington to George Bush. With visiting hours, admission charges, travel routes. 175 photographs. Index. 160pp. 8¼ x 11. 26751-2

NEW YORK IN THE FORTIES, Andreas Feininger. 162 brilliant photographs by the well-known photographer, formerly with *Life* magazine. Commuters, shoppers, Times Square at night, much else from city at its peak. Captions by John von Hartz. 181pp. 9¼ x 10¾. 23585-8

INDIAN SIGN LANGUAGE, William Tomkins. Over 525 signs developed by Sioux and other tribes. Written instructions and diagrams. Also 290 pictographs. 111pp. 6⅛ x 9¼. 22029-X

ANATOMY: A Complete Guide for Artists, Joseph Sheppard. A master of figure drawing shows artists how to render human anatomy convincingly. Over 460 illustrations. 224pp. 8⅜ x 11¼. 27279-6

MEDIEVAL CALLIGRAPHY: Its History and Technique, Marc Drogin. Spirited history, comprehensive instruction manual covers 13 styles (ca. 4th century through 15th). Excellent photographs; directions for duplicating medieval techniques with modern tools. 224pp. 8⅜ x 11¼. 26142-5

DRIED FLOWERS: How to Prepare Them, Sarah Whitlock and Martha Rankin. Complete instructions on how to use silica gel, meal and borax, perlite aggregate, sand and borax, glycerine and water to create attractive permanent flower arrangements. 12 illustrations. 32pp. 5⅜ x 8½. 21802-3

EASY-TO-MAKE BIRD FEEDERS FOR WOODWORKERS, Scott D. Campbell. Detailed, simple-to-use guide for designing, constructing, caring for and using feeders. Text, illustrations for 12 classic and contemporary designs. 96pp. 5⅜ x 8½. 25847-5

SCOTTISH WONDER TALES FROM MYTH AND LEGEND, Donald A. Mackenzie. 16 lively tales tell of giants rumbling down mountainsides, of a magic wand that turns stone pillars into warriors, of gods and goddesses, evil hags, powerful forces and more. 240pp. 5⅜ x 8½. 29677-6

THE HISTORY OF UNDERCLOTHES, C. Willett Cunnington and Phyllis Cunnington. Fascinating, well-documented survey covering six centuries of English undergarments, enhanced with over 100 illustrations: 12th-century laced-up bodice, footed long drawers (1795), 19th-century bustles, 19th-century corsets for men, Victorian "bust improvers," much more. 272pp. 5⅝ x 8¼. 27124-2

ARTS AND CRAFTS FURNITURE: The Complete Brooks Catalog of 1912, Brooks Manufacturing Co. Photos and detailed descriptions of more than 150 now very collectible furniture designs from the Arts and Crafts movement depict davenports, settees, buffets, desks, tables, chairs, bedsteads, dressers and more, all built of solid, quarter-sawed oak. Invaluable for students and enthusiasts of antiques, Americana and the decorative arts. 80pp. 6½ x 9¼. 27471-3

WILBUR AND ORVILLE: A Biography of the Wright Brothers, Fred Howard. Definitive, crisply written study tells the full story of the brothers' lives and work. A vividly written biography, unparalleled in scope and color, that also captures the spirit of an extraordinary era. 560pp. 6⅛ x 9¼. 40297-5

THE ARTS OF THE SAILOR: Knotting, Splicing and Ropework, Hervey Garrett Smith. Indispensable shipboard reference covers tools, basic knots and useful hitches; handsewing and canvas work, more. Over 100 illustrations. Delightful reading for sea lovers. 256pp. 5⅜ x 8½. 26440-8

FRANK LLOYD WRIGHT'S FALLINGWATER: The House and Its History, Second, Revised Edition, Donald Hoffmann. A total revision–both in text and illustrations–of the standard document on Fallingwater, the boldest, most personal architectural statement of Wright's mature years, updated with valuable new material from the recently opened Frank Lloyd Wright Archives. "Fascinating"–*The New York Times.* 116 illustrations. 128pp. 9¼ x 10¾. 27430-6

PHOTOGRAPHIC SKETCHBOOK OF THE CIVIL WAR, Alexander Gardner. 100 photos taken on field during the Civil War. Famous shots of Manassas Harper's Ferry, Lincoln, Richmond, slave pens, etc. 244pp. 10⅝ x 8¼.   22731-6

FIVE ACRES AND INDEPENDENCE, Maurice G. Kains. Great back-to-the-land classic explains basics of self-sufficient farming. The one book to get. 95 illustrations. 397pp. 5⅜ x 8½.   20974-1

SONGS OF EASTERN BIRDS, Dr. Donald J. Borror. Songs and calls of 60 species most common to eastern U.S.: warblers, woodpeckers, flycatchers, thrushes, larks, many more in high-quality recording.   Cassette and manual 99912-2

A MODERN HERBAL, Margaret Grieve. Much the fullest, most exact, most useful compilation of herbal material. Gigantic alphabetical encyclopedia, from aconite to zedoary, gives botanical information, medical properties, folklore, economic uses, much else. Indispensable to serious reader. 161 illustrations. 888pp. 6½ x 9¼. 2-vol. set. (Available in U.S. only.)   Vol. I: 22798-7
                                                                Vol. II: 22799-5

HIDDEN TREASURE MAZE BOOK, Dave Phillips. Solve 34 challenging mazes accompanied by heroic tales of adventure. Evil dragons, people-eating plants, blood-thirsty giants, many more dangerous adversaries lurk at every twist and turn. 34 mazes, stories, solutions. 48pp. 8¼ x 11.   24566-7

LETTERS OF W. A. MOZART, Wolfgang A. Mozart. Remarkable letters show bawdy wit, humor, imagination, musical insights, contemporary musical world; includes some letters from Leopold Mozart. 276pp. 5⅜ x 8½.   22859-2

BASIC PRINCIPLES OF CLASSICAL BALLET, Agrippina Vaganova. Great Russian theoretician, teacher explains methods for teaching classical ballet. 118 illus-trations. 175pp. 5⅜ x 8½.   22036-2

THE JUMPING FROG, Mark Twain. Revenge edition. The original story of The Celebrated Jumping Frog of Calaveras County, a hapless French translation, and Twain's hilarious "retranslation" from the French. 12 illustrations. 66pp. 5⅜ x 8½.
                                                                22686-7

BEST REMEMBERED POEMS, Martin Gardner (ed.). The 126 poems in this superb collection of 19th- and 20th-century British and American verse range from Shelley's "To a Skylark" to the impassioned "Renascence" of Edna St. Vincent Millay and to Edward Lear's whimsical "The Owl and the Pussycat." 224pp. 5⅜ x 8½.
                                                                27165-X

COMPLETE SONNETS, William Shakespeare. Over 150 exquisite poems deal with love, friendship, the tyranny of time, beauty's evanescence, death and other themes in language of remarkable power, precision and beauty. Glossary of archaic terms. 80pp. 5³⁄₁₆ x 8¼.   26686-9

THE BATTLES THAT CHANGED HISTORY, Fletcher Pratt. Eminent historian profiles 16 crucial conflicts, ancient to modern, that changed the course of civiliza-tion. 352pp. 5⅜ x 8½.   41129-X

THE WIT AND HUMOR OF OSCAR WILDE, Alvin Redman (ed.). More than 1,000 ripostes, paradoxes, wisecracks: Work is the curse of the drinking classes; I can resist everything except temptation; etc. 258pp. 5⅜ x 8½.　　　　20602-5

SHAKESPEARE LEXICON AND QUOTATION DICTIONARY, Alexander Schmidt. Full definitions, locations, shades of meaning in every word in plays and poems. More than 50,000 exact quotations. 1,485pp. 6½ x 9¼. 2-vol. set.
Vol. 1: 22726-X
Vol. 2: 22727-8

SELECTED POEMS, Emily Dickinson. Over 100 best-known, best-loved poems by one of America's foremost poets, reprinted from authoritative early editions. No comparable edition at this price. Index of first lines. 64pp. 5³⁄₁₆ x 8¼.　　　26466-1

THE INSIDIOUS DR. FU-MANCHU, Sax Rohmer. The first of the popular mystery series introduces a pair of English detectives to their archnemesis, the diabolical Dr. Fu-Manchu. Flavorful atmosphere, fast-paced action, and colorful characters enliven this classic of the genre. 208pp. 5³⁄₁₆ x 8¼.　　　29898-1

THE MALLEUS MALEFICARUM OF KRAMER AND SPRENGER, translated by Montague Summers. Full text of most important witchhunter's "bible," used by both Catholics and Protestants. 278pp. 6⅝ x 10.　　　　22802-9

SPANISH STORIES/CUENTOS ESPAÑOLES: A Dual-Language Book, Angel Flores (ed.). Unique format offers 13 great stories in Spanish by Cervantes, Borges, others. Faithful English translations on facing pages. 352pp. 5⅜ x 8½.　　　25399-6

GARDEN CITY, LONG ISLAND, IN EARLY PHOTOGRAPHS, 1869–1919, Mildred H. Smith. Handsome treasury of 118 vintage pictures, accompanied by carefully researched captions, document the Garden City Hotel fire (1899), the Vanderbilt Cup Race (1908), the first airmail flight departing from the Nassau Boulevard Aerodrome (1911), and much more. 96pp. 8⅞ x 11¾.　　　　40669-5

OLD QUEENS, N.Y., IN EARLY PHOTOGRAPHS, Vincent F. Seyfried and William Asadorian. Over 160 rare photographs of Maspeth, Jamaica, Jackson Heights, and other areas. Vintage views of DeWitt Clinton mansion, 1939 World's Fair and more. Captions. 192pp. 8⅞ x 11.　　　26358-4

CAPTURED BY THE INDIANS: 15 Firsthand Accounts, 1750-1870, Frederick Drimmer. Astounding true historical accounts of grisly torture, bloody conflicts, relentless pursuits, miraculous escapes and more, by people who lived to tell the tale. 384pp. 5⅜ x 8½.　　　　24901-8

THE WORLD'S GREAT SPEECHES (Fourth Enlarged Edition), Lewis Copeland, Lawrence W. Lamm, and Stephen J. McKenna. Nearly 300 speeches provide public speakers with a wealth of updated quotes and inspiration–from Pericles' funeral oration and William Jennings Bryan's "Cross of Gold Speech" to Malcolm X's powerful words on the Black Revolution and Earl of Spenser's tribute to his sister, Diana, Princess of Wales. 944pp. 5⅜ x 8⅜.　　　　40903-1

THE BOOK OF THE SWORD, Sir Richard F. Burton. Great Victorian scholar/adventurer's eloquent, erudite history of the "queen of weapons"–from prehistory to early Roman Empire. Evolution and development of early swords, variations (sabre, broadsword, cutlass, scimitar, etc.), much more. 336pp. 6⅛ x 9¼.　　　25434-8

CATALOG OF DOVER BOOKS

AUTOBIOGRAPHY: The Story of My Experiments with Truth, Mohandas K. Gandhi. Boyhood, legal studies, purification, the growth of the Satyagraha (nonviolent protest) movement. Critical, inspiring work of the man responsible for the freedom of India. 480pp. 5⅜ x 8½. (Available in U.S. only.) 24593-4

CELTIC MYTHS AND LEGENDS, T. W. Rolleston. Masterful retelling of Irish and Welsh stories and tales. Cuchulain, King Arthur, Deirdre, the Grail, many more. First paperback edition. 58 full-page illustrations. 512pp. 5⅜ x 8½. 26507-2

THE PRINCIPLES OF PSYCHOLOGY, William James. Famous long course complete, unabridged. Stream of thought, time perception, memory, experimental methods; great work decades ahead of its time. 94 figures. 1,391pp. 5⅜ x 8½. 2-vol. set.
Vol. I: 20381-6 Vol. II: 20382-4

THE WORLD AS WILL AND REPRESENTATION, Arthur Schopenhauer. Definitive English translation of Schopenhauer's life work, correcting more than 1,000 errors, omissions in earlier translations. Translated by E. F. J. Payne. Total of 1,269pp. 5⅜ x 8½. 2-vol. set. Vol. 1: 21761-2 Vol. 2: 21762-0

MAGIC AND MYSTERY IN TIBET, Madame Alexandra David-Neel. Experiences among lamas, magicians, sages, sorcerers, Bonpa wizards. A true psychic discovery. 32 illustrations. 321pp. 5⅜ x 8½. (Available in U.S. only.) 22682-4

THE EGYPTIAN BOOK OF THE DEAD, E. A. Wallis Budge. Complete reproduction of Ani's papyrus, finest ever found. Full hieroglyphic text, interlinear transliteration, word-for-word translation, smooth translation. 533pp. 6½ x 9¼. 21866-X

MATHEMATICS FOR THE NONMATHEMATICIAN, Morris Kline. Detailed, college-level treatment of mathematics in cultural and historical context, with numerous exercises. Recommended Reading Lists. Tables. Numerous figures. 641pp. 5⅜ x 8½.
24823-2

PROBABILISTIC METHODS IN THE THEORY OF STRUCTURES, Isaac Elishakoff. Well-written introduction covers the elements of the theory of probability from two or more random variables, the reliability of such multivariable structures, the theory of random function, Monte Carlo methods of treating problems incapable of exact solution, and more. Examples. 502pp. 5⅜ x 8½. 40691-1

THE RIME OF THE ANCIENT MARINER, Gustave Doré, S. T. Coleridge. Doré's finest work; 34 plates capture moods, subtleties of poem. Flawless full-size reproductions printed on facing pages with authoritative text of poem. "Beautiful. Simply beautiful."–*Publisher's Weekly.* 77pp. 9¼ x 12. 22305-1

NORTH AMERICAN INDIAN DESIGNS FOR ARTISTS AND CRAFTSPEOPLE, Eva Wilson. Over 360 authentic copyright-free designs adapted from Navajo blankets, Hopi pottery, Sioux buffalo hides, more. Geometrics, symbolic figures, plant and animal motifs, etc. 128pp. 8⅜ x 11. (Not for sale in the United Kingdom.) 25341-4

SCULPTURE: Principles and Practice, Louis Slobodkin. Step-by-step approach to clay, plaster, metals, stone; classical and modern. 253 drawings, photos. 255pp. 8⅛ x 11.
22960-2

THE INFLUENCE OF SEA POWER UPON HISTORY, 1660–1783, A. T. Mahan. Influential classic of naval history and tactics still used as text in war colleges. First paperback edition. 4 maps. 24 battle plans. 640pp. 5⅜ x 8½. 25509-3

THE STORY OF THE TITANIC AS TOLD BY ITS SURVIVORS, Jack Winocour (ed.). What it was really like. Panic, despair, shocking inefficiency, and a little heroism. More thrilling than any fictional account. 26 illustrations. 320pp. 5⅜ x 8½.

20610-6

FAIRY AND FOLK TALES OF THE IRISH PEASANTRY, William Butler Yeats (ed.). Treasury of 64 tales from the twilight world of Celtic myth and legend: "The Soul Cages," "The Kildare Pooka," "King O'Toole and his Goose," many more. Introduction and Notes by W. B. Yeats. 352pp. 5⅜ x 8½.

26941-8

BUDDHIST MAHAYANA TEXTS, E. B. Cowell and others (eds.). Superb, accurate translations of basic documents in Mahayana Buddhism, highly important in history of religions. The Buddha-karita of Asvaghosha, Larger Sukhavativyuha, more. 448pp. 5⅜ x 8½.

25552-2

ONE TWO THREE . . . INFINITY: Facts and Speculations of Science, George Gamow. Great physicist's fascinating, readable overview of contemporary science: number theory, relativity, fourth dimension, entropy, genes, atomic structure, much more. 128 illustrations. Index. 352pp. 5⅜ x 8½.

25664-2

EXPERIMENTATION AND MEASUREMENT, W. J. Youden. Introductory manual explains laws of measurement in simple terms and offers tips for achieving accuracy and minimizing errors. Mathematics of measurement, use of instruments, experimenting with machines. 1994 edition. Foreword. Preface. Introduction. Epilogue. Selected Readings. Glossary. Index. Tables and figures. 128pp. 5⅜ x 8½.

40451-X

DALÍ ON MODERN ART: The Cuckolds of Antiquated Modern Art, Salvador Dalí. Influential painter skewers modern art and its practitioners. Outrageous evaluations of Picasso, Cézanne, Turner, more. 15 renderings of paintings discussed. 44 calligraphic decorations by Dalí. 96pp. 5⅜ x 8½. (Available in U.S. only.)

29220-7

ANTIQUE PLAYING CARDS: A Pictorial History, Henry René D'Allemagne. Over 900 elaborate, decorative images from rare playing cards (14th–20th centuries): Bacchus, death, dancing dogs, hunting scenes, royal coats of arms, players cheating, much more. 96pp. 9¼ x 12¼.

29265-7

MAKING FURNITURE MASTERPIECES: 30 Projects with Measured Drawings, Franklin H. Gottshall. Step-by-step instructions, illustrations for constructing handsome, useful pieces, among them a Sheraton desk, Chippendale chair, Spanish desk, Queen Anne table and a William and Mary dressing mirror. 224pp. 8⅛ x 11¼.

29338-6

THE FOSSIL BOOK: A Record of Prehistoric Life, Patricia V. Rich et al. Profusely illustrated definitive guide covers everything from single-celled organisms and dinosaurs to birds and mammals and the interplay between climate and man. Over 1,500 illustrations. 760pp. 7½ x 10⅛.

29371-8